Edible Fungi
Chemical Composition, Nutrition and Health Effects

Food Chemistry, Function and Analysis

Series editors:
Gary Williamson, *Monash University, Australia*
Alejandro G. Marangoni, *University of Guelph, Canada*
Graham A. Bonwick, *AgriFoodX Limited, UK*
Catherine S. Birch, *AgriFoodX Limited, UK*

Titles in the series:

1: Food Biosensors
2: Sensing Techniques for Food Safety and Quality Control
3: Edible Oil Structuring: Concepts, Methods and Applications
4: Food Irradiation Technologies: Concepts, Applications and Outcomes
5: Non-extractable Polyphenols and Carotenoids: Importance in Human Nutrition and Health
6: Cereal Grain-based Functional Foods: Carbohydrate and Phytochemical Components
7: Steviol Glycosides: Cultivation, Processing, Analysis and Applications in Food
8: Legumes: Nutritional Quality, Processing and Potential Health Benefits
9: Tomato Chemistry, Industrial Processing and Product Development
10: Food Contact Materials Analysis: Mass Spectrometry Techniques
11: Vitamin E: Chemistry and Nutritional Benefits
12: Anthocyanins from Natural Sources: Exploiting Targeted Delivery for Improved Health
13: Carotenoid Esters in Foods: Physical, Chemical and Biological Properties
14: Eggs as Functional Foods and Nutraceuticals for Human Health
15: Rapid Antibody-based Technologies in Food Analysis
16: DNA Techniques to Verify Food Authenticity: Applications in Food Fraud
17: Advanced Gas Chromatography in Food Analysis
18: Handbook of Food Structure Development
19: Mitigating Contamination from Food Processing
20: Biogenic Amines in Food: Analysis, Occurrence and Toxicity
21: Nutrition and Cancer Prevention: From Molecular Mechanisms to Dietary Recommendations
22: Health Claims and Food Labelling
23: Nutraceuticals and Human Health: The Food-to-supplement Paradigm
24: Nutritional Signalling Pathway Activities in Obesity and Diabetes
25: The Chemistry and Bioactive Components of Turmeric
26: Foodomics
27: Food Proteins and Peptides: Emerging Biofunctions, Food and Biomaterial Applications
28: Handbook of Antioxidant Methodology: Approaches to Activity Determination
29: Fats and Associated Compounds: Consumption and Human Health

30: Oral Processing and Consumer Perception: Biophysics, Food Microstructures and Health
31: Development of Trans-free Lipid Systems and their Use in Food Products
32: Advanced Spectroscopic Techniques for Food Quality
33: Berries and Berry Bioactive Compounds in Promoting Health
34: Metabolism of Nutrients by Gut Microbiota
35: Bio- and Nano-sensing Technologies for Food Processing and Packaging
36: Edible Fungi: Chemical Composition, Nutrition and Health Effects

How to obtain future titles on publication:
A standing order plan is available for this series. A standing order will bring delivery of each new volume immediately on publication.

For further information please contact:
Book Sales Department, Royal Society of Chemistry, Thomas Graham House, Science Park, Milton Road, Cambridge, CB4 0WF, UK
Telephone: +44 (0)1223 420066, Fax: +44 (0)1223 420247,
Email: booksales@rsc.org
Visit our website at www.rsc.org/books

Edible Fungi
Chemical Composition, Nutrition and Health Effects

Edited by

Dejan Stojković
University of Belgrade, Serbia
Email: dejanbio@ibiss.bg.ac.rs

and

Lillian Barros
Instituto Politécnico de Bragança, Portugal
Email: lillian@ipb.pt

ROYAL SOCIETY OF **CHEMISTRY**

Food Chemistry, Function and Analysis No. 36

Print ISBN: 978-1-83916-401-9
PDF ISBN: 978-1-83916-752-2
EPUB ISBN: 978-1-83916-753-9
Print ISSN: 2398-0656
Electronic ISSN: 2398-0664

A catalogue record for this book is available from the British Library

The Royal Society of Chemistry is a charity, registered in England and Wales, Number 207890, and a company incorporated in England by Royal Charter (Registered No. RC000524), registered office: Burlington House, Piccadilly, London W1J 0BA, UK, Telephone: +44 (0) 20 7437 8656.

Visit our website at www.rsc.org/books

Printed in the United Kingdom by CPI Group (UK) Ltd, Croydon, CR0 4YY, UK

Dedication

To the decade of Bragança—Belgrade friendship and collaboration.

Food Chemistry, Function and Analysis No. 36
Edible Fungi: Chemical Composition, Nutrition and Health Effects
Edited by Dejan Stojković and Lillian Barros

Published by the Royal Society of Chemistry, www.rsc.org

Preface

There are many living ethnological stories on mushrooms dating back to ancient times, which have been interlaced with the supernatural in literature and art, used for human longevity, and as essential food sources. Working together in mushroom science for many past ages, investigating their chemistry, their bioactive and food preserving properties, made us somehow unintentionally feel great love and appreciation for the astonishing realm of edible mushrooms. As scientists and researchers in the modern era, we needed the matching representative for our mushroom love to transfuse the positive vibes and desire for absolute knowledge about these food diamonds: a book with the current accomplishments in myco-science of edible fungi. All the chapters summarized in this book result from enormous dedication and work, over the last few decades, of the chapter authors and book editors, who were crossing known scientific frontiers and dictating modern trends in myco-research.

The book is organized into 11 chapters with discussions of new insights into mushroom research. Chapter 1 discusses the differences between poisonous and edible mushrooms, as sometimes it is not easy to distinguish them without proper identification of the individual collected mushroom. In Chapter 2, the nutritional properties of wild-growing edible fungi, which are consumed in different parts of the world, are explored. Recent findings have presented edible fungi as a very good source of unsaturated fatty acids, to which many health beneficial properties are attributed, a field in which Chapter 3 reviews recent findings. Phenolic compounds are a diverse group of natural compounds that can be found in edible fungi. Many studies dealing with the exploration of phenolic compounds and their health-improving properties are discussed in Chapter 4. Edible mushrooms and their isolated compounds show promising antimicrobial, antioxidant, immunomodulatory, and anticancer potentials *in vitro* and *in vivo*, with

Food Chemistry, Function and Analysis No. 36
Edible Fungi: Chemical Composition, Nutrition and Health Effects
Edited by Dejan Stojković and Lillian Barros

Published by the Royal Society of Chemistry, www.rsc.org

different mechanisms of action included, and these are continuously showing novel pharmacological properties, as discussed in Chapter 5. In Chapter 6, various protein structures, lectins, and β-glucans that are found in edible mushrooms are reviewed, as they play a pivotal role in many biological effects. The characterization of terpenes and steroids from various mushroom sources and their biological activity and chemical structures are presented in Chapter 7. Meanwhile, Chapter 8 discusses the long history of mushroom cultivation but points out that reliable techniques for cultivation of fruiting bodies are known only for a limited number of species. In recent years, mushroom production has increased to a new level, especially in countries with growing economies, such as China, Poland, Hungary, and India. Chapter 9 reviews the cultural use of mushrooms in different regions of the world. Moreover, as described in Chapter 10, edible fungi do not serve only as nutraceuticals and pharmaceuticals, the uses of their extracts and compounds are described in food products as convenient natural alternatives for synthetic food additives, food fortifiers, and food enhancers. Chapter 11 reviews the most recent supplements available on the market based on mushrooms and official drugs derived from edible fungi.

Dejan Stojković
Lillian Barros

Contents

Food Chemistry, Function and Analysis No. 36
Edible Fungi: Chemical Composition, Nutrition and Health Effects
Edited by Dejan Stojković and Lillian Barros

Published by the Royal Society of Chemistry, www.rsc.org

CHAPTER 1

Diversity of the Fungi Kingdom: Molecular Tools to Distinguish Mushrooms Considered Safe and Unsafe for Human Health

A. LALLAWMSANGA[a] AND B. JAIME CARRASCO*[b,c]

[a] BioNEST Incubation Centre Mizoram University Aizawl, Mizoram-796004, India; [b] Centro Tecnológico de Investigación del Champiñón de La Rioja (CTICH), 26560 Autol, la Rioja, Spain; [c] Department of Biology, University of Oxford, South Parks Road, Oxford OX1 3RB, UK
*Email: carraco.jaime@gmail.com

1.1 Introduction

Mushroom is a term that represents a short reproductive stage in the life-cycle of a macrofungi, being generally used to describe the fruiting body of the phyla Ascomycota and Basidiomycota.[1] Basidiomycetes are one of the four branches of Eumycota (true fungi), representing 35% of the total fungal species currently described, many of them having both ecological and industrial importance.[2] Mushrooms can be either epigeous or hypogeous, also varying according to the ecological impact resulting from their lifestyles which can be saprotrophs, symbionts, or even pathogenic to fungi, plants and humans.[3] Over the years, several investigations have demonstrated that mushrooms hold unique

Food Chemistry, Function and Analysis No. 36
Edible Fungi: Chemical Composition, Nutrition and Health Effects
Edited by Dejan Stojković and Lillian Barros

Published by the Royal Society of Chemistry, www.rsc.org

properties that have played major roles in human history, religion, and culture. These highly adaptive organisms survive by feeding mainly on matter from decomposing/death of other organisms and, occasionally, on living matter. These organisms are the most visible and colorful of the fungi kingdom, being recognized for their value as food and delicacies.[4] Since mushroom fruiting is dependent on climatic conditions, the annual yields in the wild may vary from one year to another, depending on variables such as the average temperature or the precipitation rate.[5] It is therefore predictable and already proved that domestication of some mushroom species provides stable productivity independent of climate events.[6]

Wild mushroom picking is generally a leisure activity that is very common in southern Europe, providing the seasonal income of some professional pickers.[7] In Spain, for example, there are regulated mycological areas that require the purchase of picking licenses issued by the managing authority to ensure sustainable management of this resource.[8] A good approach for the conservation of fungal resources requires undertaking management and regulatory processes for edible mushrooms in picking areas, thus quantifying and controlling the pressure on the most appreciated species.[9] Northeast India is populated by over 45 million people, covers an area of 0.26 million km^2, and comprised of eight states: Arunachal Pradesh, Assam, Manipur, Meghalaya, Mizoram, Nagaland, Sikkim, and Tripura.[10] This region is rich in biodiversity, and characteristic for its heavy precipitation and high seismicity.[11,12] In Table 1.1 are presented 112 edible, medicinal, or poisonous mushrooms collected from all eight states of northeast India. The edible mushrooms foraged from northeast India include *Lentinula edodes*,[13,14] *Lentinus sajor-caju*,[15–17] *Pleurotus djamor*,[17] *Cantharellus cibarius*,[18,19] *Agaricus augustus*,[1,20] and *Lactarius deliciosus*.[20] The poisonous mushrooms found in the same areas are *Gyromitra infula*,[1] *Amanita phalloides*,[20] and *Polyozellus multiplex*;[13] and the medicinal ones are *Ganoderma lucidum*,[13,15,16] *Pleurotus pulmonarius*,[17] *Ganoderma applanatum*,[21] and *Microporus xanthopus*.[13]

The public demand for edible mushrooms continues to increase every year due to their rich nutritional value and massive industrial production. Mushroom growers increasingly require good-quality varieties with a fast maturity period, improved resistance to both pests and diseases, and a high production yield. Thus, the molecular characterization of fungi up to their species level is an essential task in mushroom research for more applications in different fields such as for agronomic and medicinal purposes.[22]

There is a growing industry of edible mushroom production based on a solid fermentation process of pasteurized or sterilized substrates inoculated with the appropriate spawn that proceeds under controlled temperature and humidity conditions.[6] The market has been pushing to improve the yield, quality and quantity of mushrooms currently produced, propelling research aimed at developing breeding programs for edible fungi and to formulate appropriate substrates and culture conditions for new species. These proposals are based on three main factors: (i) the economic value of some highly demanded fungal species; (ii) their use to produce enzymes or chemicals

Table 1.1 Edible, medicinal, and poisonous mushrooms collected from all eight states of northeast India.

Sl. no.	Name of mushroom	Family	Common name	Edibility	Location	Ref.
1	*Lentinula edodes*	Marasmiaceae	Shiitake	Edible, medicinal	Tripura, Mizoram, Meghalaya, Assam, Manipur	13,14
2	*Grifola frondosa*	Meripilaceae	Hen of the wood	Edible	Tripura, Mizoram, Meghalaya, Assam	13
3	*Ganoderma lucidum*	Ganodermataceae	Lingzhi/Reishi	Medicinal	Tripura, Mizoram, Meghalaya, Assam, Arunachal Pradesh	13, 15 and 16
4	*Pleurotus ostreatus*	Pleurotaceae	Oyster mushroom	Edible	Tripura, Mizoram, Meghalaya, Assam, Manipur	13
5	*Agaricus bisporus*	Agaricaceae	Button mushroom	Edible	Tripura, Mizoram, Meghalaya, Assam	13 and 15
6	*Hericium erinaceus*	Hericiaceae	Lion's mane	Edible	Tripura, Mizoram, Meghalaya, Assam	13
7	*Cordyceps sinensis*	Cordyceps sinesis	Caterpillar fungus	Medicinal	Tripura, Mizoram, Meghalaya, Assam, Arunachal Pradesh	13 and 16
8	*Stropharia aeruginosa*	Strophariaceae	Verdigris agaric	Poisonous	Tripura, Mizoram, Meghalaya, Assam	13
9	*Trametes versicolor*	Polyporaceae	Turkey tail	Medicinal	Tripura, Mizoram, Meghalaya, Assam	13
10	*Agaricus subrufescens*	Agaricaceae	Almond mushroom, mushroom of the sun, God's mushroom, mushroom of life, royal sun agaricus, jisongrong, or himematsutake	Edible	Tripura, Mizoram, Meghalaya, Assam	13
11	*Inonotus obliquus*	Hymenochaetaceae	Chaga	Medicinal		13

Table 1.1 (*Continued*)

Sl. no.	Name of mushroom	Family	Common name	Edibility	Location	Ref.
					Tripura, Mizoram, Meghalaya, Assam	
12	*Flammulina velutipes*	Physalacriaceae	Enoki, Velvet shank	Edible	Tripura, Mizoram, Meghalaya, Assam, Nagaland	13 and 20
13	*Fomitopsis betulina*	Fomitopsidaceae	Birch polypore, birch bracket, razor strop	Edible	Tripura, Mizoram, Meghalaya, Assam	13
14	*Polyozellus multiplex*	Thelephoraceae	Blue chanterelle	Poisonous	Tripura, Mizoram, Meghalaya, Assam	13
15	*Sparassis crispa*	Sparassidaceae	Cauliflower fungus	Edible when young	Sikkim, Tripura, Mizoram, Meghalaya, Assam, Nagaland	1, 13 and 20
16	*Pleurotus eryngii*	Pleurotaceae	King trumpet mushroom, French horn mushroom, king oyster mushroom, king brown mushroom, boletus of the steppes, trumpet royale, ali'i oyster	Edible	Tripura, Mizoram, Meghalaya, Assam	13
17	*Agrocybe aegerita*	Strophariaceae	Black poplar mushroom	Edible	Tripura, Mizoram, Meghalaya, Assam	13
18	*Tremella mesenterica*	Tremellaceae	Yellow brain, golden jelly fungus, yellow trembler, and 'witches' butter	Poisonous	Sikkim, Tripura, Mizoram, Meghalaya, Assam	1 and 13
19	*Polyporus umbellatus*	Thelephoracea	Lumpy bracket and umbrella polypore	Edible, medicinal	Tripura, Mizoram, Meghalaya, Assam	13
20	*Amauroderma* sp.	Ganodermataceae	—	Medicinal	Tripura, Mizoram, Meghalaya, Assam	13
21	*Auricularia* spp.	Auriculariaceae	Wood ear mushroom	Poisonous	Tripura, Mizoram, Meghalaya, Assam	13
22	*Microporus xanthopus*	Polyporaceae	Yellow-footed polypore	Medicinal		13 and 17

					Tripura, Mizoram, Meghalaya, Assam	
23	*Pycnoporus sanguineus*	Polyporaceae	Tropical cinnabar bracket	Poisonous	Tripura, Mizoram, Meghalaya, Assam	13
24	*Xylaria polymorpha*	Xylariaceae	Dead man's fingers	Poisonous	Tripura, Mizoram, Meghalaya, Assam	13
25	*Schizophyllum commune*	Schizophyllaceae	Split gills mushroom	Edible	Mizoram, Arunachal Pradesh	16 and 17
26	*Trametes hirsuta*	Polyporaceae	Hairy bracket	Fungal plant pathogen	Mizoram	17
27	*Marasmiellus palmivorus*	Marasmiaceae	Oil palm bunch rot	Medicinal	Mizoram	17
28	*Pholiota adiposa*	Strophariaceae	Chestnut mushroom	Edible	Mizoram	17
29	*Pleurotus pulmonarius*	Pleurotaceae	Indian oyster, Italian oyster, Phoenix mushroom, or the lung oyster	Edible, medicinal	Mizoram, Arunchal Pradesh	16 and 17
30	*Auricularia polytricha*	Auriculariaceae	Cloud ear fungus	Edible	Mizoram, Assam, Nagaland	15, 17 and 20
31	*Bjerkandera adusta*	Meruliaceae	Smoky polypore	Fungal plant pathogen	Mizoram	17
32	*Hymenopellis chiangmaiae*	Physalacriaceae	Porcelain fungus	Medicinal	Mizoram	17
33	*Trametes elegans*	Polyporaceae	Polypore	Medicinal	Mizoram	17
34	*Panus giganteus*	Polyporaceae	—	Medicinal	Mizoram	17
35	*Xylaria feejensis*	Xylariaceae	—	—	Mizoram	17
37	*Gymnopus menehune*	Omphalotaceae	—	—	Mizoram	17
38	*Lentinus sajor-caju*	Polyporaceae	White-rot fungus	Edible	Mizoram, Assam, Arunachal Pradesh	15–17
39	*Polyporus arcularius*	Polyporaceae	Spring polypore	Edible, medicinal	Mizoram	17
40	*Pleurotus djamor*	Pleurotaceae	Pink oyster mushroom	Edible, medicinal	Mizoram	17

Table 1.1 (*Continued*)

Sl. no.	Name of mushroom	Family	Common name	Edibility	Location	Ref.
41	*Calvatia gigantea*	Agaricaceae	Giant puffball	Edible when young	Meghalaya	15 and 18
42	*Cantharellus cibarius*	Cantharellaceae	Girolle/chanterelle	Edible	Meghalaya, Mizoram	18 and 19
43	*Russula integra*	Russulaceae	Entire russula	Edible	Meghalaya	18
44	*Turbinellus floccosus*	Gomphaceae	Scaly vase, the shaggy, scaly, or woolly chanterelle	Edible	Meghalaya	18
45	*Lactarius quieticolor*	Russulaceae	Milk cap	Edible	Meghalaya	18
46	*Clavulina cinerea*	Clavulinaceae	Coral fungus	Edible	Meghalaya	18
47	*Ramaria brevispora*	Gomphaceae	Coral mushroom	Edible	Meghalaya	18
48	*Lactifluus corrugis*	Russulaceae	Corrugated-cap milky	Edible	Mizoram	19
49	*Auricularia auricula-judae*	Auriculariaceae	Wood ear or black wood ear	Edible	Mizoram, Assam, Arunachal Pradesh, Nagaland	15, 16, 19 and 20
50	*Termitomyces heimii*	Lyophyllaceae	Termite mushroom	Edible, medicinal	Mizoram	19
51	*Auricularia delicata*	Auriculariaceae	Wood ear mushroom	Edible	Manipur, Mizoram, Assam, Arunachal Pradesh	14, 15 and 21
52	*Termitomyces microcarpus*	Lyophyllaceae	Termite mushroom	Edible, medicinal	Assam, Mizoram, Arunachal Pradesh	15 and 21
53	*Lactifluus volemus*	Russulaceae	Tawny milkcap	Edible	Manipur	14
54	*Pleurotus citrinopileatus*	Pleurotaceae	Golden oyster	Edible	Manipur	14
55	*Termitomyces eurhizus*	Lyophyllaceae	Termite mushroom	Edible	Sikkim, Manipur	1 and 14
57	*Agaricus campestris*	Agaricaceae	Field/meadow mushroom	Edible	Assam	15
58	*Agaricus silvaticus*	Agaricaceae	Scaly wood mushroom, blushing wood mushroom, or pinewood mushroom	Edible	Assam	15
59	*Boletus edulis*	Boletaceae	King bolete	Edible	Assam, Arunachal Pradesh	15 and 16

60	*Suillus luteus*	Suillaceae	Slippery jack or sticky bun	Edible	Sikkim, Assam, Nagaland	1, 15 and 20
61	*Marasmius oreades*	Marasmiaceae	Fairy ring mushroom or fairy ring champignon	Edible	Assam	15
62	*Termitomyces clypeatus*	Lyophyllaceae	Termite mushroom	Edible	Assam	15
63	*Termitomyces robustus*	Lyophyllaceae	Termite mushroom	Edible	Assam, Arunachal Pradesh	15 and 16
65	*Tricholoma terreum*	Tricholomataceae	Grey knight or dirty tricholoma	Edible	Assam	15
67	*Laetiporus sulphureus*	Polyporaceae	Crab-of-the-woods, sulfur polypore, sulfur shelf, and chicken-of-the-woods	Edible, medicinal	Assam	15
68	*Apioperdon pyriforme*	Lycoperdaceae	Pear-shaped puffball or stump puffball	Edible when young	Assam	15
69	*Morchella esculenta*	Morchellaceae	Common morel, morel, yellow morel, true morel, morel mushroom, and sponge morel	Edible	Assam	15
70	*Trametes betulina*	Polyporaceae	Gilled polypore, birch mazegill or multicolor gill polypore	Inedible	Assam	15
71	*Aleuria aurantia*	Pyronemataceae	Orange peel fungus	Edible	Arunachal Pradesh, Sikkim	1 and 16
72	*Polyporus squamosus*	Polyporaceae	Dryad's saddle, pheasant's back mushroom	Edible when young	Arunachal Pradesh	16
74	*Tremella fuciformis*	Tremellaceae	Snow fungus, snow ear, silver ear fungus, white jelly mushroom	Edible/ medicinal	Arunachal Pradesh	16
75	*Macrocybe lobayensis*	Tricholomataceae	—	Edible	Arunachal Pradesh	16
76	*Volvariella bombycina*	Plutaceae		Edible	Arunachal Pradesh	16

Table 1.1 (*Continued*)

Sl. no.	Name of mushroom	Family	Common name	Edibility	Location	Ref.
			Silky sheath, silky rosegill, silver-silk straw mushroom, tree mushroom			
77	*Volvariella volvacea*	Pluteaceae	Paddy straw mushroom, straw mushroom	Medicinal	Arunachal Pradesh	16
78	*Ganoderma applanatum*	Ganodermataceae	Artist's bracket, artist's conk, bear bread	Medicinal	Arunachal Pradesh, Mizoram	21
79	*Ganoderma tsugae*	Ganodermataceae	Hemlock varnish shelf	Medicinal	Mizoram	21
80	*Agaricus augustus*	Agaricaceae	The prince	Edible	Sikkim, Nagaland	1 and 20
81	*Agaricus moelleri*	Agaricaceae	Inky mushroom, dark scaled mushroom	Poisonous	Nagaland	20
82	*Agaricus xanthodermus*	Agaricaceae	Yellow-staining mushroom, the yellow-stainer	Poisonous	Nagaland, Sikkim	1 and 20
83	*Coprinus comatus*	Agaricaceae	Shaggy ink cap, lawyer's wig, or shaggy mane	Edible	Nagaland	20
84	*Lepiota felina*	Agaricaceae	Cat dapperling	Poisonous	Nagaland	20
85	*Lepiota lilacea*	Agaricaceae	Lepitoid mushroom	Poisonous	Nagaland	20
86	*Lycoperdon perlatum*	Agaricaceae	Common puffball, warted puffball, gem-studded puffball, wolf farts, or the devil's snuff-box	Edible when young	Nagaland	20
87	*Lycoperdon pyriforme*	Agaricaceae	Pear-shaped puffball or stump puffball	Edible	Nagaland	20
88	*Macrolepiota albuminosa*	Agaricaceae	Termite mushroom	Edible	Nagaland	20
89	*Lactarius deliciosus*	Russulaceae	Saffron milk cap and red pine mushroom	Edible	Nagaland	20
90	*Lactarius deterrimus*	Russulaceae	False saffron milkcap or orange milkcap	Edible	Nagaland	20
91	*Lactifluus piperatus*	Russulaceae	Blancaccio	Edible	Nagaland	20
92	*Russula nobilis*	Russulaceae	Beechwood sickener	Poisonous	Nagaland	20

93	*Amanita cokeri*	Amanitaceae	Coker's amanita, solitary lepidella	Poisonous	Nagaland	20
94	*Amanita flavoconia*	Amanitaceae	Yellow patches, yellow wart, orange amanita, yellow-dust amanita or the American yellow dust amanita	Poisonous	Sikkim, Nagaland	1 and 20
95	*Amanita phalloides*	Amanitaceae	Death cap or the death cap amanita	Poisonous	Nagaland	20
96	*Amanita virosa*	Amanitaceae	Destroying angel or the European destroying angel amanita	Poisonous	Nagaland	20
97	*Helvella crispa*	Helvellaceae	White saddle, elfin saddle or common helvel	Poisonous	Sikkim, Nagaland	1 and 20
98	*Suillus pictus*	Suillaceae	Painted slipperycap, the painted suillus or the red and yellow suillus	Edible	Nagaland	20
99	*Tricholoma imbricatum*	Tricholomataceae	Matt knight	Edible	Nagaland	20
100	*Phallus indusiatus*	Phallaceae	Bamboo mushrooms, bamboo pith, long net stinkhorn, crinoline stinkhorn, or veiled lady	Edible	Nagaland	20
101	*Gyromitra infula*	Discinaceae	Hooded false morel or the elfin saddle	Poisonous	Sikkim	1
104	*Amanita caesarea*	Amanitaceae	Caesar's mushroom	Edible	Sikkim	1
105	*Amanita franchetii*	Amanitaceae	Franchet's amanita	Edible	Sikkim	1
106	*Armillaria mellea*	Physalacriaceae	Honey fungus	Edible	Sikkim	1
107	*Hymenopellis radicata*	Physalacriaceae	Rooting shank	Edible	Sikkim	1
108	*Termitomyces medius*	Lyophyllaceae	Termite mushroom	Edible	Sikkim	1
110	*Strobilomyces floccopus*	Boletaceae	Old man of the woods	Edible	Sikkim	1
111	*Scleroderma verrucosum*	Sclerodermataceae	Common earthball, poison puffball, scaly earthball	Poisonous	Sikkim	1
112	*Scleroderma cepa*	Sclerodermataceae	Common earthball, poison puffball, scaly earthball	Poisonous	Sikkim	1

that are useful in industry or pharmacy; and (iii) their application in processes aimed at recycling industrial and/or agricultural wastes.

1.2 Good Practices for Mushroom Picking

Wild mushroom picking is a very healthful activity that requires specific knowledge about the mycological habitats and tools to discriminate among edible, toxic, and poisonous species.

Among the general rules for sustainable mushroom picking are:[23] (a) mushrooms must be harvested in an appropriate developmental stage to avoid misperception with toxic species and to take the best advantage of their culinary properties; (b) excessively trampled areas should be avoided, using only tracks and paths for vehicles to access the area; (c) collect the complete carpophores to facilitate identification; (d) clean the mushrooms immediately after harvesting to lengthen the shelf-life of the carpophores; (e) do not alter the surface of the ground by removing the leaf litter; (f) it is highly advisable to use a wicker basket instead of plastic bags to ensure spore dispersal while walking through the woods; (g) avoid mixing edible mushrooms with toxic species in the same basket and do not damage species that we are not interested in harvesting; and (h) as mushrooms can bioaccumulate toxic compounds, collection in contaminated areas should be avoided.

1.2.1 Mycological Habitats

Depending on different variables, such as the chemical composition of the soil, flora, pluviometry, elevation above sea level, and temperature, different species associated with a particular mycological habitat can be found. Among others, in Spain, different habitats, such as pine trees on alkaline soils, pine trees on siliceous soils, forests of birches, gall oaks, Pyrenean oaks, holm oaks, junipers, different pines, riverside woods, peatlands, meadows, and grasslands are inhabited by countless mycological species.[23] The characterization and understanding of the mycological habitat can facilitate the identification of the different species. Figure 1.1 shows different types of mushrooms growing in multiple habitats.

1.2.2 Morphological Identification of Mushrooms

Fungi identification requires two independent methodologies: morphological and molecular characterization. Mushrooms are macroscopic structures that conform to the sexual organs of the fungi. To prevent poisoning, mushrooms must be correctly identified before consumption. Of the visual characteristics, color, stipe, pileus, and hymenium morphologies stand out.[24] Novel tools such as mining data from mushroom shapes with tailor-made algorithms can be useful to identify edible and toxic species.[25,26] Additional characteristics such as the spore footprint, texture, and flavor are

Figure 1.1 Different mushrooms (edible, toxic, poisonous, and those with no culinary value) from Mediterranean habitats with multiple trophic behaviors, including symbionts (parasitism on living hosts) and saprotrophs (feeding on decaying organic matter): (a) *Boletus edulis* (symbiont, edible mushroom); (b) *Ganoderma pfeifferi* (saprotroph, medicinal mushroom); (c) *Trametes versicolor* (saprotroph, medicinal mushroom); (d) *Lycoperdon perlatum* (saprotroph, with no culinary value); (e) *Coprinopsis micaceous* (saprotroph, fresh and decaying mushrooms with no culinary value); (f) *Lepiota cristata* (saprotroph, toxic); (g) *Amanita phalloides* (symbiont, poisonous and lethal).

also useful tools to identify different mushroom species. It is very important that, in case of doubt, the mushroom is rejected or identified by an expert mycologist.

1.3 Identification of Mushrooms Through Molecular Tools

A variety of molecular genetic tools have been employed for the identification of fungi.[27] Molecular markers allow the selection of desired traits based on genotype rather than phenotype and can hence complement and accelerate plant breeding programs.[28] Here, we describe different types of molecular markers used for mushroom identification.

1.3.1 Random Amplified Polymorphic DNA (RAPD)

Random amplified polymorphic DNA (RAPD) is used to distinguish the genetic diversity of mushrooms[29] and to study the genetic and phylogenetic relationship between them[30] (Table 1.2). This is a simple methodology as it requires fewer DNA templates, is not susceptible to any hazardous contamination, and embraces a cheap technology.[31] Apart from the genetic characterization of different organisms, RAPD analysis is also employed for the identification and study of genetic diversity in synthetic or newly found species with agricultural and industrial importance.[32,33] Even though the traditional RAPD technique has its limitations, such as low production and poor reproducibility, the new and improved RAPD technique known as RAMP-PCR (where the ramp time of traditional RAPD-PCR is increased from annealing to extension) is found to be more accurate and more consistent as compared to the traditional RAPD analysis.[31]

1.3.2 Inter-simple Sequence Repeat (ISSR)

Inter-simple sequence repeat (ISSR) is an economical and reliable methodology that has been widely employed for the identification of different strains, populations, and genetic diversity of numerous fungi[34] (Table 1.2). ISSR markers are semi-arbitrary markers amplified by PCR in the presence of another primer corresponding to a target microsatellite.[35] This methodology has been successfully employed in the study of the genetic diversity of several mushrooms such as *Agaricus* species,[36] *Auricularia auricula* cultivars,[37] *Lentinula edodes* strains,[38] *Pleurotus citrinopileatus*,[39] *Pleurotus eryngii* cultivars,[40] and *Volvariella volvacea*.[41]

1.3.3 Sequence-related Amplified Polymorphism (SRAP)

Li and Qurios[42] first introduced a novel molecular marker named SRAP (sequence-related amplified polymorphism), which has been widely used in genetic linkage mapping, genetic diversity analysis, gene tagging, and fungal genetic diversity studies[43] (Table 1.2). This approach has been used consecutively for the evaluation of genetic diversity in *Agaricus bisporus*,[44] *Ganoderma* population,[45] *Lentinula edodes* strains,[46] *Pleurotus pulmonarius* strains,[47] and *Tricholoma matsutake*.[48]

1.3.4 Simple Sequence Repeats (SSRs)

Simple sequence repeats, also known as microsatellites, consist of only one or a few tandemly repeated nucleotides.[49] The SSR marker is a PCR-based molecular methodology with several advantages such as genetic co-dominance, high reproducibility, multi-allelic variation, and easy amplification by PCR.[50] These microsatellites are universally found in eukaryotic and prokaryotic genomes.[51] SSR found in species with little or no DNA sequence information usually involves the construction and screening of partial genomic libraries and the sequencing of SSR-positive clones.[52] SSRs have been used to evaluate the genetic diversity of several mushrooms

Table 1.2 Molecular genetic tools employed as molecular markers for the identification of mushrooms.

Sl.	RAPD[a]		ISSR		SRAP		SSRs		AFLP		SCAR		RFLPs		ITS	
no.	Application	Ref.	Application	Ref.	Application	Ref.	Application	Ref.	Application	Ref.	Application	Ref.	Application	Ref.	Application	Ref.
1	Genotyping of *Ganoderma* species	31	Genetic diversity of wild *Auricularia auricula-judae*	34	Genetic diversity of *Pleurotus pulmonarius*	47	Using SSR markers to evaluate the genetic diversity of natural germplasm of *Lentinula edodes* in China	53	Genetic diversity of the edible mushroom *Pleurotus* sp.	58	Identification of degenerate nuclei and development of a SCAR marker for *Flammulina velutipes*	95	Differentiation of commercial strains of *Agaricus species* in China (IGS-RFLP)	70	Comparative studies on the diversity of the edible mushroom *Pleurotus eryngii*	27
2	Genetic diversity of *Pleurotus pulmonarius*	47	Genotyping of *Ganoderma* species	31	Evaluation of genetic diversity in *Lentinula edodes* strains	38	Development and characterization of SSR markers for the mushroom *Flammulina velutipes*	50	Development of AFLP markers and phylogenetic analysis in *Hipsizygus marmoreus*	59	Development of strain-specific SCAR markers for authentication of *Ganoderma lucidum*	64	Differentiating homothallic *Volvariella* mushrooms by RFLPs and AP-PCR	99	Vegetative growth and phylogenetic relationship of commercially cultivated strains of *Pleurotus eryngii*	79
3	Evaluation of genetic diversity in *Lentinula edodes* strains	38	Genetic diversity of *Pleurotus pulmonarius*	47	Analysis of genetic diversity among Chinese *Auricularia auricula* cultivars	37	Development of SSR markers for typing cultivars in the mushroom *Auricularia auricula-judae*	54	Genotypic diversity of *Armillaria gallica* from mixed oak forests in Massachusetts	60	Development of SCAR markers to determine the mating types of *Lepista nuda* protoplast monokaryons	63	Genetic homogeneity of cultivated strains of shiitake (*Lentinula edodes*) used in China as revealed by the polymerase chain reaction	100	Three nonorthologous ITS1 types are present in a polypore fungus *Trichaptuma bietinum*	106
4	Evaluation of genetic diversity among Chinese *Pleurotus eryngii* cultivars	40	Evaluation of genetic diversity in *Lentinula edodes* strains	38	Analysis of genetic diversity among Chinese *Pleurotus citrinopileatus* Singer cultivars	37	Novel microsatellite markers suitable for genetic studies in the white button mushroom *Agaricus bisporus*	88	A genetic linkage map of *Lentinula edodes* (shiitake) based on AFLP markers	91	Evaluation of the use of SCAR markers for screening genetic diversity of *Lentinula edodes* strains	46	Phylogenetic analysis of oyster mushrooms (*Pleurotus* spp.) based on restriction fragment length polymorphisms of the 5 portion of 26S rDNA	101	Phylogenetic and biogeographic relationships of Eastern Asian and Eastern North American disjunct *Suillus* species (fungi) as inferred from nuclear ribosomal RNA ITS sequences	75
5	Morphological and molecular characterization of oyster mushroom (*Pleurotus* spp.)	30	Analysis of genetic diversity among Chinese *Auricularia auricula* cultivars	37	DNA fingerprinting of genetic diversity of *Agaricus bisporus*	44	Genome-wide analysis of simple sequence repeats in the model medicinal mushroom *Ganoderma lucidum*	89	Glycosyl linkage characteristics and classifications of exo-polysaccharides of some regionally	92	SCAR markers and multiplex PCR-based rapid molecular typing of *Lentinula edodes* strains	96	DNA-fingerprinting (AFLP and RFLP) for genotypic identification in species of the *Pleurotus eryngii* complex	69	Molecular phylogeny, biogeography, and speciation of the mushroom species *Pleurotus cystidiosus* and allied taxa	107

Table 1.2 *(Continued)*

Sl.	RAPD[a]		ISSR		SRAP		SSRs		AFLP		SCAR		RFLPs		ITS	
no.	Application	Ref.	Application	Ref.	Application	Ref.	Application	Ref.	Application	Ref.	Application	Ref.	Application	Ref.	Application	Ref.
									different strains of *Lentinula edodes*							
6	Genetic variation in cultivated strains of *Agaricus blazei*	77	Evaluation of genetic diversity among Chinese *Pleurotus eryngii* cultivars	40	Molecular diversity of *Auricularia polytricha*	84	An expanded genetic linkage map of an intervarietal *Agaricus bisporus* var. *bisporus* x *A. bisporus* var. *burnettii* hybrid	61	Assessment of genetic relationships among shiitake (*Lentinula* ssp.) mushrooms	62	In distinguishing a commercial strain of the superior type from other strains of *Lentinula edodes* in China	97	Genetic polymorphism of ferula mushroom growing on *Ferula sinkiangensis*	102	Internal transcribed spacer sequences from 38 recognized species of *Suillus sensulato*: phylogenetic and taxonomic implications	108
7	RAPD molecular differentiation of the cultivated strains of the jelly mushrooms, *Auricularia auricula* and *A. polytricha*	78	Analysis of genetic diversity among Chinese *Pleurotus citrinopileatus* Singer cultivars	37	In the genetic diversity of *Tricholoma matsutake*	48	Genetic diversity and population structure of *Armillaria luteovirens* (Physalacriaceae) in Qinghai-Tibet Plateau revealed by SSR markers	55	Genetic diversity and strain-typing in cultivated strains of *Lentinula edodes* (the shiitake mushroom) in Japan by AFLP analysis	91	New available SCAR markers: potentially useful in distinguishing a commercial strain of the superior type from other strains of *Lentinula edodes* in China	97	DNA polymorphisms in *Lentinula edodes*, the Shiitake mushroom	103	Differentiation and grouping of isolates of the *Ganoderma lucidum* complex by random amplified polymorphic DNA-PCR compared with grouping on the basis of internal transcribed spacer sequences	73
8	Vegetative growth and phylogenetic relationship of commercially cultivated strains of *Pleurotus eryngii*	79	Strain-typing of *Lentinula edodes* in China	81	Evaluation of the use of SCAR markers for screening genetic diversity of *Lentinula edodes* strains	46	Development of EST-SSR for preliminary analysis of genetic diversity of *Cordyceps militaris*	56	Comparative linkage mapping in the white button mushroom *Agaricus bisporus* provides foundation for breeding management	61	Development of IRAP-SCAR marker for strain identification in *Lentinula edodes*	65	Mitochondrial DNA restriction fragment length polymorphisms and phonetic relationships in natural populations of the oyster mushroom, *Pleurotus ostreatus*	104	DNA polymorphism in morels: complete sequences of the internal transcribed spacer of genes coding for rRNA in *Morchella esculenta* (yellow morel) and *Morchella conica* (black morel)	109
9	Genetic polymorphism and taxonomic infrastructure of the	80	Differentiation of commercial strains of	70	Genetic diversity of wild *Auricularia*	43	Survey and analysis of simple sequence repeats	90	Molecular differentiation of sexually	93	Mapping of a SCAR marker tightly linked to the B	98	Restriction fragment length polymorphisms in	66	Phylogenetic analysis of *Antrodia* species and *Antrodia*	110

	Pleurotus eryngii species complex		*Agaricus* species in China		*polytricha* in Yunnan Province of South-western China revealed by SRAP analysis		in the *Laccaria bicolor* genome, with development of microsatellite markers		incompatible strains of *Agaricus bitorquis* using RAPD and AFLP markers		mating type locus of *Lentinula edodes* (shiitake mushroom) by tetrad analysis		the mushrooms *Agaricus brunnescens* and *Agaricus bitorquis*		*camphorata* inferred from internal transcribed spacer region	
10	Comparative studies on the diversity of the edible mushroom *Pleurotus eryngii.*	27	DNA fingerprinting of genetic diversity of *Agaricus bisporus*	44	Analysis of genetic diversity in *Ganoderma* population with a novel molecular marker SRAP.	45	Using SSR markers to evaluate the genetic diversity of *Lentinula edodes* natural germplasm in China	53	Phylogeography and evolution in matsutake and close allies inferred by analyses of ITS sequences and AFLPs	94			Use of a culture independent method to analyze the diversity of soil fungi surrounding *Chroogomphus rutilus* in the Beijing region of China	105	Homology search of genus *Pleurotus* using an internal transcribed spacer region	111
11			Use of ISSR markers for strain identification in the button mushroom, *Agaricus bisporus*	82											Molecular phylogeny of Japanese *Amanita* species based on nucleotide sequences of the internal transcribed spacer region of nuclear ribosomal DNA	76
12			ISSR as new markers for identification of homokaryotic protoclones of *Agaricus bisporus*	83											Phylogenetic diversity of wild edible *Russula* from Northeastern Thailand on the basis of internal transcribed spacer sequence	71
13			Molecular diversity of *Auricularia polytricha*	84												
14			Use of inter-simple sequence repeat markers to develop strain-specific SCAR markers for *Flammulina velutipes*	85												

Table 1.2 *(Continued)*

Sl.	RAPD[a]		ISSR		SRAP		SSRs		AFLP		SCAR		RFLPs		ITS	
no.	Application	Ref.	Application	Ref.	Application	Ref.	Application	Ref.	Application	Ref.	Application	Ref.	Application	Ref.	Application	Ref.
15			Evaluation of growth characteristics and genetic diversity of commercial and stored lines of *Hypsizygus marmoreus*	86												
16			Phylogenetic analysis of high-yielding strain of paddy straw mushroom (*Volvariella volvacea*)	41												
17			Differentiation of homokaryons and heterokaryons of *Agaricus bisporus*	87												

[a]RAPD: random amplified polymorphic DNA; ISSR: inter-simple sequence repeat; SRAP: sequence-related amplified polymorphism; SSRs: simple sequence repeat; AFLP: amplified fragment length polymorphism; SCAR: sequence characterized amplified region; RFLPs: DNA restriction fragment length polymorphisms; ITS: internal transcribed spacer.

(Table 1.2) such as *Lentinula edodes*,[53] *Auricularia auricula-judae*,[54] *Armillaria luteovirens*[55] and *Cordyceps militaris*.[56]

1.3.5 Amplified Fragment Length Polymorphism (AFLP)

Amplified fragment length polymorphism (AFLP) is a molecular technique for DNA of any origin and complexity fingerprinting, including animal, microbe, and vegetal DNA.[57] The main advantages of the AFLP technique include its high reproducibility, reliability, and specificity. This technique has been employed to differentiate several types of mushrooms (Table 1.2) including *Pleurotus eryngii* strains,[58] *Hypsizygus marmoreus*,[59] *Armillaria gallica*,[60] and *Agaricus bisporus*.[61] Mukhopadhyay *et al.*[62] successfully studied 30 *Lentinula* sp. (including *L. edodes*, *L. lateritia*, and *L. novae-zelandiae*) using 32 AFLP primer combinations.

1.3.6 Sequence Characterized Amplified Region (SCAR)

The sequence characterized amplified region (SCAR) methodology was initially developed in 1991 and has been widely used since 1996 in studies on various aspects of basidiomycetes including breeding, population genetics, and genetic analyses.[46] SCAR markers have been employed for the identification of several mushrooms (Table 1.2) including *Flammulina velutipes*, *Lentinula edodes*, and *Laccaria bicolor*.[63] It also has been employed for the authentication and strain identification of *Ganoderma lucidum*[64] and *Lentinula edodes*.[65]

1.3.7 Restriction Fragment Length Polymorphisms (RFLPs)

DNA restriction fragment length polymorphisms (RFLPs) have been used as genetic markers in a wide variety of organisms, including humans, plants, fungi, and protozoa.[66] The most widely used RFLP method is a DNA–DNA hybridization technique that involves cutting genomic DNA with restriction endonuclease(s), followed by Southern hybridization with labeled specific probes.[67,68] This method yields highly reproducible fingerprint patterns, which are useful markers for genetic diversity studies among species[69] (Table 1.2). RFLP has been used to differentiate strains of *Agaricus* sp.,[70] phylogenetic analysis of *Pleurotus* spp., and DNA fingerprinting of *Pleurotus eryngii* complex.

1.3.8 Internal Transcribed Spacer (ITS)

The internal transcribed spacer (ITS) is a region of the nuclear ribosomal RNA which has been extensively used in molecular systematics.[71] Variation between individual ribosomal DNA (rDNA) repeats can be observed within the ITS due to its higher degree of variation than that of the small subunit (SSU) and large subunit (LSU) of rRNA genes.[72,73] Phylogenetic analysis based on DNA sequencing, ITS1-5.8S-ITS2 of rDNA sequence analysis, is normally used for fungal identification.[74] ITS has been used to study the diversity of edible mushroom *Pleurotus eryngii*,[27] phylogenetic and biogeographic relationships of *Suillus* sp.,[75] molecular phylogeny of Japanese *Amanita* species,[76] and phylogenetic diversity of wild edible *Russula* genus[71] (Table 1.2).

References

1. K. Das, Diversity and conservation of wild mushrooms in Sikkim with special reference to Barsey Rhododendron Sanctuary, *NeBIO*, 2010, **1**(2), 1.
2. L. Ramírez, L. M. Larraya and A. G. Pisabarro, Molecular tools for breeding basidiomycetes, *Int. Microbiol.*, 2000, **3**(3), 147.
3. S. T. Chang and P. G. Miles, Mushroom biology – a new discipline, *Mycologist*, 1992, **6**, 64.
4. L. Robles-Hernández, A. C. González-Franco, J. M. Soto-Parra and F. Montes-Domínguez, Review of agricultural and medicinal applications of basidiomycete mushrooms, *Tecnociencia Chihuahua*, 2008, **2**(2), 95.
5. J. G. Alday, J. A. Bonet, J. A. Oria-de-Rueda, J. Martínez-de-Aragón, J. Aldea, P. Martín-Pinto and F. Martínez-Peña, Record breaking mushroom yields in Spain, *Fungal Ecol.*, 2017, **26**, 144–146.
6. J. Carrasco, D. C. Zied, M. J. Navarro, F. J. Gea and A. Pardo-Giménez, Commercial cultivation techniques of mushrooms, in *Advances in Macrofungi*, CRC Press, 2021.
7. R. Voces, L. Diaz-Balteiro and Ó. Alfranca, Demand for wild edible mushrooms. The case of Lactarius deliciosus in Barcelona (Spain), *J. For. Econ.*, 2012, **18**(1), 47–60.
8. P. de Frutos, B. Rodríguez-Prado, J. Latorre and F. Martínez-Peña, Environmental valuation and management of wild edible mushroom picking in Spain, *For. Policy Econ.*, 2019, **100**, 177.
9. P. Ortega-Martínez and F. Martínez-Peña, A sampling method for estimating sporocarps production of wild edible mushrooms of social and economic interest, *Invest. Agrar. Sist. Recur. For.*, 2008, **17**(3), 228.
10. A. I. Ali and I. Das, Tribal situation in north east India, *Stud. Tribes Tribals*, 2003, **1**(2), 141.
11. S. K. Jain, V. Kumar and M. Saharia, Analysis of rainfall and temperature trends in northeast India, *Int. J. Biometeorol.*, 2013, **33**(4), 968.
12. K. H. Procter, K. Haridasan and G. W. Smith, How far does lowland tropical rainforest go, *Global Ecol. Biogeogr.*, 1998, **7**, 141.
13. J. Bhattacharjee, D. Bhattacharjee, T. Paul, A. Kumar and S. Chowdhury, Diversity of mushrooms in indo-Bangladesh region of north-East India, *J. Andaman Sci. Assoc.*, 2015, **19**(1), 75.

14. K. Apshahana and A. Sharma, Wild edible mushrooms traded by ethnic communities of Senapati and Kangpokpi district of Manipur, India, *J. Pharmacogn. Phytochem.*, 2018, 7(1), 2303.
15. T. C. Sarma, I. Sarma and B. N. Patiri, Wild edible mushrooms used by some ethnic tribes of Western Assam, *Bioscan*, 2010, **3**, 613.
16. S. Deb and R. K. Singh, Ethnomycological studies of wild mushrooms in Arunachal Pradesh, in *Bioresources and Traditional Knowledge of Northeast India*, ed. K. K. Singh, K. C. Das and H. Lalruatsanga, Post Graduate Science Society, Mizo, 2013.
17. Lallawmsanga, A. K. Passari, V. K. Mishra, V. V. Leo, B. P. Singh, G. Valliammai Meyyappan, V. K. Gupta, S. Uthandi and R. C. Upadhyay, Antimicrobial potential, identification and phylogenetic affiliation of wild mushrooms from two sub-tropical semi-evergreen indian forest ecosystems, *PLoS One*, 2016, **11**, e0166368.
18. D. Agrahar-Murugkar and G. J. F. C. Subbulakshmi, Nutritional value of edible wild mushrooms collected from the Khasi hills of Meghalaya, *Food Chem.*, 2005, **89**(4), 599.
19. H. Lalrinawmi, J. M. C. Vabeikhokhei and J. Zothanzama, Edible mushrooms of Mizoram, *Sci. Vision*, 2017, **17**(3), 172.
20. T. Ao and C. R. Deb, Wild Mushrooms of Nagaland, India–An Important Bioresource, *Stud. Fungi.*, 2019, **4**(1), 61–78.
21. R. Singh, R. Bhagawati, P. K. Sharma and Y. Ramakrishna, Wild Edible Fungal Resources: An Alternate Source of Food for Mizoram and Arunachal Pradesh, *Environ. Ecol.*, 2015, **33**, 1936.
22. E. Lieckfeldt, C. M. Kullnig, C. P. Kubicek, G. J. Samuels and T. Börner, *Trichoderma aureoviride*: phylogenetic position and characterization, *Mycol. Res.*, 2001, **105**(3), 313.
23. Ó. Garcia and J. M. Garcia, *Guía de setas de la provincia de Cuenca*, Ediciones de la Universidad de Castilla La Mancha, 2019.
24. U. Kües and M. Navarro-Gonzalez, How do Agaricomycetes shape their fruiting bodies? 1. Morphological aspects of development, *Fungal Biol. Rev.*, 2015, **29**(2), 63.
25. A. Wibowo, Y. Rahayu, A. Riyanto and T. Hidayatulloh, Classification algorithm for edible mushroom identification, in *2018 International Conference on Information and Communications Technology (ICOIACT)*, 2018.
26. E. S. Alkronz, K. A. Moghayer, M. Meimeh, M. Gazzaz, B. S. Abu-Nasser and S. S. Abu-Naser, Prediction of whether mushroom is edible or poisonous using back-propagation neural network, *IJAAR*, 2019, **3**(2), 1.
27. H. S. Ro, S. S. Kim, J. San Ryu, C. O. Jeon, T. S. Lee and H. S. Lee, Comparative studies on the diversity of the edible mushroom *Pleurotus eryngii*: ITS sequence analysis, RAPD fingerprinting, and physiological characteristics, *Mycol. Res.*, 2007, **111**(6), 710.
28. R. Kölliker, E. S. Jones, M. C. Drayton, M. P. Dupal and J. W. Forster, Development and characterisation of simple sequence repeat (SSR) markers for white clover (*Trifolium repens* L.), *Theor. Appl. Genet.*, 2001, **102**, 416.

29. S. M. Khan, A. Nawaz, W. Malik, N. Javed, T. Yasmin, M. ur Rehman and A. A. Khan, Morphological and molecular characterization of Oyster mushroom (*Pleurotus* spp.), *Afr. J. Biotechnol.*, 2011, **10**(14), 2638.
30. P. Gepts, The use of molecular and biochemical markers in crop evolution studies, *Evol. Biol.*, 1993, 51.
31. Z. Mei, L. Yang, M. A. Khan, M. Yang, C. Wei, W. Yang and J. Fu, Genotyping of Ganoderma species by improved random amplified polymorphic DNA (RAPD) and inter-simple sequence repeat (ISSR) analysis, *Biochem. Syst. Ecol.*, 2014, **56**, 40.
32. A. A. Bhat, M. A. Haniffa, P. R. Divya, A. Gopalakrishnan, M. J. Milton, R. Kumar and B. A. Paray, Molecular characterization of eight Indian Snakehead species (Pisces: *Perciformes Channidae*) using RAPD markers, *Mol. Biol. Rep.*, 2012, **39**(4), 4267.
33. M. Shakeel, M. Ilyas and M. Kazi, Evaluation of synthetic hexaploid wheats (derivative of durum wheats and Aegilops tauschii accessions) for studying genetic diversity using randomly amplified polymorphic DNA (RAPD) markers, *Mol. Biol. Rep.*, 2013, **40**(1), 21.
34. P. Du, B. K. Cui, C. F. Zhang and Y. C. Dai, Genetic diversity of wild *Auricularia auricula-judae* revealed by ISSR analysis, *Biochem. Syst. Ecol.*, 2013, **48**, 199.
35. B. Sikdar, M. Bhattacharya, A. Mukherjee, A. Banerjee, E. Ghosh, B. Ghosh and S. C. Roy, Genetic diversity in important members of Cucurbitaceae using isozyme, RAPD and ISSR markers, *Biol. Plant.*, 2010, **54**(1), 135.
36. X. J. Guan, L. Xu, Y. C. Shao, Z. R. Wang, F. S. Chen and X. C. Luo, Differentiation of commercial strains of *Agaricus* species in China with inter-simple sequence repeat marker, *World J. Microbiol. Biotechnol.*, 2008, **24**(8), 1617.
37. L. Tang, Y. Xiao, L. Li, Q. Guo and Y. Bian, Analysis of genetic diversity among Chinese *Auricularia auricula* cultivars using combined ISSR and SRAP markers, *Curr. Microbiol.*, 2010, **61**(2), 132.
38. L. Z. Fu, H. Y. Zhang, X. Q. Wu, H. B. Li, H. L. Wei, Q. Q. Wu and L. A. Wang, Evaluation of genetic diversity in *Lentinula edodes* strains using RAPD, ISSR and SRAP markers, *World J. Microbiol. Biotechnol.*, 2010, **26**(4), 709.
39. Q. S. Zhang, B. L. Xu, L. D. Liu, Q. Q. Yuan, H. X. Dong, X. H. Cheng and D. L. Lin, Analysis of genetic diversity among Chinese *Pleurotus citrinopileatus* Singer cultivars using two molecular marker systems (ISSRs and SRAPs) and morphological traits, *World J. Microbiol. Biotechnol.*, 2012, **28**(5), 2237.
40. S. Wang, Y. Yin, Y. Liu and F. Xu, Evaluation of genetic diversity among Chinese *Pleurotus eryngii* cultivars by combined RAPD/ISSR marker, *Curr. Microbiol.*, 2012, **65**(4), 424.
41. A. K. Sahoo, K. B. Mohapatra, B. B. Behera, K. R. Jadhao and G. R. Rout, Phylogenetic analysis of high yielding strain of paddy straw mushroom

(*Volvariella volvacea*) by using ISSR markers, *Middle-East J. Sci. Res.*, 2014, **21**(8), 1197.

42. G. Li and C. F. Quiros, Sequence-related amplified polymorphism (SRAP), a new marker system based on a simple PCR reaction: its application to mapping and gene tagging in Brassica, *Theor. Appl. Genet.*, 2001, **103**(2), 455–461.
43. P. Du, B. K. Cui and Y. C. Dai, Genetic diversity of wild Auricularia polytricha in Yunnan Province of South-western China revealed by sequence-related amplified polymorphism (SRAP) analysis, *J. Med. Plants Res.*, 2011, **5**(8), 1374.
44. Z. Wang, M. Chen, Z. Cai, J. Liao, H. Li, Z. Guo and Z. Lu, DNA fingerprinting of genetic diversity of Agaricus bisporus, in *Mushroom biology and mushroom products. Proceedings of the 7th International Conference on Mushroom Biology and Mushroom Products*, Institut National de la Recherche Agronomique (INRA), Arcachon, France, 4–7 October, 2011, vol. 2, p. 1.
45. S. J. Sun, W. Gao, S. Q. Lin, J. Zhu, B. G. Xie and Z. B. Lin, Analysis of genetic diversity in *Ganoderma* population with a novel molecular marker SRAP, *Appl. Microbiol. Biotechnol.*, 2006, **72**(3), 537.
46. J. Y. Liu, Z. H. Ying, F. Liu, X. R. Liu and B. G. Xie, Evaluation of the use of SCAR markers for screening genetic diversity of *Lentinula edodes* strains, *Curr. Microbiol.*, 2012, **64**(4), 317.
47. Y. Yin, Y. Liu, H. Li, S. Zhao, S. Wang, Y. Liu and F. Xu, Genetic diversity of *Pleurotus pulmonarius* revealed by RAPD, ISSR, and SRAP fingerprinting, *Curr. Microbiol.*, 2014, **68**(3), 397.
48. D. Ma, G. Yang, L. Mu and Y. Song, Application of SRAP in the genetic diversity of *Tricholoma matsutake* in northeastern China, *Afr. J. Biotechnol.*, 2010, **9**(38), 6244.
49. D. Tautz and M. Renz, Simple sequences are ubiquitous repetitive components of eukaryotic genomes, *Nucleic Acids Res.*, 1984, **12**(10), 4127.
50. R. Zhang, D. Hu, J. Zhang, X. Zuo, R. Jiang, H. Wang and T. B. Ng, Development and characterization of simple sequence repeat (SSR) markers for the mushroom *Flammulina velutipes*, *J. Biosci. Bioeng.*, 2010, **110**(3), 273–275.
51. J. Qian, H. Xu, J. Song, J. Xu, Y. Zhu and S. Chen, Genome-wide analysis of simple sequence repeats in the model medicinal mushroom *Ganoderma lucidum*, *Gene*, 2013, **512**(2), 331.
52. J. A. Rafalski, M. Morgante, W. Powell, J. M. Vogel and S. V. Tingey, Generating and using DNA markers in plants, in *Analysis of Nonmanmaian Genomes: A practical Guide*, ed. B. Birren and E. Lai, Academic Press, Boca Raton, FL, 1996, pp. 75–134.
53. Y. Xiao, W. Liu, Y. Dai, C. Fu and Y. Bian, Using SSR markers to evaluate the genetic diversity of 'Lentinula edodes' natural germplasm in China, *World J. Microbiol. Biotechnol.*, 2010, **26**(3), 527.
54. R. Y. Zhang, D. D. Hu, J. G. Gu, Q. X. Hu, X. M. Zuo and H. X. Wang, Development of SSR markers for typing cultivars in

the mushroom *Auricularia auricula-judae*, *Mycol. Prog.*, 2012, **11**(2), 587–592.

55. R. Xing, Q. B. Gao, F. Q. Zhang, Y. H. Li, P. C. Fu, J. H. Zhang and S. L. Chen, Genetic diversity and population structure of *Armillaria luteo-virens* (Physalacriaceae) in Qinghai-Tibet Plateau revealed by SSR markers, *Biochem. Syst. Ecol.*, 2014, **56**, 1.
56. F. Yuan, H. Yu, S. Zuo and A. Adams, Development of EST-SSR for preliminary analysis of genetic diversity of Cordyceps militaris, *Biochem. Syst. Ecol.*, 2015, **58**, 126–131.
57. M. J. Blears, S. A. De Grandis, H. Lee and J. T. Trevors, Amplified fragment length polymorphism (AFLP): a review of the procedure and its applications, *J. Ind. Microbiol. Biotechnol.*, 1998, **21**(3), 99.
58. A. Pawlik, G. Janusz, J. Koszerny, W. Małek and J. Rogalski, Genetic diversity of the edible mushroom *Pleurotus* sp. by amplified fragment length polymorphism, *Curr. Microbiol.*, 2012, **65**(4), 438.
59. L. Wang, X. Hu, Z. Feng and Y. Pan, Development of AFLP markers and phylogenetic analysis in Hypsizygus marmoreus, *J. Gen. Appl. Microbiol.*, 2009, **55**(1), 9.
60. N. J. Brazee, R. E. Marra and R. L. Wick, Genotypic diversity of Armillaria gallica from mixed oak forests in Massachusetts, *Mycologia*, 2012, **104**(1), 53.
61. M. Foulongne-Oriol, R. Dufourcq, C. Spataro, C. Devesse, A. Broly, A. Rodier and J. M. Savoie, Comparative linkage mapping in the white button mushroom *Agaricus bisporus* provides foundation for breeding management, *Curr. Genet.*, 2011, **57**(1), 39.
62. K. Mukhopadhyay, I. Haque, R. Bandopadhyay, S. Covert and D. Porter, AFLP based assessment of genetic relationships among shiitake (Lentinula ssp.) mushrooms, *Mol. Biol. Rep.*, 2012, **39**(5), 6059–6065.
63. D. Li, Y. Liu, P. Wang, Y. Ma, S. Wang, S. Zhao and F. Xu, Development of SCAR markers to determine the mating types of *Lepista nuda* protoplast monokaryons, *Curr. Microbiol.*, 2014, **68**(4), 536.
64. H. Su, L. Wang, Y. Ge, E. Feng, J. Sun and L. Liu, Development of strain-specific SCAR markers for authentication of *Ganoderma lucidum*, *World J. Microbiol. Biotechnol.*, 2008, **24**(7), 1223.
65. Y. Xiao, Y. Dai, Y. Lu, W. Liu, Z. Wang and Y. Bian, Development of IRAP-SCAR marker for strain identification in *Lentinula edodes*, *World J. Microbiol. Biotechnol.*, 2011, **27**(7), 1731.
66. A. J. Castle, P. A. Horgen and J. B. Anderson, Restriction fragment length polymorphisms in the mushrooms *Agaricus brunnescens* and *Agaricus bitorquis*, *Appl. Environ. Microbiol.*, 1987, **53**(4), 816.
67. J. Xu, R. W. Kerrigan, P. Callac, P. A. Horgen and J. B. Anderson, Genetic structure of natural populations of Agaricus bisporus, the commercial button mushroom, *J. Hered.*, 1997, **88**(6), 482.
68. J. Xu, R. W. Kerrigan, A. S. Sonnenberg, P. Callac, P. A. Horgen and J. B. Anderson, Mitochondrial DNA variation in natural populations of the mushroom *Agaricus bisporus*, *Mol. Ecol.*, 1998, 7(1), 19.

69. S. Urbanelli, V. D. Rosa, F. Punelli, D. Porretta, M. Reverberi, A. A. Fabbri and C. Fanelli, DNA-fingerprinting (AFLP and RFLP) for genotypic identification in species of the *Pleurotus eryngii* complex, *Appl. Microbiol. Biotechnol.*, 2007, **74**(3), 592.
70. X. J. Guan, L. Xu, Y. C. Shao, Z. R. Wang, F. S. Chen and X. C. Luo, Differentiation of commercial strains of *Agaricus* species in China with inter-simple sequence repeat marker, *World J. Microbiol. Biotechnol.*, 2008, **24**(8), 1617.
71. M. Manassila, T. Sooksa-Nguan, N. Boonkerd, S. Rodtong and N. Teaumroong, Phylogenetic diversity of wild edible *Russula* from northeastern Thailand on the basis of internal transcribed spacer sequence, *Sci. Asia*, 2005, **31**, 323.
72. T. J. White, T. Bruns, S. J. W. T. Lee and J. Taylor, Amplification and direct sequencing of fungal ribosomal RNA genes for phylogenetics, *PCR Protoc.: Guide Methods Appl.*, 1990, **18**(1), 315.
73. R. S. Hseu, H. H. Wang, H. F. Wang and J. M. Moncalvo, Differentiation and grouping of isolates of the *Ganoderma lucidum* complex by random amplified polymorphic DNA-PCR compared with grouping on the basis of internal transcribed spacer sequences, *Appl. Environ. Microbiol.*, 1996, **62**(4), 1354.
74. D. S. Tuckwell, M. J. Nicholson, C. S. McSweeney, M. K. Theodorou and J. L. Brookman, The rapid assignment of ruminal fungi to presumptive genera using ITS1 and ITS2 RNA secondary structures to produce group-specific fingerprints, *Microbiology*, 2005, **151**(5), 1557.
75. Q. X. Wu, G. M. Mueller, F. M. Lutzoni, Y. Q. Huang and S. Y. Guo, Phylogenetic and biogeographic relationships of eastern Asian and eastern North American disjunct Suillus species (Fungi) as inferred from nuclear ribosomal RNA ITS sequences, *Mol. Phylogenet. Evol.*, 2000, **17**(1), 37.
76. T. Oda, C. Tanaka and M. Tsuda, Molecular phylogeny of Japanese Amanita species based on nucleotide sequences of the internal transcribed spacer region of nuclear ribosomal DNA, *Mycoscience*, 1999, **40**(1), 57.
77. M. Fukuda, S. Ohno and M. Kato, Genetic variation in cultivated strains of *Agaricus blazei*, *Mycoscience*, 2003, **44**, 431–436.
78. P. S. Yan, X. C. Luo and Q. Zhou, RAPD molecular differentiation of the cultivated strains of the jelly mushrooms, *Auricularia auricula* and *A. polytricha*, *World J. Microbiol. Biotechnol.*, 2004, **20**, 795–799.
79. N. Alam, M. J. Shim, M. W. Lee, P. G. Shin, Y. B. Yoo and T. S. Lee, Vegetative Growth and Phylogenetic Relationship of Commercially Cultivated Strains of *Pleurotus eryngii*, *Mycobiology*, 2009, **37**(4), 258–266.
80. G. I. Zervakis, G. Venturella and K. Papadopoulou, Genetic polymorphism and taxonomic infrastructure of the *Pleurotus eryngii* species complex as determined by RAPD analysis, isozyme profiles and eco-morphological characters, *Microbiology*, 2001, **147**, 3183–3194.

81. R. Zhang, C. Huang, S. Zheng, J. Zhang, T. B. Ng, R. Jiang, X. Zuo and H. Wang, Strain-typing of *Lentinula edodes* in China with inter simple sequence repeat markers, *Appl. Microbiol. Biotechnol.*, 2007, **74**, 140–145.
82. K. Malekzadeh, B. J. M. Shahri and E. Mohsenifard, Use of ISSR markers for strain identification in the button mushroom, Agaricusbisporus, in *Proceedings of the 7th International Conference on Mushroom Biology and Mushroom Products (ICMBMP7)*, 2011.
83. M. I. Nazrul and Y. B. Bian, ISSR as New Markers for Identification of Homokaryotic Protoclones of *Agaricus bisporus*, *Curr. Microbiol.*, 2010, **60**, 92–98.
84. M. Yu, B. Ma, X. Luo, L. Zheng, X. Xu and Z. Yang, Molecular Diversity of *Auricularia polytricha* revealed by Inter-Simple Sequence Repeat and Sequence-Related Amplified Polymorphism Markers, *Curr. Microbiol.*, 2008, **56**, 240–245.
85. H. Su, L. Wang, L. Liu, X. Chi and Y. Zhang, Use of inter-simple sequence repeat markers to develop strain-specific SCAR markers for *Flammulina velutipes*, *J. Appl. Genet.*, 2008, **49**(3), 233–235.
86. C. Qiu, W. Yan, P. Li1, W. Deng, B. Song and T. Li1, Evaluation of Growth Characteristics and Genetic Diversity of Commercial and Stored Lines of *Hypsizygus marmoreus*, *Int. J. Agric. Biol.*, 2013, **15**(3), 479–485.
87. M. I. Nazrul and Y. B. Bian, Differentiation of homokaryons and heterokaryons of *Agaricus bisporus* with inter-simple sequence repeat markers, *Microbiol. Res.*, 2011, **166**, 226–236.
88. M. Foulongne-Oriol, C. Spataro and J. M. Savoie, Novel microsatellite markers suitable for genetic studies in the white button mushroom *Agaricus bisporus*, *Appl. Microbiol. Biotechnol.*, 2009, **84**(6), 1125–1135.
89. J. Qian, H. Xu, J. Song, J. Xu, Y. Zhu and S. Chen, Genome-wide analysis of simple sequence repeats in the model medicinal mushroom *Ganoderma lucidum*, *Gene*, 2013, **512**, 331–336.
90. J. Labbé, C. Murat, E. Morin, F. L. Tacon and F. Martin, Survey and analysis of simple sequence repeats in the *Laccaria bicolor* genome, with development of microsatellite markers, *Curr. Genet.*, 2011, **57**, 75–88.
91. K. Terashima, T. Matsumoto, E. Hayashi and Y. F. Nakai, A genetic linkage map of *Lentinula edodes* (shiitake) based on AFLP markers, *Mycol. Res.*, 2002, **106**(8), 911–917.
92. T. C. T. Loa, M. W. Kanga, B. C. Wang and C. A. Changa, Glycosyl linkage characteristics and classifications of exo-polysaccharides of some regionally different strains of *Lentinula edodes*, *Anal. Chim. Acta*, 2007, **592**, 146–153.
93. M. C. Yadav, R. K. Sharma, S. K. Singh and T. Mohapatra, Molecular Differentiation of Sexually Incompatible Strains of *Agaricus bitorquis* Using RAPD and AFLP Markers, *J. Plant Biochem. Biotechnol.*, 2006, **15**, 109–116.

94. I. H. Chapela and M. Garbelotto, Phylogeography and evolution in matsutake and close allies inferred by analyses of ITS sequences and AFLPs, *Mycologia*, 2004, **96**(4), 730–741.
95. S. Y. Kim, K. H. Kim, C. H. Im, A. Ali and C. Y. Lee, Identification of Degenerate Nuclei and Development of a SCAR Marker for *Flammulina velutipes*, *PLoS One*, 2014, **9**(9), e107207.
96. X. Wu, H. Li, W. Zhao, L. Fu, H. Peng, L. He, J. Cheng, H. Wei and Q. Wu, SCAR Makers and Multiplex PCR-Based Rapid Molecular Typing of *Lentinula edodes* Strains, *Curr. Microbiol.*, 2010, **61**, 381–389.
97. H. B. Li, X. Q. Wu, H. Z. Peng, L. Z. Fu, H. L. Wei, Q. Q. Wu, Q. Y. Jin and N. Li, New available SCAR markers: potentially useful in distinguishing a commercial strain of the superior type from other strains of *Lentinula edodes* in China, *Appl. Microbiol. Biotechnol.*, 2008, **81**, 303–309.
98. K. Miyazaki, A. Tanaka and S. Shiraishi, Mapping of a SCAR marker tightly linked to the B mating type locus of *Lentinula edodes* (shiitake mushroom) by tetrad analysis, *Micol. Apl. Int.*, 2008, **20**(2), 47–54.
99. S. W. Chid, M. Chen and S. T. Chang, Differentiating homothallic *Volvariella* mushrooms by RFLPs and AP-PCR, *Mycol. Res.*, 1995, **99**(3), 333–336.
100. S. W. Chiu, A. M. Maz, F. C. Lin and D. Moore, Genetic homogeneity of cultivated strains of shiitake (*Lentinula edodes*) used in China as revealed by the polymerase chain reaction, *Mycol. Res.*, 1996, **100**(11), 1393–1399.
101. D. Bao, H. Ishihara, N. Mori and Y. Kitamoto, Phylogenetic analysis of oyster mushrooms (*Pleurotus* spp.) based on restriction fragment length polymorphisms of the 5 portion of 26S rDNA, *J. Wood Sci.*, 2004, **50**, 169–176.
102. J. X. Zhang, C. Y. Huang, T. B. Ng and H. X. Wang, Genetic polymorphism of ferula mushroom growing on *Ferula sinkiangensis*, *Appl. Microbiol. Biotechnol.*, 2006, **71**, 304–309.
103. R. K. Kulkarni, DNA Polymorphisms in *Lentinula edodes*, the Shiitake Mushroom, *Appl. Environ. Microbiol.*, 1991, **6**, 1735–1739.
104. T. Matsumoto and Y. S. Nakai, Mitochondrial DNA restriction fragment length polymorphisms and phonetic relationships in natural populations of the oyster mushroom, *Pleurotus ostreatus*, *Mycol. Res.*, 1995, **99**(5), 562–566.
105. Y. Liu, S. Wang, Y. Yin, H. Jin, S. Zhao, M. Li, P. Wang, X. Geng and F. Xu, Use of a culture independent method to analyze the diversity of soil fungi surrounding *Chroogomphus rutilus* in the Beijing region of China, *Ann. Microbiol.*, 2012, **62**, 1743–1749.
106. K. S. Ko and H. S. Jung, Three nonorthologous ITS1 types are present in a polypore fungus *Trichaptum abietinum*, *Mol. Phylogenet. Evol.*, 2002, **23**, 112–122.
107. G. I. Zervakis, J. M. Moncalvo and R. Vilgalys, Molecular phylogeny, biogeography and speciation of the mushroom species *Pleurotus cystidiosus* and allied taxa, *Microbiology*, 2004, **150**, 715–726.

108. A. Kretzer, Y. Li, T. Szaro and T. D. Bruns, Internal Transcribed Spacer Sequences from 38 Recognized Species of *Suillus sensulato*: Phylogenetic and Taxonomic Implications, *Mycologia*, 1996, **88**(5), 776–785.
109. D. Wipf, J. C. Munch, B. Botton and F. Buscot, DNA Polymorphism in Morels: Complete Sequences of the Internal Transcribed Spacer of Genes Coding for rRNA in *Morchella esculenta* (Yellow Morel) and *Morchella conica* (Black Morel), *Appl. Environ. Microbiol.*, 1996, **62**(9), 3541–3543.
110. H. H. Chiu, Phylogenetic analysis of *Antrodia* species and *Antrodia camphorata* inferred from internal transcribed spacer region, *Antonie van Leeuwenhoek*, 2007, **91**, 267–276.
111. D. B. Choi, J. L. Ding and W. S. Cha, Homology search of genus *Pleurotus* using an internal transcribed spacer region, *Korean J. Chem. Eng.*, 2007, **24**(3), 408–412.

CHAPTER 2

Update on Research Data on the Nutrient Composition of Mushrooms and Their Potentials in Future Human Diets

MILENA J. RAŠETA,[a] MILANA S. RAKIĆ,[b] ELEONORA V. ČAPELJA[b] AND MAJA A. KARAMAN*[b]

[a] Department of Chemistry, Biochemistry and Environmental Protection, Faculty of Sciences, University of Novi Sad, Trg D. Obradovića 3, 21000 Novi Sad, Serbia; [b] Department of Biology and Ecology, Faculty of Sciences, University of Novi Sad, Trg D. Obradovića 2, 21000 Novi Sad, Serbia
*Email: maja.karaman@dbe.uns.ac.rs

2.1 Introduction

Mushrooms have been recognized as a precious food since ancient times, including by the Greeks and Romans, while the most popular edible species, such as *Amanita caesarea* (Caesar's mushroom), *Boletus edulis* (porcini), or *Morchella* spp. (morel), and *Tuber* (truffle) have been of the highest interest in human diets mostly due to their organoleptic, aphrodisiac, and health-promoting properties, including some species with hallucinogenic effects (*Psilocybe*). *Mycophagy* is a term that describes the human habit of mushroom consumption that was used from 1887 when the British mycologist Hay proposed two terms, *mycophilia* and *mycophobia*, which are still in use.

Food Chemistry, Function and Analysis No. 36
Edible Fungi: Chemical Composition, Nutrition and Health Effects
Edited by Dejan Stojković and Lillian Barros

Published by the Royal Society of Chemistry, www.rsc.org

Nowadays, mushrooms are considered a healthy food due to their high content of nutrients including proteins, carbohydrates, vitamins, fatty acids (FA), amino acids (AA), and mineral elements.[1] Generally, it is known that *mushrooms are low in calories, high in fiber, and contain many important vitamins and minerals*. Many also have medicinal constituents (metabolites) such as complex carbohydrates that strengthen the immune system. In this chapter we attempt to get closer to the world of mycophilic societies (some European, such as Italian, French, Serbian, Croatian, and Macedonian, some Asian societies like Chinese and Indian, as well as Mexican) referring to the peoples' appreciation of mushroom flavor, texture, and healthy properties, thus resulting in them using them in a regular diet based on traditional experiences.

Modern science relies on the huge experience of ethnomycology in respect to traditional usage of mushrooms as food, medicine, poison, and in spiritual practices across the world since 5000 BC.[2]

Here are some facts about the edibility of mushrooms which are generally recognized and can be found in many public sources on the internet, especially in terms of being useful for vegetarian diets. With respect to nutrition, mushrooms contain various metabolites: **proteins** – a high protein content (mostly 20–30% on dry weight [d.w.]), **fiber** – which helps in lowering cholesterol and improves digestive system functionality, niacin (B3) and other important **B vitamins** [thiamine (B1), riboflavin (B2), pantothenic acid (B5)] – which are reported to be good for the nervous system and can be a good supplement for vegetarians since certain B vitamins (*e.g.* cobalamin – B12) are not found in plants, **vitamin D** – essential for the absorption of calcium. Besides, their concentration of C and D vitamins reaches a 5–10 higher concentration Than the B3 level. **Copper** – aids in helping the body absorb oxygen and create red blood cells. **Selenium** – an antioxidant that neutralizes free radicals, thus preventing cell damage and reducing the risk of cancer and other diseases. Mushrooms contain more selenium than any other form of food. **Potassium** – regulates blood pressure and keeps cells properly functioning. A large portobello mushroom (*A. bisporus*) has been proven to have more potassium than a banana. **Other important minerals** – phosphorus (P), zinc (Zn), and magnesium (Mg). Furthermore, mushrooms are recognized as food with low levels of fat, calories, and sodium, and without cholesterol as they contain ergosterols in cell membranes. Among the 2000 types of edible mushrooms that are used for their nutritional properties, almost all possess various metabolites such as phenolics, flavonoids, polyketides, carotenoids, terpenoids, quinones, and steroids,[3] which play a crucial role in their antioxidant activity and account for their medicinal value.[4–6]

According to recent scientific reports, mushrooms constitute at least 12 000–16 000 species worldwide, while 3283 mushroom species have been confirmed as edible or conditionally edible, accounting for ~20% of all mushroom taxa recorded globally.[7] About 35 edible mushroom species are commercially cultivated, whereas nearly 200 wild species are used for medicinal purposes. *Agaricus bisporus* is the most cultivated mushroom worldwide followed by *Lentinula edodes* and *Pleurotus* spp.[8]

With respect to the fact that mushrooms are generally regarded as rich dietary sources, their composition varies greatly depending on the following factors: taxonomic position (genera of the mushrooms), growing habitat (wild or cultivated), geographical location, biological factors, physiological growth stages: fruiting bodies, mycelium, and sclerotium, as well as morphological factors, such as the cap, stalk, or hymenophore.[9–11] Due to the high water content (85–95%) mushrooms are extremely perishable foods, while their dry mass varies between 5–15%. Furthermore, the fruiting bodies contain approximately 70–95% moisture, carbohydrates (50–65%), proteins (19–35%), and some fats (2–6%), among which are many bioactive constituents, such as polysaccharides, proteins, essential AAs, unsaturated fatty acids (oleic and linoleic), vitamins, and dietary fiber.[12] In this respect, the protein content of mushrooms is ranked as higher than most other food sources, except meat.[11] Therefore, presented herein is an update of more than 100 scientifically proven results for the period from 2010 to 2021 dealing with the nutritional value of wild-growing mushrooms used across the globe with special emphasis on the following species: *Agrocybe aegerita*, *Agaricus bisporus*, *Boletus edulis*, *Cantharellus cibarius*, *Coprinus comatus*, *C. micaceus*, *C. truncorum*, *Laetiporus sulphureus*, *Macrolepiota procera*, *Meripilus giganteus*, *Pleurotus ostreatus*, and *Volvopluteus gloiocephalus* (Figure 2.1).

Since mushrooms are known to be valuable health foods, not only for their unique and subtle flavor, but also for their nutritional properties,[13] we first explain another important term with respect to this topic – "nutraceutical." This term was established in 1979 by Stephen DeFelice, founder and chairman of the foundation for Innovation in Medicine from Cranford (New Jersey). The word "nutra" comes from nutrition and "ceutical" from pharmaceutical refers

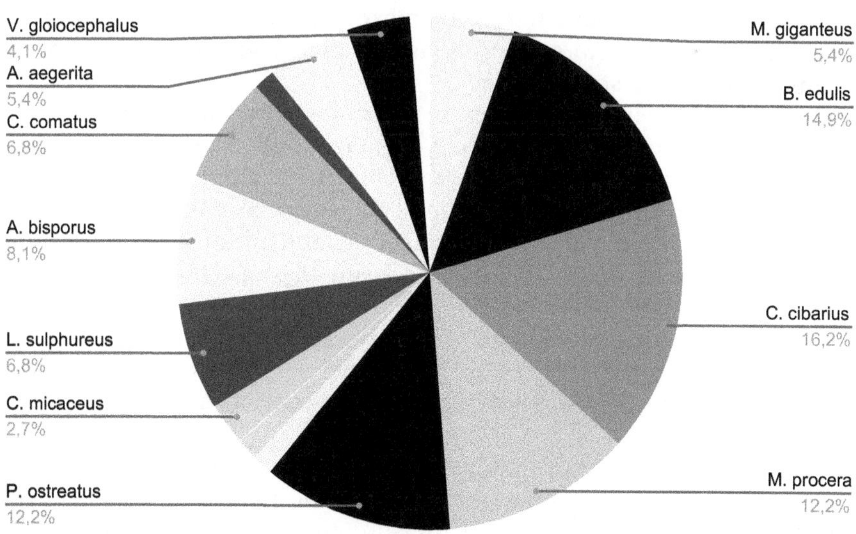

Figure 2.1 Representation percentage of species investigated in this chapter during the period 2010–2021.

to chemical compounds that may be considered a food or food supplement that may provide health benefits, including the prevention and treatment of various diseases.[14]

The results of the nutritional composition in terms of selected nutrients (proteins, AA, fats, carbohydrates, fibers, vitamins, nucleic acids, and minerals) for the 12 most common wild edible mushroom species are shown in Tables 2.1–2.5.

Some commercially available mushroom nutraceuticals include Organic ReiShi-Gen (a synergistic blend of 50% *L. edodes* and 50% *G. lucidum*) which is a daily health supplement providing immune strength. However, the production of nutraceuticals requires deep knowledge about the structure and functional relationship, an economically viable, controllable production process, and stable biologically active compound quality and quantity.[8]

This review aims to establish the nutritional properties in terms of selected nutraceuticals from 12 selected species of wild-growing edible mushrooms with a few from cultivated strains from different countries around the world (Serbia, Croatia, Portugal, Nigeria, Turkey, Macedonia, Mexico, Romania, India, Japan, China, Tunisia, Spain, Montenegro, Finland, Pakistan, and Poland) with an emphasis on updates on their nutritional value and detailed nutritional composition.

2.2 Major Nutritional Components of Selected Mushrooms

2.2.1 Proteins and Amino Acids

The nutritional value of mushrooms is primarily related to their protein and carbohydrate contents.[11,15] After the water content 70–95%, proteins are the predominant constituents in mushrooms (d.w.). Their content is dependent on environmental factors as well as on the stage of maturity of the basidiocarp and the species type.[16]

Mushroom proteins have higher nutritional quality than plant proteins as reported by the Food and Agriculture Organization,[17] however, their quality is lower than that from animal sources (for example fish or eggs).[18] Mushrooms rich in proteins contain all the essential AAs with the highest concentrations of glutamic acid, aspartic acid, and arginine.[8] Recently it has been reported that threonine (Thr) and lysine (Lys) are the major essential free AAs found in all mushrooms.

Mushroom fruiting bodies are a rich source of **ergothioneine**, an important AA that must be supplied with food. The species *A. bisporus* is one of the most famous sources of this AA with respect to its anticancer and antioxidative properties.[19]

Considerable amounts of non-essential AA such as alanine, arginine, glycine, glutamic acid, aspartic acid, and serine have been found in wild mushroom proteins.[20,21] Besides, mushrooms also contain several AAs (phenylalanine, lysine, isoleucine, leucine, valine, histidine, threonine, methionine, glutamic, and aspartic acids) that cannot be created by humans. It is interesting that glutamic and aspartic acids are the two essential AAs that give the specific taste of mushrooms.[11]

Besides these, the two unusual AA, γ-amino butyric acid (GABA) as a non-essential AA and ornithine, which are known for their peculiar physiological activities, can be also detected.[8] Lectins, which are a special type of glycoprotein in mushrooms, are of the utmost importance in the initial growth stages of fungi, having further protective effects on fungi against toxins or other microorganisms in the environment.[22] One of the lectins, ***ageritin*** from *A. aegerita*, has been reported to contain a great number of AAs – 135 – including Cys, Lys + Arg, and Asp + Glu (Table 2.1).[23] It was observed that the protein content of mushrooms depends on the composition of the species of mushrooms, size of the pileus, and harvest time.[24] Protein determination could represent a problem since different conversion factors have been calculated on the basis of nitrogen content, *e.g.*, the protein conversion factor due to the different digestible proteins present.[25] Proteins consist of over 20 AAs in variable amounts. Humans can convert some of these AAs into others, but nine are considered essential AAs (lysine, methionine, tryptophan, threonine, valine, leucine, isoleucine, histidine, and phenylalanine).[26] Generally, the protein content in mushrooms is between 20 to 30% d.w., while the concentration range of proteins in *A. bisporus* is between 33 to 43% d.w.[27] One hundred g of dried mushrooms contain proteins of 10.80% d.m. in *Armillaria mellea* to 83.40% d.m. in *Sparassis crispa* as compared to 7.3% in rice, 13.2% in wheat, 39.1% in soybean, and 25.2% in milk.[1] In this investigation, the protein content (g/100 g d.w.) in selected mushrooms varied between 1.29 in *A. bisporus* from Portugal to 239.1 from *B. edulis* and 123.7 from *C. cibarius* in Tunisia (Table 2.1).

Mushrooms are very useful for vegetarians because they contain some essential AAs which are found in animal proteins, while mushroom proteins are reported to contain all the essential AAs, some non-essential AAs, and amides.[1] The two main vital AAs, lysine and tryptophan, are absent in cereals but present in mushrooms in high amounts, while legumes usually lack methionine (Met) and tryptophan (Trp).[27] It has been already proven that mushrooms have good flavor, texture, and also all the essential AAs for humans, including lysine (Lys) and methionine (Met), which are commonly deficient in plants.[14] Proteins consist of over 20 AAs in variable amounts and the proportion of essential AAs is nutritionally favorable.

Humans can convert some of these AA into others, but nine of them are considered as essential AAs (lysine, methionine, tryptophan, threonine, valine, leucine, isoleucine, histidine, and phenylalanine). In order to simplify the calculations on nutritional values of mushrooms based on protein content, Crisan and Sands proposed in 1978 the use of a nutritional index calculated as follows:[1]

$$\text{Nutritional index} = (\text{EAA index percentage protein})/100$$

Table 2.1 Proteins and amino acids contents of mushrooms.[a]

Species	Region	Year of publication	Protein content (g/100 g d.w.)	Amino acid content (g/100 g d.w.)	Ref.
A. aegerita	Serbia	2015	6.68	—	43
	China	2020	24.79	—	78
	Italy	2021	**Ageritin**	135 amino acids, including Cys, Lys + Arg, Asp + Glu	23
	Ukraine	2021	19.62–21.78	—	79
A. bisporus	Cult. Portugal	2012	1.23	—	51
	Cult. Portugal	2012	1.29	—	51
	rev. art.	2013	23.90–34.80	—	46
	rev. art.	2015	14.10	—	80
	rev. art.	2016	23.90–34.80	Val 2.50, Leu 7.50, Ile 4.50, Thr 5.50, Met 0.90, Lys 9.10, Phe 4.20, Trp 2.00	14
	rev. art.	2017	14.08–15.43	—	8
	rev. art.	2019	26.30	—	30
	Pakistan	2019	30.00	—	81
	rev. art.	2020	—	His 2.70, Met 0.90, Phe 4.20, Thr 5.50, Lys 9.10, Trp 2.00, Val 2.50, Ile 4.50, Leu 7.50	27
	rev. art.	2020	3.09	—	11
	rev. art.	2021	24.43	—	12
	rev. art.	2021	1.41	—	82
	rev. art.	2021	3.09	—	42
	Chile	2021	18.53	—	83
	Chile	2021	17.25	—	83
B. edulis	Croatia	2011	36.91	—	21
	China	2014	28.70	Lys 0.099, Thr 0.211, Val 0.275, Ile 0.203, Leu 0.247, Met 0.0750, Phe 0.170, Arg 0.336, Ala 0.357, Tyr 0.050, Gly 0.365, Ser 0.128, Pro 0.272, His 0.208, Asp 0.286, Glu 0.493	16
	Tunisia	2016	239.10	—	84
	rev. art.	2016	29.70	—	14
	Finland	2016	—	Asp 0.285, Ser 0.153, Glu 0.262, Thr 0.090, Ala 0.667, Tyr 0.069, Val 0.054, Met 0.023, Phe 0.030, Ile 0.031, Leu 0.037, Lys 0.103, Gly 0.129, Arg 0.141, His 0.077, Trp 0.048, Asn 0.069, Gln 0.387 **Total amino acids:** 2.954	85

	Turkey	2018	9.145	**Total free amino acids:** 0.358	86
	rev. art.	2019	29.70	—	30
	rev. art.	2019	32.50	—	30
	South Africa	2020	21.90–38.00	—	61
	rev. art.	2021	21.07	—	12
C. cibarius	Croatia	2011	30.91	—	21
	India	2011	39.00	—	87
	rev. art.	2013	21.50	—	46
	Montenegro	2015	0.93	—	88
	Finland	2016	—	Asp 0.129, Ser 0.053, Glu 0.378, Thr 0.047, Ala 0.065, Tyr 0.043, Val 0.031, Met 0.014, Phe 0.025, Ile 0.023, Leu 0.046, Lys 0.100, Gly 0.021, Arg 0.447, His 0.113, Trp 0.032, Asn 0.029, Gln 0.474 **Total amino acids:** 2.199	85
	Tunisia	2016	123.70	—	84
	rev. art.	2016	26.25	—	14
	rev. art.	2020	—	Ala 0.498, Arg 0.044, Asp 0.006, Cys 0.199, Glu 2.999, Gly 0.013, His 0.315, Leu 0.021, Lys 0.574, Met 0.041, Ser 0.018, Thr 0.898, Phe 0.006, Val 0.134, Trp 0.002	89
	rev. art.	2021	21.57	—	12
C. comatus	Serbia	2013	11.84	—	90
	Serbia	2017	8.04–10.95 mg eq protein per ml sample*	—	4
	rev. art.	2019	25.40	—	30
	Pakistan	2019	31.00	—	81
	rev. art.	2020	11.80–29.50	—	64
	Com. Prep. Serbia	2020	23.07	—	91
	rev. art.	2021	15.67	—	12
	rev. art.	2021	22.70	—	12
C. micaceus	Nigeria	2012	14.00	—	92
	Pakistan	2019	16.00	—	81
	Nigeria	2019	17.05	—	93

Table 2.1 (*Continued*)

Species	Region	Year of publication	Protein content (g/100 g d.w.)	Amino acid content (g/100 g d.w.)	Ref.
C. truncorum	Serbia	2017	6.87–11.24 mg eq protein per ml sample*	—	4
L. sulphureus	Italy	2012	4.37	—	94
	Serbia	2014	15.97	—	95
	Turkey	2018	8.327	**Total free amino acids:** 0.363	86
	India	2019	22.73	—	96
	rev. art.	2019	11.90	—	30
	rev. art.	2021	22.73	—	12
M. giganteus	Turkey	2010	75.56	—	97
	Com. Prep. India	2013	21.20	—	98
	rev. art.	2019	75.56	—	30
M. procera	Croatia	2011	24.22	—	21
	Italy	2012	3.01	—	94
	Portugal	2013	19.80	—	29
	rev. art.	2019	26.35	—	30
	Serbia	2019	29.20	—	31
	Morocco	2021	—	3.57	99
	Chile	2021	8.56	—	83
P. ostreatus	Croatia	2011	24.90	Ala 0.645, Arg 0.011, Asp 0.017, Cys 0.181, Glu 4.109, Gly 0.034, His 0.442, Leu 0.029, Lys 0.465, Met 0.012, Phe 0.009, Ser 0.003, Thr 0.699, Trp 0.001, Val 0.121 **Total amino acids:** 6.778	21
	Cult. Portugal	2012	0.76	—	79
	China	2012	17.00–42.00	Leu 1.64, Val 1.05, Lys 1.13, Ile 0.99, Thr 0.94, Tyr 0.69, Met 0.27, Phe 1.11	28
	rev. art.	2013	10.50–30.40	—	46
	rev. art.	2015	7.00	—	80
	rev. art.	2016	10.50–30.40		14

				Val 0.91, Leu 1.13, Ile 0.62, Thr 1.10, Met 0.28, Lys 0.71, Phe 0.73, Trp 0.15	
	Cult. China	2017	20.46–26.13	Asp 0.86–1.32, Thr 0.42–0.67, Ser 0.47–0.71, Glu 3.43–5.96, Gly 0.48–0.79, Ala 0.56–0.90, Cys 0.01–0.03, Val 0.42–0.69, Met 0.16–0.28, Ile 0.37–0.58, Leu 0.60–0.95, Tyr 0.22–0.48, Phe 0.37–0.57, His 1.16–1.78, Lys 0.48–0.90, Arg 0.43–0.74, Pro 0.14–0.24 **Total essential AA:** 2.82–4.64 **Total AA:** 10.81–17.57	100
	rev. art.	2017	28.40	—	8
	rev. art.	2019	27.80	—	30
	rev. art.	2020	3.31	—	11
	rev. art.	2020	—	His 1.70, Met 1.50, Phe 3.70, Thr 4.60, Lys 4.50, Trp 1.30, Val 5.10, Ile 4.20, Leu 6.80	27
	Chile.	2021	18.35	—	83
	Chile	2021	20.79	—	83
	Mexico	2021	32.00	—	101
	rev. art.	2021	33.50	—	12
	Cameroon	2021	8.62	—	102
	rev. art.	2021	1.70–4.20	—	82
	Chile	2021	10.45	—	83
V. gloiocephalus	Portugal	2015	19.66	—	13

[a]d.w., dry weight; cult., cultivated mushroom species; com. prep., commercial preparation; AA, amino acids; ref., references; rev. art., review article – results are based on available and cited review articles; mg eq protein per ml sample*, proteins are expressed with different units compared to the selected unit: g/100 g d.w.

where the essential AA index is calculated according to dietary protein in terms of an essential AA pattern based on known adult human dietary requirements.

2.2.2 Fats (Lipids)

Not only do mushrooms represent a great protein source in the human diet, they are also low in calories and fats.[27] The content of total lipids (crude fat) is usually below 10% (Table 2.2), with some differences found in *A. bisporus* (up to 23%), *C. cibarius* (up to 28.5%), and *C. comatus* and *C. micaceus* (up to 18% and 20%, respectively). The lipid content in mushrooms varies depending on many environmental factors such as nutrition, oxygen, temperature, and the nature of the substrate.[28] It usually includes free fatty acids, monoglycerides, diglycerides, triglycerides, sterols, sterol esters, and phospholipids. Saponifiable lipids make up a higher percentage than nonsaponifiable lipids.[1]

Mushrooms are especially rich in polyunsaturated fatty acids (PUFA; ∼50–80%), while saturated (SFA; ∼10–30%) and monounsaturated (MUFA; ∼7–26%) fatty acids are present in lower amounts (Table 2.2). High amounts of unsaturated fatty acids (UFA) (Table 5.2), essential in the human diet, give preference to mushrooms as a healthy food choice over animal fats and various processed foods, which are often richer in saturated fatty acids, which are considered harmful to health.

Regarding the fatty acids composition, linoleic (C18:2, PUFA; ∼50–80%), palmitic (C16:0, SFA; ∼7–30%), oleic (C18:1, MUFA; ∼1–30%), and to a lesser extent stearic (c18:0, SFA; ∼0.7–10%) acids are the most prominent (Table 2.2), while other fatty acids are only found in small amounts.[29–31] Evidently, linoleic acid is the dominant fatty acid in mushrooms, while palmitic and oleic acids are often present in lesser, but similar, quantities. Interestingly, in one Turkish study, *L. sulphureus* showed surprisingly low amounts of linoleic acid (7.73%) compared to oleic, palmitic, and stearic acids, which were equally represented at 33.94%, 28.6%, and 28.55%, respectively.[32] Contrary to most known data, *Lactarius deliciosus* contains an abundant amount of stearic acid (C18:0, ∼44%).[33] Also, it should be noted that *A. bisporus* contains smaller amounts of oleic acid than those observed in other analyzed edible mushrooms (Table 2.2).

2.2.3 Vitamins

Mushrooms are a good source of vitamins, especially vitamin B complex (thiamin, riboflavin, niacin, pantothenic acid, and pyridoxine) and vitamin D,[8,11] as well as β carotene (precursor of vitamin A) and vitamin E[30] (Table 2.3).

According to some studies[34,35] mushrooms present a good source of vitamin B_{12}, but this cannot be claimed for all species investigated so far since in some, vitamin B_{12} was found only in trace amounts.[34] Despite some claims, enzyme systems for the biosynthesis of vitamin B_{12} are still not

Table 2.2 Lipids and fatty acids contents of mushrooms.[a]

Species	Region	Crude fat/total lipids (g/100 g d.w.)	FA (% of total FA) SFA	MUFA	PUFA	FA composition (g/100 g d.w.) Linoleic (c18:2n6)	Oleic (c18:1n9)	Palmitic (c16:0)	Stearic (c18:0)	Ref.
A. aegerita	China	1.18	—	—	—	—	—	—	—	78
	Serbia	2.13	17.93	3.47	78.60	23 FA determined				43
						78.40	3.03	13.07	2.13	
	India (cult.)	0.20	—	—	—	—	—	—	—	103
	Turkey		31.77	68.23 (UFA)		54.33	13.90	22.15	9.62	104
	Ukraine	2.30–2.77	—	—	—	—	—	—	—	79
A. bisporus	rev. art.	1.78–8.00	—	—	—	—	—	—	—	46
	Portugal	0.14–0.19	18.40–20.30	1.40–1.80	78.30–79.80	5 FA determined				51
						77.7–79.4	1.1–1.2	11.1–11.9	3.0–3.1	
	rev. art.	1.70–8.00	—	—	—	—	—	—	—	14
	rev. art.	1.67–2.18	—	—	—	—	—	—	—	8
	rev. art.	1.80	—	—	—	—	—	—	—	30
	Pakistan	1.25	—	—	—	—	—	—	—	81
	rev. art.	3.06	—	—	—	—	—	—	—	12
	rev. art.	0.22	—	—	—	—	—	—	—	82
	rev. art.	0.34	—	—	—	—	—	—	—	42
	Chile	2.58	—	—	—	—	—	—	—	83
B. edulis	China	4.10	—	—	—	—	—	—	—	16
	rev. art.	3.10	—	—	—	—	—	—	—	30
	rev. art.	2.45	—	—	—	—	—	—	—	12
	Croatia	2.92	—	—	—	—	—	—	—	21
	Tunisia	38.50	—	—	—	—	—	—	—	84
	S. Africa	5.00	—	—	—	—	—	—	—	62
C. cibarius	rev. art.	9.14	—	—	—	—	—	—	—	14
	rev. art.	2.88	—	—	—	—	—	—	—	12
	Croatia	1.90	—	—	—	—	—	—	—	21
	India	7.80	—	—	—	—	—	—	—	87
	Tunisia	28.20	—	—	—	—	—	—	—	84
	rev. art.	5.00	—	—	—	—	—	—	—	46

Table 2.2 (*Continued*)

Species	Region	Crude fat/total lipids (g/100 g d.w.)	FA (% of total FA) SFA	MUFA	PUFA	FA composition (g/100 g d.w.) Linoleic (c18:2n6)	Oleic (c18:1n9)	Palmitic (c16:0)	Stearic (c18:0)	Ref.
C. comatus	Serbia		16.29	7.46	76.25	28 FA determined				5
						74.19	4.32	13.32	0.78	
	Serbia (com. prep.)	2.04	15.84	7.42	76.73	8 FA determined				91
						73.39	5.82	12.93	0.97	
	rev. art.	3.30	—	—	—	—	—	—	—	30
	rev. art.	1.13–1.30	—	—	—	—	—	—	—	12
	Serbia	1.80	18.72	15.27	66.01	64.08	13.46	12.88	2.27	90
	Pakistan	1.99	—	—	—	—	—	—	—	81
	rev. art.	1.10–5.40	—	—	—	—	—	—	—	65
C. micaceus	Nigeria	0.90	—	—	—	—	—	—	—	92
	Pakistan	1.24	—	—	—	—	—	—	—	81
	Nigeria	7.14	—	—	—	—	—	—	—	93
C. truncorum	Serbia	—	14.92	11.97	73.11	28 FA determined				5
						70.09	9.08	12.39	0.92	
L. sulphureus	Serbia	2.35	20.54	15.32	64.14	24 FA determined				95
						63.27	14.52	11.68	3.17	
	Cameroon	03.07	—	—	—	—	—	—	—	101
	Italy	1.23	—	—	—	—	—	—	—	94
	Turkey		—	—	—	6 FA determined				32
			—	—	—	7.73	33.94	28.60	28.55	
M. giganteus	Turkey	0.68	—	—	—	—	—	—	—	97
	India	2.20	—	—	—	—	—	—	—	98
M. procera	Serbia	1.90	25	9.92	65.00	23 FA determined				31
						64	8.3	21	1.96	
	Morocco	6.37	—	—	—	—	—	—	—	99
	Portugal	2.20–2.90	21.40–28.00	6.90–9.30	64.00–71.60	9 FA determined				29
						63–71	5.7–7.6	18.2–22.7	1.46–2.32	
	Chile	2.10	—	—	—	—	—	—	—	83

	Croatia	0.99	—	—	—	—	—	—	—	21
	Italy	0.20	—	—	—	—	—	—	—	94
P. ostreatus	rev. art.	0.41	—	—	—	—	—	—	—	11
	China	0.50–5.00	—	—	—	—	—	—	—	28
	Mexico (cult.)	1.90	—	—	—	—	—	—	—	101
	rev. art.	1.40	—	—	—	68.9	12.3	11.2	1.6	80
	rev. art.	4.70	—	—	—	—	—	—	—	8
	rev. art.	0.56–2.20	—	—	—	—	—	—	—	14
	rev. art.	2.30	—	—	—	—	—	—	—	12
	rev. art.	0.05–0.50	—	—	—	—	—	—	—	82
	Chile	0.70–2.58	—	—	—	—	—	—	—	83
	Croatia	2.08	—	—	—	—	—	—	—	21
	Turkey	—	21.77	11.37	66.48	35 FA determined				105
		—				65.29	10.36	12.4	3.71	
	Portugal	0.15	17.00	13.60	69.40	5 FA determined				51
						68.9	12.3	11.2	1.6	
	rev. art.	1.60–2.20	—	—	—	—	—	—	—	46
V. gloiocephalus	Portugal	4.62	12.56	26.47	60.97	59.33	25.96	7.42	02.09	13

[a]FA, fatty acids; SFA, saturated fatty acids; —, not determined; ref., references; —, not determined; UFA, unsaturated fatty acids; MUFA, monosaturated fatty acids; PUFA, polysaturated fatty acids; rev. art., review article; cult., cultivated; com. prep., commercial preparation.

Table 2.3 Vitamin content of various edible mushrooms.

Species	Region	Vitamins (μg/100 g d.w.)							Vitamins (mg/100 g d.w.)									Ref.
		α-	β-	γ-	δ-vit E	TToc	Car	B_{12}	B_1	B_2	B_3	B_5	B_6	B_7	C	D_3	D_2	
A. aegerita	Serbia	2.10	8.80	86.08	3.40	100.38	—	—	—	—	—	—	—	—	—	—	—	43
Bisporus	rev. art.	—	—	—	—	—	—	0.8		0.40	3.61	1497	—	—	2.1	—	—	42
A. bisporus[a]		—	—	—	—	—	—	—	—	—	—	—	—	—	—	—	1.26	41
		—	—	—	—	—	—	—	0.08	0.40	3.61	1.50	0.10	—	—	—	—	11
	Portugal	0.23	—	—	—	—	—	—	—	—	—	—	—	—	—	—	—	51
A. cylindracea	rev. art.	—	—	—	—	—	—	—	—	—	—	—	—	—	—	—	4.27	41
C. cibarius	Montenegro	—	—	—	—	—	1945	—	—	—	—	—	—	—	100	—	—	88
	rev. art.	1.34	0.287	3	—	—	577	—	—	—	—	—	—	—	—	—	—	89
	Germany, France, Bulgaria	—	—	—	—	—	—	1.82; 1.09; 1.48	—	—	—	—	—	—	—	—	—	34
C. comatus	Serbia	13.24	—	—	31.76	45.01	—	—	—	—	—	—	—	—	—	—	—	90
C. micaceus	Nigeria	—	—	—	—	—	—	—	—	—	—	—	—	—	—	—	—	92
		—	—	—	—	—	—	—	18.80	31.25	—	—	23.40	—	200	—	—	93
C. tubaeformis	rev. art.	—	—	—	—	—	—	—	—	—	—	—	—	—	—	—	1.57	41
L. edodes[a]	Japan	—	—	—	—	—	—	5.61	—	—	—	—	—	—	—	—	—	35
	rev. art.	—	—	—	—	—	—	—	—	—	—	—	—	—	—	—	10.64	41
L. sulphureus	Serbia	109.25	—	62.07	18.42	—	—	—	—	—	—	—	—	—	—	—	—	95
	Italy	—	—	—	—	—	—	—	—	—	4.2	464	—	4.8	—	0.75	—	94
M. giganteus	Serbia	3.58	09.05	77.8	123.35	213.78	—	—	—	—	—	—	—	—	—	—	—	106
M. procera	Portugal	4.5	77	38	72	185	—	—	—	—	—	—	—	—	—	—	—	29
	Serbia	1.1	—	—	26.1	27	—	—	—	—	—	—	—	—	—	—	—	31
	Morocco	—	—	—	—	—	44	—	—	—	—	—	—	—	225	—	—	99
	Portugal	—	—	—	—	—	32	—	—	—	—	—	—	—	245	—	—	99
	Italy	—	—	—	—	—	—	—	52	0.31	4.13	410	—	4.47	—	0.5	—	94
	Portugal	2.5–4.5	51–77	9–38	65–91	146–185	—	—	—	—	—	—	—	—	—	—	—	29
P. ostreatus	rev.art.	—	—	—	—	—	—	—	0.125	0.35	4.96	1.29	0.11	—	—	—	—	11
		—	—	—	—	—	—	—	—	—	—	—	—	—	—	—	6.9	41
V. gloiocephalus	Portugal	13.83	66.06	20.20	—	—	—	—	—	—	—	—	—	—	—	—	—	13
	Turkey	—	—	—	—	—	—	—	—	—	—	—	—	—	—	—	—	107
V. volvacea	rev.art	—	—	—	—	—	—	—	—	—	—	—	—	—	—	—	0.76	41

[a]Cultivated mushrooms; Car., carotenoids; rev. art, review article; —, not detected; ref., references; d.w., dry weight; α-, β-, γ-, δ-tocopherol; TToc, total tocopherol; thiamine (B_1), riboflavin (B_2), niacin (B_3), pantothenic acid (B5), pyridoxine (B_6), biotin (B_7), folate (B_9), and cobalamin (B_{12}).

detected in mushrooms and this process can be performed only by certain bacteria.[36] Vitamin B_{12} in mushrooms is probably acquired from surface microorganisms or extracted from the substrate.[35,37] Of the six wild edible mushrooms investigated, only two species,[34] *Craterellus cornucopioides* and *Cantharellus cibarius*, contained considerable levels (1.79–2.65 μg/100 g d.w. and 1.09–1.82 μg/100 g d.w., respectively) of vitamin B_{12}, but still less than in cultivated shiitake mushrooms (1.70–12.71 μg/100 g d.w.).[35] However, mushrooms can be used as a plant-based source to prevent B_{12} deficiency in vegans and vegetarians, but cannot be relied on as the only source of this vitamin since approximately 50 g of dried mushrooms are needed to provide the recommended dietary allowance for adults (2.4 μg per day) and consumption of such a large quantity of mushrooms on a daily basis is not generally feasible.

It is noteworthy that mushrooms are the only food of non-animal origin that contains vitamin D.[10] When exposed to ultraviolet (UV) light under certain conditions, mushrooms produce vitamin D_2 in amounts much higher than that of the daily requirements for humans.[38–40] This process involves a photochemical reaction in which ergosterol, a fungal sterol, is converted to vitamin D_2 as a result of a series of photochemical and thermal reactions catalyzed by UV radiation coming from sunlight.[8] Wild mushrooms, such as *B. edulis*, are a particularly good source of vitamin D_2, which is contained in much larger quantities than in cultivated *A. bisporus*.[27,41]

2.2.4 Carbohydrate and Fiber

Mushrooms are a significant source of digestible and non-digestible carbohydrates which both present an essential part of human diet. Glucose and mannitol are the two most important digestible carbohydrate compounds[16,30,42] found in mushrooms. Based on literature data, monosaccharides (pentoses, methyl-pentoses, hexoses), as well as disaccharides, amino sugars, sugar alcohols (mannitol), sugar acids, and polysaccharides (α- and β-glucans) are an integral part of mushroom carbohydrates.[1,8,13,14,29,31,43–45]

A large amount of the carbohydrates from edible mushrooms consists of oligo- and polysaccharides, including chitin, α- and β-glucan, mannan, xylan, and galactan.[44] The indigestible polysaccharides that abound in species make edible mushrooms excellent sources of dietary fibers (DFs).[44,46] The composition of the fibrous fraction of mushrooms mainly consists of β-glucans, chitin, and hemicellulose with immunomodulatory functions.[44]

Indigestible carbohydrates form a large portion of the total carbohydrates of mushrooms, and the main compounds are oligosaccharides (trehalose)[30] and non-starch polysaccharides (chitin, β-glucans, and mannans).[16,45,47] Also, digestible carbohydrates are found in some mushrooms, including *A. bisporus*, in the usual concentration range of 5–10% of the d.w.[30,42]

In general, the carbohydrate content of mushrooms represents the majority of fruiting bodies from about 50% to 65% on a d.w. basis, while the amount of free sugars is approximately 11%[8,27] and mannitol represents nearly 80% of the total free sugars in some mushroom species.[27] This sugar alcohol is important for the volume growth and firmness of the fruiting body[30] and the highest concentration was recorded in *A. bisporus* (23.62 g/100 g d.w.), *M. procera* (12.7 g/100 g d.w.), and *P. ostreatus* (9.82 g/100 g d.w.) (Table 2.4).

A detailed overview of carbohydrate and fiber nutrients of 12 selected wild-growing mushrooms is presented in Table 2.4. The scientific papers used indicate that the carbohydrate content of the analyzed mushroom species varies between 2.26 g/100 g d.w. in *M. procera* (Italy) and 85.9 g/100 g (%) in *P. ostreatus* (Mexico). This difference could be attributed to different observed specimens or different geographical origins (Table 2.4).

Some of the selected wild edible mushrooms presented in this study have relatively high carbohydrate contents (g/100 g d.w.): *A. bisporus* 74.00, *B. edulis* 70.96, *Coprinus* spp. 76.29, and *P. ostreatus* 85.90, while only *M. giganteus* has a relatively lower carbohydrate content of 53.70.

The selected wild-growing mushrooms are comprised mostly of glucose (0.37–15.01 g/100 g d.w.), fructose (0.01–2.62 g/100 g d.w.), mannitol (0.54–23.62 g/100 g d.w.), and trehalose (0.11–12.49 g/100 g d.w.) (Table 2.4). The most common carbohydrate compounds are mannitol and trehalose, with the highest concentrations found in *A. bisporus* and *C. aegerita*, respectively.

The glucose content in edible mushrooms is usually low,[1] which correlates with most of the species analyzed in this study, with the exception of *M. procera* and *P. ostreatus*, originating from the Balkan region (Croatia), whereas the highest concentrations were 10.78 ± 0.02 g/100 g of d.w. and 15.01 ± 0.35 g/100 g of d.w., respectively.[21,22] In this regard, the level of trehalose is also usually lower in edible mushrooms, which was confirmed in most species, with the exception of *A. aegerita* and *B. edulis*: 12.49 g/100 g d.w. and 9.92 g/100 g d.w., respectively (Table 2.4). The presence of fructose, sucrose, mannose, and melezitose was also confirmed in the examined mushrooms, whereas melezitose was detected only in *M. procera*[29] (Table 2.4).

The carbohydrate content of mushrooms collected from different localities varies depending on the species, and it is interesting to note that species *M. procera*, from Europe (Croatia, Serbia, and Portugal), stands out with higher glucose (10.78 g/100 g d.w.), trehalose (6.8–7.01 g/100 g d.w.), and mannitol (11.4–12.7 g/100 g d.w.) contents (Table 2.4).

In general, the largest amounts of carbohydrates are found in *P. ostreatus*, *A. aegerita*, and *M. procera* species among others (85.9, 84.50, and 83.65 g/100 g d.w., respectively) compared to *Coprinus* sp., with *C. truncorum* as the most unexplored and *C. micaceus* with the lowest carbohydrate content (53.50–61.00 g/100 g d.w.).

Altogether, mushroom polysaccharides carry vital biological information due to their strong potential for structural variability.[48] Therefore, today's interest in characterizing the polysaccharide content from the fruiting

Table 2.4 Content of carbohydrates and fibers of selected wild edible mushrooms in publications from 2010 to 2021 (g/100 g d.w.).[a]

Species	Region	Year of publication	Carbohydrate content[b] (g/100 g d.w.)	Fiber (g/100 g d.w.)	Ref.
A. aegerita	Serbia	2015	**Carbohydrates:** 84.50 **Mannitol:** 0.93 **Trehalose:** 12.49 **Total sugars:** 13.42	—	43
	India	2016	—	**Insoluble dietary fiber (IDF):** 0.00805	103
	China	2018	**Xyl, Ara, Glc, Rha, Gal, Rib, Fuc** and **Man** are detected by GC chromatography	—	108
	China	2020	**Carbohydrates:** 26.49 **Polysaccharides:** 3.74	**Dietary fiber (DF):** 31.10	78
A. bisporus	Cult. Portugal	2012	**Carbohydrates:** 6.46 **Fructose:** 0.03 **Mannitol:** 5.60 **Trehalose:** 0.16 **Total sugars:** 5.79	—	51
	Cult. Portugal	2012	**Carbohydrates:** 5.98 **Fructose:** 0.04 **Mannitol:** 4.01 **Trehalose:** 0.22 **Total sugars:** 4.27	—	51
	India	2014	—	**Total dietary fiber (TDF):** 31.00 **Insoluble dietary fiber (IDF):** 14.00	50
	Mexico	2015	**Carbohydrates:** 74.00 **Fructose:** 0.03 **Mannitol:** 5.60 **Trehalose:** 0.16 **Total sugars:** 5.79	—	80
	rev. art.	2016	**Carbohydrates:** 51.30–62.50	**Insoluble dietary fiber (IDF):** 8.00–10.40	14
	rev. art.	2017	**Carbohydrates:** 71.54–74.00	—	8
	rev. art.	2019	**Carbohydrates:** 59.90	**Dietary fiber (DF):** 10.40	30

Table 2.4 (*Continued*)

Species	Region	Year of publication	Carbohydrate content^{b} (g/100 g d.w.)	Fiber (g/100 g d.w.)	Ref.
	rev. art.	2020	**Carbohydrates:** 3.26	**Dietary fiber (DF):** 1.00	11
	Com. Prep. Poland	2020	—	**Total glucans:** 12.60–19.50 **α-glucans:** 2.41–6.40 **β-glucans:** 8.07–13.02	50
	rev. art.	2021	**Carbohydrates:** 53.10	—	12
	rev. art.	2021	**Carbohydrates:** 7.40	—	82
	Chile	2021	**Carbohydrates:** 66.75	**Dietary fiber (DF):** 10.28	83
	Chile	2021	**Carbohydrates:** 66.33	**Dietary fiber (DF):** 11.58	83
	rev. art.	2021	**Carbohydrates:** 3.26 **Fructose:** 2.62 **Mannitol:** 23.62 **Trehalose:** 1.00–3.00	**Dietary fiber (DF):** 1.00	42
	rev. art.	2021	—	**Total glucans:** 10.10–15.00 **β-Glucans:** 8.60–12.30	45
	rev. art.	2021	—	**Total glucans:** 12.30–14.60 **β-Glucans:** 8.80–10.10	45
B. edulis	Croatia	2011	**Glucose:** 6.28 **Mannitol:** 3.72 **Trehalose:** 9.92 **Mannose:** 36.23 **Total sugars:** 64.27	—	21
	China	2014	**Carbohydrates:** 30.60	**Insoluble dietary fiber (IDF):** 15.30	16
	India	2014	—	**TDF – total dietary fiber (TDF):** 28.00 **Insoluble dietary fiber (IDF):** 13.00	49
	Tunisia	2016	**Total carbohydrates:** 57.54 **Glucose:** 0.37 **Trehalose:** 2.26 **Mannitol:** 4.10 **Total sugars:** 6.73	—	84
	Turkey	2018	**Total soluble carbohydrates** 12.95 **Glucose:** 1.95	—	86

			Fructose: 0.039 **Sucrose:** 1.57		
	rev. art.	2019	**Carbohydrates:** 46.95	—	30
	rev. art.	2019	**Carbohydrates:** 59.70	**Insoluble dietary fiber (IDF):** 8.00	30
	South Africa	2020	**Carbohydrates:** 51.70–57.20	—	62
	Poland	2020	—	**Total glucans:** 30.00–31.50 **α-Glucans:** 2.20–5.90 **β-Glucans** 25.60–29.30	50
	rev. art.	2021	**Carbohydrates:** 70.96	—	12
C. cibarius	Croatia	2011	**Glucose:** 7.98 **Mannitol:** 8.56 **Trehalose:** 6.68 **Mannose:** 28.56 **Total sugars:** 52.5	—	21
	rev. art.	2013	**Carbohydrates:** 64.90	**Insoluble dietary fiber (IDF):** 11.20	46
	India	2014	—	**Total dietary fiber (TDF):** 36.00 **insoluble dietary fiber (IDF):** 20.00	50
	Serbia	2015	**Carbohydrates:** 3.04	—	82
	rev. art.	2016	**Carbohydrates:** 18.00	**Insoluble dietary fiber (IDF):** 13.64	14
	Tunisia	2016	**Total carbohydrates:** 72.24 **Glucose:** 0.77 **Trehalose:** 12.01 **Mannitol:** 2.57 **Total sugars:** 15.35	—	84
	rev. art.	2020	**Glucose:** 7.98 **Mannitol:** 8.56 **Trehalose:** 6.68 **Mannose:** 8.56	—	89
	Poland	2020	—	**Total glucans:** 26.50–28.30 **α-Glucans:** 1.11–1.64 **β-Glucans:** 24.86–27.09	50
	rev. art.	2021	**Carbohydrates:** 66.07	—	12
	rev. art.	2021	—	**Total glucans:** 25.3–28.5 **β-Glucans:** 24.00–27.00	45

Table 2.4 (*Continued*)

Species	Region	Year of publication	Carbohydrate content[b] (g/100 g d.w.)	Fiber (g/100 g d.w.)	Ref.
C. comatus	Serbia	2013	**Carbohydrates:** 76.29 **Mannitol:** 1.84 **Trehalose:** 5.41	—	90
	Serbia	2019	mg GluE per g d.w. 8778.07* mg FruE per g d.w. 2020.44 mg XylE per g d.w. 1116.90 mg SucE per g d.w. 9.03	—	6
	rev. art.	2019	**Carbohydrates:** 58.80	**Dietary fiber (DF):** 7.30	30
	Com. Prep. Serbia	2020	**Carbohydrates:** 40.42	**Dietary fiber (DF):** 21.13	91
	rev. art.	2020	**Carbohydrates:** 49.20–76.30	**Insoluble dietary fiber (IDF):** 32.80 **Soluble dietary fiber (SDF):** 1.79	65
	rev. art.	2021	**Carbohydrates:** 70.36	—	12
	rev. art.	2021	**Carbohydrates:** 58.60	—	12
C. micaceus	Nigeria	2012	**Crude carbohydrates:** 61.00	**Insoluble dietary fiber (IDF):** 8.00	92
	Nigeria	2019	**Carbohydrates:** 53.50	**Insoluble dietary fiber (IDF):** 9.46	93
C. truncorum	Serbia	2019	mg GluE per g dw 8399.85** mg FruE per g dw 1932.25 mg XylE per g dw 1154.09 mg SucE per g dw 20.57	—	6
L. sulphureus	Italy	2012	**Carbohydrates:** 16.40	—	94
	Serbia	2014	**Carbohydrates:** 72.64 **Trehalose:** 4.00 **Mannitol:** 3.54 **Fructose:** 0.46 **Total sugars:** 8.00	—	95
	Serbia	2014	**Fructose:** 0.2–17.9 μmol Trolox per g d.e.***	—	109
	Turkey	2018	**Total soluble carbohydrates:** 26.68 **Glucose:** 4.05 **Fructose:** 0.63 **Sucrose:** 0.032	—	86

	rev. art.	2019	**Carbohydrates:** 6.49	—	30
	India	2019	**Carbohydrates:** 7.65	**Insoluble dietary fiber (IDF):** 6.09	96
	rev. art.	2021	**Carbohydrates:** 7.65	—	12
M. procera	Croatia	2011	**Glucose:** 10.78 **Mannitol:** 2.42 **Trehalose:** 0.11 **Mannose:** 11.09 **Total sugars:** 66.78	—	21
	Italy	2012	**Carbohydrates:** 2.26	—	94
	Portugal	2013	**Carbohydrates:** 6.70 **Mannitol:** 11.40 **Fructose:** 0.10 **Melezitose:** 1.02 **Trehalose:** 6.80 **Total sugars:** 19.30	—	29
	rev. art.	2019	**Carbohydrates:** 5.47	—	30
	Serbia	2019	**Total carbohydrates:** 60.30 **Mannitol:** 12.70 **Trehalose:** 7.01 **Total sugars:** 19.70	—	31
	Poland	2020	—	**Total glucans:** 10.00–11.40 **α-Glucans:** 0.23–1.25 **β-Glucans:** 8.89–12.85	50
	Morocco	2021	**Carbohydrates:** 59.00	—	99
	Portugal	2021	**Carbohydrates:** 53.71	—	99
	Chile	2021	**Carbohydrates:** 83.65	**Dietary fiber (DF):** 16.31	83
M. giganteus	Com. Prep. India	2013	**Carbohydrates:** 53.7	**Insoluble dietary fiber (IDF):** 23.46	98
P. ostreatus	Croatia	2011	**Glucose:** 15.01 **Mannitol:** 9.82 **Trehalose:** 1.79 **Mannose:** 11.13 **Total sugars:** 61.9	—	21
	China	2012	**Carbohydrates:** 37–48	**Dietary fiber (DF):** 24–31	28
	Portugal	2012	**Carbohydrates:** 9.30	—	51

Table 2.4 (*Continued*)

Species	Region	Year of publication	Carbohydrate content[b] (g/100 g d.w.)	Fiber (g/100 g d.w.)	Ref.
			Fructose: 0.01 **Mannitol:** 0.54 **Trehalose:** 4.42 **Total sugars:** 4.97		
	Mexico	2015	**Carbohydrates** 85.90 **Fructose:** 0.01 **Mannitol:** 0.54 **Trehalose:** 4.42	—	80
	rev. art.	2016	**Carbohydrates:** 57.60–81.80	**Insoluble dietary fiber (IDF):** 7.50–45.50	14
	Cult. China	2017	**Total sugar:** 38.76–40.30	**Insoluble dietary fiber (IDF):** 19.73–23.63	100
	rev. art.	2017	**Carbohydrates:** 35.40	—	8
	Com. Prep. Poland	2020	—	**Total glucans:** 45.0–47.80 **α-Glucans:** 5.09–6.29 **β-Glucans:** 39.61–41.51	50
	rev. art.	2020	**Carbohydrates:** 6.09	**Dietary fiber (DF):** 2.3	11
	Mexico	2021	**Carbohydrates:** 50.50	**Insoluble dietary fiber (IDF):** 5.60	101
	rev. art.	2021	**Carbohydrates:** 48.90	—	12
	rev. art.	2021	—	**Total glucans:** 25.60 **β-Glucans:** 24.2	45
	rev. art.	2021	**Carbohydrates:** 37.00–48.00	**Total dietary fiber (TDF):** 24–31	82
	Chile[1]	2021	**Carbohydrates:** 71.25	**Dietary fiber (DF):** 14.31	83
	Chile[2]	2021	**Carbohydrates:** 79.30	**Dietary fiber (DF):** 15.45	83
	Chile[3]	2021	**Carbohydrates:** 67.67	**Dietary fiber (DF):** 16.29	83
V. gloiocephalus	Portugal	2015	**Mannitol:** 1.45 **Trehalose:** 1.92 **Total sugars:** 3.37	**Total dietary fiber (DF):** 39.12 **Total insoluble fiber (IDF):** 37.08 **Total soluble fiber (SDF):** 2.04	13

[a]DF, dietary fiber; SDF, soluble dietary fiber; IDF, insoluble dietary fiber; com. prep., commercial preparations of mushrooms; cult., cultivated mushroom species; rev. art., results are based on available and cited review articles: GluE, glucose equivalents; FruE, fructose equivalents; XylE, xylose equivalents; SucE, sucrose equivalents; d.w., dry weight; d.e., dry extract. Chile,[1,2,3] superscript numbers signify different locations of the same mushroom species, *P. ostreatus* from Chile.
[b]*, **, ***, carbohydrates are expressed with different units compared to the selected unit: g/100 g d.w.

bodies of mushrooms has increased, and these results are presented in the following text as the content of α- and β-glucans.

Dietary fibers (DFs) were described by the American Association of Cereal Chemists (AACC) as high-molecular-weight carbohydrate polymers with 10 or more monomeric units, which are not hydrolyzed by human endogenous enzymes.[14,46] Compared to cereals, fruits, legumes, and vegetables as conventional sources of DFs, mushrooms are less utilized, but it is known that consumption of edible mushrooms as part of the daily diet can easily provide up to 25% of the recommended dietary intake of DF.[46] While plant cell walls represent the major sources of DFs, mushroom cell walls can also be considered as an important source, containing non-digestible carbohydrates (NDCs) mainly in the form of the chitin polysaccharides, with amounts of up to 80% of d.w. of the cell wall and β-glucans among others.[46] DFs exhibited great variability in different mushroom species.[30]

Cheung[47] stated that dietary fibers as a form of indigestible carbohydrate include chitin, the predominant carbohydrate in the cell wall of mushrooms, followed by mannans, hemicelluloses, and glucans which are present in significant amounts in the cell wall, and together contribute to the medicinal properties of mushrooms.[30,41] There are no data on the content of chitin in Table 2.4, however some scientists have determined that the chitin content in *A. bisporus* (9.60 g/100 g d.w.) was much higher compared to some other edible mushrooms, such as *L. edodes* and *P. ostreatus.*[42] It is important to emphasize that chitin limits the availability of other mushroom components and based on that it is indigestible to humans[30] in the form of an insoluble fiber, which promotes the functioning of immunity. The higher chitin content in *A. bisporus* contributes to its health-promoting effects.[42]

β-Glucans make up 4–13% of total fibers, with variability in the DF fractions depending on the mushroom species.[49] In this review, the total glucan content was observed in only six mushroom species (Table 2.4).

Mirończuk-Chodakowska and Witkowska[50] determined the total glucan content in *A. bisporus*, *B. edulis*, *C. cibarius*, *M. procera*, and *P. ostreatus*, where the highest content was recorded in *P. ostreatus* (39.61–41.51 g/100 g d.w.), followed by *B. edulis* (30.0–31.5 g/100 g d.w.) and *C. cibarius* (25.30–28.50 g/100 g d.w.), and the lowest in *M. procera* (10.0–11.4 g/100 g d.w.). In this regard, from *P. ostreatus* β-glucan *pleuran* was isolated, which showed a significant reduction in peripheral blood eosinophilia and stabilized the levels of immunoglobulins, suggesting a potential antiallergic effect of mushroom-derived β-glucans.[45] In the determination of glucans as polysaccharide compounds of mushrooms, β-glucan was present in higher concentrations in all examined mushroom species (8.07–41.51 g/100 g d.w.) compared to α-glucan (1.11–6.4 g/100 g d.w.), excluding *M. procera* species.[50]

In general, β-glucans are considered functional compounds with the ability to stimulate an immunomodulatory response, modulating humoral and cellular immunity.[30] Therefore, the presence of DFs with NDCs provides mushrooms with a wide range of health benefits for humans[8] and DF research has drawn attention, particularly in the growing nutraceutical industry based on their health benefits in the prevention of cancer, cardiovascular disease, weight regulation, and diabetes.[46] Related to this, feeding diabetic patients with a high-fiber diet reduces their daily insulin requirement and stabilizes their blood glucose profile, possibly by decreasing the rate of glucose absorption or delaying gastric emptying.[1,14]

Mushroom DFs mainly consist of water-insoluble, crude fiber (IDF) with chitin and β-glucans as the most representative, while the amount of water-soluble fiber (SDF) is usually less than 10% d.w. of DFs.[10,14,49] The soluble carbohydrates present in mushrooms contribute to a sweet taste, and their high content gives rise to a moderate sweet taste perception in mushrooms.[21,22] Crude fiber can improve the function of the alimentary tract and also lower cholesterol levels as well as blood glucose levels,[16] while SDF has an important function in preventing cardiac diseases by lowering the total and LDL cholesterol levels.[8,27] Previous studies on mushroom fiber content have been very limited, and literature data on the fiber content of mushroom mycelia are not available.[49] Based on the literature data, some mushrooms have low amounts of IDF, *e.g.* for *Craterellus aureus* and *Sarcodon aspratus* values about 5% d.w.,[16] for *L. edodes* varies between 6.5 to 14.7% d.w.,[49] while for many others, up to 40% d.w. (Table 2.4). The highest concentration of DF was found in *C. cibarius* (36 g/100 g d.w.), with the amount of insoluble fiber in the lowest concentration of 6.09 g/100 g d.w. and the highest concentration of 37.08 g/100 g d.w. in *L. sulphureus* and *V. gloiocephalus*, respectively. The fiber content ranges from 7.4 to 45.5 g/100 g in *P. ostreatus*[1,14,51] as compared to 16.31 g/100 g in *M. procera* and 7.3–21.13 g/100 g in *C. comatus* (Table 2.4). Although data for SDF are lacking in this overview, it is well known that DF in mushrooms shows higher levels of IDF (2.28–8.99 g/100 g d.w.) than SDF (0.32–2.20 g/100 g d.w.),[49] with the exception of *Craterellus aureus* and *Sarcodon aspratus* that have values lower than 5 g/100 g d.w.[16] Until now, the analyses of wild-growing edible mushrooms have revealed that they contain high levels of DF and IDF, thus they could be used as important ingredients in a balanced and healthy diet. The data obtained in this study on the DF and IDF contents of 11 wild-growing mushroom species (Table 2.4) could be useful in selecting appropriate mushroom species rich in SDF for inclusion in human diets to promote healthy living. Also, these data should help in identifying the correct values for available carbohydrates from the same mushroom species. For example, to date, many researchers have reported that the mycelium and fruiting body of *C. comatus* possess hypoglycemic activity based on reducing plasma glucose concentrations or on inhibiting the activity of protein tyrosine phosphatase 1B which plays a role in insulin signaling.[52] This primarily refers to polysaccharides as carbohydrate compounds from this species, whose content is 49.2–76.3 g/100 g d.w. as well as DF in the forms of IDF and SDF in the following concentrations: 32.8 g/100 g and 1.79 g/100 g, respectively. In relation to this, *C. comatus* could be the mushroom with the richest DF content, with *V. gloiocephalus* still an insufficiently researched species providing up to 30% of the recommended dietary intake of DF. Although an extraordinarily high or appreciable level of total dietary fiber (TDF) was reported for the least studied species *V. gloiocephalus* and *C. cibarius* (Table 4), herein it was *A. aegerita*, *A. bisporus*, *M. procera*, and *P. ostreatus* that gave the highest carbohydrate levels, as mentioned above. Probably, this higher carbohydrate content is due to a higher level of non-fiber carbohydrates such as monosaccharides, disaccharides, and sugar alcohols (Table 2.4).

2.2.5 Mineral Constituents

Due to the great absorption capacity of fungal mycelia, mushrooms represent a good source of mineral elements. These are taken up from the substratum by the mycelium and actively transported to the fruiting bodies. Minerals with the highest content in mushrooms (56–70%), also known as macroelements, are potassium (K), phosphorus (P), sodium (Na), calcium (Ca), and magnesium (Mg). Among these, K, as a physiologically extremely important element, accounts for nearly 45% of the total ash content.[1] Other minerals (such as Fe, Cu, Zn, Mn, Mo, Co, Ni, Se, and I), referred to as microelements or trace elements, are represented in considerably smaller percentages.[53] Nevertheless, mushrooms have the ability to uptake a relatively high amount of microelements and to manage their toxic effect, as was determined in a recent study of *C. comatus.*[54] Although these minerals may represent beneficial nutritional compounds for humans, they may also have detrimental effects if present in excessive amounts. Unfortunately, mushrooms are capable of accumulating toxic elements (Pb, Hg, Cd, Ni, As, and Cr) and radionuclides as well.[53,55] Therefore, it is of great importance to examine the metal content in wild mushrooms and consume them moderately.

The individual mineral content in mushrooms is mostly species specific (varying greatly from species to species), but it is also variable among different studies, due to the specific sampling location and environmental factors. The final mineral content in consumed edible mushrooms depends on the manner of cooking or industrial processing. Leakage of elements from the fruiting bodies into water or brine has been commonly observed when mushrooms are blanched, pickled, or conserved.[56,57]

Some of the highest values of K, Mg, Na, and Ca have recently been recorded in *C. cibarius*, *B. edulis*, *V. gloiocephalus*, *M. procera*, *C. micaceus*, and *C. comatus* (Table 2.5).

It was found that *M. procera* could serve as a good dietary source of some essential elements (Fe, Zn, Mn, Cu; Table 2.5) but the consumption of these mushrooms may pose a health risk since at the same time it shows a good Cd and Pb absorption ability (Table 2.5).[58–60]

Mushrooms are rich in one of the most important dietary elements and antioxidants, selenium. It was found that *B. edulis* may contain a high Se

Table 2.5 Mineral content of mushrooms.[a]

Species	Region	Year of publication	Macroelements ($mg\,kg^{-1}$ d.w.)	Microelements ($mg\,kg^{-1}$ d.w.)	Ref.
A. bisporus	Greece	2019	—	Pb 4.23, Cu 65.81, Mn 35.88, Zn 80.54, Fe 371.97, Ni 9.13	110
	rev. art.	2021	—	Fe 5	42
	Com. Prep. Poland	2019	—	Cu 40.8, Mn 5.91, Se 6.17, Zn 63.4	111
	Pakistan	2019	Ca 7.4, K 15.1, Mg 6.4	Mn 0.3, Zn 1.6, Se 0.09	81
	Spain	2016	—	Cd 0.08–0.39	58
	rev. art.	2020	Ca 30, Mg 90, K 3180, P 860, Na 50	Cu 3.2, Fe 5, Mn 0.5, Se 0.093, Zn 5.2	11
B. edulis	Turkey	2018	Ca 48.45, K 5471.86, Mg 17.43, Zn 275.76	Cd 0.87, Cr 4.72, Cu 109.13, Fe 148.34, Mn 33.34, Se 115.40	94
	Poland	2019	—	Cu 19, Mn 20.1, Se 13.3, Zn 170	111
	Poland	2013	—	Ni 4.24–6.94	112
	Tunisia	2016	Na 4610.24, K 16 313.49, Mg 2267.01, Ca 40.81	Cu 2.15, Fe 2.70, Zn 6.31	84
	South Africa	2020	Mg 540–860, Na 300–930, K 18 180–27 090, Ca 26 470–206 280, Pb 60–2020, Se 30 540–47 980, Si 15 570–191 170, Zn 53 630–107 070	Fe 20–130, As 0.38–1.71, Cd 0.96–1.85, Co 0.17–44.4, Cr 0.05–9.47, Cu 1.01–120, Mn 1.21–89.2, Ni 0.17–3.61	62
	Poland	2019	K 3400–50 000, Na 25.2–7900, Mg 120–2400, Ca 4.22–1500	Fe 7.81–790, Mn 1.21–89.2, Zn 22.0–500, Ba 0.24–4.82, Cu 1.01–120, Al 23.1–1000, Ag < 0.001–5.33, Sr 0.10–1.11, Cd 0.15–22.0, Pb < 0.001–33.0, Cr < 0.001–4.71, Ni 0.07–2.01	113
	Romania	2015	—	Mn 12.65–130.72, Fe 38.89–178.8, Cu 17.02–21.82, Zn 24.71–26.62	114
	Croatia	2017	—	Pb 0.99–1.42, Cd 1.73–1.93, Hg 2.35–2.56	60
	China	2018	Ca 118–572, Mg 246–748, Na 19–2720	Ba 18–182, Co 1.1–13.2, Cr 10–203, Cu 16–84, Fe 332–3568, Mn 19–71, Ni 14–76, Sr 5.4–176.9, V 4.3–60.3, Zn 40–135, Cd 3.7–91.5	61
	Poland	2019	—	Cu 19.0, Mn 20.1, Se 13.3, Zn 170	111
	Spain	2016	—	Cd 0.23–1.88	58

C. aegerita	India	2016	Na 293, K 85, Ca 273, Mg 96, P 951	Mn 58, Fe 3850, Zn 153, Cu 96, Se 7	103
C. cibarius	China	2014	—	As 0.5, Hg 0.3, Cd 0.9, Pb 4.6	16
	Poland	2019	—	Cu 69, Mn 32,5, Se 0.26, Zn 121	111
	rev. art.	2016	Mg 866.3	Cr 1.57, Mn 22.09, Fe 118.2, Co 0.05, Ni 1.07, Cu 32.49, Zn 54.29, Cd 0.38	14
	Tunisia	2016	Na 3088.97, K 20 822.94, Mg 5938.35, Ca 148.20	Cu 32.73, Fe 16.80, Zn 10.43	84
	Poland	2019	K 6200–59 000, Ca 350–2300, Mg 410–1400, Na 5.34–579	Fe 21.2–810, Zn 69.2–160, Mn 14.2–92.3, Cu 6.23–77.1, Ag < 0.001–0.82, Cd 0.31–1.11, Pb 0.15–0.57, Ni < 0.001–3.32	113
	Romania	2015	—	Mn 6.98–15.43, Fe 36.73–172.96, Cu 7.585–21.29, Zn 6.91–26.04	114
	Poland	2019	—	Cu 69.0, Mn 32.5, Se 0.26, Zn 121	113
	Spain	2016	—	Cd 0.09–0.71	58
C. comatus	Turkey	2017	Ca 2627, Mg 62.39	Fe 306.4, Mn 31.089	115
	Greece	2019	—	Pb 5.64, Cu 56.82, Mn 11.30, Zn 58.84, Fe 275.46, Ni 13.28, Cr 0.43	110
	Com. Prep. Serbia	2014	—	Pb 0.27, As 0.46, Cd 0.35, Fe 452.97, Ni 0.10, Cu 26.50, Co 0.27, Cr 3.90, Mn 14.03	116
	Pakistan	2019	Ca 5.3, K 5.2, Mg 3.4	Mn 0.18, Zn 2.5, Se 0.07	81
	Spain	2016	—	Cd 0.36–2.67	58
	Com. Prep. Serbia	2020	—	Pb 0.27, As 0.46, Cd 0.35, Fe 452.97, Ni 0.10, Cu 26.50, Co 0.27, Cr 3.90, Mn 14.03	91
C. micaceus	Turkey	2017	Ca 1794, Mg 62.34, Fe 148.4, Mn 12.91	—	115
	Nigeria	2012	Na 2890, K 5700, Ca 2050, Mg 200, P 13 200	Fe 3000	92
	Pakistan	2019	Ca 4.1, K 9.5, Mg 4.1	Mn 0.3, Zn 1.7, Se 0.06	81
	Nigeria	2019	Ca 3280, Mg 1640, P 796, K 7250	Fe 406.8	93
L. sulphureus	Turkey	2018	Ca 18.78, K 5752.54, Mg 16.86	Cd 0.41, Cr 4.03, Cu 14.35, Fe 80.62, Mn 99.49, Se 2.66, Zn 113.63	86
	Italy	2012	Ca 1010, Mg 1070, K 22 720, Na 2360	Fe 180	94

Table 2.5 (*Continued*)

Species	Region	Year of publication	Macroelements (mg kg^{-1} d.w.)	Microelements (mg kg^{-1} d.w.)	Ref.
	Turkey	2015	—	Pb 1.97, Cd 0.836, Zn 50.4, Fe 115, Mn 2.9, Cu 8.3, Cr 4.78, Ni 1.9, Co 0.993	117
	Turkey	2019	—	Fe 162.9, Cu 5, Mn 19.36, Zn 28.36	32
	Cameroon	2021	Ca 130.4, K 4336.2, Mg 138.5, P 5428.8, Na 42	Cu 11.5, Fe 86.9, Zn 26.6	102
M. giganteus	Turkey	2010	Na 93.00, K 69.33	Fe 84.33, Cu 70.17, Zn 75.63	97
	Com. Prep. India	2013	P 115, Mg 16, K 155, Ca 39	Fe 16	98
M. procera	Serbia	2016	—	Fe 89.53, Zn 53.85, Cu 109.57, Mn 9.38, Ni 0.26, Cd 6.23, Pb 1.69, Co 0.31	118
	Poland	2019	—	Cu 123, Mn 16.3, Se 2.93, Zn 84.1	111
	Serbia	2016	—	Al 29–5664.48, Ba 0.2–44.40, Cd 0.04–43.50, Co 0–12.04, Cr 0.2–13.81, Cu 28.96–382.10, Fe 29.94–4017.76, Mn 7.64–367.00, Ni 0.09–11.73, Pb 0.04–14.29, Sr 0.08–29.89, Zn 27.28–535.30	59
	Italy	2012	Ca 310, Mg 1010, K 22 880, Na 3240	Fe 150	94
	Poland	2019	K 20 000–44 000, Na 27.1–610, Ca 170–560, Mg 800–2100	Zn 39.3–180, Fe 41.4–360, Mn 24.0–100, Al 23.3–380, Cu 63.1–190, Ba 0.61–3.51, Sr 0.37–1.52	113
	Turkey	2015	—	Pb 2.84, Cd 0.542, Zn 88.4, Fe 596, Mn 53.6, Cu 68.6, Cr 5.00, Ni 9.7, Co 0.88	117
	Romania	2015	—	Mn 14.04–20.47, Fe 49.77–715.14, Cu 12.27–36.69, Zn 10.27–16.62	114
	Croatia	2017	—	Pb 1.6–1.91, Cd 2.2–2.74, Hg 1.17–1.55	60
	Poland	2019	—	Cu 123.0, Mn 16.3, Se 2.93, Zn 84.1	111
	Spain	2016	—	Cd 0.22–3.50	58

P. ostreatus	rev. art.	2020	Ca 30, Mg 180, K 4200, P 1200, Na 180	Cu 2.4, Fe 13.3, Mn 1.1, Se 0.026, Zn 7.7	11
	rev. art.	2016	Mg 37.39	Mn 0.27, Fe 1.76, Cu 0.18, Zn 1.10	14
	Romania	2015	—	Mn 1.97–4.73, Fe 22.69–31.61, Cu 5.04–5.62, Zn 20.19–34.92	114
	Africa	2018	—	Pb 0.68, Cu 9.93–12.85, Ni 0.27–6.47, Cd 0.21–0.45, Zn 29.65–31.47	119
	Com. Prep. Poland	2019	—	Cu 14.1, Mn 12.9, Se 0.19, Zn 72.4	111
	Spain	2016	—	Cd 0.19–0.81	58
V. gloiocephalus	Turkey	2017	Ca 2404, Mg 65.22	Fe 585, Mn 138.2	115
	Greece	2019	—	Pb 8.43, Cu 39.54, Mn 27.29, Zn 77.93, Fe 385.28, Ni 16.92, Cr 0.89	110
	Spain	2020	—	Zn 263, Pb 204, Ni 14.9, Mn 418, Fe 35 973, Cu 185, Cd 1.08	120
	Portugal	2015	Ca 735.2, Mg 2001.9, Na 7905.6, K 46 925.5	Fe 699.1, Cu 50.1, Mn 1.3, Zn 108.9	13

[a]d.w., dry weight; rev. art., review article.

content (~20 mg g^{-1} d.w.–20 000 mg kg^{-1} d.w.). This element is often bound to the polysaccharides in the cell walls of mushrooms which may reduce its availability for humans.[8] Unfortunately, species of the genus *Boletus* can represent a health risk for consumers since they can contain high concentrations of Hg.[60] In addition, *B. edulis* was reported to contain high amounts of As and Cd (Table 2.5).[58,61,62]

C. comatus showed good ability to absorb and concentrate various elements from its environment, especially Zn, Cu, Fe, Mn, Cr, and Ni.[63] This species may contain high levels of As (410 mg kg^{-1} d.w.), Se (410 mg kg^{-1} d.w.), and Hg (144 mg kg^{-1} d.w.).[64,65] This edible and medicinal mushroom species also successfully accumulates Cd and Pb in concentrations which may exceed permissible levels for edible mushrooms.[58,64]

2.2.6 Nucleic Acids and Their Constituents

Besides the analysis of mushroom nucleic acids, which were mostly investigated in the light of molecular identificaton[66,67] and phylogeny reconstruction,[68] there is an increase in the investigations of nucleic acid constituents (nucleotides and nucleosides) in edible mushrooms. One of the most important qualities of wild mushrooms which makes them a prized food is their unique taste. The typical mushroom flavor is constituted of both volatile compounds[69] and non-volatile compounds.[70] Among the non-volatile compounds, 5′-nucleotides are significantly attributed to mushroom taste since they are, together with some free AAs, responsible for a specific umami taste.[71] Also, the main bioactive compounds in the famous traditional medicinal mushroom *Cordyceps* are thought to be nucleosides.[72]

Since nucleosides, such as cytidine, uridine, adenosine, guanosine, thymidine, and inosine, bind to the purinergic and/or pyrimidine receptors in the human body it was determined that they are involved in the regulation and modulation of various physiological processes including enhancement of brain function, immune modulation, improvement of gastrointestinal tract repair after damage,[73] iron absorption in the gut, and had an influence on fatty acid metabolism.[74] In healthy people, nucleosides can be produced and recycled in their bodies, but dietary nucleosides can be useful in situations like illness recovery and immune system malfunction.[75] Because of this, there has been a recent increase in studies to qualitatively and quantitatively determine the distribution of nucleosides in mushrooms, which presents a remarkable reservoir of these compounds (Table 2.6). Ranogajec *et al.*,[76] Beluhan and Ranogajec,[21] and Yang *et al.*[77] investigated several wild and cultivated edible mushrooms and it was determined that the content of flavor 5′-nucleotides (GMP + IMP + XMP) ranged from 0.05 mg g^{-1} to 13.88 mg g^{-1}. It is noteworthy that the lowest and highest values of 5′-nucleotides measured in previous studies were detected in the same species of mushroom, *Craterellus cornucopioides*. This large difference in the content of flavor 5′-nucleotides can be expected since it was

Table 2.6 Nucleic acids content of mushrooms.[a]

Species	Region	Free 5′ – nucleotides (mg g^{-1} d.w.)						Ref.
		5′-AMP	5′-CMP	5′-GMP	5′-IMP	5′-UMP	5′-XMP	
B. edulis	Croatia	1.65	0.16	0.64	0.28	1.21	0.71	21
	Finland	1.39	1.87	0.6	0.35	—	—	85
C. cibarius	Croatia	0.41	0.09	0.21	0.03	0.75	0.14	21
	rev. art.	0.41	0.09	0.21	0.03	0.75	0.14	89
	Finland	0.38	0.37	0.19	0.22	0.28	—	85
M. procera	Croatia	1.03	1.39	0.12	0.16	1.61	0.24	21
P. ostreatus	Croatia	1.21	0.81	0.59	0.21	0.59	2.63	21

[a]d.w., dry weight, rev. art., review article.

determined that it is dependent on the following factors: (i) the parts of the mushroom, (ii) the maturity stages of mushrooms, (iii) the storage time, and (iv) the processing of the mushrooms, so there is no unequivocal rule to grade mushrooms based on the 5′-nucleotide content.[75]

2.3 Conclusion

Edible mushrooms are part of the human diet due to their nutritional importance based both on the primarily high content of proteins and carbohydrates (generally 20–30% d.w.) and medicinal properties. Among the large amount of available scientific data that mostly encountered crude protein and carbohydrate contents of mushrooms as sources of functional food, only some of them have investigated their specific AA composition or the individual monohydrate compounds. The richness in proteins, carbohydrates, dietary fibers, chitin, and β-glucans with potential medicinal properties, vitamins (B, A, C, and the only natural, non-animal, non-fortified edible source of vitamin D), minerals (Mg, Se, K, Fe, Cu, and P), the presence of unsaturated fatty acids, and the low fat, as well as remarkable reservoir of nucleotides have turned mushrooms into an excellent food choice to be included in low-calorie diets which are the most suitable for the modern lifestyle and vegetarian diets. What is more, all this makes them a functional human food of the future.

In summary, updates on nutritional values of mushrooms have been doubtlessly broadened during the last two decades around the world, especially in terms of species studied – the highest interest is broadening again on the most valued edible species (*B. edulis*, *C. cibarius*, *A. bisporus* and some newly investigated like *C. comatus*, *V. gloiocephalus*, and *M. procera*), but according to the average contents of the main constituents they seem to have the same results. In general, the nutritional values have been already proven but people's interest in studying mushrooms is continuously increasing, hence we can expect some new species in the future scientific data but with similar contents of the main compounds. Regardless, they will remain the first choice of functional food in future human diets.

References

1. S. T. Chang and P. G. Miles, *Mushrooms: Cultivation, nutritional value, medicinal effect and environmental impact*, CRC Press, Boca Raton, 2nd edn, 2004.
2. D. Winkler, Present and historic relevance of Yartsa Gunbu (Cordyceps sinensis). An ancient myco-medicinal in Tibet, *Fungi*, 2008, **1**, 6–7.
3. L. M. Cheung, P. C. Cheung and V. E. Ooi, Antioxidant activity and total phenolics of edible mushroom extracts, *Food Chem.*, 2003, **81**, 249–255.
4. K. Tešanović, B. Pejin, F. Sibul, M. Matavulj, M. Rašeta, L. Janjušević and M. Karaman, A comparative overview of antioxidative properties and phenolic profiles of different fungal origins: fruiting bodies and submerged cultures of *Coprinus comatus* and *Coprinellus truncorum*, *J. Food Sci. Technol.*, 2017, **54**, 430–438.
5. M. Karaman, K. Atlagić, A. R. Novaković, F. Sibul, M. Ž. Živić, K. Stevanović and B. Pejin, Fatty acids predominantly affect anti-hydroxyl radical activity and FRAP value: The case study of two edible mushrooms, *Antioxidants*, 2019, **8**(10), 480.
6. B. Pejin, K. Tešanović, D. V. Jakovljević, S. Kaisarevic, F. Sibul, M. Rašeta and M. Karaman, The polysaccharide extracts from the fungi *Coprinus comatus* and *Coprinellus truncorum* do exhibit AChE inhibitory activity, *Nat. Prod. Res.*, 2019, **33**, 750–754.
7. Y. Zhang, M. Mo, L. Yang, F. Mi, Y. Cao, C. Liu, X. Tang, P. Wang and J. Xu, Exploring the species diversity of edible mushrooms in Yunnan, Southwestern China, by DNA barcoding, *J. Fungi*, 2021, 7(4), 310.
8. H. Rathore, S. Prasad and S. Sharma, Mushroom nutraceuticals for improved nutrition and better human health: a review, *PharmaNutrition*, 2017, **5**, 35–46.
9. J. R. Al-Obaidi, N. N. Jambari and E. I. Ahmad-Kamil, Myco Pharmaceuticals and nutraceuticals: promising agents to improve human well-being and life quality, *J. Fungi*, 2021, 7, 503.
10. K. Kumari, Mushrooms as source of dietary fiber and its medicinal value: a review article, *J. Pharmacogn. Phytochem.*, 2020, **9**, 2075–2078.
11. L. Ho, N. A. Zulkifli and T. Tan, Edible mushroom: nutritional properties, potential nutraceutical values, and its utilisation in food product development, in *An Introduction to Mushroom*, InTech Open, 2020, pp. 19–38.
12. A. G. Niego, S. Rapior, N. Thong Klang, O. Raspé, W. Jaidee, S. Lumyong and K. D. Hyde, Macrofungi as a nutraceutical source: promising bioactive compounds and market value, *J. Fungi*, 2021, 7, 397.
13. S. A. Heleno, L. Barros, A. Martins, P. Morales, V. Fernández-Ruiz, J. Glamočlija, M. D. Soković and I. C. Ferreira, Nutritional value, bioactive compounds, antimicrobial activity and bioaccessibility studies with wild edible mushrooms, *LWT–Food Sci. Technol.*, 2015, **63**, 799–806.

14. D. K. Rahi and D. Malik, Diversity of mushrooms and their metabolites of nutraceutical and therapeutic significance, *J. Mycol.*, 2016, 7654123.
15. T. Fernandes, C. Garrine, J. Ferrão, V. Bell and T. Varzakas, Mushroom nutrition as preventative healthcare in Sub-Saharan Africa, *Appl. Sci.*, 2021, **11**, 4221.
16. X. Wang, J. Zhang, L. Wu, Y. Zhao, T. Li, J. Li, Y. Wang and H. Liu, A mini-review of chemical composition and nutritional value of edible wild-grown mushrooms from China, *Food Chem.*, 2014, **151**, 279–285.
17. FAO, *Protein quality evaluation*, Food and Agricultural Organization of the United Nations, Rome, 1991.
18. V. N. Pathak, N. Yadav and M. Gour, *Mushroom production and processing technology*, Agrobios, Bangalore, India, 2000, pp. 1–192.
19. B. Muszyńska, K. Kała, J. Rojowski, A. Grzywacz and W. Opoka, Composition and biological properties of *Agaricus bisporus* fruiting bodies, *Pol. J. Food Nutr. Sci.*, 2017, **67**, 173–182.
20. S. A. Heleno, L. Barros, M. J. Sousa, A. Martins and I. C. F. R. Ferreira, Study and characterization of selected nutrients in wild mushrooms from Portugal by gas chromatography and high performance liquid chromatography, *Microchem. J.*, 2009, **93**, 195–199.
21. S. Beluhan and A. Ranogajec, Chemical composition and non-volatile components of Croatian wild edible mushrooms, *Food Chem.*, 2011, **124**, 1076–1082.
22. S. S. Singh, H. Wang, Y. S. Chan, W. Pan, X. Dan, C. M. Yin, O. Akkouh and T. B. Ng, Lectins from edible mushrooms, *Molecules*, 2015, **20**, 446–469.
23. S. Ragucci, N. Landi, R. Russo, M. Valletta, P. V. Pedone, A. Chambery and A. Di Maro, Ageritin from Pioppino Mushroom: the prototype of ribotoxin-like proteins, a novel family of specific ribonucleases in edible mushrooms, *Toxins*, 2021, **13**, 263.
24. Z. Bano and S. Rajarathnam, Cultivation studies on *Pleurotus sajor-caju*, *Mushroom J.*, 1982, **115**, 243–245.
25. P. Kalač, Chemical composition and nutritional value of European species of wild growing mushrooms: a review, *Food Chem.*, 2009, **113**(1), 9–16.
26. *Wild plants, mushrooms and nuts: functional food properties and applications*, ed. I. C. Ferreira, P. Morales and L. Barros, John Wiley & Sons, 2016.
27. M. Naeem, S. Rani and S. Ozgen, Emerging role of edible mushrooms in food industry and its nutritional and medicinal consequences, *Eurasian J. Food Sci. Technol.*, 2020, **4**(1), 6–23.
28. M. A. Khan and M. Tania, Nutritional and medicinal importance of *Pleurotus* mushrooms: an overview, *Food Rev. Int.*, 2012, **28**, 313–329.
29. Â. Fernandes, L. Barros, J. C. Barreira, A. L. Antonio, M. B. Oliveira, A. Martins and I. C. Ferreira, Effects of different processing technologies on chemical and antioxidant parameters of *Macrolepiota procera* wild mushroom, *LWT-Food Sci. Technol.*, 2013, **54**, 493–499.

30. K. I. Alzand, M. S. Bofaris and A. Ugiş, Chemical composition and nutritional value of edible wild growing mushrooms: a review, *World J. Pharm. Res.*, 2019, **8**, 31–46.
31. A. Ćirić, I. Kruljević, D. S. Stojković, Â. Fernandes, L. Barros, R. C. Calhelha, I. C. Ferreira, M. D. Soković and J. Glamočlija, Comparative investigation on edible mushrooms Macrolepiota mastoidea, M. rhacodes and M. procera: functional foods with diverse biological activities, *Food Funct.*, 2019, **10**(12), 480.
32. A. S. Bengu, Some elements and fatty acid profiles of three different wild edible mushrooms from Tokat province in Turkey, *Prog. Nutr.*, 2019, **21**(1), 189–193.
33. L. Barros, P. Baptista, D. Correia, J. Morais and I. Ferreira, Effects of conservation treatment and cooking on the chemical composition and antioxidant activity of Portuguese wild edible mushrooms, *J. Agric. Food Chem.*, 2007, **55**, 4781–4788.
34. F. Watanabe, J. Schwartz, S. Takenaka, E. Miyamoto, N. Ohishi, E. Nelle, R. Nochstrasser and Y. Yabuta, Characterization of Vitamin B_{12} compounds in the wild edible mushrooms Black Trumpet (*Craterellus cornucopioides*) and Golden Chanterelle (*Cantharellus cibarius*), *J. Nutr. Sci. Vitaminol.*, 2012, **58**, 438–441.
35. T. Bito, F. Teng, N. Ohishi, S. Takenaka, E. Miyamoto, E. Sakuno, K. Terashima, Y. Yabuta and F. Watanabe, Characterization of vitamin B_{12} compounds in the fruiting bodies of shiitake mushroom (*Lentinula edodes*) and bed logs after fruiting of the mushroom, *Mycoscience*, 2014, **55**(6), 462–468.
36. Z. Schneider Biosynthesis of vitamin B12, in *Comprehensive B_{12}*, ed. Z. Scheider and A. Stroinski, Walter de Gruyter, Berlin, 1987, pp. 93–110.
37. P. Mattila, K. Könkö, M. Eurola, J. M. Pihlava, J. Astola, L. Vahteristo, V. Hietaniemi, J. Kumpulainen, M. Valtonen and V. Piironen, Contents of vitamins, mineral elements, and some phenolic compounds in cultivated mushrooms, *J. Agric. Food Chem.*, 2001, **49**(5), 2343–2348.
38. K. M. Phillips, D. M. Ruggio, R. L. Horst, B. Minor, R. R. Simon and M. J. Feeney, Vitamin D and sterol composition of 10 types of mushrooms from retail suppliers in the United States, *J. Agric. Food Chem.*, 2011, **59**(14), 7841–7853.
39. J. S. Roberts, A. Teichert and T. H. McHugh, Vitamin D_2 formation from postharvest UV-B treatment of mushrooms (*Agaricus bisporus*) and retention during storage, *J. Agric. Food Chem.*, 2008, **56**, 4541–4544.
40. S. R. Koyyalamudi, S. Jeong, K. Y. Cho and G. Pang, Vitamin B_{12} is the active corrinoid produced in cultivated white button mushrooms (*Agaricus bisporus*), *J. Agric. Food Chem.*, 2009, **57**, 6327–6333.
41. Q. Jiang, M. Zhang and A. S. Mujumdar, UV induced conversion during drying of ergosterol to vitamin D in various mushrooms: effect of different drying conditions, *Trends Food Sci. Technol.*, 2020, **105**, 200–210.

42. M. Usman, G. Murtaza and A. Ditta, Nutritional, medicinal, and cosmetic value of bioactive compounds in button mushroom (*Agaricus bisporus*): a review, *Appl. Sci.*, 2021, **11**, 5943.
43. J. Petrović, J. Glamočlija, D. S. Stojković, A. Ćirić, L. Barros, I. C. Ferreira and M. D. Soković, Nutritional value, chemical composition, antioxidant activity and enrichment of cream cheese with chestnut mushroom *Agrocybe aegerita* (Brig.) Sing, *J. Food Sci. Technol.*, 2015, **52**, 6711–6718.
44. J. Rodrigues Barbosa, M. M. dos Santos Freitas, L. H. da Silva Martins and R. N. de Carvalho, Polysaccharides of mushroom Pleurotus spp.: New extraction techniques, biological activities and development of new technologies, *Carbohydr. Polym.*, 2020, **229**, 115550.
45. C. Cerletti, S. Esposito and L. Iacoviello, Edible mushrooms and beta-glucans: impact on human health, *Nutrients*, 2021, **13**(7), 2195.
46. P. C. Cheung, Mini-review on edible mushrooms as source of dietary fiber: Preparation and health benefits, *Food Sci. Hum. Wellness*, 2013, **2**, 162–166.
47. P. C. Cheung, The nutritional and health benefits of mushrooms, *Nutr. Bull.*, 2010, **35**, 292–299.
48. G. Anusiya, U. Gowthama Prabu, N. V. Yamini, N. Sivarajasekar, K. Rambabu, G. Bharath and F. Banat, A review of the therapeutic and biological effects of edible and wild mushrooms, *Bioengineered*, 2021, **12**, 11239–11268.
49. S. H. Nile and S. W. Park, Total, soluble, and insoluble dietary fiber contents of wild growing edible mushrooms, *Czech J. Food Sci.*, 2014, **32**, 302–307.
50. I. Mirończuk-Chodakowska and A. M. Witkowska, Evaluation of Polish wild mushrooms as beta-glucan sources, *Int. J. Environ. Res. Public Health*, 2020, **17**, 7299.
51. F. S. Reis, L. Barros, A. Martins and I. C. Ferreira, Chemical composition and nutritional value of the most widely appreciated cultivated mushrooms: an inter-species comparative study, *Food Chem. Toxicol.*, 2012, **50**(2), 191–197.
52. S. Zhou, Y. Liu, Y. Yang, Q. Tang and J. Zhang, Hypoglycemic activity of polysaccharide from fruiting bodies of the Shaggy Ink Cap medicinal mushroom, Coprinus comatus (Higher Basidiomycetes), on mice induced by alloxan and its potential mechanism, *Int. J. Med. Mushrooms*, 2015, **17**(10), 957–964.
53. M. Rašeta, M. Karaman, M. Jakšić, F. Sibul, M. Kebert, A. R. Novaković and M. Popović, Mineral composition, antioxidant and cytotoxic biopotentials of wild-growing *Ganoderma* species (Serbia): *G. lucidum* (Curtis) P. Karst vs. *G. applanatum* (Pers.) Pat, *Int. J. Food Sci. Technol.*, 2016, **51**, 2583–2590.
54. M. Žižić, J. Zakrzewska, K. Tešanović, E. Bošković, M. Nešović and M. Karaman, Effects of vanadate on the mycelium of edible fungus *Coprinus comatus*, *J. Trace Elem. Med. Biol.*, 2018, **50**, 320–326.

55. M. Rakić, M. Karaman, S. Forkapić, J. Hansman, M. Kebert, K. Bikit and D. Mrdja, Radionuclides in some edible and medicinal macrofungal species from a Tara Mountain, Serbia, *Environ. Sci. Pollut. Res.*, 2014, **21**(19), 11283–11292.
56. M. Drewnowska, J. Falandysz, M. Chudzińska, A. Hanć, M. Saba and D. Barałkiewicz, Leaching of arsenic and sixteen metallic elements from *Amanita fulva* mushrooms after food processing, *LWT–Food Sci. Technol.*, 2017, **84**, 861–866.
57. S. Pankavec, A. Hanć, D. Barałkiewicz, A. Dryżałowska, J. Zhang and J. Falandysz, Mineral constituents of conserved white button mushrooms: similarities and differences, *Rocz. Panstw. Zakl. Hig.*, 2019, **70**(1), 15–25.
58. M. J. Melgar, J. V. Alonso and M. Á. García, Cadmium in edible mushrooms from NW Spain: Bioconcentration factors and consumer health implications, *Food Chem. Toxicol.*, 2016, **88**, 13–20.
59. V. M. Stefanovic, J. Trifković, J. Mutić and Ž. Tešić, Metal accumulation capacity of parasol mushroom (*Macrolepiota procera*) from Rasina region (Serbia), *Environ. Sci. Pollut. Res.*, 2016, **23**, 13178–13190.
60. I. Širić, A. Kasap, D. Bedeković and J. Falandysz, Lead, cadmium and mercury contents and bioaccumulation potential of wild edible saprophytic and ectomycorrhizal mushrooms, Croatia, *J. Environ. Sci. Health, Part B*, 2017, **52**, 156–165.
61. J. Su, J. Zhang, J. Li, T. Li, H. Liu and Y. Wang, Determination of mineral contents of wild *Boletus edulis* mushroom and its edible safety assessment, *J. Environ. Sci. Health, Part B*, 2018, **53**, 454–463.
62. M. Rasalanavho, R. Moodley and S. B. Jonnalagadda, Elemental bioaccumulation and nutritional value of five species of wild growing mushrooms from South Africa, *Food Chem.*, 2020, **319**, 126596.
63. M. Karaman, E. Čapelja, M. Rašeta and M. Rakić, Diversity, Chemistry and Environmental Contamination of Wild Growing Medicinal Mushroom Species as Sources of Biologically Active Substances (Antioxidants, Antidiabetics and AChE Inhibitors), in *A B C of Medicinal Mushrooms for Human Health and Beauty*, ed. A. Arya and K. Rusevska, Springer, Singapore, 2022, 8, 203–257.
64. P. Niedzielski, M. Mleczek, A. Budka, P. Rzymski, M. Siwulski, A. Jasińska, M. Gąsecka and S. Budzyńska, A screening study of elemental composition in 12 marketable mushroom species accessible in Poland, *Eur. Food Res. Technol.*, 2017, **243**, 1759–1771.
65. P. Nowakowski, S. K. Naliwajko, R. Markiewicz-Żukowska, M. H. Borawska and K. Socha, The two faces of *Coprinus comatus*—Functional properties and potential hazards, *Phytother. Res.*, 2020, **34**, 2932–2944.
66. E. Čapelja, N. Stević, V. Galović, M. Novaković and M. Karaman, Molecular determination of autochtonous fungal species from Serbia,

based on analysis of ITS region of rDNA, *Genetika*, 2014, **46**(1), 33–42.

67. N. Stević, E. Čapelja, V. Galović, M. Novaković and M. Karaman, Molecular characterisation of autochtonous fungal culture collection by analysis of ITS1 sequences of rDNA, *Genetika*, 2014, **46**(1), 235–242.
68. V. Galović, M. Marković, P. Pap, M. Mulett, M. Rakić, A. Vasiljević and S. Pekeč, Molecular taxonomy and phylogenetics of *Daedaleopsis confragosa* (Bolt.: Fr.) J. Schröt. from wild cherry in Serbia, *Genetika*, 2018, **50**(2), 519–532.
69. J. A. Maga, Mushroom flavor, *J. Agric. Food Chem.*, 1981, **29**, 1–4.
70. J. H. Litchfield, Morel mushroom mycelium as a food-flavoring material, *Biotechnol. Bioeng.*, 1967, **9**, 289–304.
71. P. A. Temussi, Sweet, bitter and umami receptors: a complex relationship, *Trends Biochem. Sci.*, 2009, **34**(6), 296–302.
72. M. G. Shashidhar, P. Giridhar, K. Udaya Sankar and B. Manohar, Bioactive principles from *Cordyceps sinensis*: A potent food supplement – A review, *J. Funct. Foods*, 2013, **5**(3), 1013–1030.
73. S. Yamamoto, M. F. Wang, A. A. Adjei and C. K. Ameho, Role of nucleosides and nucleotides in the immune system, gut reparation after injury, and brain function, *Nutrition*, 1997, **13**(4), 372–374.
74. E. Schlimme, D. Martin and H. Meisel, Nucleosides and nucleotides: natural bioactive substances in milk and colostrum, *Br. J. Nutr.*, 2000, **84**(S1), 59–68.
75. C. Phan, J. Wang, S. Cheah, M. Naidu, P. David and V. Sabaratnam, A review on the nucleic acid constituents in mushrooms: nucleobases, nucleosides and nucleotides, *Crit. Rev. Biotechnol.*, 2018, **38**, 762–777.
76. A. Ranogajec, S. Beluhan and Z. Šmit, Analysis of nucleosides and monophosphate nucleotides from mushrooms with reversed-phase HPLC, *J. Sep. Sci.*, 2010, **33**(8), 1024–1033.
77. F. Yang, R. Lv, Y. Zhang and Z. Xia, Comparison study on nucleosides and nucleotides in edible mushroom species by capillary zone electrophoresis, *Anal. Methods*, 2012, **4**, 546–549.
78. Q. Yu, M. Guo, B. Zhang, H. Wu, Y. Zhang and L. Zhang, Analysis of nutritional composition in 23 kinds of edible fungi, *J. Food Qual.*, 2020, **2020**, 1–9.
79. I. I. Bandura, A. S. Kulyk, S. V. Makohon, O. V. Khareba and V. V. Khareba, Influence of the substrate composition on the yield and nutritional value of the fruiting bodies of the edible mushrooms *Pleurotus citrinopileatus* and *Cyclocybe aegerita*, *Plant Var. Stud. Prot.*, 2021, **17**, 130–138.
80. M. E. Valverde, T. Hernández-Pérez and O. Paredes-López, Edible mushrooms: improving human health and promoting quality life, *Int. J. Microbiol.*, 2015, **2015**, 376387.
81. M. Manzoor, M. Asrar, S. K. Leghari and Z. Ahmed, Nutritional value of wild mushrooms from Kharan district of Balochistan, Pakistan, *FUUAST J. Biol.*, 2019, **9**(2), 209–214.

82. D. Chatterjee, D. Halder and S. Das, Varieties of mushrooms and their nutraceutical importance: a systematic review, *J. Clin. Diagn. Res.*, 2021, **15**(3), OE01–OE06.
83. B. Jacinto-Azevedo, N. Valderrama, K. Henríquez, M. Aranda and P. M. Aqueveque, Nutritional value and biological properties of Chilean wild and commercial edible mushrooms, *Food Chem.*, 2021, **356**, 129651.
84. I. K. Jedidi, I. K. Ayoub, P. Thonart and N. Bouzouita, Chemical composition and non-volatile components of three wild edible mushrooms collected from northwest Tunisia, *Mediterr. J. Chem.*, 2016, **5**(3), 434–441.
85. H. Manninen, M. Rotola-Pukkila, H. Aisala, A. I. Hopia and T. Laaksonen, Free amino acids and 5′-nucleotides in Finnish forest mushrooms, *Food Chem.*, 2018, **247**, 23–28.
86. N. Turfan, A. U. Pekşen, B. Kibar and S. S. Ünal, Determination of nutritional and bioactive properties in some selected wild growing and cultivated mushrooms from Turkey, *Acta Sci. Pol. Hortorum Cultus*, 2018, **17**(3), 57–72.
87. D. S. Kumari, M. S. Reddy and R. C. Upadhyay, Nutritional composition and antioxidant activities of 18 different wild *Cantharellus* mushrooms of Northwestern Himalayas, *Food Sci. Technol. Int.*, 2011, **17**, 557–567.
88. M. Kozarski, A. Klaus, J. Vunduk, Z. Zizak, M. Nikšić, D. V. Jakovljević, M. M. Vrvić and L. J. Van Griensven, Nutraceutical properties of the methanolic extract of edible mushroom Cantharellus cibarius (Fries): primary mechanisms, *Food Funct.*, 2015, **6**(6), 1875–1886.
89. B. Muszyńska, K. Kała, A. Firlej and K. Sułkowska-Ziaja, *Cantharellus cibarius* – culinary-medicinal mushroom content and biological activity, *Acta Pol. Pharm.*, 2016, **73**(3), 589–598.
90. D. S. Stojković, F. S. Reis, L. Barros, J. Glamočlija, A. Ćirić, L. J. van Griensven, M. D. Soković and I. C. Ferreira, Nutrients and non-nutrients composition and bioactivity of wild and cultivated *Coprinus comatus* (O.F.Müll.) Pers, *Food Chem. Toxicol.*, 2013, **59**, 289–296.
91. N. Stilinović, I. Capo, S. Vukmirović, A. Rašković, A. Tomas, M. Popović and A. Sabo, Chemical composition, nutritional profile and *in vivo* antioxidant properties of the cultivated mushroom *Coprinus comatus*, *R. Soc. Open Sci.*, 2020, **7**(9), 200900.
92. I. Okoro and F. I. Achuba, Proximate and mineral analysis of some wild edible mushrooms, *Afr. J. Biotechnol.*, 2012, **11**(30), 7720–7724.
93. E. M. Tersoo-Abiem, S. T. Gbaa and S. Sule, Effect of mushroom (*Coprinellus micaceus*) flour addition on the quality characteristics of Millet-Based Ibyer, *Res. J. Food Nutr.*, 2019, **3**(4), 1–5.
94. E. Palazzolo, M. Letizia Gargano and G. Venturella, The nutritional composition of selected wild edible mushrooms from Sicily (southern Italy), *Int. J. Food Sci. Nutr.*, 2012, **63**, 79–83.

95. J. Petrović, D. S. Stojković, F. S. Reis, L. Barros, J. Glamočlija, A. Ćirić, I. C. Ferreira and M. D. Soković, Study on chemical, bioactive and food preserving properties of *Laetiporus sulphureus* (Bull.: Fr.) Murr., *Food Funct.*, 2014, **5**(7), 1441–1451.
96. T. Ao and C. R. Deb, Nutritional and antioxidant potential of some wild edible mushrooms of Nagaland, India, *J. Food Sci. Technol.*, 2019, **56**(2), 1084–1089.
97. F. Kalyoncu, B. Ergönül, H. Yıldız, E. Kalmış and H. M. Solak, Chemical composition of four wild edible mushroom species collected from Southwest Anatolia, *Gazi Univ. J. Sci.*, 2010, **23**, 375–379.
98. K. Acharya and M. Rai, Proximate composition, free radical scavenging and NOS activation properties of a wild edible mushroom, *Int. J. Pharm. Pharm. Sci.*, 2013, **5**(1), 67–72.
99. E. H. Erbiai, L. P. da Silva, R. Saidi, Z. Lamrani, J. C. Esteves da Silva and A. Maouni, Chemical composition, bioactive compounds, and antioxidant activity of two wild edible mushrooms *Armillaria mellea* and *Macrolepiota procera* from two countries (Morocco and Portugal), *Biomolecules*, 2021, **11**, 575.
100. H. Li, Z. Zhang, M. Li, X. Li and Z. W. Sun, Yield, size, nutritional value, and antioxidant activity of oyster mushrooms grown on perilla stalks, *Saudi J. Biol. Sci.*, 2017, **24**, 347–354.
101. A. González, C. Nobre, L. S. Simões, M. Cruz, A. Loredo, R. M. Rodríguez-Jasso, J. C. Contreras, J. A. Texeira and R. Belmares, Evaluation of functional and nutritional potential of a protein concentrate from *Pleurotus ostreatus* mushroom, *Food Chem.*, 2020, **346**, 128884.
102. N. T. Ache, E. B. Manju, M. N. Lawrence and R. K. Tonjock, Nutrient and mineral components of wild edible mushrooms from the Kilum-Ijim forest, Cameroon, *Afr. J. Food Sci.*, 2021, **15**, 152–161.
103. N. Muthu and K. Shanmugasundaram, Proximate and mineral compositions of edible mushroom *Agrocybe aegerita*, *J. Pharmacogn. Phytochem.*, 2016, **5**, 116–119.
104. H. Ç. Yilmaz, H. Işık, A. Ş. Bengü and İ. Türkekul, Fatty acid contents of two edible mushroom species (Cyclocybe aegerita and Hygrophorus eburneus) collected from Tokat region, *Middle East J. Sci.*, 2020, **6**(1), 37–43.
105. P. Günç Ergönül, I. Akata, F. Kalyoncu and B. Ergönül, Fatty acid compositions of six wild edible mushroom species, *Sci. World J.*, 2013, **2013**, DOI: 10.1155/2013/163964.
106. D. S. Stojković, N. Kovacevic-Grujicic, F. S. Reis, S. Davidović, L. Barros, J. Popović, I. P. Petrović, A. Pavic, J. Glamočlija, A. Ćirić, M. Stevanović, I. C. Ferreira and M. D. Soković, Chemical composition of the mushroom *Meripilus giganteus* Karst. and bioactive properties of its methanolic extract, *LWT–Food Sci. Technol.*, 2017, **79**, 454–462.
107. G. Sadi, A. Kaya, H. A. Yalcin, B. Emsen, A. Kocabaş, D. I. Kartal and A. Altay, Wild edible mushrooms from Turkey as possible anticancer

agents on HepG2 cells together with their antioxidant and antimicrobial properties, *Int. J. Med. Mushrooms*, 2016, **18**(1), 83–95.
108. H. Jing, J. Li, J. Zhang, W. Wang, S. Li, Z. Ren, Z. Gao, X. Song, X. Wang and L. Jia, The antioxidative and anti-aging effects of acidic- and alkalic-extractable mycelium polysaccharides by *Agrocybe aegerita* (Brig.) Sing, *Int. J. Biol. Macromol.*, 2018, **106**, 1270–1278.
109. J. Petrović, M. A. Papandreou, J. Glamočlija, A. Ćirić, C. Baskakis, C. Proestos, F. N. Lamari, P. Zoumpoulakis and M. D. Soković, Different extraction methodologies and their influence on the bioactivity of the wild edible mushroom *Laetiporus sulphureus* (Bull.) Murrill, *Food Funct.*, 2014, **5**(11), 2948–2960.
110. V. Kokkoris, I. Massas, E. Polemis, G. Koutrotsios and G. I. Zervakis, Accumulation of heavy metals by wild edible mushrooms with respect to soil substrates in the Athens metropolitan area (Greece), *Sci. Total Environ.*, 2019, **685**, 280–296.
111. I. Mirończuk-Chodakowska, K. Socha, M. E. Zujko, K. M. Terlikowska, M. H. Borawska and A. M. Witkowska, Copper, manganese, selenium and zinc in wild-growing edible mushrooms from the Eastern Territory of "Green Lungs of Poland": nutritional and toxicological implications, *Int. J. Environ. Res. Public Health*, 2019, **16**, 3614.
112. M. Mleczek, Z. Magdziak, P. Goliński, M. Siwulski and K. Stuper-Szablewska, Concentrations of minerals in selected edible mushroom species growing in Poland and their effect on human health, *Acta Sci. Pol., Technol. Aliment.*, 2013, **12**(2), 203–214.
113. J. Brzezicha-Cirocka, M. Grembecka, I. Grochowska, J. Falandysz and P. Szefer, Elemental composition of selected species of mushrooms based on a chemometric evaluation, *Ecotoxicol. Environ. Saf.*, 2019, **173**, 353–365.
114. A. A. Georgescu, F. D. Danet, C. Radulescu, C. Stihi, I. D. Dulama and D. E. Chelarescu, Determination of several elements in edible mushrooms using ICP-MS, *Rom. Journ. Phys.*, 2016, **61**(5–6), 1087–1097.
115. A. Kaya, F. Kiliçel, H. S. Karapınar and Y. Uzun, Mineral contents of some wild edible mushrooms, *The Journal of Fungus*, 2017, **8**(2), 178–183.
116. N. Stilinović, B. D. Škrbić, J. Živančev, N. Mrmoš, N. Pavlović and S. Vukmirović, The level of elements and antioxidant activity of commercial dietary supplement formulations based on edible mushrooms, *Food Funct.*, 2014, **5**, 3170–3178.
117. C. SariKurkcu, B. Tepe, M. S. Kocak and M. C. Uren, Metal concentration and antioxidant activity of edible mushrooms from Turkey, *Food Chem.*, 2015, **175**, 549–555.
118. M. Kosanić, B. Ranković, A. Rančić and T. P. Stanojković, Evaluation of metal concentration and antioxidant, antimicrobial, and anticancer potentials of two edible mushrooms *Lactarius deliciosus* and *Macrolepiota procera*, *J. Food Drug Anal.*, 2016, **24**(3), 477–484.
119. S. Igbiri, N. A. Udowelle, O. C. Ekhator, R. N. Asomugha, Z. N. Igweze and O. E. Orisakwe, Edible mushrooms from Niger Delta, Nigeria with

heavy metal levels of public health concern: a human health risk assessment, *Recent Pat. Food, Nutr. Agric.*, 2018, **9**(1), 31–41.

120. M. Gil-Martínez, C. M. Navarro-Fernández, J. M. Murillo, M. T. Domínguez and T. Marañón, Trace elements and C and N isotope composition in two mushroom species from a mine-spill contaminated site, *Sci. Rep.*, 2020, **10**, 6434.

CHAPTER 3

Updates on Fatty Acids in Mushrooms: Content, Characterization, and Biological Effects

G. TEL-ÇAYAN,*[a] E. DEVECI[b] AND F. ÇAYAN[a]

[a] Department of Chemistry and Chemical Processing Technologies, Muğla Vocational School, Muğla Sıtkı Koçman University, 48000, Muğla, Turkey;
[b] Chemistry and Chemical Processing Technology Department, Technical Sciences Vocational School, Konya Technical University, 42250 Konya, Turkey
*Emails: gulsentel@mu.edu.tr; lillian@ipb.pt

3.1 Introduction

Mushrooms are described as healthy foods with ingredients of carbohydrates, protein, vitamins, fibers, minerals, valuable fatty acids, and fat that cannot be synthesized in the human body.[1] In addition to their nutritional importance, mushrooms have gained great popularity due to their delicious taste and aroma diversity.[2] Besides nutritional and gastronomic attractiveness, mushrooms have been pronounced as new generation biotherapeutics due to their described more than 100 medicinal functions, among which the most basic medicinal properties are their anticancer, immunomodulating, antidiabetic, antiallergic, anticholesterolemic, cardiovascular-protective, hepatoprotective, antibacterial, antioxidant, antifungal, antiparasitic, detoxification, and antiviral effects. They also protect against inflammation and tumor development. Also, many bioactive

Food Chemistry, Function and Analysis No. 36
Edible Fungi: Chemical Composition, Nutrition and Health Effects
Edited by Dejan Stojković and Lillian Barros

Published by the Royal Society of Chemistry, www.rsc.org

compounds (fats, lectins, proteins, alkaloids, minerals, glycosides, tocopherols, terpenoids, phenolics, folates, ascorbic, and organic acids) have been identified that are related to these valuable medicinal properties of mushrooms.[3] Mushrooms are also characterized as functional foods with these important nutritional values and excellent medicinal properties.[4] As a result of these properties, mushrooms are becoming increasingly important every day.

From a nutritional point of view, lipid synthesis in the body and dietary intake of lipids are important as they have various functions, such as structural membrane composition, the energy reserve in living organisms, protection against impacts and thermal insulation.[5] Fatty acids are rich in energy and have simple structures containing carboxylic acids derived from between 4 and 26 carbon hydrocarbon chains among the different types of lipids. Fatty acids contain an even number of carbon atoms due to the condensation of two carbon (acetate) units in an unbranched chain of 12 and 24 carbons during their biosynthesis. These fatty acids are divided into two classes: saturated fatty acids (SFAs) (only simple bonds) and unsaturated fatty acids (UFAs) [mono (MUFAs) or poly double bonds (PUFAs) between carbon atoms]. The counting of carbon atoms in fatty acids is done in two ways. In the first method, the carbon of the carboxyl group is numbered and the locations of the double bonds are indicated by superscripts of (Δ) with a number relative to the carbon. In the second, the carbon furthest from the carboxyl group, that is, the carbon of the methyl group, is called the carbon (omega-ω) that gets the number one. The most important PUFAs for human health are omega-3 (ω-3), omega-6 (ω-6), and omega-9 (ω-9) fatty acids.[6,7]

Although mushrooms contain low amounts of lipids, they have essential fatty acids such as oleic, linoleic, and linolenic acids in their lipid composition as the main components.[8] In general, the ratio of unsaturated fatty acids (UFAs) is higher than the ratio of saturated fatty acids (SFAs). For these reasons, mushrooms have the advantage of having a high percentage of unsaturated fatty acids (UFAs) in comparison to other foods of vegetable and animal origin. The percentage of these fatty acids in mushrooms varies and is usually in the range of 0.0–81.1% for linoleic acid, 1.0–60.3% for oleic acid and 0.0–28.8% for linolenic acid.[7] In addition to their energy properties, fatty acids affect many biochemical and physiological reactions related to secretory, cardiovascular and immune functions.[9] Due to the valuable properties of fatty acids, the human diet must be supplemented with foods containing fatty acids. As the consumption of certain foods has restrictions due to their ingredients such as allergy, high cholesterol, and high-calorie value, there is increasing interest in the search for new and natural sources of essential fatty acids. In this regard, mushrooms contain high amounts of protein and fiber, have low-calorie value and no cholesterol, and are easily accessible, allowing them to be considered as an advantageous source for the intake of fatty acids. In this chapter, reports on fatty acids from mushroom species that have been studied for their contents together with their biological effects in the last two decades are evaluated.

3.2 Fatty Acids Content in Mushrooms

Twenty-one fatty acids from *Lentinus crinitus* basidiocarp pileus and stipe were identified using GC analyses by Berteli *et al.*[10] The main fatty acids in pileus and stipes of *L. crinitus* were found to be linoleic (59.9, 61.80%), oleic (13.5, 12.65%), and palmitic (13.40, 12.80%) acids. The predominant class of fatty acids was PUFAs (~63%), followed by SFAs (~23%), and MUFAs (~4%). The fatty acid contents of the pileus and the stipe were similar. The major fatty acid contents of mushroom species studied including PUFAs, MUFAs and SFAs are given in Table 3.1.

Eleven different fatty acids with carbon chain lengths of 14–24 were detected from *Agaricus silvicola*, *Hydnum repandum*, and *Pleurotus eryngii* by GC-MS. The major fatty acids were oleic acid in *P. eryngii* (39.74%) and linoleic acid in *H. repandum* (60.03%) and *A. silvicola* (35.60%) mushrooms. The amount of palmitic acid was higher than the other fatty acids in all studied mushrooms.[11]

Kostic *et al.*[12] characterized fatty acid compositions of three *Russula* species (*R. rosea*, *R. nigricans*, *R. integra*) using transesterification procedures and a GC-FID technique. Twenty-one fatty acids were identified in *R. integra*, while there were 10 fatty acids in *R. rosea* and 20 in *R. nigricans*. The major fatty acids were found to be linoleic (13.39–52.0%), oleic (21.85–35.74%), stearic (6.57–36.3%) and palmitic (8.8–16.6%) acids. *R. rosea* possessed the highest percentage of SFA (50.9%), while *R. nigricans* and *R. integra* had the largest amounts of PUFAs (52.3% and 43.8%, respectively).

The analysis of the fatty acid profiles of nine mushroom species (*Bovista plumbea*, *Boletus edulis*, *Agaricus bisporus*, *Cantharellus cibarius*, *Helvella lacunose*, *Lactarius delicious*, *Morchella elata*, *Pleurotus ostreatus*, *Macrolepiota procera*) showed that linoleic (10.78–51.15%), oleic (3.46–45.32%), palmitic (10.26–29.57%) and stearic (9.83–32.67%) acids were the main fatty acids. The UFAs of the mushrooms studied in this study were in higher amounts than SFAs.[13]

Differences in the fatty acid compositions of the caps and stipes of two different *Leccinum* species (*L. aurantiacum* and *L. versipelle*) collected from three different regions of Poland (northern, central and south) were determined using GC-FID. Linoleic (49.05–66.15%), palmitic (18.44–26.22%), oleic (5.83–23.74%) and pentadecanoic (4.01–5.54%) acids were the major fatty acids in the caps of *Leccinum* species. Linoleic (41.12–59.72%), oleic (6.55–32.99%), palmitic (15.31–24.10%) and pentadecanoic (3.06–5.29%) acids were the major fatty acids in the stipes of *Leccinum* species.[14]

The most abundant fatty acids of *Pleurotus sajor-caju* were detected as linoleic (58.3%), palmitic (17.0%), oleic (16.0%) and stearic (5.2%) acids by GC. The high percentage of PUFAs (up to 85%) reported was to be considered an important contributor to the health value of mushroom species.[15]

Fatty acid composition of seventeen mushrooms namely, *Chondrostereum purpureum*, *Coprinus comatus*, *Gloeophyllum odoratum*, *Daedalea quercina*, *Gloeophyllum trabeum*, *Gloeophyllum sepiarium*, *Hydnum repandum*, *Inonotus radiatus*, *Omphalotus olearius*, *Macrolepiota procera*, *Phaeolus schweinitzii*,

Table 3.1 The contents of major fatty acids in mushrooms.[a]

Species	C18:2 (%)	C18:1 (%)	C18:3 (%)	C16:0 (%)	C18:0 (%)	SFAs (%)	MUFAs (%)	PUFAs (%)	Country	Ref.
Lentinus crinitus	59.9	13.5	2.04	13.40	2.94	23.23	14.3	62.5	Culture	10
Agaricus silvicola	60.30	4.21	—	21.62	8.23	32.81	7.16	60.03	Turkey	11
Hydnum repandum	35.60	31.14	1.15	21.50	7.09	29.58	33.66	36.75		
Pleurotus eryngii	33.57	39.74	—	19.45	3.72	25.51	40.92	33.57		
Russula integra	38.2	21.85	3.27	16.6	6.57	32.4	23.81	43.8	Serbia	12
Russula nigricans	52.0	27.1	—	8.8	7.8	19.4	28.3	52.3		
Russula rosea	13.39	35.74	—	12.0	36.3	50.9	35.74	13.39		
Agaricus bisporus	49.74	3.46	—	20.97	11.61	46.24	4.02	49.74	Turkey	13
Boletus edulis	41.99	23.21	—	17.61	9.83	32.17	25.90	41.99		
Bovista plumbea	—	51.15	—	5.31	8.52	32.56	13.83	53.61		
Cantharellus cibarius	38.64	28.62	—	14.86	11.86	29.13	32.33	38.54		
Helvella lacunosa	22.49	45.32	—	15.10	13.82	30.56	46.94	22.49		
Lactarius delicious	15.54	38.21	—	10.26	32.67	44.91	39.54	15.54		
Macrolepiota procera	48.85	8.93	—	23.32	10.95	38.33	12.23	48.55		
Morchella elata	10.78	17.65	—	29.57	14.01	48.74	22.19	29.08		
Pleurotus ostreatus	40.31	12.57	—	20.54	16.05	42.77	24.13	40.31		
Leccinum aurantiacum caps	57.21	12.18	0.10	21.16	0.98	28.96	15.67	57.43	Poland	14
Leccinum aurantiacum stipes	55.39	16.91	0.11	18.89	0.83	25.98	18.43	55.66		
Leccinum versipelle caps	57.92	7.81	0.12	24.09	1.00	32.24	9.47	58.31		
Leccinum versipelle stipes	53.01	20.39	0.09	18.35	0.75	24.79	21.89	53.34		
Pleurotus sajor-caju	58.3	16.0	—	17.0	5.2	—	—	—	Culture	15
Chondrostereum purpureum	61.63	10.66	—	19.59	2.05	23.05	15.27	61.63	Turkey	16
Coprinus comatus	33.72	25.08	—	22.20	3.18	36.67	29.55	33.72		
Daedalea quercina	6.45	18.31	—	21.68	5.65	74.31	19.16	6.45		
Gloeophyllum odoratum	41.51	25.09	—	13.51	2.93	28.22	30.21	41.51		
Gloeophyllum sepiarium	28.35	35.72	—	14.89	5.15	33.16	38.42	28.35		
Gloeophyllum trabeum	20.30	25.16	—	10.85	3.70	51.30	28.35	20.30		
Hydnum repandum	28.61	47.85	—	14.75	3.41	22.46	48.89	28.61		
Inonotus radiatus	35.62	31.18	—	17.20	—	29.98	34.34	35.62		
Macrolepiota procera	47.26	12.69	—	7.84	8.39	30.38	22.29	47.26		

Table 3.1 (*Continued*)

Species	C18:2 (%)	C18:1 (%)	C18:3 (%)	C16:0 (%)	C18:0 (%)	SFAs (%)	MUFAs (%)	PUFAs (%)	Country	Ref.
Omphalotus olearius	48.27	6.76	—	28.09	7.42	43.16	8.52	48.27		
Phellinus igniarius	41.31	14.91	—	9.66	19.77	42.89	15.74	41.31	Turkey	16
Phaeolus schweinitzii	51.47	24.42	—	16.50	1.76	23.01	25.46	51.47		
Schizophyllum commune	34.37	12.86	—	25.12	8.70	50.09	15.47	34.37		
Trametes bicolor	29.49	35.02	—	17.08	4.91	30.83	37.59	31.48		
Trametes pubescens	30.77	59.25	—	3.57	0.90	9.51	59.72	30.77		
Trametes suaveolens	31.18	43.43	—	13.89	5.30	23.04	45.72	31.18		
Trametes versicolor	16.82	42.11	—	18.92	7.02	38.27	43.74	17.92		
Morchella esculenta	57.3	7.4	0.2	17.5	12.6	32.3	9.0	58.7	Bulgaria	17
Flammulina velutipes	11.065	3.006	0.869	2.876	3.765	16.683	7.564	51.58	Pakistan	18
Ganoderma lucidum	10.257	3.544	0.0202	2.671	0.413	24.65	18.26	54.06		
Helvella crispa	1.421	1.999	3.398	3.675	1.893	11.375	19.689	15.792		
Laccaria bicolor	0.452	3.658	0.783	1.907	0.872	14.592	11.694	17.593		
Lactarius rubidus	12.985	4.868	0.957	1.985	0.859	15.859	8.647	48.64		
Mycena vitilis	16.013	2.0103	0.036	3.809	0.979	20.77	7.77	52.51		
Russula brevipes	1.246	1.063	1.006	2.758	2.985	16.784	3.967	25.867		
Russula nobilis	0.258	3.0678	3.978	3.768	1.869	12.987	17.689	22.683		
Xerocomellus chrysenteron	15.690	2.965	0.256	3.098	1.487	21.805	15.805	49.970		
Auricularia cornea	43.54	29.83	1.52	13.25	7.83	—	—	—	China	19
Coprinus comatus	73.39	5.82	1.06	12.93	0.97	15.84	7.42	76.73	Culture	20
Lycoperdon utriforme	15.47	*trace*	—	43.98	29.51	—	—	—	Turkey	21
Agaricus blazei	51.04	2.66	1.47	20.6	5.58	42.2	5.0	52.71	Market	22
Coprinus comatus	74.19	4.32	—	13.32	0.78	16.29	7.46	76.25	Serbia	23
Coprinus truncorum	70.09	9.08	0.15	12.39	0.92	14.92	11.97	73.11		
Cantharellus cinerus	40.13	16.89	—	30.95	9.3	—	—	—	Croatia	24
Clavariadelphus pistillaris	27.07	46.36	—	21.18	5.39	—	—	—		
Clitocybe nebularis	43.39	31.44	—	19.88	1.55	—	—	—		
Hygrocybe punicea	20.46	30.65	—	25.36	3.58	—	—	—		
Lactarius delicious	29.49	48.37	—	5.17	16.96	—	—	—	China	25
Agaricus bisporus (gamma irradiated)	79.7	—	—	8.1	3.4	19.2	0.8	80.1	Market	26
Agaricus bisporus (electron-beam irradiated)	78.6	—	—	8.2	4.1	20.0	1.0	79.0		

Macrolepiota mastoidea	66.59	11.10	0.120	15.84	0.88	19.16	12.91	67.92	Serbia	27
Macrolepiota rhacodes	78.62	6.06	0.081	10.00	1.16	13.97	6.79	79.2		
Macrolepiota procera	64	8.3	0.17	21	1.96	25	9.92	65		
Stereum rugosum	19.64	34.23	—	20.07	9.44	45.96	34.38	19.64	Turkey	28
Stereum sanguinolentum	12.73	20.39	—	39.14	17.78	66.21	21.03	12.73		
Pholiota aurivella	56.7	23.9	—	16.4	—	—	—	—	Turkey	29
Laetiporus sulphureus	7.73	33.94	—	28.60	28.55	—	—	—	Turkey	30
Suillus luteus	29.05	24.60	—	25.97	17.09	—	—	—		
Coprinus atramentarius	33.17	29.81	3.73	21.24	10.49	—	—	—		
Tuber indicum (Gongshan, Yunnan region)	76.52	17.94	—	9.89	2.61	13.85	18.38	67.77	China	31
Tuber indicum (Tibet region)	64.61	19.70	—	8.40	3.84	14.76	20.32	64.92		
Tuber indicum (Hebei region)	85.25	—	—	8.34	1.98	13.71	0.77	85.53		
Tuber indicum (Liaoning region)	70.72	17.87	—	6.13	1.61	10.63	18.42	70.94		
Tuber indicum (Qujing, Yunnan region)	56.05	15.52	—	6.49	2.52	15.93	19.87	64.20		
Boletus appendiculatus	72	0.11	—	15	0.81	16.92	2.35	71.55	Serbia	33
Boletus edulis	40	41	—	12.2	2.9	16.43	40.06	41.73		
Boletus impolitus	37	39	—	17	2.2	20.80	41.65	37.10		
Boletus purpureus	44	25	—	22.09	2.4	25.91	26.41	44.11		
Boletus regius	46	32	—	15	1.8	18.24	34.47	46.20		
Boletus rhodoxanthus	48	23	—	21	0.94	23.46	27.12	47.53		
Lactarius deliciosus	25	23	—	8.0	41	50.82	23.95	24.66		
Lactarius sanguifluus	20	16	—	5.9	34	63.63	16.17	19.83		
Lactarius semisanguifluus	28	64	—	10.9	57	46.41	25.27	27.88		
Lactarius volemus	19.5	25	0.64	6.3	4.9	13.26	64.39	20.15		
Xerocomellus chrysenteron	53	30	—	12.6	0.97	14.39	31.48	52.83		
Tricholoma equestre	28.29	28.59	—	28.84	7.58	40.59	31.12	28.29	Poland	34
Lenzites quercina	7.91	38.50	17.49	4.40	11.96	—	—	—	Nigeria	35
Polyporus squamosus	56.55	22.27	0.441	13.40	1.20	17.68	24.96	57.37	Romania	36
Agaricus arvensis	54.03	—	—	11.57	—	—	—	—	Turkey	37
Cantharellus cibarius	22.7	35.0	0.2	23.8	35.0	39.7	35.4	24.9	Bulgaria	38
Agaricus bisporus	71.41	1.80	0.10	12.73	5.00	25.69	2.55	71.75	Romania	39

Table 3.1 (*Continued*)

Species	C18:2 (%)	C18:1 (%)	C18:3 (%)	C16:0 (%)	C18:0 (%)	SFAs (%)	MUFAs (%)	PUFAs (%)	Country	Ref.
Boletus edulis	59.88	22.06	0.05	9.93	3.32	15.26	24.59	60.14		
Cantharellus cibarius	11.72	36.29	0.03	5.35	43.23	51.43	36.78	11.79		
Lactarius piperatus	44.51	15.96	0.12	15.11	9.01	25.93	29.44	44.63		
Pleurotus ostreatus	63.68	9.84	0.12	12.40	2.18	18.38	17.58	64.04		
Chondrogaster pachysporus	53.03	14.55	—	16.01	4.50	27.51	19.46	53.03	Turkey	40
Setchelliogaster tenuipes	5.35	35.80	—	38.60	12.26	59.85	34.80	5.35		
Lentinus tuberregium	—	5.98	7.44	4.55	6.75	—	—	—	Culture	45
Armillaria tabescens	27.64	35.92	—	19.40	5.37	31.90	40.41	27.64	Turkey	46
Leucopaxillus gentianeus	40.92	39.78	—	7.95	8.98	18.91	40.17	40.92		
Pleurotus eryngii	68.24	10.05	—	14.04	1.29	19.72	12.04	68.24		
Suillus granulatus	47.64	35.62	—	14.43	2.28	16.71	35.62	47.64		
Fomes fomentarius	42.1	9.1	8.9	20.1	8.9	31.1	—	—	Turkey	47
Funalia trogii	16.9	17.0	10.5	19.6	4.3	53.8	—	—		
Ganoderma adspersum	28.2	29.2	—	18.1	2.7	38.5	—	—		
Ganoderma applanatum	36.6	17.9	—	21.5	2.2	42.2	—	—		
Ganoderma lucidum	57.4	12.2	2.8	13.3	2.6	25.2	—	—		
Gyromitra esculenta	18.3	44.8	—	36.9	—	36.9	—	—		
Lyophyllum decastes	25.2	42.8	2.5	4.1	22.2	27.2	—	—		
Pleurotus ostreatus	56.2	12.6	2.3	15.9	3.6	25.2	—	—		
Russula delica	34.9	38.4	—	8.9	7.1	22.3	—	—		
Armillaria mellea	63.80	15.47	0.09	12.63	2.83	18.42	17.47	64.12	Serbia	48
Boletus edulis	52	32	—	10	2.4	14	33	53	Market	49
Ganoderma sp. Strain LS1	44.27	21.78	0.247	14.65	5.48	31.9	23.0	45.1	Ghana	50
Ganoderma sp. Strain LS2	40.09	25.97	0.23	15.69	6.09	31.8	27.3	40.9		
Ganoderma sp. Strain LS3	21.83	39.44	0.153	21.73	7.63	36.98	40.54	22.9		
Ganoderma sp. Strain LS4	9.74	35.03	0.167	26.80	12.27	51.6	37.6	10.8		
Ganoderma sp. Strain LS5	9.80	35.25	0.168	26.96	12.35	51.25	37.87	10.88		
Ganoderma sp. Strain LS6	17.53	42.28	0.140	21.52	8.58	38.9	43.2	17.9		
Ganoderma sp. Strain LS7	40.96	26.19	0.30	16.37	6.39	31.3	27.0	41.8		
Ganoderma sp. Strain LS8	7.57	45.80	0.210	23.76	8.89	44.4	47.3	8.31		

Ganoderma sp. Strain LS9	7.48	48.76	0.43	23.72	8.56	41.3	50.3	8.41		
Ganoderma sp. Strain DS1	30.55	24.82	0.33	19.08	9.91	42.5	26.0	31.5		
Ganoderma sp. Strain HS1	17.12	22.55	0.491	23.55	16.03	57	24.0	19.0		
Ganoderma sp. Strain AM1	40.43	28.81	0.34	15.89	5.70	29.1	29.64	41.2		
Meripilus giganteus	64.86	13.61	—	14.96	1.94	20.09	14.36	65.55	Serbia	51
Clavaria amoena	7.12	1.79	—	1.19	—	1.92	3.03	10.32	India	52
Clavaria coralloides	7.18	1.99	—	1.98	—	2.77	3.45	11.11		
Clavaria fragilis	8.18	2.46	—	2.13	—	3.05	4.03	12.33		
Clavaria purpurea	7.23	1.58	—	1.89	—	2.73	3.02	11.08		
Clavaria rosea	6.28	1.44	—	1.68	—	2.39	2.7	9.71		
Clavaria vermicularis	7.68	1.98	—	1.78	—	2.63	3.37	11.67		
Ramaria botrytis	8.28	1.48	—	2.19	—	3.13	4.08	12.48	India	52
Ramaria aurea	4.94	1.25	—	1.69	—	2.07	2.28	7.72		
Ramaria flava	5.79	1.72	—	1.95	—	2.34	3.16	8.89		
Ramaria flavescens	5.28	1.68	—	1.88	—	2.57	3.01	8.18		
Ramaria rubripermanens	6.45	1.95	—	2.11	—	2.82	3.46	9.72		
Ramaria stricta	3.93	1.12	—	1.5	—	1.76	2.20	6.32		
Rhizopogon luteolus	45.8	16.2	—	6.6	23.7	31.8	—	—	Turkey	55
Leucopaxillus candidus	49	37	—	8	2.7	13	37	50	Portugal	56
Leccinum molle	43.49	38.61	—	11.23	1.79	16.94	39.29	43.77	Portugal	57
Leccinum vulpinum	53.60	27.06	—	12.25	2.77	17.09	25.59	54.32		
Boletus pinophilus	41.0	45.8	0.70	7.3	1.9	10.3	47.5	42.2	Portugal	58
Clitocybe subconnexa	37.3	47.9	0.48	8.7	2.1	12.7	49.2	38.0		
Boletus edulis, caps	40.71	34.51	0.90	15.96	2.69	21.48	36.64	41.87	Poland	59
Boletus edulis, stems	32.68	20.69	0.48	35.95	2.40	43.01	23.64	33.34		
Aleurodiscus vitellinus	19.28	52.08	2.02	18.25	3.69	34.67	53.92	21.42	Argentina	60
Cortinarius magellanicus	63.75	9.04	0.10	19.21	3.30	25.12	10.09	64.79		
Cyttaria hariotii	34.68	7.93	11.56	18.23	3.72	35.85	16.58	47.57		
Fistulina endoxantha	43.63	23.53	0.13	20.17	4.52	30.16	25.76	44.08		
Fistulina antarctica	42.97	26.63	0.24	16.62	7.37	28.58	28.06	43.37		
Grifola gargal	26.55	45.28	0.25	17.44	7.31	26.75	46.27	26.98		
Hydropus dusenii	42.69	33.73	0.10	14.46	3.95	21.40	35.45	43.16		
Lepista nuda	71.32	9.81	0.39	12.45	1.95	16.60	10.71	72.69		

Table 3.1 (*Continued*)

Species	C18:2 (%)	C18:1 (%)	C18:3 (%)	C16:0 (%)	C18:0 (%)	SFAs (%)	MUFAs (%)	PUFAs (%)	Country	Ref.
Ramaria patagonica	45.35	29.58	0.07	13.44	3.54	23.20	30.47	46.33		
Morchella conica	63	21	—	11	2.4	14	22	64	Portugal	61
Morchella conica	67	19	—	10	1.5	13	20	67	Serbia	
Yanqi black mushroom	87.3	—	—	3.27	3.59	—	—	—	China	62
Laetiporus sulphureus (MAEH)	57.00	13.91	0.56	10.80	1.63	20.64	17.02	62.34	Serbia	64
Laetiporus sulphureus (MAEC)	56.12	13.92	0.33	11.38	1.95	22.18	17.85	59.98		
Laetiporus sulphureus (UAEH)	56.10	14.22	0.48	11.07	1.89	22.26	17.35	60.39		
Laetiporus sulphureus (UAEC)	56.45	14.38	0.42	10.74	2.04	21.37	18.12	60.51		
Hericium erinaceus	25.10	26.11	—	37.57	7.61	47.57	26.80	25.63	Portugal	65
Hericium coralloides	30.90	33.74	—	23.34	6.78	33.51	34.59	31.90		
Pleurotus sajor-caju	53.90	9.31	—	24.16	6.46	—	—	—	Culture	66
Agaricus arvensis	3.24	0.18	—	1.86	—	—	—	—	Himalayas	67
Agaricus campestris	2.04	0.35	—	1.19	—	—	—	—		
Agaricus comtulus	1.73	0.38	—	1.17	—	—	—	—		
Agaricus silvicola	1.54	0.15	—	1.56	—	—	—	—		
Amanita caesarea	2.19	0.31	—	1.96	—	—	—	—		
Amanita citrina	1.59	0.31	—	0.91	—	—	—	—		
Amanita fulva	1.64	0.15	—	0.10	—	—	—	—		
Cantharellus cibarius	2.94	0.37	—	1.10	—	—	—	—		
Conocybe tenera	0.14	0.35	—	0.07	—	—	—	—		
Gymnopilus junonius	0.49	0.88	—	1.87	—	—	—	—		
Hygrocybe coccinea	0.24	0.17	—	0.33	—	—	—	—		
Hygrocybe nivea	0.04	0.53	—	0.05	—	—	—	—		
Inocybe splendens	0.10	0.18	—	0.89	—	—	—	—		
Lactarius pubescens	3.44	0.26	—	0.75	—	—	—	—		
Laccaria laccata	2.84	0.92	—	1.01	—	—	—	—		
Lepista nuda	2.04	0.20	—	0.43	—	—	—	—		
Lentinus cladopus	2.45	0.32	—	1.46	—	—	—	—		
Pleurotus cystidiosus	2.04	0.98	—	1.35	—	—	—	—		
Russula lepida	2.04	0.49	—	033	—	—	—	—		

Russula mairei	2.01	0.63	—	0.26	—	—	—	—		
Volvopluteus gloiocephalus	59.33	25.96	—	7.42	2.09	12.56	26.47	60.97	Portugal	68
Clitocybe subconnexa	44.88	42.50	—	7.31	2.31	11.00	43.45	45.49		
Boletus edulis	57.16	30.41	0.11	7.56	1.93	10.59	31.90	57.51	Poland	69
Lentinus edodes	82.55	1.87	0.13	9.40	1.51	14.52	2.62	83.26		
Xerocomus badius	42.22	38.15	0.12	15.47	1.96	18.52	39.03	42.46		
Agrocybe aegerita	78.40	3.03	0.07	13.07	2.13	17.93	3.47	78.60	Serbia	70
Boletus aereus	40.02	41.78	—	13.17	1.84	17.23	42.54	40.23	Serbia	71
Cynomorium coccineum	19.46	37.03	8.13	15.75	2.17	33.65	44.21	24.13	Italy	72
Termitomycetes microcarpus	—	0.82	1.61	1.265	0.69	—	—	—	India	73
Termitomycetes heimii	—	1.113	2.054	1.129	0.92	—	—	—		
Ganoderma tsugae var. *jannieae*	9.75	18.1	—	11.4	1.02	—	—	—	China	74
Chroogomphus rutilus	41.6	35.9	—	18.5	1.82	20.3	—	—	Turkey	79
Laetiporus sulphureus	63.27	14.52	0.17	11.68	3.17	20.54	15.32	64.14	Serbia	80
Suillus granulatus	57.14	24.64	0.39	9.64	3.19	15.44	26.55	58.01	Portugal	81
Suillus granulatus	63.97	20.08	0.20	9.62	2.65	14.30	21.30	64.40	Serbia	81
Lactarius delicious	22.85	25.94	—	9.7	41.41	51.20	—	—	Turkey	82
Ramaria flava	9.78	47.12	—	14.43	6.68	39.12	—	—		
Phellinus linteus	25.3	13.5	1.4	26.8	12.3	58.3	14.6	27.1	Culture	83
Ganoderma lucidum	33.94	47.24	0.38	12.01	1.33	15.67	49.63	34.70	Serbia	84
Ganoderma lucidum	39.80	24.06	2.23	18.54	7.34	32.39	25.19	42.42	China	
Lactarius citriolens	22.46	25.00	—	5.35	40.58	51.85	25.42	22.74	Portugal	85
Lactarius turpis	48.55	26.29	—	8.02	12.60	23.73	27.18	49.09		
Agaricus bisporus	43.87	20.58	3.25	14.02	6.54	30.31	21.74	47.96	Culture	86
Agaricus brasiliensis	74.48	1.04	0.20	11.34	3.10	22.12	1.47	76.41		
Macrolepiota procera	56	7.6	—	27	3.4	34	9.5	56	Portugal	87
Boletus edulis	77.2	5.7	1.76	11.0	0.92	13.2	7.4	79.3	Portugal	88
Russula delica	67.5	16.3	—	12.2	1.50	15.0	17.2	67.9		
Macrolepiota procera (fresh, frozen, dried)	65,67,68	7,7,6	—	21,20,20	1.6,2.2,2.1	25,24,24	9,8,7	66,67,69	Portugal	89
Agaricus bisporus	24.1	54.8	5.0	8.7	5.9	—	—	—	Culture	95
Pleurotus ostreatus	61.9	18.4	0.7	11.2	1.5	—	—	—		
Boletus edulis	51.7	34.1	—	7.6	2.89	11.6	35.9	52.5	Portugal	96
Hydnum repandum	39.3	37.2	—	13.1	2.24	16.6	43.2	40.2		

Table 3.1 (*Continued*)

Species	C18:2 (%)	C18:1 (%)	C18:3 (%)	C16:0 (%)	C18:0 (%)	SFAs (%)	MUFAs (%)	PUFAs (%)	Country	Ref.
Amanita ceciliae	31.81	42.74	0.01	16.42	3.58	21.99	44.07	33.95	Turkey	97
Armillaria mellea	48.56	22.68	—	12.59	3.25	23.80	26.47	49.18		
Cantharellus cibarius	49.62	21.12	0.02	17.05	5.56	26.03	22.85	51.22		
Chlorophyllum rhacodes	47.77	9.06	0.02	19.08	7.72	34.49	15.07	49.75		
Rhizopogon roseolus	57.70	19.72	0.02	8.53	2.38	14.41	23.80	63.73		
Lactarius deliciosus	31.78	21.82	0.63	9.14	29.83	42.59	22.34	32.70	Greece	98
Lactarius sanguifluus	34.57	18.84	0.70	7.11	34.35	44.04	19.16	35.52		
Lactarius semisanguifluus	31.61	24.55	1.00	8.56	28.39	40.32	24.99	32.85		
Russula delica	44.32	13.66	2.50	7.22	27.10	35.87	14.26	47.19		
Suillus bellinii	36.73	39.48	0.24	12.94	2.75	17.31	40.94	37.52		
Lactarius deliciosus	20.1	8	—	5.6	62	72	8	20	Portugal	99
Macrolepiota procera	65.0	7.6	—	21.4	1.77	25.2	9.3	65.5	Portugal	100
Lignosus rhinocerotis	—	33.47	12.02	33.78	4.93	—	—	—	Malaysia	101
Lepista nuda (commercial)	63.88	16.05	—	14.86	0.88	18.90	16.69	64.41	Market	102
Tirmania pinoyi	29.72	32.29	0.20	27.87	2.68	35.82	33.50	30.67	Libya	103
Amanita caesarea	19.02	58	0.03	15	6.09	21.46	59.07	19.48	Turkey	104
Boletus reticulatus	36.60	37.61	—	11.26	5.84	21.14	40.34	38.40	Turkey	105
Flammulina velutipes	40.87	16.37	0.01	14.56	3.56	20.67	18.58	60.65		
Lactarius salmonicolor	59.44	18.98	—	7.76	7.27	18.96	19.56	61.57		
Pleurotus ostreatus	65.29	10.36	0.03	12.40	3.71	21.77	11.37	66.48		
Polyporus squamosus	38.91	33.02	0.27	17.21	3.28	25.19	34.27	40.64		
Russula anthracina	22.39	52.11	0.01	9.16	12.96	23.72	53.30	22.94		
Agaricus bitorquis	49.55	17.21	—	18.94	—	22.89	3.26	49.55	Turkey	106
Cantharellus cibarius	59.79	13.57	—	12.81	—	20.21	17.92	59.79		
Clitocybe odora	63.39	2.24	—	14.81	—	27.37	6.01	63.93		
Craterellus cornucopioides	51.01	17.68	—	10.56	—	23.79	22.82	51.93		
Lepiota excoriata	—	8.81	—	45.06	—	68.59	26.58	—		
Macrolepiota procera	62.44	17.40	—	10.95	—	15.90	19.51	62.44		
Pholiota sp.	55.83	14.34	—	12.13	—	22.64	17.64	55.83		
Pleurotus eryngii	26.60	47.82	—	9.55	—	22.47	48.73	26.60		

Polyporus squamosus	54.64	9.57	—	14.07	—	3.89	12.07	54.64		
Tricholoma terreum	60.29	16.61	—	10.50	—	17.04	18.71	62.00		
Amanita ovoidea	18	58.8	0.05	15.8	5.4	21.8	59.4	18.8	Turkey	107
Morchella esculenta	71.81	12.43	0.20	9.54	2.63	13.73	13.82	72.45	Portugal	108
Morchella esculenta	68.22	9.70	7.21	10.12	1.60	13.14	10.92	75.94	Serbia	
Agaricus blazei	72.42	1.85	—	11.38	2.80	24.39	2.03	73.58	Ltd. (Brazil)	109
Lentinus edodes	78.59	3.28	0.59	11.78	1.09	16.72	3.45	79.84		
Cordyceps militaris	68.00	7.29	0.41	12.76	5.71	23.40	7.73	68.87	Culture	110
Amanita crocea	20.14	54.46	—	17.66	3.68	24.21	55.42	20.37	Portugal	111
Amanita mairei	22.02	53.02	—	18.08	3.47	23.64	53.94	22.42		
Boletus poliporus	41.90	34.02	—	15.32	2.93	21.54	36.31	42.15		
Boletus regius	56.11	21.84	—	15.94	1.63	19.50	23.96	56.55		
Gyromitra esculenta	55.30	13.44	—	19.29	1.64	25.76	17.86	56.38		
Helvella lacunosa	12.21	43.82	—	29.05	5.51	39.87	46.44	13.70		
Russula aurea	40.32	40.63	—	9.28	5.29	17.25	42.09	40.66		
Russula virescens	29.18	40.27	—	17.31	7.16	28.78	41.51	29.71		
Cordyceps sinensis	52.26	24.15	1.08	16.43	2.98	21.77	24.89	53.34	Culture	113
Pleurotus ostreatus	67.41	10.47	0.48	16.70	1.60	20.53	11.58	67.89		
Schizophyllum commune	55.75	21.13	1.22	17.61	1.35	21.38	21.65	56.97		
Lactarius deliciosus	36.29	41.38	0.46	8.53	8.27	20.75	41.99	37.32	Turkey	114
Suillus collinitus	46.10	32.59	0.30	11.79	3.37	17.53	34.38	47.42		
Agrocybe cylindracea	66.72	1.74	0.10	15.23	5.52	28.13	2.83	69.13		
Coprinus comatus	62.36	9.29	1.16	14.21	4.82	13.84	11.37	64.76		
Tricholoma myomyces	36.98	45.26	0.13	7.77	3.13	15.83	46.29	37.77		
Agaricus campestris	68.97	6.09	—	12.48	2.77	20.91	9.05	70.04	Portugal	115
Agaricus comtulus	72.88	3.50	—	12.98	2.66	22.04	4.42	73.55		
Agaricus lutosus	74.40	6.11	—	12.03	2.26	18.49	6.63	74.88		
Amanita umbrinolutea	18.81	58.82	—	15.10	3.87	21.18	59.82	19.00		
Boletus armeniacus	48.95	27.61	—	15.68	2.92	21.01	29.67	49.32		
Boletus impolitus	60.95	14.21	—	16.77	1.10	23.19	15.48	61.33		
Bovista aestivalis	41.51	12.63	—	21.43	4.32	41.80	15.53	42.68		
Bovista nigrescens	38.28	21.01	—	17.39	4.19	37.78	23.16	39.06		
Chlorophyllum rhacodes	72.61	5.68	—	16.35	1.59	20.11	6.91	72.98		

Table 3.1 (*Continued*)

Species	C18:2 (%)	C18:1 (%)	C18:3 (%)	C16:0 (%)	C18:0 (%)	SFAs (%)	MUFAs (%)	PUFAs (%)	Country	Ref.
Clavariadelphus pistillaris	24.74	49.11	—	16.76	1.99	24.86	50.11	25.03		
Clavariadelphus truncatus	29.77	47.26	—	14.80	2.11	21.43	48.31	30.26		
Clitocybe costata	34.68	37.27	—	12.76	5.99	22.34	38.02	39.64		
Clitocybe gibba	64.45	4.91	—	13.81	7.89	27.87	6.16	66.02		
Cortinarius praestans	59.95	20.76	—	13.44	1.78	17.93	21.49	60.59		
Flammulina velutipes	56.33	15.08	—	10.31	1.38	14.36	17.56	68.08		
Hygrophorus chrysodon	1.23	57.26	—	25.95	3.88	35.32	63.05	1.63		
Leucoagaricus leucothites	74.72	6.27	—	12.16	1.81	18.00	6.74	75.25		
Lycoperdon umbrinum	29.36	22.83	—	19.92	7.14	42.48	24.79	32.74		
Ramaria aurea	25.60	56.92	—	7.32	4.07	15.27	58.47	26.26		
Suillus variegates	37.44	42.00	—	12.71	3.47	18.09	44.24	37.67		
Agaricus bohusii	72.85	4.50	0.25	10.10	2.96	21.05	5.17	73.79	Serbia	116
Fistulina hepatica	26.74	49.98	—	16.05	5.90	—	—	—	Italy	117
Infundibulicybe geotropa	18.16	57.18	0.21	17.56	3.84	—	—	—		
Laetiporus sulphureus	20.62	52.35	0.64	22.29	2.24	—	—	—		
Macrolepiota procera var. *procera*	26.83	49.28	0.23	16.41	6.08	—	—	—		
Suillus granulatus	26.63	50.12	0.19	19.13	2.21	—	—	—		
Terfezia boudieri	44.34	10.37	2.12	23.72	3.99	—	—	—	Turkey	118
Gomphus clavatus	30.4	33.4	—	14.3	6.9	—	—	—	Greece	119
Agaricus bisporus brown	79.4	1.2	0.1	11.1	3.0	18.4	1.8	79.8	Market	120
Agaricus bisporus white	77.7	1.1	0.1	11.9	3.1	20.3	1.4	78.3		
Flammulina velutipes	45.4	5.7	28.8	11.0	2.0	18.5	7.2	74.3		
Lentinula edodes	81.1	2.3	0.1	10.3	1.6	15.1	2.9	82.0		
Pleurotus eryngii	68.8	12.3	0.3	12.8	1.7	17.4	13.1	69.4		
Pleurotus ostreatus	68.9	12.3	0.1	11.2	1.6	17.0	13.6	69.4		
Agaricus arvensis	59.8	2.4	—	16.7	5.4	—	—	—	Bulgaria	121
Agaricus augustus	58.5	6.6	—	15.8	4.9	—	—	—		
Agaricus campestris	59.8	9.6	0.4	14.6	4.6	—	—	—		
Agaricus silvaticus	62.3	3.2	—	16.5	5.6	—	—	—		
Agaricus xanthoderma v. *eleagris*	61.4	8.0	0.3	12.5	4.9	—	—	—		

Clitocybe clavipes	38.8	33.4	—	18.9	3.2	—	—	—		
Endoptychum agaricoides	62.9	14.5	—	13.8	1.7	—	—	—		
Hygrocybe nigrescens	18.5	42.7	—	22.0	7.5	—	—	—		
Lepista gilva	58.8	6.7	—	16.1	6.6	—	—	—		
Lepista inversa	58.6	19.6	—	15.9	0.2	—	—	—		
Lyophyllum connatum	49.0	16.4	—	18.0	5.3	—	—	—		
Lyophyllum decastes	35.4	33.4	—	16.3	8.1	—	—	—		
Marasmius oreades	30.9	16.9	—	23.1	20.3	—	—	—		
Pleurotus dryinus	56.7	15.4	—	17.9	1.8	—	—	—		
Pluteus atricapillus	22.8	38.7	—	16.3	5.3	—	—	—		
Catathelasma ventricosum	73.92	15.45	0.98	5.12	2.17	7.29	16.28	76.42	China	122
Clitocybe maxima	28.34	46.82	0.07	12.64	5.44	18.81	51.61	29.58		
Craterellus cornucopioides	28.35	54.02	1.26	10.15	5.37	15.76	54.45	29.78		
Laccaria amethystea	74.39	13.74	1.14	6.94	3.34	10.33	13.74	75.92		
Stropharia rugoso-annulata	45.42	39.32	0.35	9.46	3.95	13.56	39.99	46.45		
Agaricus bisporus	67.29	6.07	1.52	13.55	3.72	20.28	—	—	Turkey	124
Agaricus bitorquis	64.38	6.87	2.03	12.67	3.94	22.56	—	—		
Agaricus essettei	61.82	8.11	2.45	14.71	3.97	22.50	—	—		
Armillaria mellea	38.89	30.58	—	13.51	3.02	—	—	—	Croatia	125
Collybia platyphylla	34.20	36.43	—	20.03	4.05	—	—	—		
Lactarius deterrimus	21.96	22.77	0.43	4.85	46.82	—	—	—		
Macrolepiota procera	52.61	9.86	0.24	23.00	2.49	—	—	—		
Ganoderma lucidum	*tr*	48	—	15	7	—	—	—	Ltd. (China)	126
Amanita caesarea	25.74	53.78	0.16	12.42	4.88	19.16	54.81	26.03	Portugal	127
Amanita muscaria	21.87	60.08	0.05	7.88	6.13	16.46	61.38	22.16		
Amanita pantherina	21.72	53.06	0.03	9.41	10.41	23.51	54.17	22.31		
Chroogomphus fulmineus	31.54	47.09	—	12.59	2.47	16.89	49.77	33.33		
Cortinarius anomalus	44.39	37.74	0.04	12.12	2.23	15.93	39.40	44.68		
Cortinarius collinitus	35.89	35.21	0.10	19.03	2.94	24.58	38.54	36.87		
Cortinarius violaceus	66.22	14.59	0.15	14.02	1.67	17.46	15.96	66.58		
Lactarius quietus	35.15	24.55	4.42	13.71	12.76	33.68	26.47	39.86		
Lactarius volemus	34.38	39.27	0.01	12.19	6.44	23.92	40.80	35.28		
Russula sardonia	16.93	63.19	1.71	7.84	6.56	15.71	65.12	19.17		

Table 3.1 (*Continued*)

Species	C18:2 (%)	C18:1 (%)	C18:3 (%)	C16:0 (%)	C18:0 (%)	SFAs (%)	MUFAs (%)	PUFAs (%)	Country	Ref.
Suillus luteus	52.31	31.24	0.29	10.57	2.06	14.32	32.93	52.75		
Tricholoma ustale	59.29	23.46	0.07	9.72	3.46	16.22	24.37	59.41		
Agaricus arvensis	78.0	2.53	—	11.1	2.83	16.6	3.61	78.8	Turkey	128
Boletus edulis	75.6	8.22	1.73	12.7	0.67	13.4	10.5	76.1		
Cantharellus cibarius	45.6	8.44	—	13.7	3.34	17.1	37.3	45.6		
Cantharellus tubaeformis	19.8	57.8	—	15.2	6.19	21.7	58.5	19.8		
Clavulina rugose	35.5	41.7	—	15.2	5.06	20.3	44.3	35.5		
Hydnum repandum	42.5	42.4	—	10.9	1.71	12.9	44.4	42.7		
Laccaria laccata	31.6	45.9	—	20.0	1.84	21.9	46.6	31.6		
Lepista nuda	53.6	30.8	—	11.9	0.89	13	32.7	57.3		
Leucopaxillus giganteus	38.3	43.3	—	13.9	3.73	18	43.8	38.3		
Russula rosea	23.2	31.5	—	9.07	34.6	44.3	38.1	23.2		
Tricholoma var. *saponaceum*	43.1	38.8	—	12.6	2.96	16.0	40.9	43.1		
Boletus erythropus	48.76	14.74	—	21.33	4.15	33.62	16.06	50.32	Portugal	129
Boletus fragrans	56.89	19.80	—	14.91	2.39	20.75	21.86	57.38		
Calvatia utriformis	70.29	6.00	—	13.54	2.43	22.37	6.47	71.16		
Clitopilus prunulus	59.92	20.42	0.04	10.11	2.47	16.45	21.51	62.04		
Hygrophorus pustulatus	34.74	51.53	—	6.96	3.96	13.18	51.85	34.98		
Lycoperdon echinatum	69.92	9.16	0.06	10.82	2.08	19.45	9.75	70.80		
Lyophyllum decastes	32.03	47.43	0.08	10.95	4.88	19.31	47.94	32.75		
Macrolepiota excoriata	66.19	10.21	—	17.83	1.94	22.49	11.10	66.41		
Russula cyanoxantha	43.65	28.39	—	12.95	11.10	26.90	29.06	44.04		
Russula olivacea	50.20	25.99	—	16.13	2.78	21.75	27.40	50.85		
Agrocybe chaxingu	69.3	5.0	—	13.7	5.4	25.7	5.0	69.3	Culture	130
Pleurotus ostreatus	76.5	4.8	—	11.0	2.6	17.4	6.1	76.5		
Flammulina velutipes	63.7	5.6	—	18.6	3.1	30.7	5.6	63.7	Market	
Armillaria mellea	27.71	47.74	0.04	11.04	3.53	17.23	55.01	27.76	Portugal	131
Calocybe gambosa	43.88	32.54	0.93	13.57	3.24	21.54	33.38	45.07		
Clitocybe odora	34.90	46.07	0.06	12.46	3.46	18.57	46.39	35.04		
Coprinus comatus	74.86	6.27	1.90	10.56	1.90	15.42	7.12	77.46		

Boletus aereus	43.83	36.72	0.07	12.47	3.80	18.08	37.78	44.15	Portugal	132
Boletus edulis	41.32	42.05	0.03	9.57	3.11	14.77	43.53	41.71		
Boletus reticulatus	32.83	47.20	0.09	11.00	4.25	17.95	48.50	33.55		
Boletus purpureus	51.22	24.18	0.32	17.49	1.83	21.27	26.84	51.89		
Boletus rhodoxanthus	62.34	7.21	—	17.41	3.39	26.15	10.83	63.03		
Boletus satanas	32.18	44.53	0.05	15.24	3.03	20.57	46.85	32.58		
Amanita porphyria	71.22	5.45	—	11.99	2.23	21.65	6.32	72.03	Portugal	133
Collybia fusipes	55.30	13.47	—	18.49	5.22	30.05	14.08	55.87		
Fomitopsis pinicola	65.07	16.17	—	11.81	3.09	18.15	16.49	65.36		
Hebeloma sinapizans	69.89	7.92	—	14.61	3.11	19.87	8.93	71.20		
Inocybe splendens	29.90	53.07	—	10.27	2.20	14.84	54.48	30.68		
Lactarius hepaticus	44.99	31.24	—	10.00	5.82	21.24	32.34	46.42		
Lentinus tigrinus	46.89	28.85	—	12.25	1.57	17.16	34.06	48.78		
Piptoporus betulinis	60.94	8.67	—	11.96	6.17	29.10	8.92	61.98		
Pluteus murinus	34.74	51.53	—	6.96	3.96	13.18	51.84	34.98		
Russula emetic	18.47	65.44	—	6.91	5.02	14.19	65.90	19.91		
Boletus aestivalis	57.6	19.1	1.0	8.3	3.5	—	—	—	Israel	136
Boletus aereus	51.7	21.7	1.2	11.1	3.0	—	—	—		
Boletus appendiculatus	50.9	28.4	2.1	9.7	2.2	—	—	—		
Boletus badius	38.2	36.5	1.6	15.1	2.1	—	—	—		
Boletus crocipodius	47.3	16.5	3.1	17.0	2.0	—	—	—		
Boletus edulis	51.7	29.4	1.2	8.9	3.0	—	—	—		
Boletus granulatus	58.3	17.1	0.8	12.1	2.3	—	—	—		
Boletus impolitus	55.9	20.2	0.8	9.4	3.4	—	—	—		
Boletus luridus	53.4	24.1	0.9	10.2	2.4	—	—	—		
Boletus luteus	44.1	37.8	0.7	7.6	4.1	—	—	—		
Boletus pinicola	39.7	42.6	2.1	7.0	4.4	—	—	—		
Boletus sp.	39.4	35.2	0.8	11.4	2.1	—	—	—		
Boletus queletii	52.1	17.1	2.5	13.4	1.9	—	—	—		
Boletus scaber	45.8	31.7	3.1	9.7	9.7	—	—	—		
Boletus versipellis	55.2	15.2	1.6	14.1	14.1	—	—	—		
Tricholoma matsutake	26.7	57.6	—	9.1	1.8	12.4	—	—	China	137
Russula griseocarnosa pileus	28.40	47.92	1.02	19.30	—	19.30	—	—	China	138

Table 3.1 (*Continued*)

Species	C18:2 (%)	C18:1 (%)	C18:3 (%)	C16:0 (%)	C18:0 (%)	SFAs (%)	MUFAs (%)	PUFAs (%)	Country	Ref.
Cortinarius glaucopus	54.99	24.01	—	12.05	2.98	19.60	25.12	55.28	Portugal	139
Fistulina hepatica	52.37	31.51	0.16	10.42	2.54	15.11	32.19	52.70		
Hygrophoropsis aurantiaca	55.45	17.82	8.16	9.97	0.92	14.75	20.76	64.49		
Hypholoma capnoides	35.67	16.98	0.55	16.43	4.10	44.37	19.11	36.53		
Laccaria laccata	20.45	60.68	0.39	11.64	2.02	15.44	63.64	21.03		
Lactarius salmonicolor	26.44	18.45	0.47	7.35	40.13	54.00	18.93	27.07		
Lepista inversa	44.58	28.78	4.64	16.36	1.71	21.30	29.14	49.56		
Russula delica	27.15	41.20	0.39	12.02	10.34	27.20	44.89	27.91		
Suillus mediterraneensis	43.72	36.42	0.02	11.93	3.56	18.04	37.92	44.03		
Tricholoma imbricatum	33.03	51.53	0.19	7.44	4.10	14.91	51.80	33.29		
Cordyceps gunnii	42.577	30.238	—	13.859	6.123	21.06	30.67	46.98	Culture	140
Cordyceps jiangxiensis	35.951	36.443	—	12.777	6.763	46.98	40.37	38.10		
Cordyceps taii	38.431	24.703	—	22.263	8.736	32.44	25.22	37.72		
Agaricus bisporus	75.72	1.30	0.06	9.97	4.08	22.08	1.52	76.41	Portugal	144
Agaricus silvaticus	74.78	6.67	0.10	11.74	1.41	17.10	7.67	75.23		
Agaricus silvicola	76.50	3.49	0.05	9.96	2.64	18.80	4.25	76.95		
Boletus edulis	44.32	39.72	0.07	10.03	2.75	14.52	40.91	44.56	Market	
Calocybe gambosa	57.75	18.10	0.45	15.16	2.14	22.51	19.05	58.42		
Cantharellus cibarius	53.59	10.78	0.08	13.08	6.49	22.63	23.29	54.08		
Craterellus cornucopioides	23.67	51.85	0.07	6.66	7.83	16.36	59.85	23.79		
Marasmius oreades	50.66	28.52	0.07	13.82	1.65	18.88	30.16	50.96		
Cantharellus cibarius	50.01	8.13	0.10	7.19	3.34	12.04	37.54	50.42	Portugal	145
Lepista nuda	51.48	29.53	0.21	11.77	2.98	17.58	30.32	52.10		
Lycoperdon molle	64.15	8.58	0.06	13.74	2.35	25.04	9.04	65.92		
Lycoperdon perlatum	70.69	4.59	0.22	12.90	2.95	23.57	4.91	71.52		
Ramaria botrytis	38.32	43.93	0.02	9.91	2.36	16.38	44.69	38.91		
Cordyceps militaris	61.3	6.0	—	24.5	5.8	—	—	—	Market	146
Agaricus bisporus	44.19	21.47	—	4.34	1.79	—	—	—	Pakistan	147
Agaricus bitorquis	31.03	40.13	11.27	4.28	2.03	—	—	—		
Lentinula edodes	35.61	—	0.33	17.75	2.75	—	—	—	Market	148

Auricularia polytricha	29.5	27.1	3.6	11.2	10.2	39.8	27.1	33.1	India	149
Boletus edulis	33.8	31.1	1.7	21.6	9.1	33.4	31.1	35.5		
Cantharellus cibarius	17.3	35.4	—	18.3	6.0	39.0	43.7	17.3		
Cantharellus clavatus	20.8	25.9	—	24.7	8.1	48.3	30.9	20.8		
Geastrum arinarius	5.4	55.1	1.1	19.9	8.3	33.7	59.8	6.5		
Helvella crispa	62.7	22.7	0.3	10.6	—	12.8	24.2	63.0		
Hydnum repandum	27.2	26.4	20.3	15.7	0.9	24.6	27.9	47.5		
Lentinus sajor-caju	54.9	23.5	—	15.4	—	20.0	25.1	54.9		
Lentinus squarrulosus	61.4	5.8	1.2	16.6	1.4	30.6	6.8	62.6		
Lactarius deliciosus	37.1	33.0	—	16.3	6.1	28.5	34.4	37.1		
Lactarius sanguifluus	35.1	32.4	—	23.1	4.9	31.2	33.7	35.1		
Macrolepiota procera	47.0	17.2	15.6	4.6	—	20.2	17.2	62.6		
Morchella angusticeps	64.6	10.7	0.2	8.3	6.0	22.1	13.1	64.8		
Morchella conica	68.6	11.3	—	8.5	5.4	18.5	12.9	68.6		
Pleurotus djamor	45.5	28.8	—	15.8	—	25.1	29.4	45.5		
Pleurotus sajor-caju	53.8	16.4	—	13.9	3.4	27.1	19.1	53.8		
Russula brevipes	33.9	39.2	—	8.0	4.5	21.7	44.4	33.9		
Sparassis crispa	31.3	49.0	—	10.4	1.7	19.0	49.7	31.3		
Termitomyces heimii	35.2	28.1	—	19.7	4.8	35.9	28.9	35.2		
Termitomyces mummiformis	58.0	18.2	—	16.0	4.9	22.2	19.8	58.0		
Termitomyces microcarpus	31.9	34.2	—	21.8	7.7	33.2	34.9	31.9		
Termitomyces schimperi	34.5	25.7	—	14.6	10.4	39	26.5	34.5		
Agaricus arvensis	58.7	4.48	0.13	20.2	5.50	33.4	7.27	59.3	Canada	150
Agaricus campestris	68.8	3.99	0.17	16.5	3.04	24.5	6.12	69.4		
Amanita rubescens	19.0	58.0	0.09	14.5	4.47	21.8	59.0	19.3		
Amanita vaginata	16.9	60.3	0.14	15.2	3.72	21.3	61.5	17.3		
Boletus edulis	42.7	36.2	0.18	10.5	3.17	16.4	39.2	43.3		
Boletus subglabripes	39.4	39.2	0.22	12.5	3.65	18.8	41.2	40.0		
Coprinus comatus	64.5	7.45	0.47	18.9	1.84	23.8	10.4	65.3		
Ganoderma applanatum	17.9	37.4	0.27	19.0	4.60	34.6	46.1	18.4		
Lycoperdon pyriforme	50.7	5.12	0.19	22.6	7.27	39.2	8.33	51.4		
Suillus granulatus	44.6	33.7	0.28	13.0	3.03	19.2	35.9	45.2		
Boletus aestivalis	57.65	19.11	0.97	8.33	3.53	12.47	23.74	58.62	Israel	151

Table 3.1 (*Continued*)

Species	C18:2 (%)	C18:1 (%)	C18:3 (%)	C16:0 (%)	C18:0 (%)	SFAs (%)	MUFAs (%)	PUFAs (%)	Country	Ref.
Boletus aereus	51.68	21.67	1.23	11.12	2.95	18.71	24.15	52.91		
Boletus appendiculatus	50.93	28.35	2.13	9.75	2.19	13.50	30.46	53.06		
Boletus badius	38.22	36.57	1.65	15.11	2.11	18.60	38.91	41.05		
Boletus crocipodius	47.32	16.55	3.11	17.03	1.98	23.12	20.24	50.43	Israel	151
Boletus edulis	51.69	29.44	1.23	8.96	3.06	13.85	31.45	52.92		
Boletus granulatus	58.35	17.11	0.78	12.14	2.33	17.68	20.76	59.13		
Boletus impolitus	55.94	20.18	0.84	9.38	3.35	15.33	22.71	56.78		
Boletus luridus	53.48	24.08	0.92	10.24	2.44	14.92	27.20	54.40		
Boletus luteus	44.13	37.87	0.65	7.64	4.11	12.52	38.75	44.78		
Boletus pinicola	39.71	42.61	0.77	6.98	4.39	12.00	44.32	41.02		
Boletus sp.	39.40	35.22	2.15	11.44	2.07	16.80	37.57	42.77		
Boletus queletii	53.15	17.14	2.51	13.41	1.88	19.23	22.02	56.98		
Boletus scaber	45.84	31.67	3.08	9.76	3.42	14.44	33.02	48.92		
Boletus versipellis	55.15	15.21	1.58	14.12	1.76	20.98	20.33	57.25		
Bovista plumbea	51.9	14.0	0.4	14.1	3.4	23.6	—	—	Bulgaria	152
Calvatia utriformis	42.4	23.0	0.8	12.2	3.6	22.8	—	—		
Lycoperdon pyriforme	64.8	6.8	0.1	18.1	2.0	23.4	—	—		
Lycoperdon perlatum	37.6	24.1	3.9	14.5	6.4	28.6	—	—		
Agaricus arvensis	56.11	15.46	0.19	14.55	3.37	23.47	19.85	56.68	Portugal	153
Lactarius deliciosus	17.06	41.26	0.26	12.08	25.33	40.14	42.28	17.59		
Leucopaxillus giganteus	46.18	21.09	0.09	13.46	2.11	19.25	34.08	46.67		
Sarcodon imbricatus	35.38	45.06	0.16	11.14	3.65	18.08	46.20	35.73		
Tricholoma portentosum	30.88	58.36	0.40	5.60	2.33	9.57	59.03	31.40		
Lactarius deliciosus	25.37	22.93	0.15	7.02	37.82	50.60	23.32	26.06	Portugal	154
Lactarius piperatus	8.57	26.99	0.01	5.15	54.53	64.00	27.33	8.64		
Paecilomyces tenuipes	12.35	38.35	28.3	17.08	3.16	—	—	—	Culture	155
Boletus edulis	42.2	36.1	0.17	9.84	2.72	43.1	—	—	Canada	157

Boletus erythropus	63.0	18.0	0.20	11.2	1.29	64.2	—	—		
Boletus piperatus	47.4	29.8	0.25	7.37	1.78	52.1	—	—		
Boletus subglabripes	42.2	35.0	0.27	12.6	2.97	43.6	—	—		
Boletus subtomentosus	42.2	31.7	0.29	16.3	1.68	43.2	—	—		
Boletus variipes	34.2	49.2	0.05	7.62	2.72	35.2	—	—		
Leccinum aurantiacum	52.9	21.1	0.16	16.5	0.90	53.7	—	—		
Leccinum scabrum	42.3	37.8	0.13	13.8	1.11	43.1	—	—		
Suillus granulatus	46.6	24.2	0.22	12.0	3.75	47.1	—	—		
Suillus grevillei	40.1	43.9	0.63	8.90	1.49	41.9	—	—		
Tylopilus felleus	54.4	25.2	0.08	16.0	1.72	54.8	—	—		
Agaricus bisporus	33.3	10.7	—	1.93	2.11	27.6	47.2	34.5	Turkey	158
Agaricus campestris	44.0	4.17	—	1.47	5.11	26.4	51.6	45.4		
Armillaria mellea	49.6	4.55	—	1.04	4.05	20.6	56.1	50.6	Turkey	158
Boletus edulis	33.6	30.2	—	0.942	2.88	17.7	64.7	33.6		
Coprinus comatus	25.8	6.17	—	0.199	3.46	20.0	32.3	26.0		
Oudemansiella radicata	41.9	9.79	—	2.22	5.60	21.2	54.7	42.7		
Pleurotus ostreatus	42.0	20.80	—	0.961	3.58	20.2	63.9	42.1		
Termitomyces aurantiacus	55.8	14.3	0.3	15.9	4.0	24.9	17.5	56.1	Market	159
Termitomyces le testui	57.3	14.7	—	14.1	0.8	17.6	16.0	57.3		
Termitomyces mammiformis	57.9	14.3	—	15.0	4.8	21.8	16.2	57.9		
Termitomyces mboudaeina sp. nov.	59.0	12.2	—	16.1	4.2	25.3	14.7	59.0		
Termitomyces schimperi	45.6	23.0	—	15.5	6.7	24.1	26.3	45.6		
Termitomyces subclypeatus forme bisporus sp. nov.	65.1	9.0	—	16.6	3.9	23.2	11.7	65.1		
Tricholoma acerbum	65.6	5.8	—	16.2	4.9	—	—	—	Spain	160
Tricholoma batschii	58.5	7.1	—	23.0	5.3	—	—	—		
Tricholoma nudum	63.9	4.8	—	21.2	1.9	—	—	—		
Tricholoma pardinum	61.4	4.0	—	18.2	10.1	—	—	—		
Tricholoma portentosum	27.9	58.0	—	7.6	3.4	—	—	—		
Tricholoma terreum	29.7	56.7	—	10.1	1.8	—	—	—		
Lentinus edodes	68.8	8.3	0.6	19.2	2.7	22.3	—	—	Market	161
Schizophyllum commune	61.3	10.4	4.8	20.8	2.5	23.5	—	—		

Table 3.1 (*Continued*)

Species	C18:2 (mg kg^{-1})	C18:1 (mg kg^{-1})	C18:3 (mg kg^{-1})	C16:0 (mg kg^{-1})	C18:0 (mg kg^{-1})	SFAs (mg kg^{-1})	MUFAs (mg kg^{-1})	PUFAs (mg kg^{-1})	Country	Ref.
Suillus bellini	124.578	191.046	—	160.702	76.787	325.294	222.648	128.753	Portugal	141
Suillus luteus	650.694	1097.004	—	11.209	3.107	18.138	1100.903	651.864		
Suillus granulatus	663.008	1177.661	—	6.926	6.690	24.794	1187.203	663.473		
Hygrophorus agathosmus	76.380	159.028	—	19.045	4.481	31.348	163.922	77.892		
Amanita rubescens	19.365	69.810	—	39.761	27.844	80.791	71.945	19.984		
Russula cyanoxantha	976.316	2165.499	0.022	16.015	6.941	30.345	2167.730	976.961		
Boletus edulis	7.536	11.583	—	5.285	3.084	9.930	12.756	7.840		
Tricholoma equestre	203.032	244.048	0.019	7.312	2.787	12.691	244.367	203.051		
Fistulina hepatica	18.607	8.448	—	4.857	1.782	7.801	8.577	18.648		
Cantharellus cibarius	654.706	148.168	—	15.660	7.017	26.953	148.493	655.176		
Amanita caesarea	310.698	1282.709	—	14.101	7.111	24.732	1284.685	311.271		
Hydnum rufescens	7.885	11.798	—	9.933	4.066	16.305	16.985	7.962		

Species	C18:2 (mg g^{-1})	C18:1 (mg g^{-1})	C18:3 (mg g^{-1})	C16:0 (mg g^{-1})	C18:0 (mg g^{-1})	SFAs (mg g^{-1})	MUFAs (mg g^{-1})	PUFAs (mg g^{-1})	Country	Ref.
Chanterelle cibarius	5.75	9.46	1.24	2.90	6.56	21.02	11.32	15.18	Shop	43
Chanterelle cinnabarinus schwein	0.96	5.17	1.06	3.17	8.37	22.47	5.59	4.57		
Chanterelle tubaeformis	2.49	4.95	—	2.42	7.31	15.37	5.26	2.49		
Pleurotus ostreatus	10.257	3.544	0.0202	2.671	0.413	24.65	18.26	54.06	America	44
Pleurotus sajor-caju	16.013	2.0103	0.036	3.809	0.979	20.77	7.77	52.51	Canada	
Pleurotus sapidus	1.421	1.170	0.031	1.246	0.258	48.7	5.85	30.5	UK	
Pleurotus columbinus	12.547	6.075	0.196	4.229	1.193	30.33	28.64	47.31	UK	
Handkea utriformis	25.68	2.58	—	5.59	0.53	—	—	—	Serbia	63
Handkea excipulliformis	7.44	0.33	—	0.20	0.47	—	—	—		
Vascellum pratense	11.74	0.84	—	4.40	0.72	—	—	—		
Lactarius tenuicystidiatus	17.7	21.8	—	3.7	2.5	8.2	22.0	17.7	China	90
Lactarius deterrimus	15.9	1.8	0.1	2.9	37.9	41.2	1.9	16.0		
Lactarius hatsudake	10.7	6.8	0.2	2.6	12.1	15.4	6.9	10.9		
Lactarius volemus sensu lato	9.7	15.0	—	1.2	0.5	2.1	15.0	9.9		

Agaricus brasiliensis	65.8	5.75	1.10	13.6	6.20	—	—	—	Ltd. (Israel)	91
Lentinus edodes	75.8	3.46	0.28	13.7	1.88	—	—	—		
Ophiocordyceps sinensis	32.9	36.3	2.10	20.7	5.20	—	—	—		
Coprinus militaris	58.7	11.1	0.30	18.0	4.60	—	—	—		
Grifola frondosa	35.1	44.1	—	16.8	2.34	—	—	—		
Trametes versicolor	13.7	16.2	1.70	27.2	18.9	—	—	—		
Pleurotus ostreatus	14.7	17.1	6.60	25.8	17.7	—	—	—		
Ganoderma lucidum	11.6	52.4	1.70	20.1	7.80	—	—	—		
Hericium erinaceus	26.9	33.7	0.40	24.0	8.10	—	—	—		
Coprinus comatus	71.1	5.70	1.10	17.5	2.00	—	—	—		
Flammulina velutipes	51.2	10.7	13.0	15.9	2.80	—	—	—		
Tremella fuciformis	17.2	51.8	0.70	19.0	4.90	—	—	—		
Grifola frondosa	142.0	448.9	—	61.41	9.28	—	—	—	Ltd. (Taiwan)	112
Hericium erinaceum	111.2	151.4	—	136.6	66.93	—	—	—		
Agaricus blazei	228.6	4.54	—	39.10	15.72	—	—	—		
Ganoderma lucidum	43.70	16.1	—	29.01	5.39	—	—	—		
Coriolus versicolor	61.92	28.16	5.30	31.80	3.56	—	—	—		
Phellinus linteus	24.99	13.61	—	18.65	4.17	—	—	—		
Phellinus linteus (Zhouyulin Edible Mushroom Institute of Wuhan)	1.3486	0.09857	—	1.0798	0.2774	—	—	—	China	134
Phellinus linteus (Culture Center of Huazhong Agricultural University)	0.4449	0.04101	—	0.6283	0.3040	—	—	—		
Phellinus linteus (Xinyu Edible Mushroom Institute of Wuhan)	0.9397	0.07742	—	0.8191	0.2256	—	—	—		
Phellinus linteus (Agricultural Culture Collection of China)	0.4136	0.03369	—	0.657	0.3391	—	—	—		
Phellinus linteus (Microbiological Research Center of Sichuan Academy of Agricultural Science)	0.8667	0.04813	—	0.8459	0.2562	—	—	—		
Phellinus linteus (Microbiological Research Center of Sichuan Academy of Agricultural Science)	0.5941	0.06317	—	0.9894	0.2998	—	—	—		
Phellinus linteus (Agricultural Culture Collection of China)	1.6483	0.10889	—	1.0196	0.3336	—	—	—	—	

Table 3.1 (*Continued*)

Species	C18:2 (mg g^{-1})	C18:1 (mg g^{-1})	C18:3 (mg g^{-1})	C16:0 (mg g^{-1})	C18:0 (mg g^{-1})	SFAs (mg g^{-1})	MUFAs (mg g^{-1})	PUFAs (mg g^{-1})	Country	Ref.
Phellinus baumii (China General Microbiological Culture Collection Center)	0.7721	0.0762	—	0.4871	0.1464	—	—	—	—	
Phellinus baumii (Guangda Edible Mushroom Center of Jining in Shandong)	1.144	0.06089	—	1.1034	0.2345	—	—	—	—	
Phellinus baumii (Xinyu Edible Mushroom Institute of Wuhan)	0.7316	0.05094	—	0.6312	0.2910	—	—	—		
Phellinus baumii (Fungi Institute of Jinxiang in Shandong)	0.5181	003801	—	0.9778	0.2839	—	—	—		
Phellinus baumii (Edible Mushroom Institute of Yichun in Heilongjiang)	1.0969	0.09553	—	1.2403	0.3821	—	—	—		
Phellinus pomaceus	0.2982	0.02074	—	0.3952	0.2292	—	—	—		
Phellinus pini	0.1869	0.03654	—	0.4405	0.2984	—	—	—		
Phellinus robustus	0.1629	0.02732	—	0.5223	0.2591	—	—	—		
Phellinus sp.	0.4638	0.01644	—	0.7253	0.3242	—	—	—		
Antrodia camphorata	12.69	10.87	—	14.74	14.34	—	—	—	Ltd. (Taiwan)	156
Amanita rubescens	21.745	69.304	—	11.301	6.254	—	—	—	Mexico	162
Boletus frostii	16.847	19.527	—	6.167	1.754	—	—	—		
Lactarius indigo	13.276	10.757	—	3.491	32.057	—	—	—		
Ramaria flava	15.715	12.032	—	4.041	0.910	—	—	—		
Clitocybe aurantiaca	661	58	8	172	63	—	—	—	Italy	163
Clitocybe candida	626	59	7	198	72	—	—	—		
Clitocybe cinerascens	684	51	10	164	49	—	—	—		
Clitocybe geotropa	639	65	12	169	67	—	—	—		
Clitocybe nebularis	652	53	6	188	58	—	—	—		
Hygrophorus conicus	482	220	2	157	72	—	—	—	Italy	164
Hygrophorus lucorum	491	244	*tr*	140	67	—	—	—		
Hygrophorus lucorum	486	228	3	145	78	—	—	—		
Lactarius controversus	425	288	*tr*	173	50	—	—	—		
Lactarius sanguifluus	418	264	3	169	46	—	—	—		
Lactarius torminosus	404	297	3	176	55	—	—	—		
Lactarius uvidus	432	283	*tr*	167	49	—	—	—		

Table 3.1 (*Continued*)

Species	C18:2 (mg/100 g)	C18:1 (mg/100 g)	C18:3 (mg/100 g)	C16:0 (mg/100 g)	C18:0 (mg/100 g)	SFAs (mg/100 g)	MUFAs (mg/100 g)	PUFAs (mg/100 g)	Country	Ref.
Agaricus bisporus	2759.4	45.3	1.75	472.8	132.5	736.5	66.0	2774.8	Culture	75
Lentinus edodes	1044.5	43.8	5.59	223.7	13.5	281.3	57.8	1055.6		
Pleurotus ostreatus	1663.2	323.4	—	310.1	38.8	420.8	356.3	1676.6		
Agaricus campestris	2370.1	94.1	1.87	439.1	109.9	667.6	127.5	2391.4	Ethiopia	
Laetiporus sulphureus	673.7	337.8	1.02	189.9	32.0	277.1	358.6	698.2		
Termitomyces clypeatus	1625.3	434.6	1.41	537.4	113.9	752.2	543.9	1654.8		
Termitomyces microcarpus	1831.7	1558.8	0.89	1081.5	374.1	1600.4	1693.1	1861.5		
Termitomyces le testui	2222.8	783.9	3.64	852.6	267.6	1226.4	855.1	2249.7		
Termitomyces spp.	2056.5	670.5	1.77	649.4	169.7	981.9	766.5	2069.8		
Agrocybe aegerita	6.42	24.4	—	94.5	14.3	—	—	—	Portugal	92
Amanita caesarea	5.42	41.2	—	238	132	—	—	—		
Amanita muscaria	5.82	1172	—	178	131	—	—	—		
Agaricus silvicola	12.1	—	—	93.5	22.5	—	—	—		
Amanita vaginata	5.88	—	—	191	67.8	—	—	—		
Boletus edulis	5.03	—	—	28.3	4.86	—	—	—		
Collybia butyracea	10.8	26.9	—	60.8	39.5	—	—	—		
Clitocybe dealbata	6.39	19.2	—	48.2	17.7	—	—	—		
Hebeloma sinapizans	21.4	85.6	—	76.3	33.3	—	—	—		
Lactarius aurantiacus	8.04	50.9	—	50.7	64.0	—	—	—		
Lactarius controversus	8.44	270	—	286	457	—	—	—		
Lycoperdon perlatum	7.19	10.1	—	49.5	14.7	—	—	—		
Macrolepiota procera	13.7	44.5	—	140	29.1	—	—	—		
Mycena rosea	9.54	53.6	—	107	60.8	—	—	—		
Russula cyanoxantha	5.79	345	—	220	95.7	—	—	—		
Russula delica	5.22	335	—	188	200	—	—	—		
Rhizopogon luteolus	7.16	25.7	—	51.2	11.6	—	—	—		
Suillus bovinus	5.56	29.0	—	44.7	17.0	—	—	—		
Sarcodon imbricatus	6.54	32.8	—	68.6	22.7	—	—	—		
Tricholoma acerbum	5.58	190	—	104	70.9	—	—	—		
Tricholoma equestre	12.8	84.2	—	72.4	43.1	—	—	—		
Tricholoma portentosum	5.51	39.1	—	65.9	26.1	—	—	—		

Table 3.1 (*Continued*)

Species	C18:2 (g/100 g)	C18:1 (g/100 g)	C18:3 (g/100 g)	C16:0 (g/100 g)	C18:0 (g/100 g)	SFAs (g/100 g)	MUFAs (g/100 g)	PUFAs (g/100 g)	Country	Ref.
Amanita hemibapha	18.641	38.818	—	10.626	15.830	22.75	—	—	India	53
Trogia cantharelloides	34.556	18.529	0.10	10.233	2.683	16.19	—	—		
Lentinus polychrous	25.30	18.50	6.70	20.00	4.70	—	—	—	Karnataka	76
Hericium erinaceus	38.69	27.36	0.05	18.60	6.15	27.56	32.94	39.50	Portugal	77
Pholiota nameko	62.22	17.82	0.08	10.29	1.64	14.93	22.51	62.56		
Pleurotus citrinopileatus var. *cornucopiae*	69.09	5.47	—	16.52	1.79	21.76	6.32	71.93		
Pleurotus eryngii	53.90	27.13	0.09	12.49	1.66	16.36	29.04	54.59		
Pleurotus salmoneo stramineus	78.33	3.71	—	12.45	1.85	16.51	4.56	78.93		
Agaricus bisporus	81.2	1.2	0.2	13.5	3.7	—	—	—	Market	142
Lentinus edodes	80.7	2.7	0.2	14.5	1.2	—	—	—		
Pleurotus ostreatus	70.8	14.8	0.05	11.4	2.2	—	—	—		

Species	C18:2 (μg g^{-1})	C18:1 (μg g^{-1})	C18:3 (μg g^{-1})	C16:0 (μg g^{-1})	C18:0 (μg g^{-1})	SFAs (μg g^{-1})	MUFAs (μg g^{-1})	PUFAs (μg g^{-1})	Country	Ref.
Grifola frondosa	19.0	13.1	—	36.3	—	—	—	—	Culture	42
Astraeus spp.	4146	7052	—	1639	108	—	—	—	Japan	143
Astraeus spp.	92.5	29.8	—	29.9	5.07	—	—	—		

Species	C18:2 (g kg^{-1})	C18:1 (g kg^{-1})	C18:3 (g kg^{-1})	C16:0 (g kg^{-1})	C18:0 (g kg^{-1})	SFAs (g kg^{-1})	MUFAs (g kg^{-1})	PUFAs (g kg^{-1})	Country	Ref.
Agrocybe aegerita	0.618	0.019	0.003	0.107	0.046	0.153	—	—	Italy	54
Agaricus bisporus	0.858	0.007	0.243	0.103	0.054	0.157	—	—		
Pleurotus giganteus (commercial strain)	5.0	10.3	—	3.8	1.1	—	—	—	Malaysia	93
Pleurotus giganteus (wild strain)	4.5	8.9	—	3.3	0.8	—	—	—		
Agaricus abruptibulbus	0.71	1.40	—	7.02	2.20	12.30	—	—	India	94
Termitomyces globulus	—	16.95	—	10.58	—	10.91	—	—		

Species	C18:2 (mg/10 g)	C18:1 (mg/10 g)	C18:3 (mg/10 g)	C16:0 (mg/10 g)	C18:0 (mg/10 g)	SFAs (mg/10 g)	MUFAs (mg/10 g)	PUFAs (mg/10 g)	Country	Ref.
Armillaria luteo-virens	47.14	22.15	0.26	8.29	3.00	—	—	—	China	123
Agrocybe aegerita	33.31	4.08	0.16	6.93	2.20	—	—	—		
Copyinds comatus	49.34	9.41	0.77	14.88	1.68	—	—	—	Market	
Flammulina velutipes	40.23	5.73	18.02	9.69	2.16	—	—	—		
Pleurotus eryngii	27.17	5.17	0.20	6.49	1.15	—	—		—	

Species	C18:2 (mg/10 mg)	C18:1 (mg/10 mg)	C18:3 (mg/10 mg)	C16:0 (mg/10 mg)	C18:0 (mg/10 mg)	SFAs (mg/10 mg)	MUFAs (mg/10 mg)	PUFAs (mg/10 mg)	Country	Ref.
Auricularia polytricha	0.679	0.665	—	0.364	0.368	—	—	—	Thailand	32

Species	C18:2 (mg mL^{-1})	C18:1 (mg mL^{-1})	C18:3 (mg mL^{-1})	C16:0 (mg mL^{-1})	C18:0 (mg mL^{-1})	SFAs (mg mL^{-1})	MUFAs (mg mL^{-1})	PUFAs (mg mL^{-1})	Country	Ref.
Grifola frondosa	327	388	—	71.3	5.71	—	—	—	Ltd. (Taiwan)	78

[a]MAEH: microwave-assisted extraction with hexane; MAEC: microwave-assisted extraction with hexane; UAEH: ultrasound-assisted extraction with hexane; UAEC: ultrasound-assisted extraction with hexane.

Schizophyllum commune, *Phellinus igniarius*, *Trametes pubescens*, *Trametes versicolor*, *Trametes bicolor* and *Trametes suaveolens* were identified using GC and GC-MS. Linoleic (6.45–61.63%), oleic (6.76–59.25%), palmitic (3.57–28.09%) and stearic (0.90–19.77%) acids were detected as the most abundant fatty acids in all studied mushroom species. Also, chemometric analysis of the fatty acids of the mushroom species was conducted using principal component analyses (PCA) and hierarchical clustering analyses (HCA), and the main fatty acids and their concentrations found in the mushrooms have been reported in distinguishing mushroom species.[16]

In the study of Papazav *et al.*,[17] linoleic (57.3%), palmitic (17.5%), stearic (12.6%) and oleic (7.4%) acids were detected as the main fatty acids of *Morchella esculenta* by GC-FID. The amount of SFA was 32.3%, while the amount of UFA was 67.7%, in the oil from the mushroom.

Fatty acid profiling of *Flammulina velutipes*, *Helvella crispa*, *Ganoderma lucidum*, *Mycena vitilis*, *Lactarius rubidus*, *Laccaria bicolor*, *Russula brevipes*, *Russula nobilis* and *Xerocomellus chrysenteron* were reported as palmitic (1.907–3.768%), stearic (0.413–3.765%), oleic (1.063–4.868%), linoleic (0.258–15.690%), nonadecanoic (0.432–12.547%) and linolenic (0.0202–3.978%) acids through GC-MS. The amounts of PUFAs of *F. velutipes*, *G. lucidum*, *L. rubidus*, *M. vitilis* and *X. chrysenteron* mushrooms were higher than for other studied mushrooms.[18]

GC-MS analysis led to analysis of fatty acid compounds of *Auricularia cornea* with a predominance of linoleic (43.54%), oleic (29.83%), palmitic (13.25%) and stearic (7.83%) acids.[19]

As a result of transesterification and GC-FID analysis, linoleic (73.39%), oleic (12.93%) and palmitic (5.82%) acids were recorded as the major fatty acids in *Coprinus comatus*. The concentrations of α-linolenic, eicosatrienoic and eicosapentaenoic acids, which are ω-3 fatty acids, were also found to be greater than 0.5%.[20]

Fatty acid compositions consisting of palmitic acid (40.54–43.98%), stearic acid (28.14–33.94%), linoleic acid (2.06–15.47%), 2,4-di-*tert*-butylphenol (7.61–13.89%) and oleic acid (*trace*) in water, methanol, acetone, ethyl acetate, chloroform and hexane extracts of *Lycoperdon utriforme* were reported using GC-MS.[21]

The most abundant fatty acids were found to be linoleic (51.04%) and palmitic (20.6%) acids in *Agaricus blazei* among the 22 detected fatty acids with trans-esterification and GC-MS analysis of the mushroom. Regarding the fatty acids classes, PUFAs dominated, followed by SFAs and MUFAs.[22]

The fatty acid profiles of *Coprinellus truncorum* and *Coprinus comatus* from nature (fruiting body) and submerged cultivation (cultured mycelium and fermentation broth) were investigated using GC-MS. Linoleic (51.41–74.19%), oleic (4.38–20.28%) and palmitic (13.32–16.15%) acids were reported as the major fatty acids of *C. comatus* and linoleic (16.94–70.09%), palmitic (12.39–28.58%) and oleic (6.55–9.08%) acids of *C. truncorum*.[23]

Fatty acid compositions of *Cantharellus cinerus*, *Clavariadelphus pistillaris*, *Clitocybe nebularis* and *Hygrocybe punicea* including linoleic (20.46–43.39%), oleic (16.89–46.36%), palmitic (19.88–30.95%) and stearic (1.55–9.53%) acids were revealed by Dimitrijevic *et al.*[24] The levels of linoleic acid were high in all studied mushrooms with the highest value of 43.39% in *C. nebularis*. The concentration of oleic acid, which is a monounsaturated fatty acid, was observed in higher percentages in *C. pistillaris* (46.36%) followed by *H. punicea* and *C. nebularis* mushrooms (30.65 and 31.44%), respectively.

Oleic (48.37%), linoleic (29.49%), stearic (16.96%) and palmitic (5.17%) acids were detected as the fatty acids of *Lactarius deliciosus* by trans-esterification and GC-MS analysis. The amounts of UFAs (C18:1 and C18:2) were high in *L. deliciosus*.[25]

Different radiation [gamma (0, 1, 2, 5 kGy) and electron beam (0, 1, 2, 5 kGy)] doses and storage times (0, 4, 8 days) effects on the fatty acid composition of *Agaricus bisporus* were studied by Cardoso *et al.*[26] The levels of the main fatty acid, linoleic acid, were reported as 79.7, 79.6, 78.7, 78.6, 79.4, 79.5 and 78.6%, respectively, for gamma irradiated (0, 1, 2, 5 kGy) and stored (0, 4, 8 days), and 78.6, 78.9, 78.0, 78.2, 78.5, 78.6 and 78.1%, respectively, for electron-beam irradiated (0, 1, 2, 5 kGy) and stored (0, 4, 8 days).

Ciric *et al.*[27] determined the major fatty acids of three *Macrolepiota* species (*M. rhacodes*, *M. mastoidea*, *M. procera*) as consisting of linoleic (64–78.62%), palmitic (10.00–21%) and oleic (6.06–11.10%) acids. PUFAs were detected as dominant in all mushrooms studied, while SFAs and MUFAs were identified in lower amounts.

The dominant fatty acids were identified by using GC and GC-MS as palmitic (20.07, 39.14%), oleic (34.23, 20.39%), linoleic (19.64, 12.73%) and stearic (9.44, 17.78%) acids in two *Stereum* species (*S. rugosum* and *S. sanguinolentum)* among the 30 identified fatty acids. Also, the linoleic/oleic acid ratio was calculated as 0.57 for *S. rugosum* and 0.62 for *S. sanguinolentum*.[28]

Linoleic (56.7%), oleic (23.9%) and palmitic (16.4%) acids were found to be the main fatty acids of *Pholiota aurivella* in GC-MS analysis.[29]

In the study by Bengu,[30] oleic, linoleic, palmitic and stearic acid levels of *Suillus luteus*, *Laetiporus sulphureus* and *Coprinus atramentarius* were given as 24.60–33.94, 7.73–33.17, 21.24–28.60 and 10.49–28.55%, respectively. Also, seven different fatty acids (C14:0, C15:0, C16:1, C18:3, C22:0, C20:3 and C20:4) were investigated in this study in addition to these fatty acids. The highest concentration of linoleic and linolenic acids was identified in *C. atramentarius* while the highest concentrations of oleic, myristic, stearic and palmitic acids were found in *L. sulphureus*. Behenic and palmitoleic acids were richest in *S. luteus* in this report.

Fatty acids detected in *Tuber indicum* collected from five geographical regions of China (Gongshan, Tibet, Liaoning, Qujing, Hebei) were reported as linoleic (56.05–76.52%), oleic (nd–19.70%), palmitic (6.13–9.89%) and stearic (1.61–3.84%) acids. The major fatty acid classes in the studied truffles from the different geographical regions were PUFAs (64.20–85.53%), followed by MUFAs (0.77–20.32%) and SFAs (10.63–15.93%). As a result of the highest amount of linoleic acid, the PUFA content (85.53%) was highest in *T. indicum* collected from Hebei.[31]

Linoleic (0.679 mg/10 mg), oleic (0.665 mg/10 mg), palmitic (0.364 mg/10 mg) and stearic (0.368 mg/10 mg) acids were detected as the fatty acids of *Auricularia polytricha* in a quantitative analysis using GC-MS.[32]

Eleven mushroom species (*Boletus appendiculatus*, *Boletus regius*, *Boletus edulis*, *Boletus purpureus*, *Boletus impolitus*, *Boletus rhodoxanthus*, *Lactarius deliciosus*, *Lactarius sanguifluus*, *Lactarius semisanguifluus*, *Lactarius volemus*, *Xerocomellus chrysenteron*) were examined by GC and GC-MS analysis for the presence of fatty acids. Linoleic (19.5–72%), oleic (0.11–64%), stearic (0.81–57%) and palmitic (5.9–22%) acids were reported as the main fatty acids among the 26 determined fatty acids. Palmitoleic, α-linolenic, eicosenoic, erucic and nervonic acids were also present in lower percentages, but not in all species.[33]

Palmitic (28.84%), oleic (28.59%) and linoleic (28.29%) acids were determined in the highest amounts of the total content in *Tricholoma equestre* by GC and GC-MS. The UFA content was found to be higher than the SFA content.[34]

Eighteen fatty acids with carbon chain lengths of 4–24 were detected in raw and fermented *Lenzites quercina* by using GC. The main fatty acids of raw and fermented *L. quercina* were identified as oleic (38.50, 37.92%), lauric (22.70, 18.71%), linolenic (17.40, 22.50%) and stearic (11.96, 16.99%) acids.[35]

Twenty-three fatty acids, saturated and unsaturated, were identified in *Polyporus squamosus*. The major fatty acids determined in *Polyporus squamosus* in high percentages were reported as linoleic (56.55%), oleic (22.27%) and palmitic (13.40%) acids in the study of Mocan *et al.*[36] Furthermore, PUFAs (57.37%), which are known to have several beneficial health effects, were found to be dominant, followed by MUFAs (24.96%) and SFAs (17.68%).

Linoleic (54.03%) and palmitic (11.57%) acids were characterized as the most abundant fatty acids of *Agaricus arvensis* by GC-MS.[37]

Dospatliev[38] identified the major fatty acids found in *Cantharellus cibarius* as oleic (35%), palmitic (23.8%), linoleic (22.7%) and stearic (12.3%) acids. The UFA content was 60.3% while the MUFA content was 39.7%.

Linoleic (11.72–71.41%), stearic (2.18–43.23%), oleic (1.80–36.29%) and palmitic (5.35–12.73%) acids were identified as the major fatty acids of *Lactarius piperatus*, *Agaricus bisporus*, *Cantharellus cibarius*, *Pleurotus ostreatus* and *Boletus edulis* through GC-MS. Besides the four major fatty acids, 20 fatty acids were identified and quantified in total. PUFAs were the predominant group of fatty acids in all species except *C. cibarius*. The highest PUFA content (71.75%) was in *A. bisporus* due to the higher concentration of linoleic acid.[39]

Linoleic (53.03, 5.35%), oleic (14.55, 35.80%), palmitic (16.01, 38.60%) and stearic (4.50, 12.26%) acids were noted as the most abundant fatty acids in *Chondrogaster pachysporus* and *Setchelliogaster tenuipes*. Also, the linoleic/oleic acid ratios were calculated as 3.64 and 0.15.[40]

Fatty acids of four different *Ganoderma* species (*G. lucidum*, *G. adspersum*, *G. resinaceum*, *G. applanatum*) collected from four different host trees were extracted using three different techniques (maceration, soxhlet and ultrasonic) and the changes in the fatty acid compositions were investigated. Linoleic (9.80–3467 $\mu g\,g^{-1}$), oleic (16.9–3356 $\mu g\,g^{-1}$), palmitic (15.5–1272 $\mu g\,g^{-1}$), tricosanoic (1.11–406 $\mu g\,g^{-1}$), tetracosanoic (1.11–408 $\mu g\,g^{-1}$), lauric (2.17–314 $\mu g\,g^{-1}$) and stearic (1.11–286 $\mu g\,g^{-1}$) acids were reported as the most abundant fatty acids in all *Ganoderma* species. The oleic acid/linoleic acid ratio ranged from 0.18 to 2.88 in all samples.[41]

Palmitic (36.3 $\mu g\,mg^{-1}$), linoleic (19 $\mu g\,mg^{-1}$) and oleic (13.1 $\mu g\,mg^{-1}$) acids were presented as the fatty acids of *Grifola frondosa* by GC-MS.[42]

Li *et al.*[43] reported the main fatty acids of three *Chanterelle* species (*C. cibarius*, *C. cinnabarinus schwein*, *C. tubaeformis*) as oleic (4.95–9.46 $mg\,g^{-1}$), stearic (6.56–8.37 $mg\,g^{-1}$), linoleic (0.96–5.75 $mg\,g^{-1}$), and palmitic (2.42–3.17 $mg\,g^{-1}$) acids.

Fatty acid profiles by GC-MS revealed the main fatty acids in four *Pleurotus* species (*P. sapidus*, *P. sajor-caju*, *P. columbinus*, *P. ostreatus*) to be linoleic (1.421–16.013 $mg\,g^{-1}$), oleic (1.170–6.075 $mg\,g^{-1}$) and palmitic (1.246–4.229 $mg\,g^{-1}$) acids. UFAs were found in higher amounts than SFAs.[44]

The main fatty acids were recorded as linolenic (7.44%), stearic (6.75%), oleic (5.98%) and palmitic (4.55%) acids in *Lentinus tuberregium* by GC analysis.[45]

The analysis of the fatty acid compounds of *Suillus granulatus*, *Pleurotus eryngii*, *Armillaria tabescens* and *Leucopaxillus gentianeus* revealed linoleic (27.64–68.24%), oleic (10.05–39.78%) and palmitic (7.95–19.40%) acids as the main fatty acids. In addition, the linoleic/oleic acid ratios were calculated as 0.77, 1.03, 6.79 and 1.34, respectively.[46]

Linoleic (16.9–57.4%), oleic (9.1–44.8%) and palmitic (4.1–36.9%) acids were found to be the major fatty acids of *Russula delica*, *Funalia trogii*, *Ganoderma adspersum*, *Fomes fomentarius*, *Ganoderma applanatum*, *Pleurotus ostreatus*, *Gyromitra esculenta*, *Ganoderma lucidum* and *Lyophyllum decastes*. The linoleic/oleic acid ratios were calculated as 4.63, 0.99, 0.97, 2.04, 4.70, 0.41, 0.59, 4.46 and 0.91, respectively.[47]

A total of 25 fatty acids were detected and quantified in *Armillaria mellea* by GC-FID. Among the identified fatty acids, linoleic, oleic and palmitic acids were found to be in the highest levels of 63.80, 15.47 and 12.63%, respectively. PUFAs were found to be predominant in *A. mellea* with 64.12%, followed by SFAs (18.42%) and MUFAs (17.47%).[48]

The fatty acid profile of *Boletus edulis* treated by different gamma radiation doses (0, 2, 6, 10 kGy) and storage times (0, 6, 12 months) was researched and the amounts of the main fatty acids consisting of linoleic, oleic and palmitic acids were reported to be in the ranges of 50–54, 31–33 and 10–11%, respectively.[49]

The fatty acid profiles of 12 wild strains of *Ganoderma* sp. (DS1, HS1, AM1, LS1–LS9) were screened by Obodai *et al.*[50] and oleic (21.78–48.76%), linoleic (9.74–44.27%), and palmitic (14.65–26.96%) acids were found to be the main fatty acids. In general, a higher content of SFA over PUFA was observed, except for strains AM1, LS1, LS2 and LS7, where PUFAs were predominant.

A total of 24 fatty acids were identified in *Meripilus giganteus*, linoleic acid (64.86%) being the most abundant fatty acid, followed by palmitic (14.96%) and oleic (13.61%) acids. PUFAs predominated over SFAs, representing 65.55% of total fatty acids.[51]

The predominant fatty acids of 12 coral mushrooms (*Clavaria vermicularis*, *Clavaria fragilis*, *Clavaria purpurea*, *Clavaria amoena*, *Clavaria rosea*, *Clavaria coralloides*, *Ramaria botrytis*, *Ramaria flava*, *Ramaria rubripermanens*, *Ramaria flavescens*, *Ramaria stricta*, *Ramaria aurea*) were identified as linoleic (3.93–8.28%), paullinic (2.39–4.20%) and palmitic (1.19–2.19%) acids. In general, PUFAs were the main class of fatty acids in all studied species. *C. fragilis* and *R. botrytis* included lower concentrations of SFA and higher PUFAs as compared to other mushrooms.[52]

The main fatty acids of *Amanita hemibapha* and *Trogia cantharelloides* were found to be oleic (38.818, 18.529 g/100 g), linoleic (18.641, 34.556 g/100 g), stearic (15.830, 2.683 g/100 g) and palmitic (10.626, 10.233 g/100 g) acids among the 20 identified fatty acids by Ravikrishnan *et al.*[53] There are about 17 fatty acids detected in *A. hemibapha* and about 18 fatty acids in *T. cantharelloides*.

Landi *et al.*[54] reported linoleic (0.618 $g\,kg^{-1}$) and palmitic (0.107 $g\,kg^{-1}$) acids as the main fatty acids in *Agaricus aegerita*; and linoleic (0.858 $g\,kg^{-1}$), γ-linolenic (0.243 $g\,kg^{-1}$) and palmitic (0.103 $g\,kg^{-1}$) acids in *Agaricus bisporus*. Generally, the fatty acid contents of *A. aegerita* and *A. bisporus* were richer in UFAs than in SFAs.

The fatty acid profile of *Rhizopogon luteolus* was found to be rich in linoleic (45.8%), stearic (23.7%), oleic (16.2%) and palmitic (6.6%) acids using GC and GC-MS.[55]

The fatty acids with the highest percentages in *Leucopaxillus candidus* were found to be linoleic (49%), oleic acid (37%), palmitic (8.0%) and stearic (2.7%) acids by Vieira *et al.*[56] Consequently, *L. candidus* was rich in PUFAs (50%), followed by MUFAs (37%) and SFAs (13%).

Leccinum molle and *Leccinum vulpinum* revealed a similar profile, with linoleic (43.49, 53.60%), oleic (38.61, 27.06%), palmitic (11.25, 12.23%) and stearic (1.79, 2.77%) acids being the main fatty acids. PUFAs were present in higher quantities in both species compared to SFAs.[57]

The effects of 2 kGy gamma irradiation on the fatty acid profile of *Boletus pinophilus* and *Clitocybe subconnexa* were discussed by Fernandes *et al.*[58] When there was no irradiation the amounts of the most abundant fatty acids, the percentages of linoleic, oleic and palmitic acids were found to be 41.0, 37.3%; 45.8, 47.9%; and 7.3, 8.7%, respectively. Oleic (45.1, 52.5%), linoleic (42.3, 35.1%) and palmitic (6.9, 7.5%) acids were determined to be in the highest amounts after 2 kGy of gamma irradiation.

The fatty acid composition of *Boletus edulis* originating from four different regions (A, B, C and D) of Poland was identified using GC-MS. The most abundant fatty acids were found to be linoleic (32.78–50.62%), oleic (19.56–44.75%) and palmitic (8.39–28.84%) acids. Also, the highest amount of linoleic (40.71, 32.68%), oleic (34.51, 20.69%) and palmitic (15.96,

35.95%) acids were found in the caps and stems of *B. edulis* in this study. The concentration of other fatty acids of *B. edulis* ranged between 0.10–2.69%. SFAs were about twice as rich in caps (43.01%) as in the stems (21.48%). The caps were quantified by a higher content of MUFAs (36.64%) and PUFAs (41.87%) compared to the stems (23.64% and 33.34%, respectively).[59]

The maximum amounts of linoleic (19.28–63.75%), oleic (7.93–52.08%) and palmitic (12.45–20.17%) acids were recorded among the 24 detected fatty acids in *Aleurodiscus vitellinus*, *Cyttaria hariotii*, *Cortinarius magellanicus*, *Fistulina antarctica*, *Fistulina endoxantha*, *Grifola gargal*, *Lepista nuda*, *Hydropus dusenii* and *Ramaria patagonica*. PUFAs, due to the high content of linoleic acid (72%), were the predominant group of fatty acids in *C. hariotii*, *H. dusenii*, *R. patagonica*, *F. endoxantha*, *L. nuda*, *C. magellanicus* and *F. antarctica* mushrooms.[60]

Fatty acid compositions of *Morchella conica* species collected from two different localities were reported as linoleic (63, 67%), oleic (21, 19%), palmitic (11, 10%) and stearic (2.4, 1.5%) acids. PUFAs were the main group studied in both samples.[61]

The main fatty acids of yanqi black mushroom were given as linoleic (87.3%), stearic (3.59%) and palmitic (3.27%) acids as a result of GC-MS analysis.[62]

Linoleic (7.44–25.68 $mg\,g^{-1}$), palmitic (0.20–5.59 $mg\,g^{-1}$) and oleic (0.33–2.58 $mg\,g^{-1}$) acids were reported as the main fatty acids of *Handkea utriformis*, *Handkea excipulliformis* and *Vascellum pratense*.[63]

The fatty acid profiles of four extracts of *Laetiporus sulphureus* obtained by microwave and ultrasound-assisted extraction with hexane and chloroform were determined in the study of Sinanoglou *et al.*[64] Linoleic (56.10–57.00%), oleic (13.91–14.38%) and palmitic (10.74–11.38%) acids were found to be the main fatty acids among the 31 identified fatty acids. PUFAs were observed in the highest amounts, followed by SFAs and MUFAs. The linoleic acid/oleic acid ratio was found to be from 3.93 to 4.34.

Proximate fatty acid compositions of *Hericium erinaceus* and *Hericium coralloides* were reported to consist of oleic (26.11, 33.74%), linoleic (25.10, 30.90%), palmitic (37.57, 23.34%) and stearic (7.61, 6.78%) acids. SFAs predominated over PUFAs and MUFAs, due to the high concentration of palmitic and stearic acids for *H. erinaceus*. The concentrations of SFA, PUFA and MUFA were close to each other for *H. coralloides*.[65]

The fatty acid contents of *Pleurotus sajor-caju* exhibited the presence of saturated and unsaturated fatty acids with 12–18 carbons. Linoleic (53.50%) and palmitic (24.16%) acids were recorded as the most abundant fatty acids in *P. sajor-caju* grown in whey.[66]

Linoleic (0.04–3.44%) and palmitic (0.05–1.96%) acids were detected as the major fatty acids in *Agaricus arvensis*, *Agaricus comtulus*, *Agaricus campestris*, *Agaricus silvicola*, *Amanita citrine*, *Amanita fulva*, *Amanita caesarea*, *Cantharellus cibarius*, *Conocybe tenera*, *Gymnopilus junonius*, *Hygrocybe coccinea*, *Hygrocybe nivea*, *Inocybe splendens*, *Lactarius pubescens*, *Laccaria laccata*, *Lepista nuda*, *Lentinus cladopus*, *Pleurotus cystidiosus*, *Russula lepida* and *Russula mairei*. Besides these three major fatty acids, six more were identified and characterized. PUFAs were the main class of fatty acids reported in all these mushrooms. However, UFAs predominated over SFAs in all the studied mushrooms, accounting for between 65% and 70%.[67]

The main fatty acids in *Volvopluteus gloiocephalus* and *Clitocybe subconnexa* were identified as linoleic (59.33, 44.88%), oleic (25.96, 42.50%), palmitic (7.52, 7.31%) and stearic (2.09, 2.15%) acids. The concentrations of palmitic and stearic acids were similar in both species, but the concentration of oleic acid was significantly higher in *C. subconnexa*, contributing to its higher amounts of MUFA. In addition, the concentration of linoleic acid was higher in *V. gloiocephalus*, contributing to its higher amounts of PUFA.[68]

As a result of the measurements of *Boletus edulis*, *Lentinus edodes* and *Xerocomus badius* fatty acids based on GC-MS, linoleic (42.22–82.55%), oleic (1.87–38.15%), palmitic (7.56–15.47%) and stearic (1.51–1.96%) acids were determined to be the main fatty acids. *L. edodes* presented a significant linoleic acid content compared with *B. edulis* and *X. badius,* but a much lower amount of oleic acid.[69]

Twenty-four fatty acids were identified in *Agrocybe aegerita* by GC-FID and linoleic, palmitic, oleic and stearic acids were determined to be in the highest amounts of 78.40, 13.07, 3.03 and 2.13%, respectively. Other fatty acids presented with a share of less than 1.00%.[70]

Quantitative analysis of *Boletus aereus* revealed the presence of oleic (41.78%), linoleic (40.02%), palmitic (13.17%) and stearic (1.84%) acids as the main fatty acids. Twenty additional fatty acids were quantified in trace amounts.[71]

The major fatty acids of *Cynomorium coccineum* oil were reported to be oleic (37.03%), linoleic (19.46%) and palmitic (15.75%) acids by GC, and oleic (286.5 mg g^{-1}), linoleic (149.4 mg g^{-1}) and linolenic (72.6 mg g^{-1}) acids by HPLC.[72]

The predominant fatty acids in *Termitomycetes microcarpus* and *Termitomycetes heimii* were recorded as linolenic (1.61, 2.054%), palmitic (1.265, 1.129%), α-linolenic (1.069, 0.98%) and oleic (0.82, 1.113%) acids by GC-MS. Linoleic acid was the principal unsaturated fatty acid, while palmitic acid was the principal saturated fatty acid present in both species.[73]

Chan *et al.*[74] reported linoleic (62.0%), palmitic (16.87%) and oleic (13.87%) acids as the main fatty acids in the mycelium of *Ganoderma tsugae* var. *jannieae*; oleic (43.0%), with elaidic (28.13%) and palmitic (11.4%) acids being the main ones in the fruiting body of *G. tsugae* var. *jannieae.*

Woldegiorgis *et al.*[75] reported the main fatty acids of *Agaricus campestris*, *Agaricus bisporus*, *Laetiporus sulphureus*, *Lentinus edodes*, *Pleurotus ostreatus*, *Termitomyces microcarpus*, *Termitomyces clypeatus*, *Termitomyces le-testui* and *Termitomyces* spp. as linoleic (673–2759 mg/100 g), oleic (43.8–1558.8 mg/100 g), palmitic (189.9–1081.5 mg/100 g) and stearic (13.5–374.1 mg/100 g) acids among 25 identified fatty acids. Also, the linoleic/oleic acid ratios were reported to be 5.14, 23.9, 60.9, 25.2, 1.99, 3.74, 1.18, 2.84 and 3.07, respectively.

Linoleic (25.30 g/100 g), palmitic (20.0 g/100 g) and oleic (18.50 g/100 g) acids were determined to be in the highest concentrations in the *Lentinus polychrous* fatty acid profile. UFA contents were higher than SFAs.[76]

Rodrigues *et al.*[77] identified the main fatty acids of *Pleurotus eryngii*, *Pholiota nameko*, *Pleurotus citrinopileatus* var. *cornucopiae*, *Hericium erinaceus* and *Pleurotus salmoneo stramineus* as linoleic (38.69–78.33 g/100 g), oleic (3.71–27.36 g/100 g) and palmitic (10.29–18.60 g/100 g) acids. The fatty acid contents of the mushrooms were identified by a much higher content of MUFAs and PUFAs than of SFAs.

Oleic (388 $mg\,mL^{-1}$), linoleic (327 $mg\,mL^{-1}$) and palmitic (71.3 $mg\,mL^{-1}$) acids were detected as the major fatty acids of *Grifola frondosa* by GC-MS analysis.[78]

The methylated extract of *Chroogomphus rutilus* was reported to possess 13 fatty acids by GC and GC-MS, and linoleic (41.61%) and oleic acids (35.87%) were the most abundant fatty acids. The linoleic acid/oleic acid (L/O) ratio was found to be 1.16.[79]

As a result of transesterification and GC-FID analysis, linoleic (63.27%), oleic (14.52%) and palmitic (11.68%) acids were found to be the major fatty acids in *Laetiporus sulphureus*. Polyunsaturated fatty acids (64.14%) in *L. sulphureus* had a higher content, while SFA (20.54%) and MUFA (15.32%) fatty acids had lower contents.[80]

Linoleic (57.14, 63.97%), oleic (24.64, 20.08%), palmitic (9.64, 9.62%) and stearic (3.19, 2.65%) acids were identified as the major fatty acids in the fatty acid contents of two *Suillus granulatus* originating from two different locations. In general, MUFAs and PUFAs were the prevailing fatty acids in both samples (21.30–64.40%). The Portuguese sample showed a higher concentration of MUFA (26.55%), whereas the Serbian sample revealed a higher concentration of PUFA (64.40%).[81]

Unbaked and baked *Lactarius delicious* and *Ramaria flava* species were investigated for fatty acid compositions. Oleic (25.94, 47.12%), linoleic (22.85, 9.78%), palmitic (9.7, 14.43%) and stearic (1.41, 6.68%) acids were found to be the main fatty acids in unbaked mushrooms; meanwhile, oleic (18.07, 45.13%), stearic (49.94, 6.23%), linoleic (23.36, 9.25%) and palmitic (7.92, 19.12%) acids were highest in baked mushrooms. The oleic and palmitic acid contents of *L. delicious* decreased when the mushroom was baked; however, the linoleic and stearic acids contents increased. Only the palmitic acid content of *R. flava* increased after baking.[82]

The main fatty acids detected in *Phellinus linteus* were reported as palmitic (26.8%), linoleic (25.3%), oleic (13.5%), and stearic (12.3%) acids. SFAs dominated (58.3%) over UFAs (MUFAs: 14.6%; PUFAs: 27.1%).[83]

Ganoderma lucidum from two different regions revealed a similar profile, with linoleic (33.94, 39.80%), oleic (47.24, 24.06%), palmitic (12.01, 18.54%), stearic (1.33, 7.34%) and linolenic (0.38, 2.23%) acids being the fatty acids present.[84]

Vieira *et al.*[85] characterized the fatty acid compositions of two *Lactarius* species (*L. citriolens* and *L. turpis*) by using transesterification procedures and GC-FID technique and linoleic (22.46, 48.55%), stearic (40.58, 12.60%), oleic (25.00, 26.29%) and palmitic (5.35, 8.02%) acids were found to be the main fatty acids.

In the study by Stojkovic *et al.*,[86] while linoleic (43.87%), oleic (20.58%) and palmitic (14.02%) acids were detected as the main fatty acids of *Agaricus bisporus*, linoleic (74.48%) and palmitic (11.34%) acids were reported to be the main fatty acids of *Agaricus brasiliensis.* PUFAs predominated over SFAs and MUFAs.

The fatty acid profiles of *Macrolepiota procera* treated by different electron beam radiation doses (0, 0.5, 1, 6 kGy) and storage times (0, 6, 12 months) were researched, and the amounts of the main fatty acids, consisting of linoleic, palmitic and oleic acids, were reported to be in the range of 56–62, 22–27 and 7.6–9.4%, respectively.[87]

The variations in the fatty acid compositions of *Russula delica* and *Boletus edulis* submitted to different electron-beam irradiation (0, 2, 6, 10 kGy) doses were studied. At 0, 2, 6 and 10 kGy irradiation, the percentages of the major fatty acids, linoleic, oleic and palmitic acids, were found to be in the range of 72.2–77.9, 3.8–12.5 and 9.0–11.0%, respectively, for *B. edulis*, and 63.5–67.5, 16.3–19.4 and 12.1–13.3%, respectively, for *R. delica.*[88]

The main fatty acids irradiated with 0, 0.5, 1.0 kGy doses and fresh, frozen and dried *Macrolepiota procera* were identified as linoleic (65–68%), palmitic (20–21%) and oleic (6–7%) acids.[89]

A total of 14 fatty acids were identified in a derivatized methyl ester of four *Lactarius* species (*L. deterrimus*, *L. volemus* sensu lato, *L. hatsudake*, *L. tenuicystidiatus*) using GC. Stearic (0.5–37.9 $mg\,g^{-1}$), oleic (1.8–21.8 $mg\,g^{-1}$), linoleic (9.7–17.7 $mg\,g^{-1}$) and palmitic (1.2–3.7 $mg\,g^{-1}$) acids were found to be the major fatty acids.[90]

The fatty acid contents of mycelium and fruiting bodies of 12 mushrooms (*Agaricus brasiliensis*, *Coprinus comatus*, *Coprinus militaris*, *Grifola frondosa*, *Ganoderma lucidum*, *Flammulina velutipes*, *Hericium erinaceus*, *Ophiocordyceps sinensis*, *Lentinus edodes*, *Pleurotus ostreatus*, *Tremella fuciformis*, *Trametes versicolor*) were researched. Linoleic, oleic, palmitic and stearic acids were determined to be in the highest amounts in fruiting bodies (11.6–75.8, 3.46–51.8, 13.65–25.8, 1.88–17.7 $mg\,g^{-1}$, respectively) and mycelium (3.60–42.4, 16.2–45.7, 14.0–27.5, 3.70–18.9 $mg\,g^{-1}$, respectively).[91]

The fatty acid compositions of 22 mushroom species were identified by GC-MS by Carvalho *et al.*[92] Palmitic acid was recorded as the main fatty acid in *Agrocybe aegerita* (94.5 mg/100 g), *Amanita caesarea* (238 mg/100 g), *Agaricus silvicola* (93.5 mg/100 g), *Amanita vaginata* (191 mg/100 g), *Boletus edulis* (28.3 mg/100 g), *Collybia butyracea* (60.8 mg/100 g), *Clitocybe dealbata* (48.2 mg/100 g), *Lycoperdon perlatum* (49.5 mg/100 g), *Macrolepiota procera* (140 mg/100 g), *Mycena rosea* (107 mg/100 g), *Rhizopogon luteolus* (51.2 mg/100 g), *Suillus bovinus* (44.7 mg/100 g), *Sarcodon imbricatus* (68.5 mg/100 g); oleic acid in *Amanita muscaria* (1172 mg/100 g), *Hebeloma sinapizans* (85.6 mg/100 g), *Lactarius aurantiacus* (50.9 mg/100 g), *Russula cyanoxantha* (345 mg/100 g), *Russula delica* (335 mg/100 g), *Tricholoma acerbum* (190 mg/100 g), *Tricholoma equestre* (84.2 mg/100 g), *Tricholoma portentosum* (65.9 mg/100 g); and stearic acid in *Lactarius controversus* (457 mg/100 g).

Oleic (10.3, 8.9 g kg^{-1}), linoleic (5.0, 4.5 g kg^{-1}), palmitic (3.8, 3.3 g kg^{-1}) and caprylic (1.3, 1.1 g kg^{-1}) acids were determined as the dominant fatty acids in wild and commercial strains of *Pleurotus giganteus*.[93]

The main fatty acids were reported to be palmitic (7.02 g kg^{-1}) and stearic (2.20 g kg^{-1}) acids in uncooked *Agaricus abruptibulbus*; oleic (3.06 g kg^{-1}) and palmitic (1.98 g kg^{-1}) acids in cooked *Agaricus abruptibulbus*; oleic (16.65 g kg^{-1}), elaidic (11.42 g kg^{-1}) and palmitic (10.58 g kg^{-1}) acids in uncooked *Termitomyces globulus*; and stearic (12.28 g kg^{-1}), palmitic (8.88 g kg^{-1}) and oleic (7.50 g kg^{-1}) acids in cooked *Termitomyces globulus*.[94]

The fatty acid contents of caps and stipes of *Agaricus bisporus* and *Pleurotus ostreatus* were examined and linoleic (24.1, 61.9%), oleic (54.8, 18.4%), palmitic (8.7, 11.2%) and stearic (5.9, 1.5%) acids were found to be as the prominent fatty acids in caps; and linoleic (41.9, 36.6%), oleic (38.7, 17.6%), palmitic (7.6, 7.1%) and stearic (9.1, 3.2%) acids in stipes.[95]

The fatty acid compositions of *Boletus edulis* and *Hydnum repandum* irradiated by different gamma radiation doses (0, 1, 2 kGy) were investigated. The levels of the main fatty acids, consisting of linoleic, oleic and palmitic acids, were reported to be in the range of 48.1–51.7, 34.1–37.4 and 7.6-–8.5%, respectively, in *B. edulis* and 33–39.3, 37.2–40.9 and 13.1–17%, respectively, in *H. repandum*.[96]

Fatty acid analysis of five mushroom species (*Rhizopogon roseolus*, *Amanita ceciliae*, *Cantharellus cibarius*, *Armillaria mellea*, *Chlorophyllum rhacodes*) showed the presence of linoleic (31.81–57.70%), oleic (9.06–42.74%), palmitic (8.53–19.08%) and stearic (2.38–7.72%) acids as the major ones present. The carbon chain lengths of fatty acids of mushroom samples ranged from 4 to 24. Generally, UFA contents were at higher percentages than SFA contents.[97]

The highest percentages of linoleic (31.61–44.32%), oleic (13.66–39.48%), palmitic (7.11–12.94%) and stearic (2.75–34.35%) acids were recorded among the 26 identified fatty acids in *Russula delica*, *Suillus bellinii*, *Lactarius sanguifluus*, *Lactarius deliciosus* and *Lactarius semisanguifluus*. PUFAs dominated in *R. delica* and MUFAs in *S. bellinii*, while SFAs dominated in the three *Lactarius* species.[98]

The fatty acid profiles of *Lactarius deliciosus* treated by different gamma radiation doses (0.0, 0.5, 1.0 kGy) and storage times (0, 4, 8 days) were examined, and the amounts of the main fatty acids consisting of stearic (60–67%), linoleic (13–20%), oleic (8–10%) and palmitic (5.6–7%) acids were reported.[99]

The effects of different processing technologies consisting of drying, freezing and gamma irradiation (0.1 kGy) on *Macrolepiota procera* fatty acid composition were investigated. Linoleic (62–71.0%), palmitic (18.2–22.7%), oleic (5.7–8.3%) and stearic (1.46–2.32%) acids were recorded as the main fatty acids, and gamma irradiation was reported as the processing technology with the highest ability to preserve the fatty profile of fresh mushrooms.[100]

Determination of the fatty acid profiles of the pileus, stipe, sclerotium and mycelium parts of *Lignosus rhinocerotis* showed oleic (17.02–34.44%), palmitic (13.89–36.21%), stearic (4.51–5.34%), γ-linolenic (0.00–12.02%), lauric (1.05–3.71%) and α-linolenic (0.59–7.34%) acids as the most abundant fatty acids. In general, *L. rhinocerotis* possessed higher contents of UFAs than SFAs, with a high percentage of essential fatty acids.[101]

The highest levels of linoleic (37.63–63.88%), oleic (16.05–35.74%), palmitic (14.86–26.97%) and stearic (0.88–2.77%) acids were found in the fatty acid profiles of *Lepista nuda* fruiting bodies from three different habitats (wild oak forest, wild pine forest, commercial). The commercial sample had the highest PUFA content (64.41%) due to the higher concentration of linoleic acid, while the wild sample from oak forest had the highest MUFA levels (36.74%) due to the higher oleic acid content. The wild sample from the pine forest exhibited the highest SFA content (34.43%) due to the contribution of palmitic and stearic acids.[102]

Oleic (32.29%), linoleic (29.72%) and palmitic (27.97%) acids were recorded as the main fatty acids of *Tirmania pinoyi* using transesterification and GC-FID analysis. SFAs were the most dominant (35.82%), followed by MUFAs (33.50%) and PUFAs (30.67%).[103]

Doğan and Akbaş[104] characterized the main fatty acids of *Amanita caesarea* as oleic (58%), linoleic (19.02%), palmitic (15%) and stearic (6.09%) acids using transesterification procedures and GC-FID technique. A total of 37 fatty acids were identified with a carbon chain length ranging from C6 to C22. In addition, MUFAs accounted for 59.07% of the total fatty acid composition and were more dominant than SFAs (21.46%) and PUFAs (19.48%).

The main fatty acids in *Flammulina velutipes* var. *velutipes*, *Boletus reticulatus*, *Lactarius salmonicolor*, *Russula anthracina*, *Polyporus squamosus* and *Pleurotus ostreatus* were reported to consist of linoleic (22.39–65.29%), oleic (10.36–52.11%), palmitic (7.76–17.21%) and stearic (3.28–12.96%) acids.[105]

The dominant fatty acids were found to be linoleic (26.60–63.39%), oleic (2.24–47.82%) and palmitic (9.55–45.06%) acids in *Agaricus bitorquis*, *Macrolepiota procera*, *Cantharellus cibarius*, *Craterellus cornucopioides*, *Lepiota excoriata*, *Pleurotus eryngii*, *Polyporus squamosus*, *Clitoceybe odora*, *Pholiota* sp. and *Tricholoma terreum* among the 13 detected fatty acids. SFAs were higher than UFAs only for *L. procera*. However, UFAs were higher in the other studied species. The most dominant MUFA percentage (48.73%) was in *P. eryngii*.[106]

A total of 37 different fatty acids were identified via GC in *Amanita ovoidea*. These fatty acids differed in length between C6 and C24. Oleic (58.8%), linoleic (18%), palmitic (15.8%) and stearic (5.4%) acids were detected as the dominant fatty acids.[107]

The main fatty acids in *Morchella esculenta* species harvested from two different localities (Portugal and Serbia) were reported as linoleic (71.81, 68.22%), oleic (12.43, 9.70%), palmitic (9.54, 10.12%) and stearic (2.63, 1.60%) acids. PUFAs dominated over SFAs and MUFAs.[108]

Linoleic (72.42, 78.59%), and oleic (11.38, 11.78%) acids were observed as the most abundant fatty acids in *Agaricus blazei* and *Lentinus edodes* among the 26 qualified fatty acids by GC-FID.[109]

The main fatty acids found in *Cordyceps militaris* were reported to be linoleic (68.00%), palmitic (12.76%), oleic acid (7.29%) and stearic (5.71%) acids. PUFAs (68.87%) predominated over SFAs (23.40%) and MUFAs (7.73%), which was related with the higher contents of linoleic acid.[110]

The relative percentages of linoleic, oleic, palmitic and stearic acids of *Amanita crocea*, *Gyromitra esculenta*, *Amanita mairei*, *Boletus regius*, *Boletus poliporus*, *Helvella lacunosa*, *Russula virescens* and *Russula aurea* were demonstrated to be 12.21–56.11, 13.44–54.46, 9.28–29.05 and 1.63–7.16%, respectively. Up to 23 fatty acids were identified in almost all of the studied mushrooms. *G. esculenta* and *B. regius* showed the highest contents of PUFA (~56%) while *Amanita* species exhibited the highest content of MUFA (~54%).[111]

Su *et al.*[112] demonstrated oleic (4.54–448.9 $mg\,g^{-1}$), linoleic (24.99–228.6 $mg\,g^{-1}$), palmitic (18.65–136.6 $mg\,g^{-1}$) and stearic (3.56–66.93 $mg\,g^{-1}$) acids as the main fatty acids of *Agaricus blazei*, *Coriolus versicolor*, *Ganoderma lucidum*, *Grifola frondosa*, *Hericium erinaceum* and *Phellinus linteus* based on GC-MS analysis.

The main fatty acid contents of *Cordyceps sinensis*, *Pleurotus ostreatus* and *Schizophyllum commune* consisting of linoleic, oleic and palmitic acid were identified in the range of 52.26–67.41, 10.47–24.15, 16.43–17.61% in mycelium, and 42.43–61.09, 16.32–32.54, 17.30–20.14% in culture broth. PUFAs which were predominant in the studied mushrooms were from 53.34 to 67.89% in culture mycelium and from 42.65 to 61.41% in the culture broth.[113]

Linoleic (36.29–66.72%), oleic (1.74–45.26%), palmitic (7.77–15.23%) and stearic (3.13–8.27%) acids were recorded as the main fatty acids of *Coprinus comatus*, *Lactarius deliciosus*, *Agrocybe cylindracea*, *Tricholoma myomyces* and *Suillus collinitus*. UFAs has higher contents than SFAs in the mushroom species studied.[114]

The main fatty acids found in *Agaricus comtulus*, *Agaricus lutosus*, *Agaricus campestris*, *Amanita umbrinolutea*, *Boletus armeniacus*, *Boletus impolitus*, *Bovista nigrescens*, *Bovista aestivalis*, *Clavariadelphus pistillaris*, *Clavariadelphus truncates*, *Clitocybe gibba*, *Chlorophyllum rhacodes*, *Clitocybe costata*, *Cortinarius praestans*, *Hygrophorus chrysodon*, *Flammulina velutipes*, *Leucoagaricus leucothites*, *Lycoperdon umbrinum*, *Ramaria aurea* and *Suillus variegates* were detected as linoleic (1.23–74.72%), oleic (3.50–58.82%), palmitic (7.32–25.95%) and stearic (1.10–7.89%) acids. *A. lutosus*, *L. leucothites*, *A. comtulus* and *C. rhacodes* possessed the highest concentrations of PUFA (72–75%), while *H. chrysodon* had the highest concentration of MUFA (63%).[115]

Linoleic (72.8%), palmitic (10.10%), oleic (4.5%), stearic (2.96%) and arachidic (2.46%) acids were detected as fatty acids of *Agaricus bohusii* by transesterification and GC-MS analysis. PUFAs (73.79%) predominated over SFAs (21.05%) and MUFAs (5.17%).[116]

GC-MS analysis of *Fistulina hepatica*, *Macrolepiota procera* var. *procera*, *Suillus granulatus*, *Laetiporus sulphureus* and *Infundibulicybe geotropa* revealed the presence of oleic (49.28–57.18%), linoleic (18.16–26.83%), esadecanoic (16.05–22.29%) and stearic (2.21–6.08%) acids as the major fatty acids.[117]

Fatty acid profiling of *Terfezia boudieri* exhibited linoleic (44.34%), palmitic (23.72%), oleic (10.37%) and decanoic (10.36%) acids as the main fatty acids.[118]

The major fatty acids identified were reported as oleic (33.4%), linoleic (30.4%), palmitic (14.3%) and stearic (6.9%) acids in the fraction of *Gomphus clavatus* by GC-MS.[119]

The fatty acid compositions of *Agaricus bisporus* (brown and white), *Flammulina velutipes*, *Pleurotus ostreatus*, *Lentinula edodes* and *Pleurotus eryngii* were identified as rich in linoleic (45.4–81.1%), palmitic (10.3–12.8%), stearic (1.6–3.1%), oleic (1.1–3.1%) and linolenic (0.1–28.8%) acids. PUFAs were the most abundant fatty acid in all studied mushrooms.[120]

Applying GC-MS to *Agaricus augustus*, *Agaricus arvensis*, *Agaricus silvaticus*, *Agaricus campestris*, *Agaricus xanthoderma* var. *meleagris*, *Endoptychum agaricoides*, *Marasmius oreades*, *Hygrocybe nigrescens*, *Clitocybe clavipes*, *Lyophyllum connatum*, *Lyophyllum decastes*, *Lepista gilva*, *Pluteus atricapillus* and *Pleurotus dryinus* resulted in the identification of linoleic (18.5–62.3%), oleic (2.4–42.7%) and palmitic (14.6–23.1%) acids as the major fatty acids. It was demonstrated in this study that the fatty acid profile was effective in the classification of species using chemometry. The linoleic acid/oleic acid ratio varied from 1.1 to 25.3. The lowest value was found in *L. decastes*, whereas the highest value was observed in *A. arvensis*.[121]

The dominant fatty acid concentrations of *Catathelasma ventricosum*, *Clitocybe maxima*, *Craterellus cornucopioides*, *Laccaria amethystea* and *Stropharia rugoso-annulata* were reported to vary between 28.34% and 74.39% for linoleic acid, 13.74% and 54.02% for oleic acid, 5.12% and 12.64% for palmitic acid. UFAs predominated over SFAs in all studied mushrooms. PUFAs were the main class of fatty acids in *C. ventricosum*, *L. amethystea* and *S. rugoso-annulata*, while MUFAs were the main class in *C. maxima* and *C. cornucopioides*.[122]

Fatty acids of *Agrocybe aegerita*, *Armillaria luteo-virens*, *Copyinds comatus*, *Flammulina velutiper* and *Pleurotus eryngii* were derivatized by 2-(12-oxobenzo[*b*]acridin-5(12*H*)-yl)-ethyl-4-toluenesulfonate (BAETS) and identified by HPLC-FLD-MS/MS. Among 18 detected fatty acids, linoleic (27.17–49.34 mg/10 g), oleic (4.08–22.15 mg/10 g), palmitic (6.49–14.88 mg/10 g) and stearic (1.15–3.00 mg/10 g) acids were characterized as the main fatty acids.[123]

The fatty acids of *Agaricus bitorquis*, *Agaricus essettei* and *Agaricus bisporus* were identified using GC and GC-MS. Linoleic (61.82–67.29%) and palmitic (12.67–14.71%) acids were reported as the dominant fatty acids amongst the 13 fatty acids identified in the oils. Other fatty acids consisting of C8:0, C10:0, C12:0, C14:0, C15:0, C17:0, C18:3 and C20:0 were also observed in all the studied *Agaricus* species, but all were in lower amounts. All studied species were identified as having a high content of UFA greater than 75% of the total fatty acid concentration. The linoleic acid/oleic acid (L/O) ratio was 11.09 for *A. bisporus*, 9.37 for *A. bitorquis* and 7.62 for *A. essettei.*[124]

GC-FID was used for quantitative evaluation of fatty acid compositions of *Armillaria mellea*, *Lactarius deterrimus*, *Collybia platyphylla* and *Macrolepiota procera* and linoleic (21.96–52.61%), oleic (9.86–36.43%), palmitic (4.85–23.00%) and stearic (2.49–46.82%) acids were identified as the dominant fatty acids.[125]

The fatty acid profile of *Ganoderma lucidum* was found to be rich in oleic (48%), palmitic (15%), stearic (7%) and palmitoleic (5%) acids by GC-MS identification. Linoleic, pentadecanoic and heptadecanoic acids were in small quantities. The UFA concentration was about 68% of the total oil.[126]

Linoleic (16.93–66.22%), oleic (14.59–63.19%), palmitic (7.84–19.03%) and stearic (1.67–12.76%) acids were found to be the most abundant fatty acids in *Amanita muscaria*, *Amanita caesarea*, *Amanita pantherina*, *Cortinarius anomalus*, *Chroogomphus fulmineus*, *Russula sardonia*, *Cortinarius violaceus*, *Suillus luteus*, *Cortinarius collinitus*, *Lactarius volemus*, *Lactarius quietus* and *Tricholoma ustale.* UFA contents were higher than for SFAs in all studied mushrooms. *R. sardonia* possessed the highest contents of MUFAs (65.12%) with the highest concentration of oleic acid; *C. violaceus* had the highest PUFA content (66.58%) with the highest concentration of linoleic acid.[127]

Agaricus arvensis, *Boletus edulis*, *Cantharellus cibarius*, *Cantharellus tubaeformis*, *Clavulina rugosa*, *Laccaria laccata*, *Hydnum repandum*, *Lepista nuda*, *Russula rosea*, *Tricholoma* var. *saponaceum* and *Leucopaxillus giganteus* were researched for their fatty acid profiles using GC-MS, and linoleic (19.8–78.0%), oleic (2.53–57.8%), palmitic (9.07–20.0%) and stearic (0.89–34.6%) acids were detected as the predominant fatty acids. In addition to these four main fatty acids, seven others were quantified: C15:0, C18:0, C20:0, C16:1, C18:1, C20:0 and C18:3. MUFAs were the main group for *C. tubaeformis*, *C. rugosa*, *R. rosea*, *H. repandum*, *L. giganteus* and *L. laccata*, while PUFAs were the main group for *A. arvensis*, *B. edulis*, *C. cibarius*, *L. nuda and T. saponaceum*. The linoleic acid/oleic acid ratio ranged from 0.34 to 30.8.[128]

Fatty acid compositions of 10 saprotrophic and mycorrhizal wild mushrooms (*Boletus erythropus*, *Boletus fragrans*, *Calvatia utriformis*, *Clitopilus prunulus*, *Hygrophorus pustulatus*, *Lyophyllum decastes*, *Lycoperdon echinatum*, *Russula cyanoxantha*, *Macrolepiota excoriate* and *Russula olivacea*) were determined by Grangeia *et al.*[129] Linoleic acid (32.03–70.29%) was found to be the most abundant fatty acid, followed by oleic (6.0–51.53%), palmitic (6.96–21.33%) and stearic (1.94–11.10%) acids.

Linoleic (63.7–76.5%), palmitic (11.0–18.6%), oleic (4.8–5.6%), and stearic (2.6–5.4%) acids were reported to be the main fatty acids in *Agrocybe chaxingu*, *Flammulina velutipes* and *Pleurotus ostreatus*. PUFA contents predominated with values of 69.3%, 76.5% and 63.7%, respectively.[130]

The main fatty acid contents of *Armillaria mellea*, *Coprinus comatus*, *Clitocybe odora* and *Calocybe gambosa* comprising of linoleic, oleic palmitic and stearic acids were identified in the ranges of 27.71–74.86, 6.27–47.74, 10.56–13.57 and 1.90–3.53%. Up to 22 fatty acids were identified and quantified.[131]

Heleno *et al.*[132] reported the main fatty acids of six *Boletus* species (*B. edulis*, *B. aereus*, *B. reticulatus*, *B. rhodoxanthus*, *B. purpureus*, *B. satanas*) to be linoleic (32.18–62.34%), oleic (7.21–47.20%) and palmitic (9.57–17.49%) acids. The main fatty acid was found to be oleic acid for *B. reticulatus*, *B. satanas* and *B. edulis*, and linoleic acid for *B. purpureus*, *B. aereus* and *B. rhodoxanthus*.

The major fatty acids of *Amanita porphyria*, *Collybia fusipes*, *Fomitopsis pinicola*, *Hebeloma sinapizans*, *Inocybe splendens*, *Lactarius hepaticus*, *Lentinus tigrinus*, *Piptoporus betulinis*, *Pluteus murinus* and *Russula emetica* were observed to be linoleic (18.47–71.22%), oleic (5.45–65.44%), palmitic (6.91–18.49%) and stearic (1.57–6.17%) acids by Reis *et al.*[133] PUFAs predominated for *A. porphyria*, *C. fusipes*, *F. pinicola*, *H. sinapizans*, *L. hepaticus*, *L. tigrinus* and *P. betulinis*, while MUFAs dominated for *I. splendens*, *P. murinus* and *R. emetica*.

Five *Phellinus* species (*P. linteus*, *P. baumii*, *P. pomaceus*, *P. pini*, *P. robustus*) were screened for palmitoleic (0.00461-0.10032 $mg\,g^{-1}$), linoleic (0.1629–1.6483 $mg\,g^{-1}$), oleic (0.01644–0.10889 $mg\,g^{-1}$), palmitic (0.3952–1.2403 $mg\,g^{-1}$), and stearic (0.1464–0.3821 $mg\,g^{-1}$) acids using HPLC-PDA.[134]

The UV and sunlight effects on the fatty acid profile of *Agaricus bisporus* have been investigated. The main fatty acids, linoleic, oleic, stearic and palmitic acids were identified in untreated (1.877, 0.066, 0.381, 0.096% dw, respectively), UV-treated (1.972, 0.063, 0.397, 0.104% dw, respectively), and sunlight-treated (2.042, 0.062, 0.373, 0.117% dw, respectively), mushrooms.[135]

The main fatty acids of 15 *Boletus* species (*B. aestivalis*, *B. appendiculatus*, *B. aereus*, *B. crocipodius*, *B. badius*, *B. granulatus*, *B. edulis*, *B. luridus*, *B. impolitus*, *B. pinicola*, *Boletus* sp., *B. luteus*, *B. scaber*, *B. queletii*, *B. versipellis*) were demonstrated as linoleic (38.2–57.6%), oleic (15.2–42.6%), palmitic (7.0–17.0%) and stearic (1.9–14.1%) acids.[136]

Oleic, linoleic and palmitic acids were identified as the major fatty acids of *Tricholoma matsutake* with percentages of 57.65, 26.75 and 9.1%, respectively, and the linoleic/oleic acid ratio was measured as 0.46. UFAs were identified as having a high concentration of about 65%.[137]

The examination of the fatty acid profiles of pileus and stipe of *Russula griseocarnosa* sp. *nov.* displayed linoleic (28.40, 50.22%), oleic (47.92, 14.20%), and palmitic (19.30, 30.70%) acids as the main fatty acids. UFAs were characterized with a proportion above 65% of the total fatty acids.[138]

The most abundant fatty acids of *Suillus mediterraneensis*, *Fistulina hepatica*, *Cortinarius glaucopus*, *Lactarius salmonicolor*, *Hygrophoropsis aurantiaca*, *Tricholoma imbricatum*, *Laccaria laccata*, *Hypholoma capnoides*, *Lepista inversa* and *Russula delica* were described as oleic (16.98–60.68%), linoleic (20.45–55.45%), stearic (0.92–40.13%) and palmitic (7.35–16.43%) acids. Overall, UFAs predominated over SFAs in all the studied mushrooms except *L. salmonicolor* (through the high content of stearic acid) and varied from 56% to 85%.[139]

The dominant fatty acids were identified using GC-MS as linoleic (35.951–42.577%), oleic (24.703–36.443%), palmitic (12.777–22.263%) and stearic (6.123–8.736%) acids in three *Cordyceps* species (*C. gunnii*, *C. taii* and *C. jiangxiensis*) among the 35 identified compounds. PUFAs were the main class of fatty acids in all studied *Cordyceps* species.[140]

The fatty acid compositions of *Amanita caesarea*, *Hydnum rufescens*, *Boletus edulis*, *Amanita rubescens*, *Fistulina hepatica*, *Russula cyanoxantha*, *Cantharellus cibarius*, *Hygrophorus agathosmus*, *Suillus luteus*, *Suillus granulatus*, *Tricholoma equestre* and *Suillus bellini* were examined by GC-MS. Oleic (8.448–2165.499 $mg\,kg^{-1}$), linoleic (7.536–976.316 $mg\,kg^{-1}$), palmitic (5.285–160.702 $mg\,kg^{-1}$) and stearic (1.782–76.782 $mg\,kg^{-1}$) acids were found to be the most abundant fatty acids. SFAs were found in higher amounts in *A. rubescens* and *S. bellini*. PUFAs and MUFAs were dominant in all of the remaining mushrooms, in the range of 12–79% and 18–79%, respectively.[141]

Linoleic (70.8–81.2 g/100 g) and palmitic (11.4–14.5 g/100 g) acids were identified as the major fatty acids of *Agaricus bisporus*, *Lentinus edodes* and *Pleurotus ostreatus* by GC-FID.[142]

Oleic acid (7052, 3167 $\mu g\,g^{-1}$) was found to be the most abundant fatty acid, followed by linoleic (4146, 2542 $\mu g\,g^{-1}$) and palmitic (1639, 775 $\mu g\,g^{-1}$) acids in intact and homogenized *Astraeus* spp.[143]

Barros *et al.*[144] studied the fatty acid contents of *Agaricus silvaticus*, *Agaricus silvicola*, *Agaricus bisporus*, *Boletus edulis*, *Craterellus cornucopioides*, *Cantharellus cibarius*, *Calocybe gambosa* and *Marasmius oreades* and identified the main fatty acids as linoleic (23.67–76.50%), oleic (1.30–51.85%), palmitic (6.66–15.16%) and stearic (1.41–7.83%) acids. PUFAs were the main class of fatty acids in all species excluding *C. cornucopioides*. The highest SFA contents were found in *C. cibarius* (22.63%). However, UFAs dominated over SFAs in all mushroom species, ranging from 77% to 85%. The highest UFA (85%) and the lowest SFA (15%) concentrations were found in *B. edulis*.

Linoleic (38.32–70.69%), oleic (4.59–43.93%) and palmitic (7.19–13.74%) acids were determined to be in the highest amounts in the fatty acid compositions of *Lepista nuda*, *Cantharellus cibarius*, *Lycoperdon perlatum*, *Ramaria botrytis* and *Lycoperdon molle*. PUFAs were the main class of fatty acids in all mushrooms studied except for *R. botrytis*, in which MUFAs were the main class. However, UFAs dominated over SFAs for all studied mushrooms, ranging from 75% to 88%. *L. perlatum* contained the lowest MUFA level, and the highest PUFA level, due to the higher concentration of linoleic acid. *L. molle* contained the highest SFA concentration, owing to the higher contribution of palmitic acid.[145]

Hur[146] described the main fatty acids of the fruiting body and corpus of *Cordyceps militaris* as linoleic (61.3, 33.0%), palmitic (24.5, 21.5%) and oleic

(6.0, 17.7%) acids by GC. The *C. militaris* fruiting body had the highest content of UFA. Palmitic acid was predominant regarding SFAs; its contents were 21.5% in the corpus and 24.5% in the fruiting body. Linoleic acid was predominant regarding UFAs; its contents were 33.0% in the corpus and 61.3% in the fruiting body.

GC analysis of fatty acid methyl esters in *Agaricus bisporus* and *Agaricus bitorquis* revealed that linoleic acid (44.19, 31.03%) and oleic acid (21.47, 40.13%) were reported to be the dominant fatty acids, respectively.[147]

The fatty acid composition of *Lentinula edodes* was found to consist of linoleic (35.61%), palmitic (17.75%), stearic (2.75%) and linolenic (0.33%) acids.[148]

The major fatty acids of *Auricularia polytricha*, *Cantharellus cibarius*, *Hydnum repandum*, *Cantharellus clavatus*, *Boletus edulis*, *Helvella crispa*, *Geastrum arinarius*, *Lentinus squarrulosus*, *Lactarius sanguifluus*, *Lentinus sajor-caju*, *Lactarius deliciosus*, *Macrolepiota procera*, *Termitomyces tylerance*, *Morchella angusticeps*, *Pleurotus sajor-caju*, *Termitomyces microcarpus*, *Morchella conica*, *Russula brevipes*, *Termitomyces schimperi*, *Sparassis crispa*, *Termitomyces mummiformis*, *Pleurotus djamor* and *Termitomyces heimii* were reported as linoleic (5.4–68.6%), oleic (5.8–55.1%) and palmitic (4.6–24.7%) acids. UFAs predominated over SFAs for all studied mushroom species, ranging from about 50% to 80%. The linoleic/oleic acid ratios were found to be 1.08, 1.08, 0.48, 0.80, 0.09, 2.76, 1.03, 2.33, 10.58, 1.12, 1.08, 2.73, 6.03, 6.07, 1.57, 3.28, 0.86, 0.63, 1.25, 3.18, 0.93, 1.34 and 0.78, respectively.[149]

Fatty acid profiles of neutral and polar lipids revealed the main fatty acids in *Amanita vaginata*, *Agaricus arvensis*, *Amanita rubescens*, *Agaricus campestris*, *Boletus subglabripes*, *Boletus edulis*, *Coprinus comatus*, *Lycoperdon pyriforme*, *Ganoderma applanatum* and *Suillus granulatus* were presented as linoleic (17.9–68.8%), oleic (3.99–60.3%) and palmitic (10.5–22.6%) acids by GC-MS. In addition, high concentrations of UFA ranging from 62.7 to 82.3% (polar lipids) and 59.8 to 82.5% (neutral lipids) were observed in both lipid fractions.[150]

The most abundant fatty acids were determined to be linoleic (38–58%), oleic (15–42%) and palmitic (7–17%) acids in 15 *Boletus* species (*B. aereus*, *B. aestivalis*, *B. badius*, *B. appendiculatus*, *B. edulis*, *B. crocipodius*, *B. impolitus*, *B. granulatus*, *B. luteus*, *B. luridus*, *B. pinicola*, *B. queletii*, *Boletus* sp., *B. versipelli*, *B. scaber*). PUFA contents predominated over SFAs for all studied *Boletus* species in the range of 41.05–59.13%.[151]

TLC and GC-MS of fatty acid methyl esters and fatty acid 4,4-dimethyloxazoline derivatives in the determination of fatty acid compositions in *Lycoperdon perlatum*, *Calvatia utriformis*, *Lycoperdon pyriforme* and *Bovista plumbea* led to the identification of 27 fatty acids. Linoleic (37–65%), oleic (7–24%), palmitic (12–18%) and stearic (2–6%) acids were found to be the major contributors.[152]

GC-FID quantifications of fatty acid compositions of *Tricholoma portentosum*, *Agaricus arvensis*, *Leucopaxillus giganteus*, *Lactarius deliciosus* and *Sarcodon imbricatus* were carried out, with oleic (15.46–58.36%), linoleic

(17.06–56.11%), stearic (2.11–25.33%) and palmitic (5.60–14.55%) acids being found to be the main fatty acids among the 15 identified fatty acids. PUFAs were the main class of fatty acids in *L. giganteus* and *A. arvensis*, while MUFAs were the main class in *S. imbricatus*, *L. deliciosus* and *T. portentosum*. However, UFAs were predominant over SFAs for all mushrooms studied, ranging from 59.9% to 90.4%.[153]

Lactarius deliciosus and *Lactarius piperatus* fruiting bodies in different stages of maturity: immature (cap diameter less than 4.5 cm), mature (cap diameter between 4.5 and 7 cm, with immature spores), and mature (cap diameter higher than 7 cm, with mature spores) were analyzed for fatty acid compositions and stearic (37.82–81.46%), oleic (5.50–26.99%), linoleic (6.66–25.37%), palmitic (1.10–7.02%) and decanoic (0.03–5.46%) acids were reported to be the main fatty acids. Also, it was suggested that MUFA and PUFA percentages increased with the fruiting body maturity stage, while the SFA content decreased.[154]

Oleic (38.5%), linolenic (28.3%), palmitic (17.08%) and linoleic (12.35%) acids were demonstrated as the main fatty acids of *Paecilomyces tenuipes* by GC analysis. UFAs were rich in *P. tenuipes* with about 79% of the total fatty acid content.[155]

The fatty acid composition of *Antrodia camphorata* was reported as palmitic (14.74 mg g^{-1}), stearic (14.34 mg g^{-1}), linoleic (12.69 mg g^{-1}), oleic (10.78 mg g^{-1}), heptadecanoic (6.20 mg g^{-1}) and myristic (1.46 mg g^{-1}) acids.[156]

Pedneault *et al.*[157] determined the major fatty acids involving linoleic (34.2–63.0%), oleic (18.0–49.2%) and palmitic (7.37–16.5%) acids for 11 mushroom species, namely *Boletus erythropus*, *Boletus edulis*, *Boletus subglabripes*, *Boletus piperatus*, *Boletus variipes*, *Boletus subtomentosus*, *Leccinum scabrum*, *Leccinum aurantiacum*, *Suillus grevillei*, *Suillus granulatus* and *Tylopilus felleus*. UFAs accounted for about 83.3% of the total fatty acids and largely predominated over SFAs in all the studied mushrooms.

The most abundant fatty acids were identified as linoleic, oleic, palmitic and stearic acids in the fruiting bodies (25.8–49.6, 4.17–30.2, 0.199–2.22, 2.11–5.60%, respectively) and stems (2.56–59.5, 2.84–59.0, 0.828–6.09, 1.89–10.6%, respectively) of *Agaricus campestris*, *Agaricus bisporus*, *Armillaria mellea*, *Boletus edulis*, *Oudemansiella radicata*, *Coprinus comatus* and *Pleurotus ostreatus*. UFAs were richer than SFAs in all studied mushrooms.[158]

Screening fatty acid profiles of six *Termitomyces* species (*T. aurantiacus*, *T. le testui*, *T. mammiformis*, *T. mboudaeina* sp. nov., *T. schimperi* and *T. subclypeatus forme bisporus* sp. nov.) enabled the characterization of linoleic (45.6–65.1%), oleic (9.0–23.0%), palmitic (14.1–16.6%) and stearic (0.8–6.7%) acids as the main fatty acids. PUFAs had a high amount, with the content ranging from 45.6 to 65.1% in studied *Termitomyces* species.[159]

The fatty acid profiles of six *Tricholoma* species (*T. acerbum*, *T. batschii*, *T. nudum*, *T. pardinum*, *T. portentosum*, *T. terreum*) were found to be rich in linoleic (29.7–65.6%), oleic (4.0–58.0%), palmitic (7.6–23.0%) and stearic (1.8–10.1%) acids by GC-FID identification. Also, the linoleic acid/oleic acid ratios were determined to be 11.3, 8.23, 13.3, 15.35, 0.48 and 0.52, respectively.[160]

Linoleic (61.3, 68.8%), palmitic (20.8, 19.2%),and oleic (10.4, 8.3%) acids were identified as the main fatty acids of *Schizophyllum commune* and *Lentinus edodes* through transesterification and GC analysis.[161]

The major fatty acids of *Boletus frostii*, *Amanita rubescens*, *Ramaria flava* and *Lactarius indigo* were found to be oleic (10.757–69.304 mg g^{-1}), linoleic (13.276–21.745 mg g^{-1}), palmitic (3.491–11.301 mg g^{-1}) and stearic (0.910–32.057 mg g^{-1}) acids.[162]

Senatore[163] carried out GC-MS to identify the fatty acids present in five *Clitocybe* species (*C. aurantiaca*, *C. candida*, *C. cinerascens*, *C. geotropa*, *C. nebularis*) and linoleic (626–684 mg g^{-1}), palmitic (164–198 mg g^{-1}), stearic (49–72 mg g^{-1}) and oleic (51–65 mg g^{-1}) acids were revealed to be the most abundant fatty acids. Linoleic acid/oleic acid ratios of studied *Clitocybe* species were determined to range from 9.8 to 12.3.

A total of 13 fatty acids were detected in *Hygrophorus conicus*, *Hygrophorus lucorum*, *Lactarius controversus*, *Lactarius sanguifluus*, *Lactarius torminosus* and *Lactarius uvidus*. Linoleic (404–491 mg g^{-1}), oleic (220–297 mg g^{-1}), palmitic (140–176 mg g^{-1}) and stearic (49–78 mg g^{-1}) acids were noted to be the major fatty acids.[164]

3.2.1 Linoleic Acid

Linoleic acid (C18:2), which is an essential fatty acid, is very significant for human health. It is well-proven that linoleic acid is the precursor of eight carbon aromatic compounds in mushrooms, such as 1-octen-3-one, 3-octanol, 1-octen-3-ol and 3-octanone, which are the main volatile compounds in many species.[24]

In addition, linoleic acid, a polyunsaturated fatty acid and a member of the ω-6 fatty acids, is an essential fatty acid due to the deficiency of Δ12 and Δ15 desaturases required for synthesis in humans.[165] After oleic acid was synthesized *de novo* in plastids, the generation of plant-based linoleic acid is coupled to that of membrane glycerolipids, which is carried out through two pathways: the 'eukaryotic pathway' in the endoplasmic reticulum by fatty acid desaturases and the 'prokaryotic pathway' in plastids.[166] Oxygen and NADH are required for the insertion of the double bond to oleic acid, and the synthesis of animal-based linoleic acid occurs via Δ12 and Δ15 desaturases.[167] Linoleic acid is metabolized by oxidative desaturation and elongation steps principally to arachidonic acid after ingestion. The four main functions of linoleic acid can be listed as follows: (1) it can be used as a source of energy like other fatty acids. (2) It can be esterified to procedure polar and neutral lipids such as triacylglycerols, cholesterol esters and phospholipids. (3) As part of membrane phospholipids, linoleic acid serves as a structural element to maintain a certain degree of membrane fluidity of the epidermis transdermal water barrier. (4) When released from membrane phospholipids, linoleic acid can be enzymatically oxidized to derivatives of various compounds involved in cell signaling.

Although linoleic acid is an essential nutrient, there are no reported data on the concentration of linoleic acid required to correct the symptoms of PUFA deficiency. In addition, there is no upper limit for excess linoleic acid consumption, as no discomfort caused by its over-consumption has been reported.[168] The American Heart Association, based on data compiled from randomized and case-control trials, recommended that a 5–10% PUFA-induced (especially linoleic acid) energy intake is connected with a reduced risk of cardiovascular disease.[169]

In many studies, linoleic acid intake was generally inversely related to risk factors including ischemic heart disease, blood pressure, diabetes, total cholesterol, and low-density lipoprotein (LDL) cholesterol.[170] A comprehensive analysis of 20 studies involving more than 39 000 people from 10 countries revealed that higher amounts of linoleic in the blood are linked with a 43% lower relative risk for type 2 diabetes. Higher linoleic acid consumption was found to cause lower glycemia as measured by oral glucose tolerance tests in non-diabetic men.[171] In randomized controlled crossover trials with linoleic-rich oil supplementation in women with type 2 diabetes, significant improvements in high-density lipoprotein, C-reactive protein, and fasting glucose and HbA1c, resulted in glycemic control.[172,173] The lower risk or therapeutic effect of high linoleic acid status in type 2 diabetes caused by insulin sensitivity, decreased body or visceral fat mass, increased lean mass, and decreased inflammation has been discussed in a detailed review.[174] In an international epidemiological study reporting the relationship between blood pressure and linoleic acid intake in over 4000 individuals, it was explained that dietary linoleic acid supplementation contributed to the prevention and control of adverse blood pressure levels.[175]

In a study investigating the effects on liver fat in people fed a diet rich in linoleic acid, lower liver fat formation was determined after 10 weeks in overweight individuals.[176] Correspondingly, a study with D6KO mice found that fat accumulation and markers of inflammation were reduced in D6KO mice compared to wild-type mice fed linoleic acid-based high-fat diets. This study suggested that arachidonic acid produced from linoleic acid had a pro-inflammatory effect.[177]

Choi[178] demonstrated that linoleic acid suppressed AGS human gastric cancer cell proliferation by inducing apoptosis. The growth-inhibitory effects of linoleic acid were found to be linked with specific inhibition of COX-2 expression and connected with a loss of PGE2 synthesis. Linoleic acid was reported to be a potent proliferation inhibitor in HT-29 colorectal cancer cells.[179] Kuriki *et al.*[180] analyzed erythrocyte fatty acid composition in patients with colorectal cancer and reported that arachidonic acid produced from linoleic acid was related to decreased, not increased, risk of colorectal cancer. It has been proven that dietary and serum linoleic acid levels are inversely related to prostate and other types of cancer in elderly men.[181] A comprehensive review concluded that high linoleic acid consumption does not significantly increase breast, prostate or colorectal cancer risk.[182]

3.2.2 Oleic Acid

Oleic acid, a monounsaturated fatty acid and a member of the ω-9 fatty acids, is a valuable fatty acid found in plasma triglycerides and dietary fat.

In the human liver, elongation of palmitic acid to stearic acid followed by Δ-9 desaturation produces oleic acid.[167] It has been reported that diets rich in MUFA, especially oleic acid, have beneficial effects on the low prevalence of atherosclerosis and lipoprotein metabolism.[183] ω-9 fatty acids are not essential because humans can synthesize oleic acid from stearic acid via the Δ9-desaturase enzyme. Oleic acid has been the focus, as it has traditionally been associated with its protection against many diseases, of the "Mediterranean diet" characterized by the consumption of olive oil rich in oleic acid.[184]

Oleic acid enhanced mitochondrial function, β-cell function, endothelial function, inflammation, glucolipotoxicity, oxidative stress, hypothalamic function, enzyme dysregulation and apoptosis.[183] In *in vitro* studies, oleic acid was reported to contribute to the prevention of insulin resistance by causing a significant increase in M2 expression in the murine macrophage cell line.[185] In a different study, oleic acid protected insulin resistance and prevented diabetes mellitus through its anti-inflammatory effect on BEND3 cells.[186] The expression of adiponectin is upregulated by oleic acid which helps to mediate its antidiabetic effect and, therefore, increases insulin sensitivity in human and murine adipocytes.[187] Oleic acid administered to mice caused a significant decrease of TNF-α, which was evidenced in an experimental study demonstrating the potential of oleic acid to decrease the activity/release of pro-inflammatory mediators and thus additional damage.[188] In a study conducted on rats, it was suggested that the negative effects of oxidative stress were overcome by suppressing endothelial activation and minimizing LDL oxidation tendency with dietary supplements rich in oleic acid.[189] In hypertensive rats, a diet rich in oleic acid has been proven to normalize blood pressure.[190] The cardioprotective and atheroprotective effects of oleic acid were well-defined as related to improved endothelial dysfunction in response to pro-inflammatory signals and reduced proliferation and apoptosis by Steffen *et al.*[191]

The administration of oleic acid in a rodent model has been confirmed to have a neuroprotective effect against delayed neuronal death in transient and persistent focal cerebral ischemia.[192] In addition, a different study showed that oleic acid supplementation caused a decrease in amyloid plaques in the brains of transgenic mice involved in the pathology of Alzheimer's disease.[193] It has been suggested by Sakurai *et al.*[194] that consumption of a diet rich in oleic acid, in particular, has a beneficial effect against cognitive decline in elderly people living in the community.

In *in vivo* and *in vitro* studies, the effect of oleic acid on cancer has been examined and it has been observed that it decreases the risk of breast, colorectal and prostate cancer. In a comprehensive review, the inhibitory proliferation effects of oleic acid on various cancer cells have been summarized.[195] It has been revealed that the mechanism of action of oleic acid on the HT29 colorectal cancer cell line occurs by inhibiting the decreased Ca^{2+} influx flow pathway. In the study by Menendez *et al.*,[196] the anticancer effect of oleic acid on SK-Br3, MDA and MB-231 breast cancer cell lines was described through downregulated Her-2/neu promoter activity. Oleic acid increased the proliferation of PC3 and DU-145 prostate cancer cell lines, showed chemoresistance and formed an aggressive phenotype via free-FA receptor-1, calcium and PI3K/Akt signaling.[197,198]

3.2.3 Linolenic Acid

Linolenic acid, a polyunsaturated fatty acid and a member of the ω-3 fatty acids is described as an essential fatty acid due to the deficiency of Δ12 and Δ15 desaturases required for synthesis in humans.[165] After oleic acid was synthesized *de novo* in plastids, the generation of plant-based linolenic acid is coupled to that of membrane glycerolipids, which is carried out in two pathways: the 'eukaryotic pathway' in the endoplasmic reticulum by fatty acid desaturases and the 'prokaryotic pathway' in plastids.[166] Oxygen and NADH are required for insertion of the double bond to oleic acid and the synthesis of animal-based linolenic acid occurs via Δ12 and Δ15 desaturases.[167] Linolenic acid does not accumulate even if taken at high dietary levels, because it mostly undergoes β-oxidation in mitochondria and turns into docosahexaenoic acid (DHA) and eicosapentaenoic acid (EPA), which are very valuable for health.[199]

Kaur *et al.*[200] reported the health benefits of ω-3 fatty acids as raising energy levels, performance and stamina, learning, calmness, improving concentration, IQ and behavior, lowering the cardiovascular risk factors, increasing insulin sensitivity, inhibiting cancer growth and metastasis, physical exertion and surgery, speeding up the healing of wounds caused by accidental injury, decreasing inflammation and joint pain, improving bone mineral metabolism, reducing the symptoms of autoimmune diseases, improving weight management and increasing fat burning and decreasing fat production.

It has been revealed that ω-3 fatty acids and linolenic acid prevent and treat systemic lupus erythematosus, rheumatoid arthritis, multiple sclerosis, neurodegenerative diseases, age-related maculopathy and type 1 diabetes, and that these effects are due to other mechanisms such as anti-inflammatory effects and regulation of mTOR activity.[201,202]

In a comprehensive review, it was stated that as a result of 47 studies conducted with 2630 participants, linolenic acid intake lowered the blood lipids including cholesterol, low-density lipoprotein cholesterol, high-density lipoprotein cholesterol and very-low-density lipoprotein.[203] In a meta-analysis study published by Pan *et al.*[204] on the risk of linolenic acid on cardiovascular diseases, it was reported that there was an opposite relationship between total cardiovascular diseases and linolenic acid, and that the risk was reduced by 10%. Considering individual subtypes of cardiovascular diseases, an increase in dietary linolenic acid has been reported to result in a 10% reduction in the risk of coronary heart disease death. In a published prospective cohort study, high intakes of linolenic acid among 3088 older adults were linked with a lower risk of type 2 diabetes.[205]

Many *in vivo* and *in vitro* studies have demonstrated that DHA and EPA, which are metabolized from linolenic acid, have protective and therapeutic properties on breast cancer and colorectal cancer.[206–208]

3.2.4 Palmitic Acid

Palmitic acid can be taken in the diet or synthesized endogenously from carbohydrates, amino acids, and other fatty acids. Palmitic acid is synthesized by the fatty acid synthase complex by adding seven malonyl CoA to acetyl CoA in the human liver.[167] Palmitic acid constitutes 20–30% of the total fatty acids in adipose triacylglycerols and membrane phospholipids. An average of 3.5 kg of a 70-kg person consists of palmitic acid.[209] Palmitic acid is a vital component of cell membranes, with key roles in palmitoylated signal molecules and protein palmitoylation.[210] In newborns, 13–15% of the body consists of fat, and 45–50% of this fat ratio is endogenously sourced palmitic acid. The baby reaches 25% body fat by 4–5 months by accumulating high amounts of adipose tissue after birth. At this time of growth of the babies, human milk provides 10% dietary energy as palmitic acid, but it is an unusual structure (known as 2 palmitate) in that this fatty acid is located in the center of the glycerol central carbon (sn-2 position). Studies have suggested that large amounts of palmitic acid supplements increase survival rates in newborns.[211]

In a study examining the lipid–peptide interaction at the molecular level, it was determined that palmitic acid, which is widely used as an important additive as a pulmonary surfactant, provides a suitable hydrophilic–lipophilic balance.[212]

Palmitic acid was reported to decrease cell viability in MCF-7 breast cancer cells. Treatment with palmitic acid improved the expression of apoptosis-related proteins including caspase-3, 9, p53 and Bax, whereas the expression of anti-apoptotic protein Bcl-2 was decreased.[213] A study in mice with diabetes suggested that palmitic acid regulated cancer cell proliferation under conditions of obesity and excessive consumption of dietary fat.[214] Zeng *et al.*[215] showed that palmitic acid regulated p21 and Bax expression in primary mouse embryonic fibroblasts and osteoblasts in Atrp53-dependent and -independent pathways. Also, the authors suggested that palmitic acid exerted some positive effects on MEF proliferation against DNA damage, and by negatively regulating the DNA damage response pathway, it may support cell turnover and contribute to drug resistance and tumor progression in precancerous/cancer cells. As a result of many extensive studies, it has been stated that the increase in palmitic acid concentration as a result of a sedentary lifestyle causes excessive accumulation of palmitic acid, resulting in hyperglycemia, dyslipidemia, increased inflammation and increased ectopic fat accumulation.[216,217]

3.2.5 Stearic Acid

Stearic acid is the second most prevalent consumed saturated fatty acid. Following the synthesis of palmitic acid by fatty acid synthase in the liver, stearic acid is synthesized as a result of the elongation of palmitic acid.[167] Stearic acid is an important fatty acid that has demolished the general belief that saturated fatty acids are harmful to human health due to its proven beneficial effects. It is well known that certain fatty acids have biological properties that trigger the risk of cardiovascular disease. In the concept of a food-based nutrition guide designed to reduce the risk of cardiovascular disease, fat and fatty acid recommendations focused on reducing trans fatty acids and saturated fatty acids and ensuring adequate intake of unsaturated fatty acids. In this context, stearic acid has been evaluated as the primary substitute for *trans* fatty acids in recent years.[218] In addition, it has been revealed in animal experiments that stearic acid esterified with glycerol at sn-1 and sn-3 positions have limited bioavailability, while it has unique properties when glycerol is located at the sn-2 position.

Many animal studies have shown that glycerides rich in stearic acid are less efficiently absorbed than lauric, myristic and palmitic acids or their triglycerides. This result partially explains why stearic acid is less cholesterolemic and atherogenic than saturated fatty acids with a shorter chain length.[219] The effects of stearic acid on serum cholesterol levels are considered to be unique. It has been proven by studies in experimental animals and humans that stearic acid intake has a neutral or cholesterol-lowering effect related to coronary heart diseases, unlike the effects of myristic, lauric and palmitic acids.[220–222] It has been explained that fat accumulation is a risk factor for both cardiovascular disease and cancer. In an *in vivo* study involving both healthy subjects and type 2 diabetic patients, it was determined that stearic acid intake caused stronger mitochondrial fusion in diabetic patients compared to controls, reducing the risk of cancer and cardiovascular diseases.[223]

Wang *et al.*[224] reported that stearic acid could preserve cortical neurons against oxidative stress by increasing the internal antioxidant enzymes. It has also been suggested that the neuroprotective effect of stearic acid is mediated by activation of PPARγ and new protein synthesis in cortical neurons.

In a study carried out with the participation of more than 25 000 adults, relationships between pre-diagnostic levels of circulating metabolites and the risks of prostate, colorectal and breast cancer were investigated and stearic acid was found to be consistently associated with lower risks of prostate, colorectal and breast cancer.[225]

Stearic acid activated apoptosis preferentially in Hs578t, MDA-MB-231, Hs578Bst and MCF-10A breast cancer cells and inhibited *de novo* protein kinase C blocked stearate-induced caspase-3 activity or diacylglycerol synthesis in the study of Evans *et al.*[226] In a case study of 5098 case-patients and 6649 control patients, it was observed that higher amounts of stearic acid intake were connected with a lower risk of total prostate cancer.[227] Stearic acid and its conjugate indicated the best potency on MDA-MB-231, MCF-7 and MDA-MB-361 breast cancer cells with inhibition of cell growth, migration and adhesion. In addition, they also induced the expression of caspase-3, an effect of apoptosis, in human breast cancer cells.[228] Stearic acid inhibited DNA synthesis and human cervical cancer (HOG-1) cell growth in a dose-dependent manner. Delayed entry of treated cells into exponential phase growth revealed that stearic acid affected the early signals leading to cell proliferation.[229]

3.2.6 Ratio of ω-6/ω-3

The omega-6/omega-3 (ω-6/ω-3) ratio was determined in rodent experiments in which a high level of dietary linoleic acid partially suppresses the conversion of dietary linolenic acid to docosahexaenoic acid (DHA) and eicosapentaenoic acid (EPA) in the body. Linoleic acid and linolenic acid are metabolized through a common enzyme system. The fact that the ω-6/ω-3 ratio is too high causes a decreasing increase in DHA/EPA levels in tissues with the competitive inhibitory effect of linoleic acid on linolenic acid. It has been found that even when the amount of linolenic acid was stabilized, lower ratios than the higher ratio afford better conversion efficiency of linolenic acid to DHA/EPA. Higher linolenic acid intake is a strategy used to moderately increase the conversion of linolenic acid to EPA, with a much lower ratio of ω-6/ω-3. The presence of linoleic acid and linolenic acid results in competition for the respective end products. Therefore, the ω-6/ω-3 ratio directly influences the type of compounds formed. This ratio is considered to be important as this competition is a factor in thromboxane's platelet aggregation that causes thrombosis. In addition, this ratio was found to be associated with arthritis, cancer, lupus and asthma in the immune/inflammatory system response and was used for controlling the synthesis of ω-6 eicosanoids.[230] Epidemiological and experimental studies have reported that an ω-6/ω-3 ratio of approximately 1 : 1–2 : 1 is a diet balanced in terms of fatty acids, which has the greatest protective effect against the development and growth of breast and colon cancers.[231]

3.3 Fatty Acids Isolated from Mushrooms

Fifty-eight fatty acid compounds isolated from 27 mushroom species (*Stropharia rugosoannulata*, *Amanita pantherine*, *Hypsizygus marmoreus*, *Amanita spissacea*, *Auricularia polytricha*, *Calvatia nipponica*, *Morchella esculenta*, *Lactarius deliciosus*, *Rhizopogon luteolus*, *Fistulina hepatica*, *Hypoxylon truncatum*, *Cantharellus cibarius*, *Gomphus clavatus*, *Boletus pseudocalopus*, *Sparassis crispa*, *Termitomyces titanicus*, *Pleurocybella porrigens*, *Hygrophorus discoxanthus*, *Filoboletus sp.*, *Agrocybe aegerita*, *Hypoxylon rubiginosum*, *Clitocybe clavipes*, *Gymnopilus spectabilis*, *Cantharellus friesii*, *Cantharellus tubaeformis*, *Lepista nebularis* and *Hericium erinaceum*) have been reported in the last 20 years. Also, the bioactivities of isolated fatty acids were evaluated in this review. The structures of fatty acids are given in Figure 3.1.

So *et al.*[232] isolated palmitic acid (**1**), (9*E*)-8-oxo-9-octadecenoic acid (**2**), (10*E*)-9-oxo-10-octadecenoic acid (**3**), (9*E*)-8-oxo-9-octadecenoate methyl ester (**4**), (9*Z*)-9-octadecenoate-(2′*S*)-2′,3′-dihydroxy propyl ester (**5**) and (9*Z*)-9-octadecenoic acid

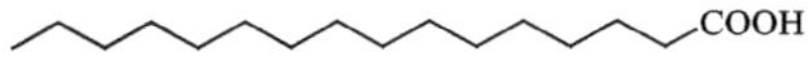

1

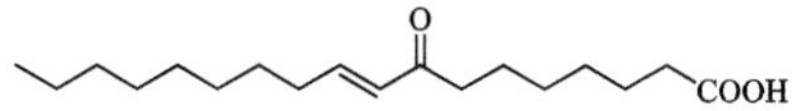

2

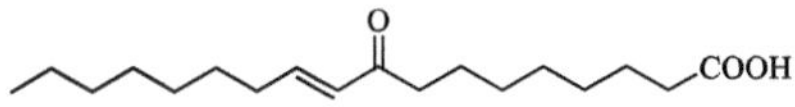

3

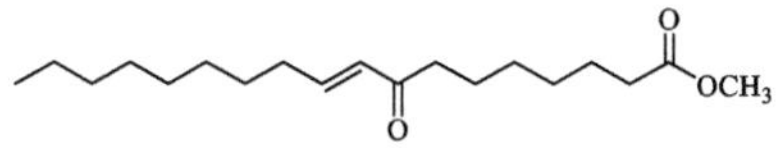

4

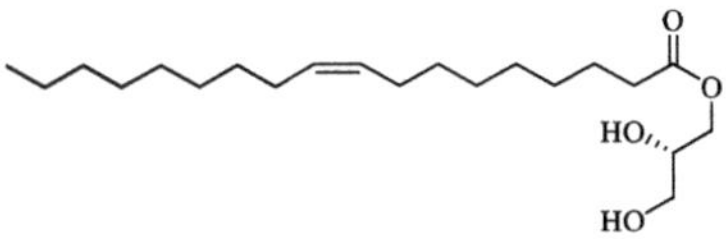

5

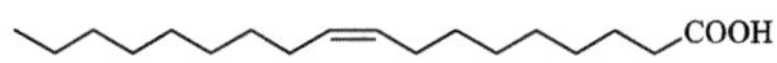

6

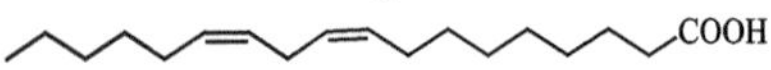

7

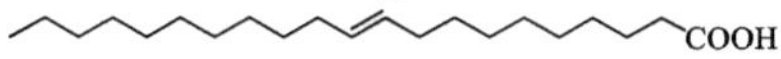

8

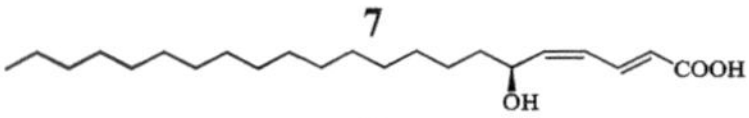

9

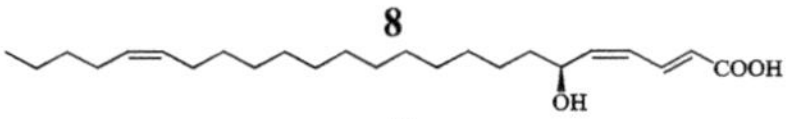

10

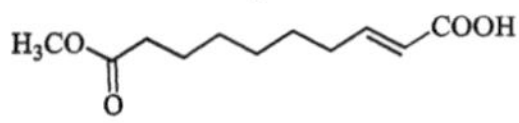

11

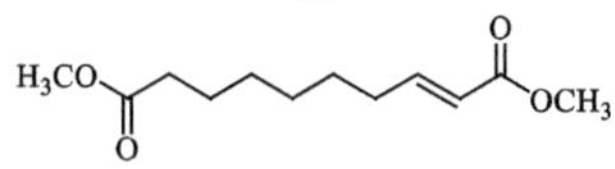

12

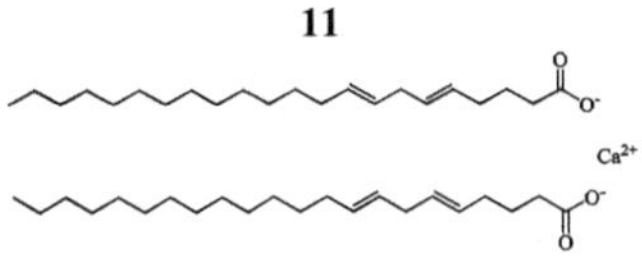

13

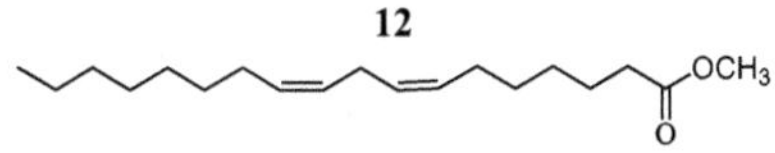

14

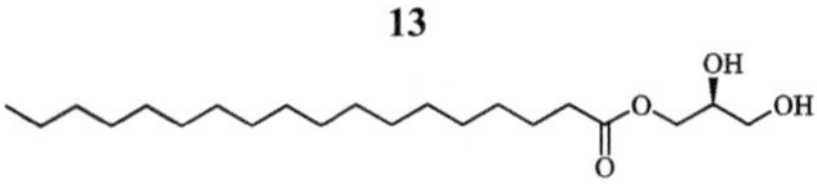

15

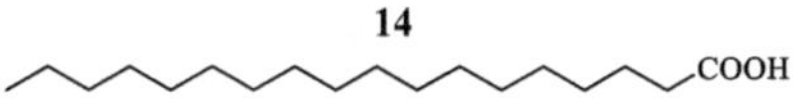

16

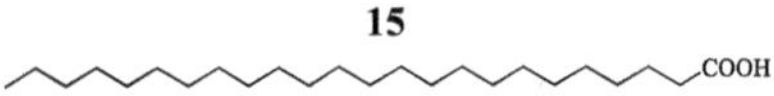

17

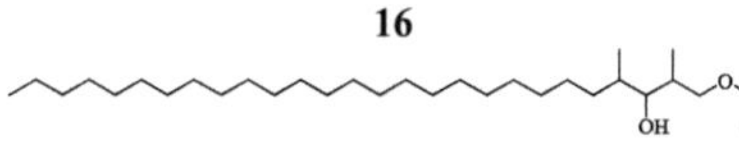

18

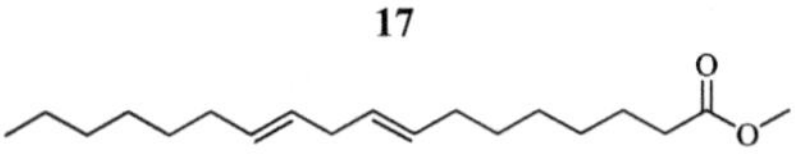

19

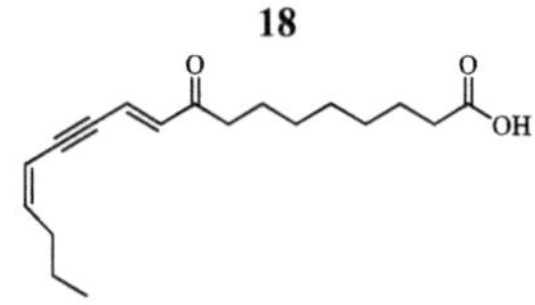

20

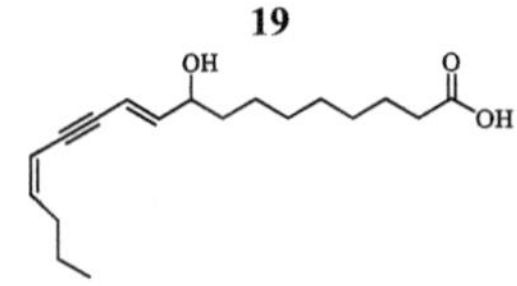

21

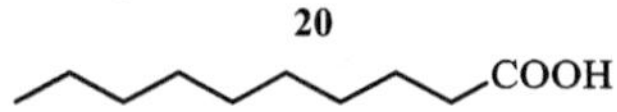

22

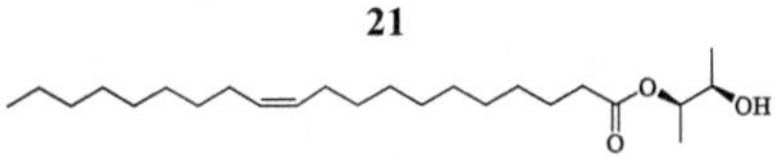

23

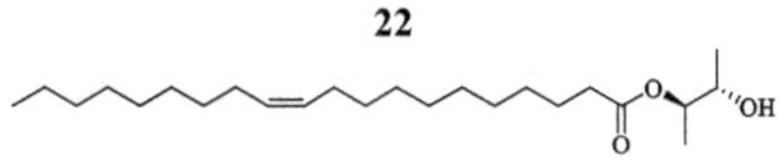

24

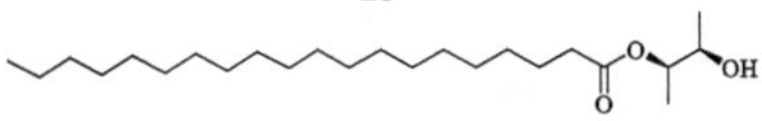

25

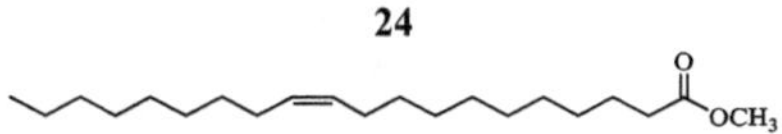

26

Figure 3.1 The structures of fatty acids isolated from 27 mushroom species. The ChemDraw Ultra 14.0 version was used to draw the structures of compounds (**1–58**) using the related papers published.

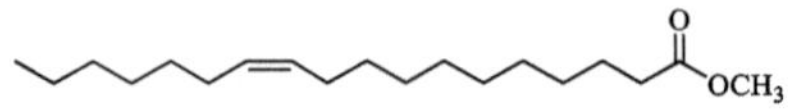

27

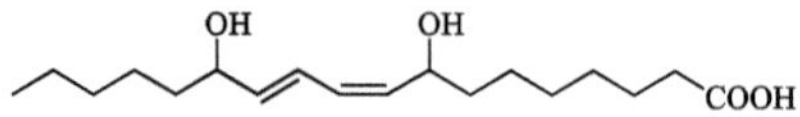

28

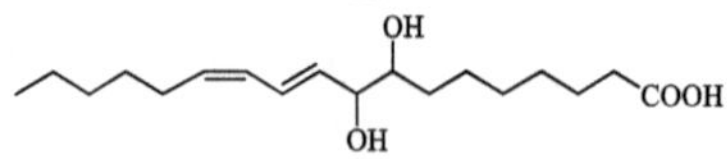

29

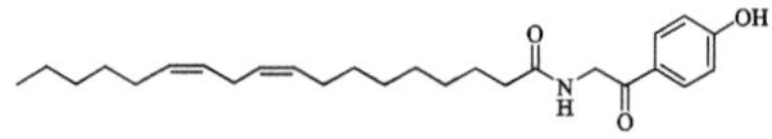

30

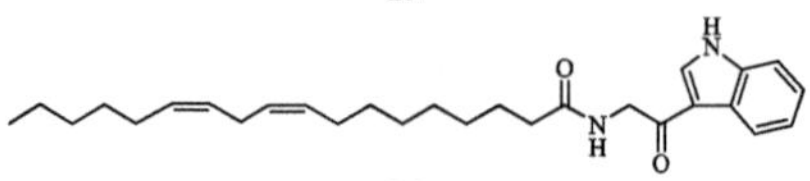

31

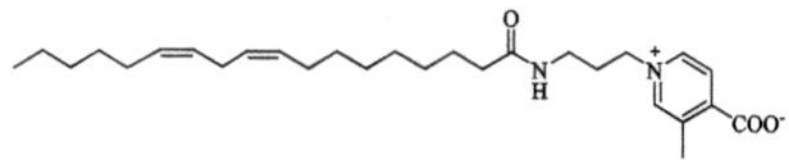

32

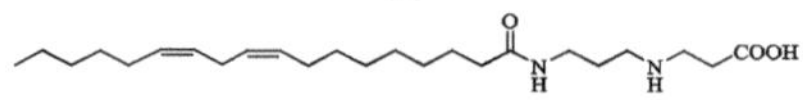

33

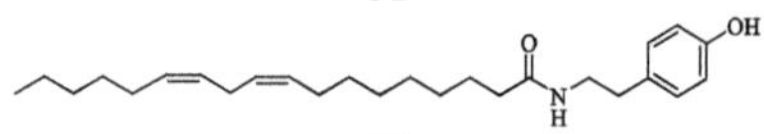

34

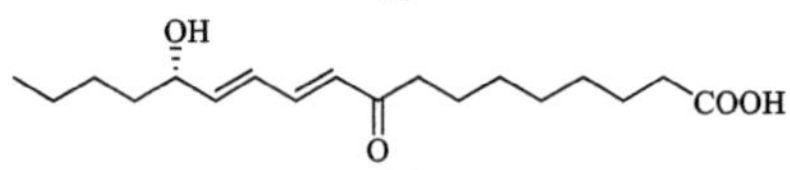

35

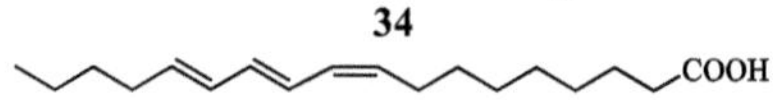

36

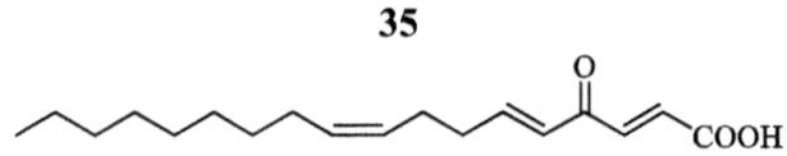

37

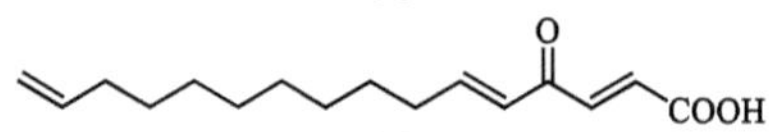

38

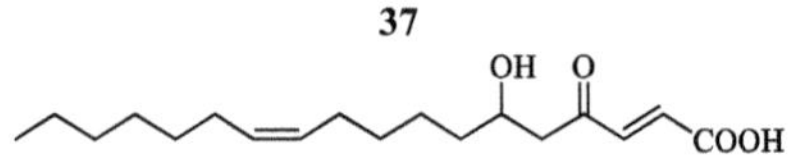

39

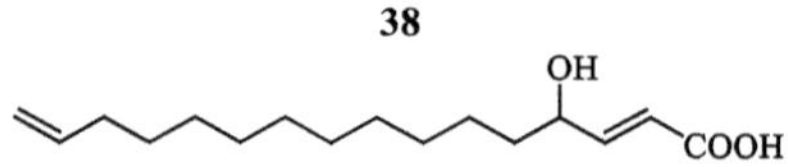

40

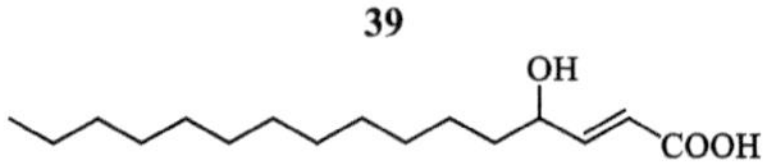

41

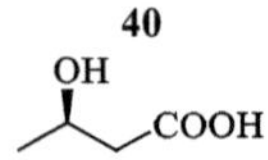

42

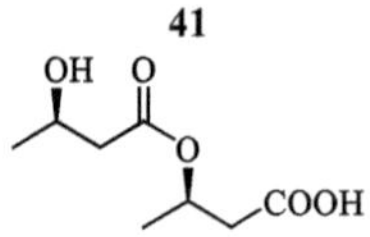

43

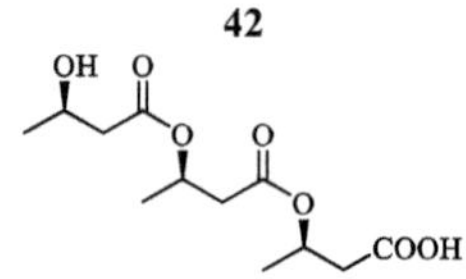

44

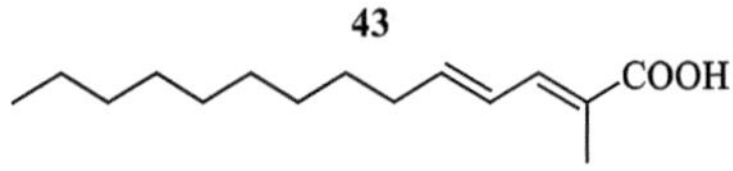

45

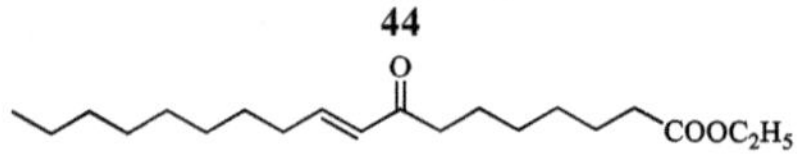

46

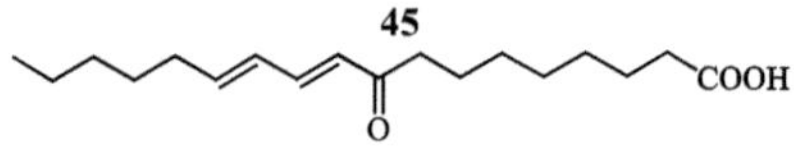

47

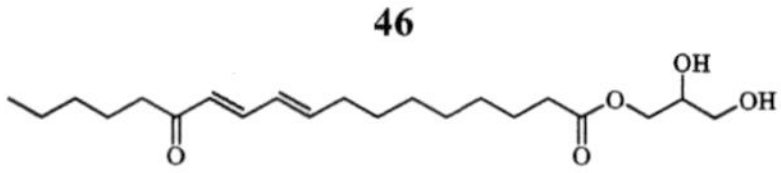

48

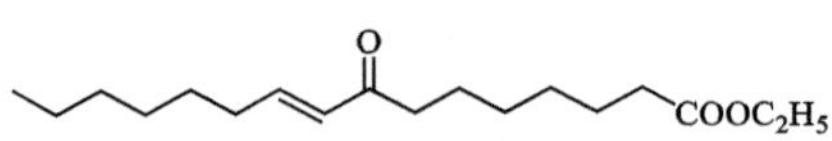

49

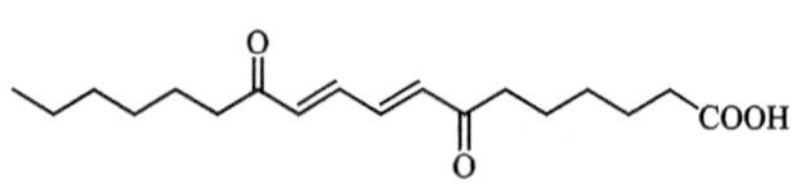

50

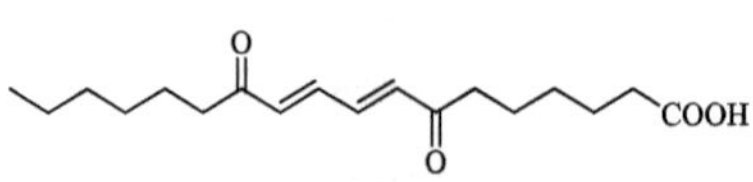

51

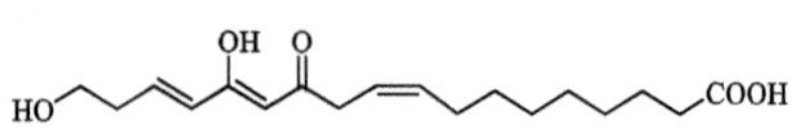

52

Figure 3.1 Continued.

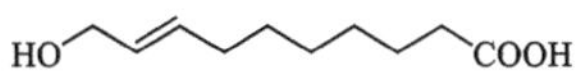

53

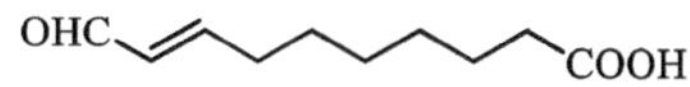

54

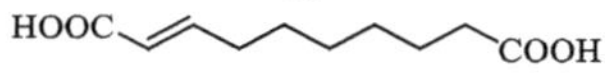

55

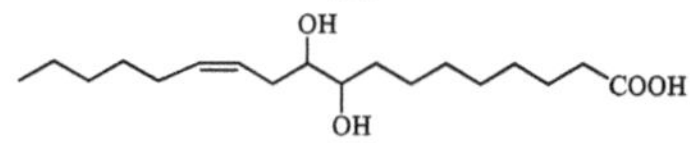

56

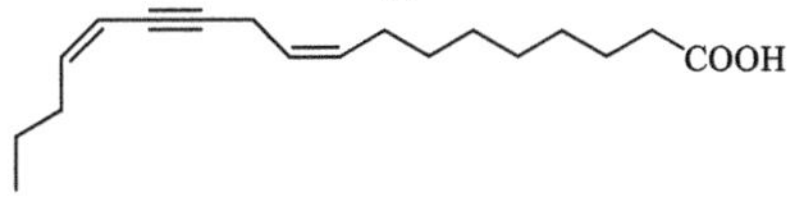

57

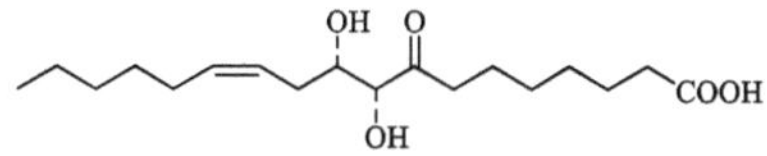

58

Figure 3.1 Continued.

(oleic acid) (**6**) from *Amanita spissacea* and searched their cytotoxic effects on human lung cancer cells (H1264, A549, Calu-6, H1299). Compounds **2** and **3** possessed the most potent cytotoxic activity against H1264, A549, Calu-6 and H1299 human lung cancer cell lines. IC_{50} values of **2** were calculated as 321.0 ± 2.7, 311.3 ± 2.0, 255.7 ± 0.4, 278.3 ± 4.0 μM, respectively, and **3** as 22.5 ± 0.9, 321.5 ± 2.9, 250.2 ± 8.8, 293.0 ± 6.0 μM, respectively. It has been demonstrated that the mechanisms of action of **2** and **3** were mediated by apoptosis associated with caspase-3 activation.

Mycochemical investigation of the sporophore of *Stropharia rugosoannulata* resulted in isolation and identification of palmitic acid (**1**), linoleic acid (**7**) and 10-eicosanoic acid (**8**).[233]

Pantheric acid A (**9**), pantheric acid B (**10**), pantheric acid C (**11**) and 1,10-dimethyl ester-2-decenedioic acid (**12**) were purified from *Amanita pantherina* and tested for their effects on lipid accumulation through adipocyte maturation. **9–11** were reported to support the expansion of lipid droplets in 3T3-L1 adipocytes and change the lipid metabolism by inhibiting lipolysis and inducing lipogenesis.[234]

A hemolytic compound, *trans*,*trans*-5,8-docosadienoic acid calcium salt (**13**) was purified from *Hypsizygus marmoreus* using the modified Bligh and Dyer method followed by a combination of chromatographic methods.[235]

Isolation and anti-HIV protease activity of linoleic acid (**7**) from *Auricularia polytricha* were studied in a previous report. Linoleic acid (**7**) exhibited important inhibition of HIV-1 PR activity (~20% inhibition at 20 μM concentration).[236]

A fatty acid methyl acid, namely (7*Z*,10*Z*)-7,10-octadecadienoic acid methyl ester (**14**), was obtained from *Calvatia nipponica*, and anti-inflammatory activity was determined by illuminating its mechanism of action. **14** was found to inhibit NO production with an IC_{50} value of 27.50 ± 0.08 μM. Also, in the experiment in which the mechanism of inhibition of **14** on NO production was elucidated using the western blot method, it was revealed that **14** (100 μM, 200 μM) suppressed iNOS and COX-2 at the protein level in a dose-dependent manner. **14** was defined as a potent anti-inflammatory compound via downregulation of iNOS and COX-2 in LPS-stimulated RAW264.7 macrophages.[237]

Cytotoxicity of (*Z*,*Z*)-9,12-octadecadienoic acid (**7**), 1-*O*-octadecanoyl-*sn*-glycerol (**15**) and 1-octadecanoic acid (stearic acid) (**16**) isolated from *Morchella esculenta* on human lung adenocarcinoma cells (H1264, A549, Calu-6, H1299) have been investigated. Compound **15** displayed the highest cytotoxic activity on human lung cancer cell lines. Calculated IC_{50} values of **2** and **15** on A549 were reported as 470.4 ± 4.6 and 207.8 ± 7.1 μM, on H1264 400.2 ± 5.0 and 171.6 ± 11.0 μM, on H1299 356.6 ± 10.3 and 266.8 ± 14.0 μM, and on Calu-6 291.6 ± 4.4 and 278.0 ± 2.7 μM. For **16**, the IC_{50} value only against the H1299 cell was calculated as 491.65 ± 1.2 μM.[238]

^{1}H NMR-guided fractionation of the extracts from *Lactarius deliciosus* allowed the characterization of stearic acid (**16**) in the study of Tala *et al.*[239]

In the study of Tel-Çayan *et al.*,[55] tetracosanoic acid (**17**) and 3-hydroxy-2,4-dimethylheptacosyl acetate (**18**) were purified from *Rhizopogon luteolus* and checked for antioxidant and anticholinesterase activities. **17** and **18** exhibited moderate antioxidant activity in β-carotene-linoleic acid (8.48 ± 1.69, 5.52 ± 1.80%), DPPH$^•$ (7.10 ± 0.47, 5.78 ± 0.19%) ABTS$^{•+}$ (20.76 ± 1.77, 27.66 ± 0.62%), $O_2^{•-}$ (2.29 ± 0.76, 7.83 ± 1.46%), CUPRAC (absorbance: 0.23 ± 0.01, 0.11 ± 0.00) and metal-chelating (20.10 ± 1.50, 11.28 ± 1.88%) assays at 200 μg mL^{-1} concentration. In addition, moderate inhibitory activities were recorded on AChE (3.18 ± 0.03% for **17**, 8.14 ± 0.20% for **18**) and BChE (16.42 ± 1.02% for **17**, 29.29 ± 1.53% for **18**) at 200 μg mL^{-1} concentration.

Fistulina hepatica afforded octadeca-8,11-dienoic acid methyl ester (**19**) and its antimicrobial property was investigated. **19** indicated no antimicrobial properties against *Staphylococcus aureus*, *Bacillus subtilis*, *Bacillus cereus*, *Candida albicans* and *Pseudomonas aeruginosa*, but very low activity (>1000 μg mL^{-1}) against *Escherichia coli* and *Proteus vulgaris*.[240]

(10*E*,14*Z*)-9-Oxooctadeca-10,14-dien-12-ynoic acid (**20**) and (10*E*,14*Z*)-9-hydroxyoctadeca-10,14-dien-12-ynoic acid (**21**) were obtained in an isolation study of *Cantharellus cibarius* and investigated for their effects on the gene expression of peroxisome proliferator-activated receptor-gamma target genes. **20** was found to activate peroxisome proliferator-activated receptor (PPAR)-γ with an EC_{50} value of 1.88 μM in the reporter gene assay when **21** was inactive (EC_{50} >20 μg mL^{-1}).[241]

A chemical investigation of *Gomphus clavatus* led to the isolation and identification of palmitic acid (**1**), linoleic acid (**7**), stearic acid (**16**) and capric acid (**22**).[119]

Bioassay-guided fractionation and chemical analysis of *Boletus pseudocalopus* resulted in the identification of calopusin A (**23**), calopusin B (**24**), calopusin C (**25**), 11-(*Z*)-eicosenoate (**26**) and methyl palmitoleate (**27**). The new compounds **23–25** exhibited significant inhibitory activity on the proliferation of the cancer cell lines tested. Cytotoxic activities of all compounds were tested and IC_{50} values of **23–27** were reported as 4.97, 6.02, 4.04, 35.30 and 44.59 μM on A549 (lung carcinoma), 3.65, 5.37, 2.77, 24.41 and 29.65 μM on Hs746T (stomach carcinoma), 12.51, 9.07, 8.26, 66.16 and 60.06 μM on H3122 (non-small cell lung carcinoma), and 30.97, 42.92, 41.71, 55.27 and 59.59 μM on HUVEC (human umbilical cord endothelial cells).[242]

Yoshikawa *et al.*[243] isolated (9*Z*,11*E*)-8,13-dihydroxyoctadeca-9,11-dienoic acid (filoboletic acid) (**28**) and 8,9-dihydroxy-10*E*,12*Z*-octadecadienoic acid (**29**) from *Sparassis crispa.*

Five fatty acid amides, termitomycamide A (**30**), termitomycamide B (**31**), termitomycamide C (**32**), termitomycamide D (**33**) and termitomycamide E (**34**), were purified from *Termitomyces titanicus* and their protective activity against endoplasmic reticulum stress-dependent cell death was described. While **31** and **34** showed significant endoplasmic reticulum stress-dependent cell death caused by tunicamycin, **32** and **33** were recorded as inactive.[244]

The isolation and structural elucidation of porrigenic acid (**35**) from *Pleurocybella porrigens* were studied by Hasegawa *et al.*[245] and **35** (IC_{50}: 46.5 $\mu g\,mL^{-1}$) was found to have cytotoxicity against cultured myeloma THP-1 cells, while no cytotoxic activity was found on murine B16F1 cells cultured with different concentrations (0.8–50 $\mu g\,mL^{-1}$).

α-Eleostearic acid (**36**), accepted as a characteristic fatty acid, was reported from *Pleurocybella porrigens.*[246]

Five oxidized 4-oxo fatty acids, namely (2*E*,5*E*,9*Z*)-4-oxo-octadeca-2,5,9-trienoic acid (**37**), (2*E*,5*E*)-4-oxo-hexadeca-2,5,15-trienoic acid (**38**), (*Z*)-6-hydroxy-4-oxo-octadeca-11-enoic acid (**39**), 6-hydroxy-4-oxo-hexadeca-15-enoic acid (**40**) and 6-hydroxy-4-oxo-hexadecanoic acid (**41**) were isolated from *Hygrophorus discoxanthus* and tested for fungicidal activity. **37** and **38** were moderately fungicidal against the phytopathogenic fungus *Cladosporium cucumerinum* Ell. & Arth.[247]

3*R*-Hydroxybutyric acid (**42**), its dimer 3*R*-(3*R*-hydroxybutyryloxy)-butyric acid (**43**), and trimer 3*R*-[3*R*-(3*R*-hydroxybutyryloxy) butyryloxy]-butyric acid (**44**), which are well-known as sex pheromones of the spider *Linyphia triangularis*, were isolated from *Hypoxylon truncatum.*[248]

Zhang *et al.*[249] isolated palmitic acid (**1**) from *Agrocybe aegerita* and investigated cyclooxygenase (COX) enzyme inhibitory and antioxidant activities. At 100 $\mu g\,mL^{-1}$ concentration, the inhibition values of **1** were reported as 45% in liposome peroxidation assay, 39% against COX-1 enzyme and 45% against COX-2 enzyme assays.

A rare fatty acid with 15 carbon atoms in nature was isolated as crystal, namely 2-methyl-2*E*,4*E*-tetradecadienoic acid (rubiginosic acid) (**45**), from *Hypoxylon rubiginosum.*[250]

Five fatty acid derivatives, namely (*E*)-8-oxo-9-oactadecenoic acid (**2**), (*E*)-8-oxo-9-oactadecenoic acid ethyl ester (**46**), (*E,E*)-9-oxooctadeca-10,12-dienoic acid (**47**), a C_{18} fatty acid having an α,β,γ,δ-dienone moiety (**48**) and an ethyl ester of a C_{16} fatty acid (**49**) were obtained from *Clitocybe clavipes* and investigated for *in vitro* aldehyde dehydrogenase inhibition effects. Compounds **2**, **46** and **49** were reported to have strong inhibitory activity on aldehyde dehydrogenase with IC_{50} values of 0.28, 0.43 and 0.17 mM, respectively; while compounds **47** and **48** indicated weak activity, with inhibition ratios lower than 50% at 0.5 mM.[251]

(*E,E*)-7,12-Dioxooctadeca-8,10-dienoic acid (ostopanic acid) (**50**) was isolated as a result of the mycochemical study of *Gymnopilus spectabilis* and *in vitro* cytotoxicity on P-388 lymphocytic leukemia cells was observed with an ED_{50} of 1.5 mg mL^{-1}.[252]

14-Hydroxy-12-oxo-10*E*,13*E*,15*E*-octadecatrienoic acid (**51**), which is a new unsaturated carboxylic acid, was obtained from *Cantharellus friesii* by gel-permeation and liquid chromatography techniques.[253]

Isolation, antimicrobial and cytotoxic activities of (9*Z*,13*Z*,15*E*)-14,18-dihydroxy-12-keto-9,13,15-octadecatrienoic acid (cibaric acid) (**52**) and 10-hydroxy-8-decenoic acid (**53**) from *Cantharellus cibarius* and *Cantharellus tubaeformis* were studied by Anke *et al.*[254] Both isolated compounds possessed weak antimicrobial activity against bacteria (*Acinetobacter calcoaceticus*, *Bacillus brevis*, *Bacillus subtilis*, *Micrococcus luteus*, *Proteus vulgaris*, *Staphylococcus aureus*, *Pseudomonas fluorescens*) and yeasts and fungi (*Mucor miehei*, *Nematos poracoiyli*, *Paecilomyces variotii*, *Penicillium nototum*, *Saccharomyces cerevisiae*, *Candida albicans*, *Nadsonia fulvescens*) with MIC values of >100 μg mL^{-1}. Also, weak cytotoxic activities of **52** and **53** on L1210 (lymphocytic leukemia cell) (IC_{50}: 50, 40 μg mL^{-1}) and BHK (baby hamster kidney fibroblasts) (IC_{50}: >50, 50 μg mL^{-1}) were reported.

(9*Z*,11*E*)-8,13-Dihydroxyoctadeca-9,11-dienoic acid (filoboletic acid) (**31**) was isolated from a *Filoboletus* sp. strain.[255]

8*E*-10-Hydroxy-8-decenoic acid (**53**), 8*E*-10-oxo-8-decenoic acid (**54**), *E*-2-decenedioic acid (**55**) and (12*Z*)-9,10-dihydroxy-12-octadecenoic acid (**56**) were obtained in the chemical investigation of *Lepista nebularis*.[256]

Isolation and structure elucidation of (9*Z*,13*Z*,15*E*)-14,18-dihydroxy-12-keto-9,13,15-octadecatrienoic acid (cibaric acid) (**52**) and (9*Z*,14*Z*)-9,14-octadecadien-12-ynoic acid (14,15-dehydrocrepenyic acid) (**57**) from *Cantharellus cibarius* have been reported.[257]

Kawagishi *et al.*[258] described the isolation of (9*R**, 10*S**, 12*Z*)-9,10-dihydroxy-8-oxo-12-octadecenoic acid (**58**), which was reported to exhibit cytotoxicity (the minimum concentration resulting in complete growth inhibition of the cells was 100 μg mL^{-1}) against HeLa cells and an inhibitory effect (the inhibition zone against growth was 5.63 mm at 250 μg mL^{-1}) on tea pollen growth from *Hericium erinaceum*.

3.4 Conclusions

This chapter has reviewed the fatty acid contents of mushroom species with attention given to the importance of fatty acids and fatty acid-derived molecules for the human diet and health. To date, the fatty acid contents of approximately 143 mushroom species from different countries, mainly *Boletus edulis*, *Agaricus bisporus*, *Pleurotus ostreatus*, *Lactarius deliciosus* and *Macrolepiota procera* species have been reported by several researchers. These reported fatty acids include C4:0, C5:0, C6:0, C7:0, C8:0, C9:0, C10:0, C11:0, C12:0, C13:0, C14:0, C14:1, C15:0, C15:1, C16:0, C16:1, C17:0, C17:1, C18:0, C18:1, C18:2, C18:3, C19:0, C20:0, C20:1, C20:2, C20:3, C20:4, C20:5, C21:0, C22:0, C22:1, C22:2, C22:3, C22:4, C22:5, C22:6, C23:0, C24:0, C24:1, C25:0, C26:0 fatty acids, with linoleic, oleic, palmitic and stearic acids predominating. The results of this review reveal that mushrooms are rich in polyunsaturated fatty acids, especially linoleic acid (18:2). Linoleic and linolenic acids, which are essential fatty acids, are necessary for human health. A lack of these essential fatty acids can cause a number of disorders, including cardiovascular and chronic diseases.

Abbreviations

A549	Non-small cell lung adenocarcinoma
ABTS	2,2′-Azino-bis (3-ethylbenzothiazoline-6-sulfonic acid)
AChE	Acetylcholinesterase
ADP	Adenosine diphosphate
AGS	Human gastric cancer cells
B16F1	Mouse melanoma
BAETS	2-(12-Oxobenzo[*b*]acridin-5(12*H*)-yl)-ethyl-4-toluenesulfonate
BChE	Butyrylcholinesterase
BEND3	Mouse brain microvascular cell line
Bax	Pro-apoptotic Bcl-2 protein
Bcl-2	B-cell lymphoma 2
BHK	Baby hamster kidney fibroblasts
COX	Cyclooxygenase
Calu-6	Human lung adenocarcinoma cell line
CoA	Coenzyme A
DHA	Docosahexaenoic acid
DPPH	2,2-Diphenyl-1-picrylhydrazyl
DU-145	Prostate cancer cell line
EPA	Eicosapentaenoic acid
GC	Gas chromatography
GC-MS	Gas chromatography mass spectrometry
GC-FID	Gas chromatography-flame ionization detector
H1264	Human lung cancer cell line
H1299	Human lung cancer cell line
H3122	Non-small cell lung carcinoma
HCA	Hierarchical clustering analyses
HeLa	Human epitheloid cervix carcinoma
HIV	Human immunodeficiency virus
HPLC	High performance liquid chromatography
HPLC-FLD-MS/MS	High-performance liquid chromatography with fluorescence detection-mass spectrometry
HPLC-PDA	High-performance liquid chromatography-photodiode-array detection
HOG-1	Human cervical cancer cell
Hs746T	Stomach carcinoma
Hs578Bst	Breast cancer cell line
Hs578t	Breast cancer cell line

HT-29	Colorectal cancer cell line
HUVEC	Human umbilical vein endothelial cells
IC50	Half maximal inhibitory concentration
INOS	Inducible nitric oxide synthase
LPS	Lipopolysaccharide
LDL	Low-density lipoprotein
MAEH	Microwave-assisted extraction with hexane
MAEC	Microwave-assisted extraction with chloroform
MB-231	Breast cancer cell line
MCF-7	Human breast cancer cell line
MCF-10A	Breast cancer cell line
MDA	Breast cancer cell lines
MDA-MB-231	Breast cancer cell line
MEFs	Mouse embryonic fibroblasts
MIC	Minimum inhibitory concentration
mTOR	Mammalian target of rapamycin
MUFA	Monounsaturated fatty acid
NADH	Nicotinamide adenine dinucleotide
NO	Nitric oxide
p21	Cyclin-dependent kinase
p-388	Lymphocytic leukemia cell line
p53	Tumor protein 53
PPARy	Peroxisome proliferator-activated receptor
(PPAR)-γ	Peroxisome proliferator-activated receptor
p53	Cellular tumor antigen
PAF	Platelet-activating factor
PCA	Principal component analyses
PC12	Rat pheochromocytoma cell line
PGE2	Prostaglandin E2
PMA	Phorbol-12-myristate-13-acetate
pRb	Retinoblastoma protein
PUFA	Polyunsaturated fatty acid
RAW 264.7	Mouse leukemic monocyte macrophage cell line
SFA	Saturated fatty acid
SK-Br3	Human breast cancer cell line
sn	Stereospecific numbering
THP-1	Human monocytic cell line
UAEH	Ultrasound-assisted extraction with hexane
UAEC	Ultrasound-assisted extraction with chloroform
UFA	Unsaturated fatty acid

References

1. W. Breene, *J. Food Prot.*, 1990, **53**(10), 883.
2. P. Kalac, *J. Sci. Food Agric.*, 2013, **93**(2), 209.

3. M. E. Valverde, T. Hernandez-Perez and O. Paredes-Lopez, *J. Microbiol.*, 2015, **2015**, 14.
4. M. Dimitrijevic, V. S. Jovanovic, J. Cvetkovic, M. Mitic, G. Petrovic, A. Dordevic and V. Mitic, *Anal. Lett.*, 2017, **50**, 1691.
5. D. J. Murphy, *Prog. Lipid Res.*, 2001, **40**, 325.
6. H. M. Roche, *Proc. Nutr. Soc.*, 1999, **58**, 397.
7. D. Sande, G. P. de Oliveira, M. A. F. Moura, B. A. Martins, M. T. N. S. Lima and J. A. Takahashi, *Food Res. Int.*, 2019, **125**, 10852.
8. E. Guillamon, A. García-Lafuente, M. Lozano, M. D'Arrigo, M. A. Rostagno, A. Villares and J. A. Martinez, *Fitoterapia*, 2010, **81**, 715.
9. J. E. Kinsella, K. S. Broughton and J. W. Whelan, *J. Nutr. Biochem.*, 1990, **1**, 123.
10. M. B. D. Berteli, O. B. Q. Oliveira Filho, J. D. S. Freitas, W. C. Bortolucci, G. R. Silva, Z. C. Gazim, F. A. R. Livero, E. C. W. Lovato, J. S. Valle, G. A. Linde, L. Barros, F. S. Reis, I. C. F. R. Ferreira, L. D. Paccola-Meirelles and N. B. Colauto, *Eur. Food Res. Technol.*, 2021, **247**, 1355.
11. H. Isik, *Chem. Nat. Compd.*, 2020, **56**, 6.
12. M. Kostic, M. Ivanov, A. Fernandes, J. Pinela, R. C. Calhelha, J. Glamoclija, L. Barros, I. C. F. R. Ferreira, M. Sokovic and A. Ciric, *Molecules*, 2020, **25**, 4336.
13. A. S. Bengu, *Prog. Nutr.*, 2020, **22**, 185.
14. R. Pietrzak-Fiecko, M. Galgowska and J. Falandysz, *Chem. Biodivers.*, 2020, **17**, e2000032.
15. G. Maheshwari, D. K. Gessner, S. Meyer, J. Ahlborn, G. Wen, R. Ringseis, H. Zorn and K. Eder, *Mol. Nutr. Food Res.*, 2020, **64**, 2000591.
16. F. Çayan, E. Deveci, G. Tel-Çayan and M. E. Duru, *Anal. Lett.*, 2020, **53**, 2784.
17. P. Papazav, D. Vassilev, N. Valchev and P. Denev, *Oxid. Commun.*, 2020, **43**, 194.
18. M. Shahid, H. Fatima, F. Anjum, M. Riaz, N. Akhter and M. A. Murtaza, *Acta Poloniae Pharmaceutican Drug Res.*, 2020, 77, 145.
19. R. J. Rebecca, Z. Jiachan, Z. Wensheng, L. Yu, K. L. Wasantha and W. Shouxian, *Prog. Nutr.*, 2020, **22**, 323.
20. N. Stilinovic, I. Capo, S. Vukmirovic, A. Raskovic, A. Tomas, M. Popovic and A. Sabo, *R. Soc. Open Sci.*, 2020, 7, 200900.
21. S. Sezgin, A. Dalar and Y. Uzun, *J. Food Sci. Technol.*, 2020, **57**(5), 1866.
22. O. Taofiq, F. Rodrigues, L. Barros, R. M. Peralta, M. F. Barreiro, I. C. F. R. Ferreira and M. B. P. P. Oliveira, *Food Funct.*, 2019, **10**, 565.
23. M. Karaman, K. Atlagic, A. Novakovic, F. Sibul, M. Zivic, K. Stevanovic and B. Pejin, *Antioxidants*, 2019, **8**, 480.
24. M. V. Dimitrijevic, V. D. Mitic, J. S. Nikolic, A. S. Djordjevic, J. J. Mutic, V. P. S. Jovanovic and G. S. Stojanovic, *Chem. Biodivers.*, 2019, **16**, e1800492.

25. Z. Xu, L. Fu, S. Feng, M. Yuan, Y. Huang, J. Liao, L. Zhou, H. Yang and C. Ding, *Molecules*, 2019, **24**, 1357.
26. R. V. C. Cardoso, A. Fernandes, J. C. M. Barreira, S. C. Verde, A. L. Antonio, A. M. Gonzalez-Paramas, L. Barros and I. C. F. R. Ferreira, *Food Chem.*, 2019, **278**, 760.
27. A. Ciric, I. Kruljevic, D. Stojkovic, A. Fernandes, L. Barros, R. C. Calhelha, I. C. F. R. Ferreira, M. Sokovic and J. Glamoclija, *Food Funct.*, 2019, **10**, 7678.
28. F. Çayan, G. Tel-Çayan, E. Deveci, M. Öztürk and M. E. Duru, *Int. J. Med. Mushrooms*, 2019, **21**(11), 1075.
29. A. Doğan, A. Dalar, A. Battal, A. Uyar, C. Sadullahoğlu, İ. Çelik and Y. Uzun, *Fresenius Environ. Bull.*, 2019, **28**, 9503.
30. A. S. Bengu, *Prog. Nutr.*, 2019, **21**, 189.
31. J. M. Li, H. Q. Liang, P. Qiao, K. M. Su, P. G. Liu, S. X. Guo and J. Chen, *Chem. Biodivers.*, 2019, **16**, e1800609.
32. C. Sillapachaiyaporn, S. Nilkhet, A. T. Ung and S. Chuchawankul, *BMC Complementary Altern. Med.*, 2019, **19**, 351.
33. M. V. Dimitrijevic, V. D. Mitic, O. P. Jovanovic, V. P. S. Jovanovic, J. S. Nikolic, G. M. Petrovic and G. S. Stojanovic, *Chem. Biodivers.*, 2018, **15**, e1700434.
34. B. Muszynska, K. Kala, J. Radovic, K. Sulkowska-Ziaja, A. Krakowska, J. Gdula-Argasinska, W. Opoka and T. Kundakovic, *Eur. Food Res. Technol.*, 2018, **244**, 2255.
35. C. O. Ogidi, V. O. Oyetayo, B. J. Akinyele, C. A. De Carvalho and M. C. M. Kasuya, *Prev. Nutr. Food Sci.*, 2018, **23**(3), 228.
36. A. Mocan, A. Fernandes, L. Barros, G. Crişan, M. Smiljkovic, M. Sokovic and I. C. F. R. Ferreira, *Food Funct.*, 2018, **9**, 160.
37. A. Dogan, A. Dalar, C. Sadullahoglu, A. Battal, Y. Uzun, I. Celik and K. Demirel, *Mol. Biol. Rep.*, 2018, **45**, 787.
38. L. Dospatliev, *C. R. Acad. Bulg.*, 2018, **71**, 1458.
39. M. Fogarasi, S. A. Socaci, F. V. Dulf, Z. M. Diaconeasa, A. C. Farcas, M. Tofana and C. A. Semeniuc, *Molecules*, 2018, **23**, 3272.
40. G. Tel-Çayan, E. Deveci, F. Çayan and M. E. Duru, *Marmara Pharm. J.*, 2018, **22**(1), 59.
41. O. Tokul-Olmez, E. Kaplaner, M. Ozturk, Z. Ullah and M. E. Duru, *Biochem. Syst. Ecol.*, 2018, **78**, 91.
42. H. Aoki, M. Hanayama, K. Mori and R. Sato, *Biosci., Biotechnol., Biochem.*, 2018, **82**, 1550.
43. X. Li, Y. Guo, Y. Zhuang, Y. Qin and L. Sun, *Int. J. Food Sci. Technol.*, 2018, **53**, 1855.
44. A. Irshad, S. Sharif, M. Riaz and F. Anjum, *J. Biol. Regul. Homeostatic Agents*, 2018, **32**(1), 107.
45. J. Manjunathan, S. S. Gowri, N. Vennila, A. Ezhilarasu and V. Kaviyarasan, *Int. J. Pharma Res. Health Sci.*, 2017, 5(5), 1886.
46. F. Çayan, G. Tel-Çayan, M. A. Özler and M. E. Duru, *Eurasian J. Anal. Chem.*, 2017, **12**(3), 257.

47. G. Tel-Cayan, M. Ozturk, M. E. Duru and A. Turkoglu, *Chem. Nat. Compd.*, 2017, **53**, 351.
48. M. Kostic, M. Smiljkovic, J. Petrovic, J. Glamoclija, L. Barros, I. C. F. R. Ferreira, A. Ciric and M. Sokovic, *Food Funct.*, 2017, **8**, 3239.
49. A. Fernandes, J. C. M Barreira, T. Günaydi, H. Alkan, A. L. Antonio, M. B. P. P. Oliveira, A. Martins and I. C. F. R. Ferreira, *Food Control*, 2017, **72**, 328.
50. M. Obodai, D. L. N. Mensah, A. Fernandes, N. K. Kortei, M. Dzomeku, M. Teegarden, S. J. Schwartz, L. Barros, J. Prempeh, R. K. Takli and I. C. F. R. Ferreira, *Molecules*, 2017, **22**, 196.
51. D. S. Stojkovic, N. Cevic-Grujicic, F. S. Reis, S. Davidovic, L. Barros, J. Popovic, I. Petrovic, A. Pavic, J. Glamoclija, A. Ciric, M. Stevanovic, I. C. F. R. Ferreira and M. Sokovic, *LWT–Food Sci. Technol.*, 2017, **79**, 454.
52. S. K. Sharma and N. Gautam, *Sci. Rep.*, 2017, 7, 46570.
53. V. Ravikrishnan, S. Ganesh and M. Rajashekhar, *Int. Food Res. J.*, 2017, **24**(2), 679.
54. N. Landi, S. Pacifico, S. Ragucci, A. M. Di Giuseppe, F. Iannuzzi, A. Zarrelli, S. Piccolella and A. Di Maro, *J. Sci. Food Agric.*, 2017, **97**, 5388.
55. G. Tel-Çayan, A. Muhammad, M. E. Duru, M. Öztürk and A. Adhikari, *Nat. Prod. Res.*, 2016, **30**(20), 2258.
56. V. Vieira, L. Barros, A. Martins and I. C. F. R. Ferreira, *Molecules*, 2016, **21**, 99.
57. F. S. Reis, L. Barros, A. Martins, M. H. Vasconcelos, P. Morales and I. C. F. R. Ferreira, *Molecules*, 2016, **21**, 246.
58. A. Fernandes, J. C. M. Barreira, A. L. Antonio, M. B. P. P. Oliveira, A. Martins and I. C. F. R. Ferreira, *LWT–Food Sci. Technol.*, 2016, **67**, 99.
59. R. Pietrzak-Fiecko, M. Galgowska and S. Bakula, *Ital. J. Food Sci.*, 2016, **28**, 402.
60. C. V. Toledo, C. Barroetavena, A. Fernandes, L. Barros and I. C. F. R. Ferreira, *Molecules*, 2016, **21**, 1201.
61. V. Vieira, A. Fernandes, L. Barros, J. Glamoclija, A. Ciric, D. Stojkovic, A. Martins, M. Sokovic and I. C. F. R. Ferreira, *J. Sci. Food Agric.*, 2016, **96**, 90.
62. Z. Fu, J. Wang and H. Aisa, *Toxicol. Environ. Chem.*, 2016, **98**, 439.
63. P. Petrovic, J. Vunduk, A. Klaus, M. Kozarski, M. Niksic, Z. Zizak, N. Vukovic, G. Sekularac, S. Drmanic and B. Bugarski, *J. Funct. Foods*, 2016, **21**, 36.
64. V. J. Sinanoglou, P. Zoumpoulakis, G. Heropoulos, C. Proestos, A. Ciric, J. Petrovic, J. Glamoclija and M. Sokovic, *J. Food Sci. Technol.*, 2015, **52**(6), 3264.
65. S. A. Heleno, L. Barros, A. Martins, M. J. R. P. Queiroz, P. Morales, V. Fernandez-Ruiz and I. C. F. R. Ferreira, *LWT–Food Sci. Technol.*, 2015, **63**, 475.
66. R. Mukhopadhyay and A. K. Guha, *LWT–Food Sci. Technol.*, 2015, **61**, 339.

67. S. K. Sharma and N. Gautam, *BioMed Res. Int.*, 2015, **2015**, 12.
68. S. A. Heleno, L. Barros, A. Martins, P. Morales, V. Fernandez-Ruiz, J. Glamoclija, M. Sokovic and I. C. F. R. Ferreira, *LWT–Food Sci. Technol.*, 2015, **63**, 799.
69. S. A. Heleno, R. Carolina Ferreira, A. L. Antonio, M. J. R. P. Queiroz, L. Barros and I. C. F. R. Ferreira, *Food Biosci.*, 2015, **11**, 48.
70. J. Petrovic, J. Glamoclija, D. Stojkovic, A. Ciric, L. Barros, I. C. F. R. Ferreira and M. Sokovic, *J. Food Sci. Technol.*, 2015, **52**(10), 6711.
71. D. S. Stojkovic, F. S. Reis, A. Ciric, L. Barros, J. Glamoclija, I. C. F. R. Ferreira and M. Sokovic, *J. Food Sci. Technol.*, 2015, **52**(11), 7385.
72. A. Rosa, M. Nieddu, A. Piras, A. Atzeri, D. Putzu and A. Rescigno, *Nutrients*, 2015, 7(2), 849.
73. J. G. Kaviyarasan, *Indo Am. J. Pharm. Sci.*, 2015, **2**, 731.
74. J. S. L. Chan, M. D. Asatiani, L. Sharvit, B. Trabelcy, G. S. Barseghyan and S. P. Wasser, *Int. J. Med. Mushrooms*, 2015, **17**(8), 735.
75. A. Z. Woldegiorgis, D. Abate, G. D. Haki, G. R. Ziegler and K. J. Harvatine, *J. Nutr. Food Sci.*, 2015, **5**(3), 5.
76. V. Ravikrishnan, P. Naik, S. Ganesh and M. Rajashekhar, *Int. J. Plant, Anim. Environ. Sci.*, 2015, **5**, 278.
77. D. M. F. Rodrigues, A. C. Freitas, T. A. P. Rocha-Santos, M. W. Vasconcelos, M. Roriz, L. M. Rodriguez-Alcala, A. M. P. Gomes and A. C. Duarte, *Food Sci. Technol.*, 2015, **52**(11), 6927.
78. K. P. Shen, C. H. Su, T. M. Lu, M. N. Lai and L. T. Ng, *Pharm. Biol.*, 2015, **53**(5), 705.
79. F. Çayan, G. Tel, M. E. Duru, M. Öztürk, A. Türkoğlu and M. Harmandar, *Food Anal. Methods*, 2014, 7, 449.
80. J. Petrovic, D. Stojkovic, F. S. Reis, L. Barros, J. Glamoclija, A. Ciric, I. C. F. R. Ferreira and M. Sokovic, *Food Funct.*, 2014, **5**(7), 1441.
81. F. S. Reis, D. Stojkovic, L. Barros, J. Glamoclija, A. Ciric, M. Sokovic, A. Martins, M. H. Vasconcelos, P. Morales and I. C. F. R. Ferreira, *Food Funct.*, 2014, **5**, 2861.
82. M. Öztürk, G. Tel, F. A. Öztürk and M. E. Duru, *Rec. Nat. Prod.*, 2014, **8**(2), 189.
83. F. S. Reis, J. C. M. Barreira, R. C. Calhelha, L. J. I. D. van Griensven, A. Ciric, J. Glamoclija, M. Sokovic and I. C. F. R. Ferreira, *LWT–Food Sci. Technol.*, 2014, **58**, 478.
84. D. S. Stojkovic, L. Barros, R. C. Calhelha, J. Glamoclija, A. Ciric, L. J. L. D. van Griensven, M. Sokovic and I. C. F. R. Ferreira, *Int. J. Food Sci. Nutr.*, 2014, **65**(1), 42.
85. V. Vieira, L. Barros, A. Martins and I. C. F. R. Ferreira, *Molecules*, 2014, **19**, 20650.
86. D. Stojkovic, F. S. Reis, J. Glamoclija, A. Ciric, L. Barros, L. J. L. D. Van Griensven, I. C. F. R. Ferreira and M. Sokovic, *Food Funct.*, 2014, **5**, 1602.

87. A. Fernandes, J. C. M. Barreira, A. L. Antonio, M. B. P. P. Oliveira, A. Martins and I. C. F. R. Ferreira, *Food Bioprocess Technol.*, 2014, **7**, 1606.
88. A. Fernandes, J. C. M. Barreira, A. L. Antonio, M. B. P. P. Oliveira, A. Martins and I. C. F. R. Ferreira, *Innovative Food Sci. Emerging Technol.*, 2014, **22**, 158.
89. A. Fernandes, J. C. M. Barreira, A. L. Antonio, M. B. P. P. Oliveira, A. Martins and I. C. F. R. Ferreira, *Food Chem.*, 2014, **15**(149), 91.
90. X. H. Chen, J. J. Li and B. Q. Zhao, *J. Food Saf.*, 2014, **25**, 161.
91. N. Cohen, J. Cohen, M. D. Asatiani, V. K. Varshney, H. T. Yu, Y. C. Yang, Y. H. Li, J. L. Mau and S. P. Wasser, *Int. J. Med. Mushrooms*, 2014, **16**(3), 273.
92. L. M. Carvalho, F. Carvalho, M. L. Bastos, P. Baptista, N. Moreira, A. R. Monforte, A. C. S. Ferreira and P. G. de Pinho, *Talanta*, 2014, **118**, 292.
93. C. W. Phan, P. David, Y. S. Tan, M. Naidu, K. H. Wong, U. R. Kuppusamy and V. Sabaratnam, *Sci. World J.*, 2014, **2014**, 10.
94. N. M. Sudheep and K. R. Sridhar, *Mycology*, 2014, **5**, 64.
95. N. Cherno, S. Osolina and A. Nikitina, *J. Fac. Food Eng.*, 2013, **12**, 291.
96. A. Fernandes, J. C. M. Barreira, A. L. Antonio, P. M. P. Santos, A. Martins, M. B. P. P. Oliveira and I. C. F. R. Ferreira, *Food Res. Int.*, 2013, **54**, 18.
97. I. Akata, P. G. Ergonul, B. Ergonul and F. Kalyoncu, *J. Pure Appl. Microbiol.*, 2013, **7**(4), 3143.
98. N. Kalogeropoulos, A. E. Yanni, G. Koutrotsios and M. Aloupi, *Food Chem. Toxicol.*, 2013, **55**, 378.
99. A. Fernandes, A. L. Antonio, J. C. M. Barreira, M. L. Botelho, M. B. P. P. Oliveira, A. Martins and I. C. F. R. Ferreira, *Food Bioprocess Technol.*, 2013, **6**, 2895.
100. A. Fernandes, L. Barros, J. C. M. Barreira, A. L. Antonio, M. B. P. P. Oliveira, A. Martins and I. C. F. R. Ferreira, *LWT–Food Sci. Technol.*, 2013, **54**, 493.
101. B. F. Lau, N. Abdullah and N. Aminudin, *J. Agric. Food Chem.*, 2013, **61**, 4890.
102. S. Pinto, L. Barros, M. J. Sousa and I. C. F. R. Ferreira, *Food Res. Int.*, 2013, **51**, 496.
103. D. Stojkovic, F. S. Reis, I. C. F. R. Ferreira, L. Barros, J. Glamoclija, A. Ciric, M. Nikolic, T. Stevic, A. Giveli and M. Sokovic, *Food Res. Int.*, 2013, **53**, 56.
104. H. H. Doğan and G. Akbaş, *Pharm. Biol.*, 2013, **51**(7), 863.
105. P. G. Ergönül, I. Akata, F. Kalyoncu and B. Ergönül, *Sci. World J.*, 2013, **2013**, 4.
106. N. Yılmaz, I. Türkekul, S. Bulut, F. Sahin and Ö. F. Bayrak, *Asian J. Chem.*, 2013, **25**, 1216.
107. H. H. Doğan, *J. Food Compos. Anal.*, 2013, **31**, 87.
108. S. A. Heleno, D. Stojkovic, L. Barros, J. Glamoclija, M. Sokovic, A. Martins, M. J. R. P. Queiroz and I. C. F. R. Ferreira, *Food Res. Int.*, 2013, **51**, 236.

109. A. A. J. Carneiro, I. C. F. R. Ferreira, M. Duenas, L. Barros, R. da Silva, E. Gomes and C. Santos-Buelga, *Food Chem.*, 2013, **138**(4), 2168.
110. F. S. Reis, L. Barros, R. C. Calhelha, A. Ciric, L. J. L. D. van Griensven, M. Sokovic and I. C. F. R. Ferreira, *Food Chem. Toxicol.*, 2013, **62**, 91.
111. A. R. Leal, L. Barros, J. C. M. Barreira, M. J. Sousa, A. Martins, C. Santos-Buelga and I. C. F. R. Ferreira, *Food Res. Int.*, 2013, **50**, 1.
112. C. H. Su, M. N. Lai and L. T. Ng, *Food Funct.*, 2013, **4**, 644.
113. T. A. Krupodorova, V. Y. Barshteyn, N. A. Bisko and T. S. Lvanova, *Int. J. Med. Mushrooms*, 2012, **14**(3), 285.
114. P. G. Ergonul, B. Ergonul, F. Kalyoncu and I. Akata, *Int. J. Pharmacol.*, 2012, **8**, 463.
115. E. Pereira, L. Barros, A. Martins and I. C. F. R. Ferreira, *Food Chem.*, 2012, **130**, 394.
116. F. S. Reis, D. Stojkovic, M. Sokovic, J. Glamoclija, A. Ciric, L. Barros and I. C. F. R. Ferreira, *Food Res. Int.*, 2012, **48**, 620.
117. E. Palazzolo, M. L. Gargano and G. Venturella, *Int. J. Food Sci. Nutr.*, 2012, **63**(1), 79.
118. A. Dundar, O. F. Yesil, H. Acay, V. Okumus, S. Ozdemir and A. Yildiz, *Food Sci. Technol. Int.*, 2012, **18**(4), 317.
119. M. Makropoulou, N. Aligiannis, Z. Gonou-Zagou, H. Pratsinis, A. L. Skaltsounis and N. Fokialakis, *J. Med. Food*, 2012, **15**(2), 216.
120. F. S. Reis, L. Barros, A. Martins and I. C. F. R. Ferreira, *Food Chem. Toxicol.*, 2012, **50**, 191.
121. I. Marekov, S. Momchilova, B. Grung and B. Nikolova-Damyanov, *J. Chromatogr. B: Anal. Technol. Biomed. Life Sci.*, 2012, **910**, 54.
122. Y. T. Liu, J. Sun, Z. Y. Luo, S. Q. Rao, Y. J. Su, R. R. Xu and Y. J. Yang, *Food Chem. Toxicol.*, 2012, **50**, 1238.
123. N. Jing, J. Shi, G. Li, Z. Sun and J. You, *Food Res. Int.*, 2012, **48**, 155.
124. M. Öztürk, M. E. Duru, Ş. Kivrak, N. Mercan-Doğan, A. Türkoglu and M. A. Özler, *Food Chem. Toxicol.*, 2011, **49**, 1353.
125. I. Mujic, Z. Zekovic, S. Vidovic, M. Radojkovic, J. Zivkovic and D. G. Cevac, *J. Med.*, 2011, **14**(11), 1330.
126. Z. Huang, F. Fang and C. W. Wong, *J. Food Biochem.*, 2011, **35**, 1508.
127. F. S. Reis, S. A. Heleno, L. Barros, M. J. Sousa, A. Martins, C. Santos-Buelga and I. C. F. R. Ferreira, *J. Food Sci.*, 2011, **76**, 824.
128. F. A. Ayaz, L. T. Chuang, H. Torun, A. Colak, E. Sesli, J. Presley, B. R. Smith and R. H. Glew, *Int. J. Food Sci. Nutr.*, 2011, **62**(4), 328.
129. C. Grangeia, S. A. Heleno, L. Barros, A. Martins and I. C. F. R. Ferreira, *Food Res. Int.*, 2011, **44**, 1029.
130. K. J. Lee, I. J. Yun, K. H. Kim, S. H. Lim, H. J. Ham, W. S. Eum and J. H. Joo, *J. Food Compos. Anal.*, 2011, **24**, 175.
131. J. A. Vaz, L. Barros, A. Martins, C. Santos-Buelga, M. H. Vasconcelos and I. C. F. R. Ferreira, *Food Chem.*, 2011, **126**, 610.
132. S. A. Heleno, L. Barros, M. J. Sousa, A. Martins, C. Santos-Buelga and I. C. F. R. Ferreira, *LWT–Food Sci. Technol.*, 2011, **44**, 1343.

133. F. S. Reis, E. Pereira, L. Barros, M. J. Sousa, A. Martins and I. C. F. R. Ferreira, *Molecules*, 2011, **16**, 4328.
134. K. Deng, Y. Zhang, Z. Ren, L. Xie, W. Peng and B. Gan, *J. Med. Plants Res.*, 2011, **5**(13), 2816.
135. R. R. Simon, K. M. Phillips, R. L. Horst and I. C. Munro, *J. Agric. Food Chem.*, 2011, **59**, 8724.
136. V. M. Dembitsky, A. O. Terentev and D. O. Levitsky, *Rec. Nat. Prod.*, 2010, **4**(4), 218.
137. G. Liu, H. Wang, B. Zhou, X. Guo and X. Hu, *J. Med. Plants Res.*, 2010, **4**(12), 1222.
138. X. H. Chen, L. X. Xia, H. B. Zhou and G. Z. Qiu, *J. Agric. Food Chem.*, 2010, **58**, 6966.
139. S. A. Heleno, L. Barros, M. J. Sousa, A. Martins and I. C. F. R. Ferreira, *Microchem. J.*, 2009, **93**, 195.
140. J. H. Xiao, D. M. Xiao, Z. H. Sun, Q. Xiong, Z. Q. Liang and J. J. Zhong, *J. Food, Agric. Environ.*, 2009, **7**(3&4), 91.
141. B. Ribeiro, P. G. de Pinho, P. B. Andrade, P. Baptista and P. Valentao, *Microchem. J.*, 2009, **93**, 29.
142. R. Vidrih, S. Filip and J. Hribar, *Czech J. Food Sci.*, 2009, **27**, 125.
143. P. Kakumyan and K. Matsui, *Biosci. Biotechnol. Biochem.*, 2009, **73**(12), 2742.
144. L. Barros, T. Cruz, P. Baptista, L. M. Estevinho and I. C. F. R. Ferreira, *Food Chem. Toxicol.*, 2008, **46**, 2742.
145. L. Barros, B. A. Venturini, P. Baptista, L. M. Estevinho and I. C. F. R. Ferreira, *J. Agric. Food Chem.*, 2008, **56**, 3856.
146. H. Hur, *Mycobiology*, 2008, **36**(4), 233.
147. S. Saiqa, N. Haq, B. Nawaz, A. H. Muhammad and M. A. A. Ata-ur-Rehman, *Iran. J. Chem. Chem. Eng.*, 2008, **27**, 151.
148. V. Molnar, F. Billes, E. Tyihak and P. G. Ott, *J. Planar Chromatogr.*, 2008, **21**(6), 423.
149. S. Kavishree, J. Hemavathy, B. R. Lokesh, M. N. Shashirekha and S. Rajarathnam, *Food Chem.*, 2008, **106**, 597.
150. K. Pedneault, P. Angers, A. Gosselin and R. J. Tweddell, *Mycol. Res.*, 2008, **112**, 1428.
151. L. O. Hanus, I. Shkrob and V. M. Dembitsky, *J. Food Lipids*, 2008, **15**, 370.
152. D. Nedelcheva, D. Antonova, S. Tsvetkova, I. Marekov, S. Momchilova, B. Nikolova-Damyanova and M. Gyosheva, *J. Liq. Chromatogr. Relat. Technol.*, 2007, **30**, 2717.
153. L. Barros, P. Baptista, D. M. Correia, S. Casal, B. Oliveira and I. C. F. R. Ferreira, *Food Chem.*, 2007, **105**, 140.
154. L. Barros, P. Baptista, L. M. Estevinho and I. C. F. R. Ferreira, *J. Agric. Food Chem.*, 2007, **55**, 8766.
155. I. P. Hong, S. H. Nam, G. B. Sung, I. M. Chung, H. Hur, M. W. Lee, M. K. Kim and S. X. Guo, *Mycobiology*, 2007, **35**(4), 215.
156. C. C. Peng, K. C. Chen, R. Y. Peng, C. C. Chyau, C. H. Su and H. M. Hsieh-Li, *J. Ethnopharmacol.*, 2007, **109**, 93.

157. K. Pedneault, P. Angers, A. Gosselin and R. J. Tweddell, *Mycol. Res.*, 2006, **110**, 1179.
158. N. Yilmaz, M. Solmaz, İ. Türkekul and M. Elmastaş, *Food Chem.*, 2006, **99**, 168.
159. G. Kansci, D. C. Mossebo, A. B. Selatsa and M. Fotso, *Nahrung/Food*, 2003, **47**, 213.
160. V. A. Diez and A. Alvarez, *Food Chem.*, 2001, **75**, 417.
161. T. Longvah and Y. G. Deosthale, *Food Chem.*, 1998, **63**, 331.
162. M. F. Leon-Guzman, I. Silva and M. G. Lopez, *J. Agric. Food Chem.*, 1997, **45**, 4329.
163. F. Senatore, *J. Agric. Food Chem.*, 1992, **58**, 499.
164. F. Senatore, *J. Agric. Food Chem.*, 1990, **51**, 91.
165. N. R. Glick and M. H. Fischer, *J. Evid. Complementary Altern. Med.*, 2013, **18**(4), 268.
166. M. He, C. X. Qin, X. Wang and N. Z. Ding, *Front. Plant Sci.*, 2020, **11**, 390.
167. S. M. Innis, *Crit. Rev. Food Sci. Nutr.*, 2016, **56**, 1952.
168. J. Whelan and K. Fritsche, *Adv. Nutr.*, 2013, **4**(3), 311.
169. W. S. Harris, D. Mozaffarian, E. Rimm, P. Kris-Etherton, L. L. Rudel, L. J. Appel, M. M. Engler, M. B. Engler and F. Sacks, *Circulation*, 2009, **119**, 902.
170. J. V. Zhao and C. M. Schooling, *BMC Med.*, 2019, **17**, 61.
171. Y. Mahendran, J. Agren, M. Uusitupa, H. Cederberg, J. Vangipurapu, A. Stancakova, U. Schwab, J. Kuusisto and M. Laakso, *Am. J. Clin. Nutr.*, 2014, **99**, 79.
172. M. Asp, A. L. Collene, L. E. Norris, R. M. Cole, M. B. Stout, S. Y. Tang, J. C. Hsu and M. A. Belury, *Clin. Nutr.*, 2011, **30**, 443.
173. L. E. Norris, A. L. Collene, M. L. Asp, J. C. Hsu, L. F. Liu, J. R. Richardson, D. Li, D. Bell, K. Osei, R. D. Jackson and M. A. Belury, *Am. J. Clin. Nutr.*, 2009, **90**(3), 468.
174. M. A. Belury, R. M. Cole, D. B. Snoke, T. Banhb and A. Angelotti, *Prostaglandins, Leukotrienes Essent. Fatty Acids*, 2018, **132**, 30.
175. K. Miura, J. Stamler, H. Nakagawa, P. Elliott, H. Ueshima, Q. Chan, I. J. Brown, I. Tzoulaki, S. Saitoh, A. R. Dyer, M. L. Daviglus, H. Kesteloot, A. Okayama, J. D. Curb, B. L. Rodriguez, P. J. Elmer, L. M. Steffen, C. Robertson and L. Zhao, *Hypertension*, 2008, **52**, 408.
176. H. Bjermo, D. Iggman, J. Kullberg, I. Dahlman, L. Johansson, L. Persson, J. Berglund, K. Pulkki, S. Basu, M. Uusitupa, M. Rudling, P. Arner, T. Cederholm, H. Ahlström and U. Riserus, *Am. J. Clin. Nutr.*, 2012, **95**, 1003.
177. J. Monteiro, F. Askarian, M. T. Nakamura, M. H. Moghadasian and D. W. L. Ma, *Can. J. Physiol. Pharmacol.*, 2013, **91**, 469.
178. Y. H. Choi, *J. Cancer Prev.*, 2014, **19**, 31.
179. X. Llor, E. Pons, A. Roca, M. Aolvarez, J. Mane, F. Fernandez-Banares and M. A. Gassull, *Clin. Nutr.*, 2003, **22**(1), 71.

180. K. Kuriki, K. Wakai, K. Hirose, K. Matsuo, H. Ito, T. Suzuki, T. Saito, Y. Kanemitsu, T. Hirai, T. Kato, M. Tatematsu and K. Tajima, *Cancer Epidemiol., Biomarkers Prev.*, 2006, **15**, 1791.
181. D. E. Laaksonen, J. A. Laukkanen, L. Niskanen, K. Nyyssönen, T. H. Rissanen, S. Voutilainen, E. Pukkala, A. Hakkarainen and J. T. Salonen, *Int. J. Cancer*, 2004, **111**, 444.
182. P. L. Zock and M. B. Katan, *Am. J. Clin. Nutr.*, 1998, **68**, 142.
183. K. Rehman, K. Haider, K. Jabeen and M. S. H. Akash, *Rev. Endocr. Metab. Disord.*, 2020, **21**, 631.
184. M. Granado-Casas and D. Mauricio, Oleic Acid in the Diet and What It Does: Implications for Diabetes and Its Complications, in *Bioactive Food as Dietary Interventions for Diabetes*, 2nd edn, 2019, ch. 14.
185. C. Camell and C. W. Smith, *PLoS One*, 2013, **8**(9), e75147.
186. K. Rogers, D. Davis and D. Kurjiaka, *FASEB J.*, 2017, **31**, lb763.
187. E. Scoditti, M. Massaro, M. A. Carluccio, M. Pellegrino, M. Wabitsch, N. Calabriso, C. Storelli and R. De Caterina, *PLoS One*, 2015, **10**(6), e0128218.
188. I. M. Medeiros-de-Moraes, C. F. Gonçalves-de-Albuquerque, A. R. Kurz, F. M. Oliveira, V. H. P. Abreu, R. C. Torres, V. F. Carvalho, V. Estato, P. T. Bozza, M. Sperandio, H. C. de Castro-Faria-Neto and A. R. Silva, *Oxid. Med. Cell. Longevity*, 2018, **2018**, 13.
189. A. Nakbi, W. Tayeb, S. Dabbou, M. Issaoui, A. K. Grissa, N. Attia and M. Hammami, *Lipids Health Dis.*, 2010, **9**(1), 89.
190. S. Teres, G. Barcelo-Coblijn, M. Benet, R. Alvarez, R. Bressani, J. Halver and P. V. Escriba, *Proc. Natl. Acad. Sci. U. S. A.*, 2008, **105**(37), 13811.
191. B. T. Steffen, D. Duprez, M. Szklo, W. Guan and M. Y. Tsai, *J. Clin. Lipidol.*, 2018, **12**(6), 1404.
192. J. Song, Y. S. Kim, D. H. Lee, S. H. Lee, H. J. Park, D. Lee and H. Kim, *Sci. Rep.*, 2019, **9**, 10732.
193. Z. Amtul, D. Westaway, D. F. Cechetto and R. F. Rozmahel, *Brain Pathol.*, 2011, **21**, 321.
194. K. Sakurai, C. Shen, I. Shiraishi, N. Inamura and T. Hisatsune, *Nutrients*, 2021, **13**, 284.
195. C. Carrillo, M. M. Cavia and S. R. Alonso-Torre, *Nutr. Hosp.*, 2012, **27**(5), 1860.
196. J. A. Menendez, A. Papadimitropoulou, L. Vellon and R. A. Lupu, *Eur. J. Cancer*, 2006, **42**, 2425.
197. A. Liotti, V. Cosimato, P. Mirra, G. Cali, D. Conza, A. Secondo, G. Luongo, D. Terracciano, P. Formisano, F. Beguinot, L. Insabato and L. Ulianich, *J. Cell. Physiol.*, 2018, **233**, 7367.
198. H. Uehara, T. Takahashi, M. Oha, H. Ogawa and K. Izumi, *Int. J. Cancer*, 2014, **135**, 2558.
199. S. Rajaram, *Am. J. Clin. Nutr.*, 2014, **100**, 443.
200. N. Kaur, V. Chugh and A. K. Gupta, *J. Food Sci. Technol.*, 2014, **51**(10), 2289.

201. J. W. Fetterman and M. M. Zdanowicz, *Am. J. Health-Syst. Pharm.*, 2009, **66**, 1169.
202. X. Li, X. Bi, S. Wang, Z. Zhang, F. Li and A. Z. Zhao, *Front. Immunol.*, 2019, **10**, 2241.
203. H. Yue, B. Qiu, M. Jia, W. Liu, X. Guo, N. Li, Z. X. Xu, F. L. Du, T. Xu and D. Li, *Crit. Rev. Food Sci. Nutr.*, 2020, **9**, 1.
204. A. Pan, M. Chen, R. Chowdhury, J. H. Y. Wu, Q. Sun, H. Campos, D. Mozaffarian and F. B. Hu, *Am. J. Clin. Nutr.*, 2012, **96**(6), 1262.
205. L. Djousse, M. L. Biggs, R. N. Lemaitre, I. B. King, X. Song, J. H. Ix, K. J. Mukamal, D. S. Siscovick and D. Mozaffarian, *Am. J. Clin. Nutr.*, 2011, **94**, 527.
206. J. S. Zheng, X. J. Hu, Y. M. Zhao, J. Yang and D. Li, *BMJ*, 2013, **346**, 3706.
207. A. J. Cockbain, G. J. Toogood and M. A. Hull, *Gut*, 2012, **61**, 135.
208. C. J. Fabian, B. F. Kimler and S. D. Hursting, *Breast Cancer Res.*, 2015, **17**, 62.
209. G. Carta, E. Murru, S. Banni and C. Manca, *Front. Physiol.*, 2017, **8**(8), 902.
210. J. B. German, *Matern. Child Nutr.*, 2011, **2**, 2.
211. C. Agostoni, L. Moreno and R. Shamir, *Crit. Rev. Food Sci. Nutr.*, 2016, **56**(12), 1941.
212. H. Nakahara, S. Lee, Y. Shoyama and O. Shibata, *Soft Matter*, 2011, 7, 11351.
213. M. Zafaryab, K. U. Fakhri, M. A. Khan, K. Hajela and M. M. A. Rizvi, *Int. J. Life Sci. Res.*, 2019, **7**, 166.
214. D. Bai, Y. Wu, P. Deol, Y. Nobumori, Q. Zhou, F. M. Sladek and X. Liu, *Cancer Lett.*, 2021, **496**, 127.
215. L. Zeng, G. Z. Wu, K. J. Goh, Y. M. Lee, C. C. Ng, A. B. You, J. Wang, D. Jia, A. Hao, Q. Yu and B. Li, *PLoS One*, 2008, **3**(6), e2329.
216. A. Mancini, E. Imperlini, E. Nigro, C. Montagnese, A. Daniele, S. Orru and P. Buono, *Molecules*, 2015, **20**, 17339.
217. M. A. van Rooijen and R. P. Mensink, *Nutrients*, 2020, **12**, 615.
218. P. M. Kris-Etherton, A. E. Griel, T. L. Psota, S. K. Gebauer, J. Zhang and T. D. Etherton, *Lipids*, 2005, **40**, 1193.
219. D. Kritchevsky, *Am. J. Clin. Nutr.*, 1994, **60**, 997.
220. C. L. Schneider, R. L. Cowles, C. L. Stuefer-Powell and T. P. Carr, *J. Nutr.*, 2000, **130**(5), 1232.
221. P. W. Siri-Tarino, Q. Sun, F. B. Hu and R. M. Krauss, *Curr. Atheroscler. Rep.*, 2010, **12**, 384.
222. J. E. Hunter, J. Zhang and P. M. Kris-Etherton, *Am. J. Clin. Nutr.*, 2010, **91**, 46.
223. D. Senyilmaz-Tiebe, D. H. Pfaff, S. Virtue, K. V. Schwarz, T. Fleming, S. Altamura, M. U. Muckenthaler, J. G. Okun, A. Vidal-Puig, P. Nawroth and A. A. Teleman, *Nat. Commun.*, 2018, **9**, 3129.
224. Z. J. Wang, C. L. Liang, G. M. Li, C. Y. Yu and M. Yin, *Acta Pharm. Sin. B*, 2007, **28**(3), 315.

225. T. Kühn, A. Floegel, D. Sookthai, T. Johnson, U. Rolle-Kampczyk, W. Otto, M. von Bergen, H. Boeing and R. Kaaks, *BMC Med.*, 2016, **14**, 13.
226. L. M. Evans, S. L. Cowey, G. P. Siegal and R. W. Hardy, *Nutr. Cancer*, 2009, **61**(5), 746.
227. F. L. Crowe, P. N. Appleby, R. C. Travis, M. Barnett, T. M. Brasky, H. B. Bueno-de-Mesquita, V. Chajes, J. E. Chavarro, M. D. Chirlaque, D. R. English, R. A. Gibson, G. G. Giles, G. E. Goodman, S. M. Henning, R. Kaaks, I. B. King, L. N. Kolonel, A. R. Kristal, M. L. Neuhouser, S. Y. Park, G. Severi, A. Siddiq, M. J. Stampfer, P. Stattin, C. M. Tangen, A. Tjonneland, D. Trichopoulos, R. Tumino, L. R. Wilkens, T. J. Key and N. E. Allen, *J. Natl. Cancer Inst.*, 2014, **106**(9), dju240.
228. A. A. Khan, A. M. Alanazi, M. Jabeen, A. Chauhan and A. S. Abdelhameed, *Anticancer Res.*, 2013, **33**, 2517.
229. J. S. Beesley, W. P. Soutter and J. O. White, in *Effect of stearic acid on human cervical cancer cell growth*, ed. S. Nigam, K. V. Honn, L. J. Marnett and T. L. Walden, Springer, Boston, MA, 1993, pp. 573–575.
230. N. Kaur, V. Chugh and A. K. Gupta, *J. Food Sci. Technol.*, 2014, **51**(10), 2289.
231. V. Chajes and P. Bougnoux, *World Rev. Nutr. Diet.*, 2003, **92**, 133.
232. H. M. So, S. Lee, K. H. Baek, H. S. Roh, S. Kim, M. S. Jo, S. C. Baek, S. Seok, R. Ryoo and K. H. Kim, *Nat. Prod. Res.*, 2021, **35**(4), 649.
233. Q. X. Yan, M. X. Huang, P. Sun, S. X. Cheng, Q. Zhang and H. Dai, *Biochem. Syst. Ecol.*, 2020, **88**, 103963.
234. S. R. Lee, S. A. Yi, K. H. Nam, R. Ryoo, J. Lee and K. H. Kim, *J. Nat. Prod.*, 2019, **82**(12), 3489.
235. Y. Oda, K. Saito and M. Nakata, *BioSci. Trends*, 2019, **13**(1), 86.
236. C. Sillapachaiyaporn, S. Nilkhet, A. T. Ung and S. Chuchawankul, *BMC Complementary Altern. Med.*, 2019, **19**, 351.
237. S. Lee, D. Lee, J. C. Lee, K. S. Kang, R. Ryoo, H. J. Park and K. H. Kim, *Chem. Biodivers.*, 2018, **15**(9), e1800203.
238. S. R. Lee, H. S. Roh, S. Lee, H. B. Park, T. S. Jang, Y. J. Ko, K. H. Baek and K. H. Kim, *J. Funct. Foods*, 2018, **40**, 249.
239. M. F. Tala, J. Qin, J. T. Ndongo and H. Laatsch, *Nat. Prod. Bioprospect.*, 2017, **7**, 269.
240. V. Ivanova, M. Kolarova, K. Aleksieva, R. Schlegel, P. Schumann and U. Graefe, *J. Mod. Med. Chem.*, 2013, **1**, 43.
241. S. S. Hong, J. H. Lee, W. Jeong, N. Kim, H. Z. Jin, B. Y. Hwang, H. J. Lee, S. J. Lee, D. S. Jang and D. Lee, *Bioorg. Med. Chem. Lett.*, 2012, **22**, 2347.
242. K. H. Kim, S. U. Choi and K. R. Lee, *Lipids*, 2012, **47**, 593.
243. K. Yoshikawa, N. Kokudo, T. Hashimoto, K. Yamamoto, T. Inose and T. Kimura, *Biol. Pharm. Bull.*, 2010, **33**(8), 1355.
244. J. H. Choi, K. Maeda, K. Nagai, E. Harada, M. Kawade, H. Hirai and H. Kawagishi, *Org. Lett.*, 2010, **12**, 5012.
245. T. Hasegawa, M. Ishibashi, T. Takata, F. Takano and T. Ohta, *Chem. Pharm. Bull.*, 2007, **55**(12), 1748.

246. Y. Amakura, K. Kondo, H. Akiyama, H. Ito, T. Hatano, T. Yoshida and T. Maitani, *Food Hyg. Saf. Sci.*, 2006, **47**, 178.
247. G. Gilardoni, M. Clericuzio, A. Marchetti, P. V. Finzi, G. Zanoni and G. Vidari, *Nat. Prod. Commun.*, 2006, **1**, 1079.
248. D. N. Quang, T. Hashimoto, M. Toyota and Y. Asakawa, *J. Nat. Prod.*, 2003, **66**, 1613.
249. Y. Zhang, G. L. Mills and M. G. Nair, *Phytomedicine*, 2003, **10**, 386.
250. D. N. Quang, T. Hashimoto, M. Stadler and Y. Asakawa, *J. Nat. Prod.*, 2004, **67**, 1152.
251. H. Kawagishi, M. Ando, T. Mizuno, H. Yokota and S. Konishi, *Agric. Biol. Chem.*, 1990, **54**(5), 1329.
252. J. Nunez-Alarcon, J. C. C. Paredes, M. T. Carmona and M. Quinones, *Bol. Soc. Chil. Quim.*, 2001, **46**, 227.
253. T. T. Kiet, H. Dörfelt, M. Ritzau, S. Heinze and U. Grafe, *J. Basic Microbiol.*, 1999, **39**, 25.
254. H. Anke, P. Morales and O. Sterner, *Planta Med.*, 1996, **62**, 181.
255. B. T. Simon, T. Anke and O. Sterner, *Phytochemistry*, 1994, **34**, 815.
256. Z. Pang, H. Anke and O. Sterner, *Acta Chem. Scand.*, 1994, **48**, 408.
257. Z. Pang and O. Sterner, *J. Org. Chem.*, 1991, **56**, 1233.
258. H. Kawagishi, M. Ando, T. Mizuno, H. Yokota and S. Konishi, *Agric. Biol. Chem.*, 1990, **54**(5), 1329.

CHAPTER 4

Bioactive Phenolic Compounds from Mushrooms

MIHAI BABOTĂ, OLEG FRUMUZACHI, ALEXANDRU NICOLESCU, IRINA IELCIU, RAMONA PĂLTINEAN, GIANINA CRIŞAN AND ANDREI MOCAN*

Department of Pharmaceutical Botany, "Iuliu Haţieganu" University of Medicine and Pharmacy, Gheorghe Marinescu street 23, 400337 Cluj-Napoca, Romania
*Email: mocan.andrei@umfcluj.ro

4.1 Introduction

Based on the huge amount of research published in recent years, mushrooms have been increasingly promoted as a promising alternative source of micro- and macronutrients with high nutritional value, but also as functional foods and medicinal species. This fact is supported by the fast demographic growth which has led to increased demand for alternative food sources. Mushrooms ensure the main dietetic requirements are met and present certain economic advantages in comparison with conventional animal and plant sources.[1] Mushrooms are also attractive alternatives for the development of new foods and bioactive compounds; their growth process capitalizes on the use of waste from the wood, paper and agricultural industries, with a positive impact from environmental and economical perspectives.[2] Beyond their well-established food and medicinal properties, fungal species still gain scientists' interest through their wide diversity of secondary metabolites

Food Chemistry, Function and Analysis No. 36
Edible Fungi: Chemical Composition, Nutrition and Health Effects
Edited by Dejan Stojković and Lillian Barros

Published by the Royal Society of Chemistry, www.rsc.org

belonging to different groups of bioactive compounds. Among them, polyphenols are one of the most studied, and their importance is directly linked with the health-related benefits of mushroom consumption.[3,4] Thousands of studies published in the last decades have assessed and confirmed the presence of different types of polyphenols in the extracts obtained from mushrooms. The link between phenolic-rich fractions of this species and their antitumor, hypoglycemic, immunostimulatory, neuroprotective, anti-inflammatory, cytotoxic, antihyperlipidemic, antiviral, hypotensive, anti-diabetic, antioxidant, and antibacterial potentials has been intensively studied and proven through *in vitro* and *in vivo* assays.[2,5–7] In this chapter, we aim to present an up-to-date, brief overview regarding the phenolic composition of the most commonly used mushrooms for food or medicinal purposes.

4.2 Overview of the Main Classes of Phenolic Compounds Found in Mushrooms

Among the organic secondary metabolites of mushrooms, phenolic compounds are one of the most representative and studied constituents. Their structural diversity in fungi regnum is reduced in comparison with plants.[8] The main criterion used in the classification of these compounds is the structural one. According to Harborne *et al.*,[9,10] taking into account the structural complexity of phenolic compounds, they can be divided into three main categories, namely C_6 compounds (which contain only one aromatic ring in their molecule, being named generically *simple phenols*), C_6–C_n compounds (which have attached to the main aromatic ring another chain containing a maximum of seven carbon atoms), and C_6–C_n–C_6 compounds (comprising compounds with two aromatic rings linked by a chain containing a varied number of carbon atoms, usually no more than seven). Each of these categories being sub-divided.

Regarding the C_6 compounds, they are rarely reported as secondary fungal metabolites. Zhuang *et al.*[11] confirmed through GC-MS the presence of several simple phenols in *Boletus* species (*B. aereus*, *B. edulis*, *B. rubellus*, *B. auripes*), their occurrence being correlated with their smoky taste (more prominent for *B. auripes*) after sensory evaluation.

Compounds belonging to the C_6–C_n class are, overall, the most cited types of phenolics found in mushrooms. They can be divided into three main categories based on the number of carbon atoms of the chain directly linked to the aromatic ring: C_6–C_1, namely *hydroxybenzoic acids*, C_6–C_2 described as *acetophenones* or *phenylacetic acids*, and C_6–C_3 known as *hydroxycinnamic acids* or *phenylpropanoids*.[12] All the above-mentioned compounds have a common biosynthetic origin, being formed from L-phenylalanine or L-tyrosine through the shikimic acid pathway.[13] Their occurrence in the different mushroom species is described in the next section. Figure 4.1. summarizes the structures of the most common phenolic compounds belonging to the C_6–C_n class.

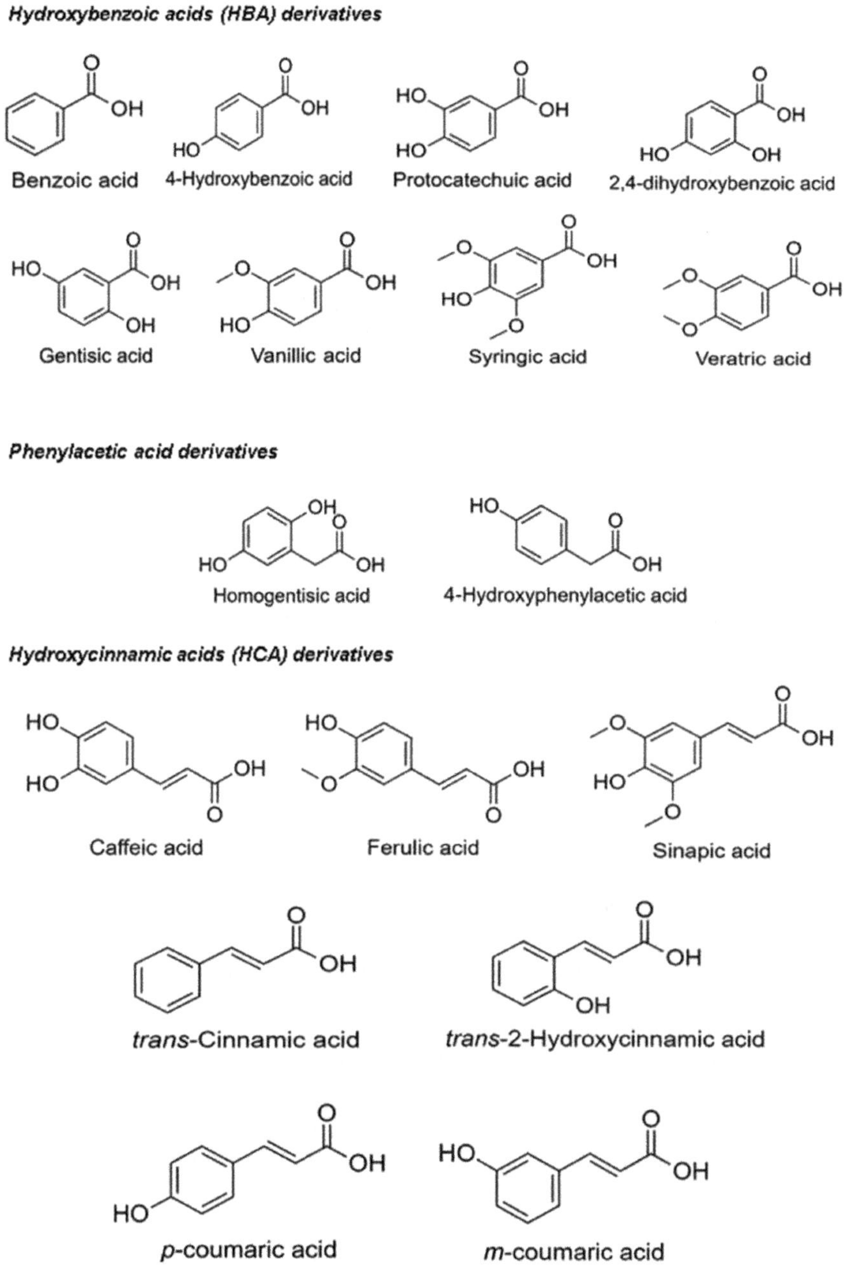

Figure 4.1 The structures of several of the most common phenolic compounds found in mushrooms.

The most heterogeneous category of phenolic compounds comprises the C_6–C_n–C_6 structural types, which include, among others, the huge class of flavonoids. The presence of flavonoids as fungal secondary metabolites is still considered controversial, with the scientific community's opinions on this topic being divergent. Although many earlier works have reported flavonoid compounds in mushrooms using modern analytical techniques (*i.e.*, high-performance liquid chromatography coupled with tandem mass spectrometry), several authors still contest this hypothesis. Gil-Ramírez *et al.*[8] published in 2016 a comprehensive study focused on showing the major error behind this theory, explaining that the metabolic pathways specific for flavonoid synthesis are incomplete due to the lack of the specific enzymes that assure this process in plants, namely chalcone synthase (CHS) and chalcone flavanone isomerase (CHI). Moreover, they showed that the absorption of flavonoids from neighboring plants or from those that establish symbiotic interactions *via* mycorrhizae is limited, as long as no significant absorption was noticed from fruiting bodies cultivated on flavonoid-enriched substrates or from mycelia grown on flavonoid-supplemented lab media. Conversely, T. K. Mohanta published in 2020 a work in which is highlighted the presence of gene/protein sequences associated with the biosynthesis of flavonoids in fungal species.[14] Considering all the factors above-mentioned, we consider that there remains a lack of knowledge on this hypothesis. That is the reason why we chose to limit our findings to the phenolic composition of mushrooms excluding all data about the flavonoids previously cited as constituents of this species.

4.3 Phenolic Composition of the Most Popular Mushrooms with Food and/or Medicinal Applications

4.3.1 Agaricaceae Family

The Agaricaceae family is comprised of more than 60 accepted genera, being considered the largest family of lamellate mushrooms.[15] Besides the well-known *Agaricus* species, this taxonomic subunit provides an important segment of edible mushrooms like *Calvatia*, *Coprinus*, *Leucoagaricus*, *Lycoperdon*, and *Macrolepiota* species. Regarding the phenolic profile of the different Agaricaceae, the scientific evidence highlights their fruiting bodies as the main sources of phenolic acids, but there are also reports about the presence of other non-specific compounds like coumarins.[16–18] Gąsecka *et al.*[19] investigated the phenolic profile of six *Agaricus* fruiting bodies (*A. arvensis*, *A. bisporus*, *A. bitorquis*, *A. brasiliensis*, *A, campestris*, and *A. silvaticus*), revealing the presence of nine phenolic acids differently distributed in each species. Gallic, caffeic, and ferulic acids were found in all samples, while *trans*-cinnamic acid (one of the most abundant phenolic compounds quantified in this study) was detected in all species except *A. campestris*. Interestingly, it was observed that wild species (*A. bitorquis*, *A, campestris*, and *A. silvaticus*) were found to be poorest in phenolic acids in comparison with the cultivated ones. Several other studies have revealed a high content of phenolic acids in *A. bisporus*, one of the most popular members of the Agaricaceae family.[20–30] Moreover, in a comparative study on edible *Agaricus* mushrooms, Glamočlija *et al.*[25] reported the presence of cinnamic acid in *A. bitorquis* and *A. macrosporus*, two less-studied species of this genus.

From *Calvatia* genus, *C. nipponica* is one of the recently reported species with a promising angiogenetic inhibitory potential. This activity was correlated with the presence of several alkaloids and phenolic compounds in this mushroom extract.[31] *Leucoagaricus leucothites* is another popular mushroom belonging to the Agaricaceae family. According to the literature, this mushroom has a higher content of phenolic acids, namely *p*-coumaric, protocatechuic, *p*-hydroxybenzoic, 2,4-dihydroxybenzoic, ellagic, and *trans*-cinnamic acids, as identified by Çayan *et al.*[32] Similarly, the phenolic profile was described for three *Lycoperdon* species (*L. molle*, *L. perlatum*, *L. utriforme*), confirming the presence of sinapic acid, vanillic acid isomers, and *p*-hydroxybenzoic acid.[33–35] *Macrolepiota* is another popular group of wild-growing edible mushrooms described as a fungal species with an important nutritional value. *M. procera*, *M. mastoidea*, and *L. rhacodes* were also cited for their phenolic acids (*p*-hydroxybenzoic, *p*-coumaric, and cinnamic acids) content.[32,36]

Concerning Agaricaceae, *Coprinus* genus is another important genus that comprises several edible species studied for their composition in bioactive phenolic compounds. Scientific evidence supports the taxonomic link between *Coprinus* and *Coprinellus* species, since several authors consider their names as synonyms.[16] In a recent study, Stilinović *et al.*[17] investigated the chemical profile of cultivated *C. comatus*, one of the most representative taxons of *Coprinus* genus, showing that *p*-hydroxybenzoic, protocatechuic, cinnamic, *p*-coumaric, caffeic, and quinic acids are the main constituents of the 70% hydroethanolic extracts obtained from the fruiting bodies of this species. Similarly, Karaman *et al.*[37] and Vamanu *et al.*[38] published similar reports, highlighting the presence of gallic and homogentisic acids as phenolic secondary metabolites of *C. comatus*. There are comparative studies cited between *C. comatus* and *Coprinellus truncorum* which revealed a variable distribution of phenolic acids in the fruiting bodies of these species, depending on their harvesting location.[16,18] Moreover, after comprehensive LC-MS/MS screening of phenolic compounds from the two above-mentioned species, Karaman *et al.*[18] identified two coumarins (esculetin and umbelliferone), reported also by Stilinović *et al.*[17] in *Coprinellus truncorum*. An overview regarding the phenolic profile of several Agaricaceae species is presented in Table 4.1.

4.3.2 Boletaceae Family

Boletus genus comprises the group of mushrooms also known as *porcini*, popular species extensively used and appreciated for their culinary applications.[39] Recent studies regarding the chemical profile of different *Boletus*

Table 4.1 Phenolic compounds from several Agaricaceae species.

Species	Phenolic compounds	Ref.
Agaricus arvensis	Gallic acid, caffeic acid, *p*-coumaric acid, ferulic acid, chlorogenic acid, *trans*-cinnamic acid	1
A. bisporus	Gallic acid, caffeic acid, *p*-hydroxybenzoic acid, *p*-coumaric acid, ferulic acid, chlorogenic acid, *trans*-cinnamic acid, protocatechuic acid	1
A. bisporus	*p*-Hydroxybenzoic acid	2
A. bisporus	Gallic acid, protocatechuic acid, *p*-hydroxybenzoic acid, syringic acid, cinnamic acid.	3
A. bisporus	Chlorogenic acid, syringic acid, *p*-coumaric acid, ferulic acid, 2-hydroxycinnamic acid, protocatechuic acid, abscisic acid, *trans*-cinnamic acid	4
A. bisporus	Gallic acid, *p*-coumaric acid, cinnamic acid	5
A. bisporus	Gallic acid, protocatechuic acid, gentisic acid, *p*-hydroxybenzoic acid, chlorogenic acid, vanillic acid, *p*-coumaric acid, vanillic acid isomer, cinnamic acid, ferulic acid	6
A. bisporus	Cinnamic acid	7 and 8
A. bisporus	Gallic acid, chlorogenic acid, vanillic acid, caffeic acid, *p*-coumaric acid	9
A. bisporus	Protocatechuic acid, *p*-hydroxybenzoic acid, syringic acid	10
A. bisporus	4-Hydroxybenzoic acid, 2,4-dihydroxybenzoic acid, 4-hydroxyphenylacetic acid, protocatechuic acid, cinnamic acid, 4-feruloylquinic acid, 5-feruloylquinic acid, 3,5-dicaffeoylquinic acid	11
A. bisporus	Gallic acid, caffeic acid, 3,4-dihydroxybenzoic acid, vanillic acid, *p*-coumaric acid, *trans*-cinnamic acid	12
A. bitorquis	Gallic acid, caffeic acid, *p*-coumaric acid, ferulic acid, chlorogenic acid, *trans*-cinnamic acid	1
A. bitorquis	*p*-Hydroxybenzoic acid, cinnamic acid	5
A. brasiliensis	Gallic acid, caffeic acid, *p*-hydroxybenzoic acid, *p*-coumaric acid, ferulic acid, chlorogenic acid, syringic acid, *trans*-cinnamic acid, protocatechuic acid	1
A. campestris	Gallic acid, caffeic acid, *p*-hydroxybenzoic acid, *p*-coumaric acid, ferulic acid	1
A. campestris	Protocatechuic acid, *p*-hydroxybenzoic acid, *p*-coumaric acid, cinnamic acid	5
A. macrosporus	Cinnamic acid	5
A. silvaticus	Gallic acid, caffeic acid, *p*-hydroxybenzoic acid, *p*-coumaric acid, ferulic acid, syringic acid, *trans*-cinnamic acid	1
Calvatia nipponica	5,7-Dihydroxy-isobenzofuran-1(3*H*)-one, 5-(hydroxymethyl)-1,3-benzenediol	13

Table 4.1 (*Continued*)

Species	Phenolic compounds	Ref.
Coprinus comatus	*p*-Hydroxybenzoic acid, protocatechuic acid, vanillic acid, gallic acid, gentisic acid, syringic acid, cinnamic acid, *p*-coumaric acid, *o*-coumaric acid, caffeic acid, ferulic acid	14
C. comatus	*p*-Hydroxybenzoic acid, protocatechuic acid, cinnamic acid, *p*-coumaric acid, caffeic acid, quinic acid	15
C. comatus	*p*-Hydroxybenzoic acid, protocatechuic acid, vanillic acid, gallic acid, gentisic acid, esculetin, umbelliferone	16
C. comatus	Gallic acid, homogentisic acid, quinic acid	17 and 18
Coprinellus truncorum	*p*-Hydroxybenzoic acid, protocatechuic acid, vanillic acid, gallic acid, gentisic acid, *p*-coumaric acid, *o*-coumaric acid, caffeic acid, esculetin, scopoletin, umbelliferone	15
C. truncorum	*p*-Hydroxybenzoic acid, protocatechuic acid, vanillic acid, gallic acid, gentisic acid, cinnamic acid, *p*-coumaric acid, *o*-coumaric acid, caffeic acid, esculetin	16
Leucoagaricus leucothites	*p*-Coumaric acid, protocatechuic acid, *p*-hydroxybenzoic acid, 2,4-dihydroxybenzoic acid, *trans*-cinnamic acid	19
Lycoperdon molle	*p*-Hydroxybenzoic acid, vanillic acid isomers	20
Lycoperdon perlatum	Cinnamic acid, protocatechuic acid	20
Lycoperdon utriforme	Protocatechuic acid, gallic acid, chlorogenic acid	21
Macrolepiota procera	*p*-Hydroxybenzoic acid, *p*-coumaric acid, gallic acid, protocatechuic acid, caffeic acid, *trans*-2-hydroxycinnamic acid, rosmarinic acid	22 and 23
M. mastoidea	*p*-Hydroxybenzoic acid, *p*-coumaric acid	22
M. rhacodes	*p*-Hydroxybenzoic acid, *p*-coumaric acid	22

species have proven the diversity of phenolic compounds found in the fruiting bodies of these species, their distribution being responsible for the organoleptic properties and biological activities of these mushrooms.[39–41] *B. edulis* is one of the most studied mushrooms of this group, reflecting its popularity as an edible species and wide distribution around the world. The latest findings on its phenolic profile have revealed the presence of hydroxybenzoic acid derivatives (*i.e.* *p*-hydroxybenzoic, 2,4-dihydroxybenzoic, protocatechuic, vanillic, and syringic acids) and hydroxycinnamic acid derivatives (*i.e.* *o*-coumaric, *p*-coumaric, ferulic, and caffeic acids) as the main constituents of this species.[27–30,39,40] A similar phenolic composition was reported for *B. aereus*, another edible mushroom belonging to this genus.[27,39]

In a comparative study focused on the evaluation of the flavor profile of five *Boletus* mushrooms, Zhuang *et al.*[39] confirmed the presence of several methylated and methoxylated phenol derivatives (*i.e.* 2,3-dimethylphenol, 2,4-dimethylphenol, 2-methoxy-4-methylphenol) in dried fruiting bodies of *B. edulis*, *B. aereus*, *B. auripes Pk* and *B. rubellus Krombh*, their occurrence being correlated with significant aroma differences for the analyzed samples. Additionally, individual phenolic acid profiles were evaluated by Taofiq *et al.*[27] for *B. flagrans*, *B. impolitus*, and *B. reticulatus*. In Table 4.2 the phenolic composition for some of the most popular *Boletus* species are summarized.

4.3.3 Cantharellaceae Family

The Chanterelle group comprises mushrooms belonging to the *Chantarellus* genus, widely recognized as one of the most desirable edible mushrooms.[42] Among this species, *C. cibarius* has been extensively studied for its bioactive potential, which was correlated with the presence of phenolic acids in the fruiting bodies of this mushroom. Kała *et al.*[20,29] and Fogarasi *et al.*[30,40] analyzed the phenolic composition of *C. cibarius* collected from Romania and Poland, emphasizing the presence of various phenolic acids, both as simple or glycosylated forms (*i.e.* protocatechuic acid-4-*O*-glucoside, 1-*O*-galloyl-β-D-glucoside, *p*-hydroxybenzoic acid-4-*O*-glucoside). Khalili *et al.*[42] reported the presence of gallic, chlorogenic, caffeic, *p*-coumaric, and ferulic acids in ethyl acetate and methanolic extracts of *C. cibarius*. Due to the high content of phenolic acids, the extracts were further tested for their ability to chelate excessive iron ions in mice. The obtained results revealed preliminary perspectives on the use of these preparations as new natural alternatives to treat diseases correlated with iron overload. Another study conducted by Çol Ayvaz *et al.*[43] was focused on the evaluation of phenolic composition of *C. cibarius* collected from Ordu, Turkey. The phytochemical analysis confirmed the presence of pyrogallol (a simple phenol) and several phenolic acids (*p*-hydroxybenzoic, caffeic, *trans*-cinnamic, benzoic, gallic, homogentisic, and *p*-coumaric); their occurrences were correlated with *in vitro* anticholinesterase and anti-tyrosinase activities of this species.

Craterellus cornucopioides is another mushroom belonging to the Cantharellaceae family, also known for its food uses and bioactive properties. Previous reports about the phytochemical profile of this mushroom highlighted the presence of various types of chemical constituents. Among them, craterellins D and E (two specific sesquiterpenoids isolated for the first time from this species), polysaccharides, and several phenolic compounds were highlighted as the main ones.[44] In their recent work, Kosanić *et al.*[44] evaluated the phenolic acids profile of an acetone extract obtained from the fruiting bodies of. *C. cornucopioides*, confirming the presence of gallic, *p*-coumaric, chlorogenic, syringic, and ferulic acids. Moreover, Vasdekis *et al.*[45] analyzed the piceatannol (a stilbene derivative with significant antioxidant, anticancer, and anti-inflammatory activity)

Table 4.2 Phenolic compounds from several Boletaceae species.

Species	Phenolic compounds	Ref.
Boletus aereus	*p*-Hydroxybenzoic acid, protocatechuic acid	24
B. aereus	Phenol, *p*-cresol, 2,6-dimethoxyphenol, 2-methoxy-4-methylphenol	25
B. aereus	*p*-Hydroxybenzoic acid, *p*-coumaric acid, cinnamic acid	8
B. auripes	Phenol, 2,3-dimethylphenol, 2,4-dimethylphenol, 3-ethylphenol, 2,6-dimethoxyphenol, *o*-cresol, 2,5-dimethylphenol, 4-ethyl-2-methoxyphenol	25
B. edulis	2,4-Dihydroxybenzoic acid, gallic acid, protocatechuic acid 4-*O*-glucoside, *p*-hydroxybenzoic acid-4-*O*-glucoside, *p*-hydroxybenzoic acid, *trans*-cinnamic acid	26
B. edulis	Phenol, 2,4-dimethylphenol, *p*-cresol, 2,6-dimethoxyphenol	25
B. edulis	Cinnamic acid, *p*-hydroxybenzoic acid, syringic acid, gallic acid	8 and 10
B. edulis	4-Hydroxybenzoic acid, 2,4-dihydroxybenzoic acid, 4-hydroxyphenylacetic acid, protocatechuic acid, *p*-coumaric acid, sinapic acid, *o*-coumaric acid, cinnamic acid, 3,5-dicaffeoylquinic acid	11
B. edulis	Gallic acid, chlorogenic acid, vanillic acid, caffeic acid, *p*-coumaric acid, ferulic acid	9
B. edulis	Protocatechuic acid, *p*-coumaric acid, *p*-hydroxybenzoic acid	27
B. flagrans	Cinnamic acid	8
B. impolitus	*p*-hydroxybenzoic acid, *p*-coumaric acid, cinnamic acid	8
B. reticulatus	Cinnamic acid	8

content in methanol extracts of 29 different wild edible mushrooms widely consumed in north-western Greece, showing its presence in high quantities for chanterelles (*Cantharellus cibarius*, *C. cinereus*, and *Craterellus cornucopioides*).

4.3.4 Hymenochaetaceae Family

Species from this family are mainly recognized as arboreal fungi responsible for fungal-related diseases of different trees, but also as a source of remedies in folk medicine. The *Inonotus* and *Phellinus* genera were intensively studied as potential sources of bioactive molecules.[46,47] In a recent study, Wei *et al.*[48] identified in *I. obliquus* sclerotia two new phenolic compounds, respectively, (±)3-hydroxy-1-(4-hydroxy-3,5-dimethoxyphenyl)-3-(4-(3-hydroxypropanoyl)-2,6-dimethoxyphenoxy)propan-1-one and inoscavin A, proving a significant neuroprotective activity against human neuroblastoma SH-SY5Y for the second compound. Similarly, Kou *et al.*[49] reported the presence of three new phenolic derivatives with hispolone structure generically named inonophenols A, B, and C in fruiting bodies of cultivated edible *I. hispidus*; all three above-mentioned molecules showed important *in vitro* anti-inflammatory and neuroprotective activities. From the same species, Ren *et al.*[50] isolated several other phenolic derivatives (hispinine, hispolone, hispidine, pinillidine, (*E*)-4-(3,4-dihydroxyphenyl)but-3-en-2-one, 3,14′-bihispidinyl). Gallic, *p*-hydroxybenzoic, *p*-coumaric, *trans*-2-hydroxycinnamic, *trans*-cinnamic, and rosmarinic acids were found as the main phenolic constituents of *I. radiatus* collected from Turkey,[51] whereas for *I. sanghuang* a high content of chlorogenic acid was reported.[52]

On the other hand, Sułkowska-Ziaja *et al.*[53] developed an LC-DAD method for comparative evaluation of phenolic constituents of four *Phellinus* species: *P. igniarius* (containing gallic, protocatechuic, and syringic acids), *P. pini* (containing 3,4-dihydroxyphenylacetic and protocatechuic acids), *P. pomaceus* (containing gallic, protocatechuic, 3,4-dihydroxyphenylacetic, and syringic acids), and *P. robustus* (for which only gallic acid was detected). Suabjakyong *et al.*[54] evidenced the presence of some special types of phenolic compounds in *P.* igniarius [3,4-dihydroxybenzaldehyde, 4-(3,4-dihydroxyphenyl-3-buten-2one, inonoblin C, phelligridin D, inoscavin C, phelligridin C, and interfungin B]. The extracts obtained from this species demonstrated good capacity in the prevention of acrolein toxicity in a mouse neuroblastoma (Neuro-2a) cell line.

4.3.5 Lyophyllaceae Family

There have been reported studies regarding some edible mushrooms belonging to Lyophyllaceae, the most recent being focused on *Calocybe* and *Hypsizygus* species. Milky white mushroom (*Calocybe indica*) is widespread in eastern Asian countries and appreciated for its pleasant taste, especially for its similarities with the button mushroom (*Agaricus bisporus*).[55] Datta *et al.*[55] evaluated the chemical composition and antibacterial activity of a methanolic extract obtained from *C. indica* fruiting bodies, proving the presence of veratric acid glucuronide for this species. In another study, Alam *et al.*[56] identified through LC-DAD six phenolic acids (gallic, homogentisic, protocatechuic, chlorogenic, caffeic, and ferulic acids) in *C. indica*, their presence being correlated with the *in vitro* antityrosinase properties exerted by the extracts obtained from this mushroom. Moreover, the antiproliferative potential of *C. indica* was evaluated by Ghosh *et al.*,[57] showing the presence of significant amounts of cinnamic and gallic acids in tested extracts.

Hypsizygus species, recognized as part of the edible *shimeji* mushrooms group (popular in Japanese cuisine), were also studied for their phenolic content. Ukaegbu *et al.*[58] evaluated the chemical composition of brown and white varieties of *H. tessellatus* (popularly named *bunapi* and *buna shimeji*), showing that the brown variant contains pyrrogallic acid and ferroxidin. Some studies were also published on *H. marmoreus*, confirming the presence of gallic, protocatechuic, gentisic, and *p*-hydroxybenzoic acids as the main phenolic constituents for this species.[59,60]

4.3.6 Polyporaceae Family

This family comprises several well-established medicinal species, including *Fomes*, *Laetiporus*, *Polyporus*, and *Trametes*. Doğan *et al.*[61] evaluated the phenolic profile of the methanolic extract of *Fomes fomentarius*, showing the presence of phenolic acids (gallic, syringic, *p*-coumaric, benzoic, cinnamic, *p*-hydroxybenzoic, vanillic, and protocatechuic) as the main compounds. Additionally, ferulic and salicylic acids were identified in ethanolic extracts of the same species by Nowacka *et al.*,[62] whereas Bal *et al.*[63] reported the presence of chlorogenic and caffeic acids in a *F. fomentarius* ethanolic extract obtained through Soxhlet method extraction.

Laetiporus sulphureus is recognized as a valuable source of biomolecules responsible for the immunomodulatory and antitumoral properties of this mushroom. The main phenolic constituents reported for this species were gallic, chlorogenic, caffeic, *p*-coumaric, *p*-hydroxybenzoic, salicylic, and protocatechuic acids.[64,65] Regarding the *Polyporus* species, Mocan *et al.*[66] and Fernandes *et al.*[67] evaluated the phenolic composition of *P. squamosus* extracts, highlighting the presence of *p*-hydroxybenzoic, cinnamic, and *p*-coumaric acids.

Trametes is another important genus belonging to Polyporaceae, its species have been intensively studied in recent years. In a comparative study conducted by Tel-Çayan *et al.*[68] four *Trametes* species (*T. bicolor*, *T. pubescens*, *T. suaveolens*, and *T. versicolor*) were evaluated for their phenolic content, and antioxidant and anticholinesterase properties, with the most abundant compounds cited in this study being fumaric acid for *T. bicolor* and *trans*-cinnamic acid for *T. pubescens*. Other studies have proved the presence of several phenolic acids[69–71] (*p*-hydroxybenzoic, protocatechuic, gentisic, vanillic, gallic, syringic, *p*-coumaric, ferulic, caffeic) or coumarins[71] (*i.e.* esculetin, scopoletin, umbeliferone) in *T. versicolor* extracts. Protocatechuic, *p*-hydroxybenzoic, chlorogenic, caffeic, benzoic, gallic, and rosmarinic acids were reported for *T. hirsuta*,[72–74] whereas in *T. gibbosa* gallic, chlorogenic, syringic, protocatechuic, *p*-hydroxybenzoic, and caffeic acid were identified.[63,73]

4.3.7 Other Mushrooms Rich in Phenolic Compounds

Due to the considerable diversity of fungal species promoted as medicinal and/or edible mushrooms, it is difficult to maintain up-to-date evidence regarding the newest findings of their chemical composition. New compounds or new species are constantly being discovered and tested for different therapeutic potentials, broadening the knowledge in this field. Hence, in Table 4.3 the latest reports about the phenolic composition of other popular mushrooms used for medicinal and food purposes are summarized.

Table 4.3 Phenolic compounds from other popular edible or/and medicinal mushrooms.

Species	Phenolic compounds	Ref.
Albatrellus confluens	Grifolin, grifolic acid	28–30
Amanita caesarea	*p*-Hydroxybenzoic acid, cinnamic acid, *p*-coumaric acid, quinic acid	31
Armillaria mellea	Cinnamic acid, ferulic acid, gallic acid, *p*-coumaric acid, *p*-hydroxybenzoic acid, protocatechuic acid, syringic acid, vanillic acid	32
Auricularia polytricha	Gallic acid, vanillic acid, protocatechuic acid	10 and 33
Chlorophyllum agaricoides	Gallic acid, *trans*-cinnamic acid	34
Chroogomphus rutilus	Protocatechuic acid, *p*-hydroxyphenylacetic acid, *p*-hydroxybenzoic acid, 4-hydroxybenzaldehyde, (4-hydroxyphenyl) acetic acid, 3-(3,4-dihydroxyphenyl)-2-propenoic acid	35 and 36
Clitocybe nebularis	Ferulic acid, vanillic acid, caffeic acid, syringic acid, *trans*-cinnamic acid, ellagic acid, *p*-coumaric acid	37
C. subconnexe	Gallic acid, *p*-coumaric acid, protocatechuic acid, cinnamic acid.	38
Cordyceps bassiana	Bassiamide A, bassiamate	39
C. militaris	Caffeic acid, vanillic acid	40
C. sinensis	Salicylic acid, *p*-methoxyphenol, *p*-methoxybenzoic acid, 4-hydroxyacetophenone, vanillic acid, methyl-*p*-hydroxyphenylacetate, syringic acid, cetovanillone, *p*-hydroxyphenylacetic acid, *p*-hydroxybenzoic acid, protocatechuic acid	41
Fistulina antarctica, *F. endoxantha*	Gallic acid	42
Fomes fomentarius	Gallic acid, syringic acid, *p*-coumaric acid, benzoic acid, cinnamic acid, *p*-hydroxybenzoic acid, vanillic acid, rutin, protocatechuic acid	43–45
Fomitopsis betulina	Syringic acid, gallic acid, 5-hydroxybenzoic acid, 3,4-dihydroxyphenylacetic acid	46 and 47
Ganoderma adspersum	Caffeic acid, 2,4-dihydroxybenzoic acid, ferulic acid, coumarin, ellagic acid, rosmarinic acid	48
G. applanatum	*p*-Hydroxybenzoic acid, protocatechuic acid, *p*-coumaric acid, vanillic acid, gallic acid, aesculetin, caffeic acid, quinic acid, syringic acid	49–51
G. carnosum	Gallic acid, protocatechuic acid, 3,4-dihydroxyphenylacetic acid, 2,5-dihydroxybenzoic acid, 4-hydroxybenzoic acid, vanillic acid, 3-hydroxybenzoic acid, chlorogenic acid, ferulic acid, rosmarinic acid	52
G. cochlear	Cochlearols N–Y, ganocochlearins B and C, cochlearoids F–K, cochlearoids N–P	53–55

G. duripora	Ganoduriporols A and B	56
G. lingzhi	Gallic acid, protocatechuic acid, syringic acid	44
G. lipsiense	4-Hydroxymethyl benzoic acid, *p*-anisic acid, salicylic acid, syringic acid	57
G. lucidum	Gallic acid, *trans*-cinnamic acid, *p*-hydroxybenzoic acid, protocatechuic acid, *p*-coumaric acid, vanillic acid, aesculetin, caffeic acid, quinic acid, syringic acid, caffeic acid 3-glucoside, ellagic acid, ferulic acid 4-*O*-glucuronide, chlorogenic acid, lingzhifuran A, lingzhilactones D–F, chizhines A–F	49, 51 and 58–64
G. pfeifferi	Gallic acid, protocatechuic acid, 3,4-dihydroxyphenylacetic acid, 2,5-dihydroxybenzoic acid, 4-hydroxybenzoic acid, vanillic acid, 3-hydroxybenzoic acid, syringic acid	51, 52 and 65
G. sinense	Ganosinensols E–J, other farnesyl phenolic enantiomers	66
G. sinense	Ganosinensols A–D	67
G. resinaceum	Protocatechuic acid, *p*-hydroxybenzoic acid, chlorogenic acid, caffeic acid, benzoic acid, rosmarinic acid, apigenin, vanillic acid, gallic acid, aesculetin, caffeic acid, quinic acid, syringic acid	50, 51 and 65
Grifola gargal	Gallic acid, *p*-hydroxybenzoic acid	42
Helvella leucopus	Protocatechuic acid	68
Hericium erinaceus	Gallic acid, *p*-hydroxybenzoic acid, *p*-coumaric acid, chlorogenic acid, quinic acid, vanillic acid, *o*-aminophenol, *o*-hydroxybenzoic acid, *p*-, *m*-, *o*-toluic acid, ferulic acid, sinapic acid, *trans*-cinnamic acid, erinacerins C–L	62 and 69–75
Hydnum repandum	Gallic acid, *p*-coumaric acid, chlorogenic acid, syringic acid, ferulic acid, gentisic acid, protocatechuic acid, vanillic acid, cinnamic acid	76 and 77
Hygrocybe acutoconica	*p*-Hydroxybenzoic acid, *p*-coumaric acid, quinic acid	31
Hypholoma fasciculare	Rosmarinic acid, *trans*-2-hydroxycinnamic acid, *trans*-cinnamic acid, gallic acid, ferulic acid, ellagic acid	78
Hypomyces perniciosus	Syringic acid, vanillic acid, cinnamic acid, 2-hydroxycinnamic acid, ferulic acid, *p*-coumaric acid	79
H. lactiflorum	Protocatechuic acid dimer	80
Lactarius aurantiacus	Cinnamic acid	81
L. deliciosus	*p*-Hydroxybenzoic acid, caffeic acid, chlorogenic acid, ferulic acid, gallic acid, gentisic acid, *p*-hydroxybenzoic acid, homogentisic acid, protocatechuic acid, pyrogallol	23
L. indigo	*p*-Hydroxybenzoic acid, *o*-coumaric acid dimer, chlorogenic acid derivative	80
L. sanguifluus	*p*-Hydroxybenzoic acid, gallic, protocatechuic, vanillic, syringic, *p*-coumaric, ferulic acid	32
Laetiporus sulphureus	Gallic acid, chlorogenic acid, caffeic acid, *p*-coumaric acid, *p*-hydroxybenzoic acid, salicylic acid, protocatechuic acid	23 and 82

Table 4.3 (*Continued*)

Species	Phenolic compounds	Ref.
Lentinula edodes	Protocatechuic acid, *p*-hydroxybenzoic acid, benzoic acid, caffeic acid, cinnamic acid, coumaric acid, ferulic acid, gallic acid, syringic acid, vanillic acid	83 and 84
Lepista nuda	Protocatechuic acid, *p*-hydroxybenzoic acid, *p*-coumaric acid, cinnamic acid, gallic acid, 2,4-dihydroxybenzoic acid, rosmarinic acid, *trans*-cinnamic acid, *trans*-2-hydroxy cinnamic acid	20 and 23
L. personata	Protocatechuic acid, gallic acid, *p*-hydroxybenzoic acid, *p*-coumaric, ferulic acid, rosmarinic acid, *trans*-cinnamic acid	23
L. sordida	Protocatechuic acid, chlorogenic acid, caffeic acid, gallic acid, vanillic acid	85
Leucopaxillus candidus	*p*-Hydroxybenzoic acid, *p*-coumaric acid	86
L. gentianeus	Gallic acid, chlorogenic acid, coumaric acid, cinnamic acid	87
L. tricolor	Protocatechuic acid, gallic acid, catechin hydrate, *p*-hydroxybenzoic acid, 2,4-dihydroxybenzoic acid, ellagic acid	23
Morchella anguisticeps	*p*-Hydroxybenzoic acid, *p*-coumaric acid	88
M. conica	Tannic acid, gallic acid, protocatechuic acid, *p*-coumaric acid	88
M. elata	Protocatechuic acid, gallic acid, catechin hydrate, *p*-hydroxybenzoic acid, caffeic acid, *p*-coumaric acid, ferulic acid, *trans*-cinnamic acid	23
M. esculenta	Protocatechuic acid, gallic acid, *p*-coumaric acid, ellagic acid, rosmarinic acid	23
M. pulchella	Caffeic acid	68
Mycena haematopus	Protocatechuic acid, cinnamic acid	81
Omphalotus olearius	Gallic, protocatechuic, ferulic acid, rosmarinic acid	89
Phaeolus schweinitzii	Gallic, protocatechuic, *p*-hydroxybenzoic acid, caffeic, *p*-coumaric, ferulic acid, *trans*-2-hydroxycinnamic acid	23
Pleurotus eryngii	Gallic acid, 3,4-dihydroxybenzoic acid, *p*-hydroxybenzoic acid, protocatechuic acid	84 and 90
P. ostreatus	Gallic acid, homogentisic acid, chlorogenic acid	84
Ramaria aurea	Caffeic acid, cinnamic acid, gallic acid	91 and 92
R. botrytis	Protocatechuic, *p*-hydroxybenzoic acid	92
R. largentii	Protocatechuic and vanillic acid	92

R. flava	Gallic acid, protocatechuic acid, *p*-coumaric acid, *trans*-cinnamic acid, caffeic acid hexoside, sinapic acid	23, 80 and 92
Russula aurora	Gallic acid, ellagic acid, rosmarinic acid, *trans*-cinnamic acid	23
R. azurea	Gallic acid, ellagic acid, rosmarinic acid, *trans*-cinnamic acid, *p*-coumaric acid, ferulic acid, 6,7-dihydroxycoumarin	23
R. chloroides	Gallic acid, ferulic acid	93
R. delica	Gallic acid, protocatechuic acid, *trans*-cinnamic acid, ferulic acid, salicylic acid, caffeic, *p*-hydroxybenzoic acid	23 and 94
R. lepida	Gallic acid, protocatechuic acid, *p*-hydroxybenzoic acid, cinnamic acid, benzoic acid	94
R. mustelina	Salicylic acid, protocatechuic acid, caffeic acid, *p*-hydroxybenzoic acid	94
R. vinosa	Gallic acid, protocatechuic acid, *trans*-cinnamic acid, *p*-coumaric acid	23
Sarcodon aspratus	Gallic acid, gentisic acid, 3,4-dihydroxybenzoic acid, vanillic acid, *p*-hydroxybenzoic acid	90
S. imbricatus	Protocatechuic, chlorogenic, gallic acid, *p*-hydroxybenzoic acid, syringic acid	80, 90 and 95
Schizophyllum commune	Gallic acid, ferulic acid, vanillic acid, *m*-hydroxybenzoic acid, *o*-hydroxybenzeneacetic acid, 3-hydroxy-5-methylbenzoic acid, 4-hydroxybenzoic acid	89 and 96
Sparassis crispa	Gallic acid, protocatechuic acid, *p*-hydroxybenzoic acid, vanillic acid, caffeic acid, syringic acid, *p*-coumaric acid, veratric acid (ether), pyrogallol	77 and 84
S. latifolia	4,6-Dihydroxy-3-methoxy-phthalic acid, methyl 2,4,6-trihydroxyphenylglyoxylate, carboxy-4-hydroxy-5-methoxybenzaldehyde, 5,6-dihydroxy-7-methoxy phthalide, 3-formylsalicylic acid	97
Suillus bellinii	Protocatechuic acid, *p*-hydroxybenzoic acid, cinnamic acid	98
S. bovis	Protocatechuic acid, *p*-hydroxybenzoic acid	10
S. luteus	Protocatechuic acid, *p*-hydroxybenzoic acid	10
Tricholoma matsutake	Gallic acid, protocatechuic acid, chlorogenic acid, gentisic acid, caffeic acid, vanillic acid, β-resorcylic acid, *p*-coumaric acid, sinapic acid, ferulic acid, veratric acid, *m*-coumaric acid, *o*-coumaric acid	99
T. virgatum	Gallic acid, chlorogenic acid, coumaric acid	100
T. terreum	*p*-Coumaric acid and gallic acid	101
T. equestre	Protocatechuic acid, *p*-hydroxybenzoic acid, syringic acid	10
Volvopluteus gloiocephalus	Gallic acid, *p*-hydroxybenzoic acid, *p*-coumaric acid, protocatechuic acid, cinnamic acid	38
Xerocomus badius	Protocatechuic acid, caffeic acid, vanillic acid, syringic acid, *p*-coumaric acid, *p*-hydroxybenzoic acid, cinnamic acid	27, 102 and 103

4.4 Conclusions

Phenolic compounds are an important class of secondary metabolites found in mushrooms. Their occurrence has been correlated with bioactive and nutraceutical potential demonstrated by these fungal species. The latest findings prove the presence of classic phenolic structures in mushrooms (*i.e.* hydroxybenzoic and hydroxycinnamic acid derivatives, chlorogenic acids, tannins, coumarins, *etc.*), emphasizing, at the same time, several new phenolic derivatives that are specific for these fungi. Due to all the above-mentioned, the research interest in this field is growing continuously, promoting new perspectives regarding the development of mushrooms as promising alternative sources for bioactive molecules.

Acknowledgements

This work was supported by a grant from the Romanian Ministry of Education and Research, CNCS–UEFISCDI, project number PN-III-P2-2.1-PED-2019-5360.

References

1. A. González, M. Cruz, C. Losoya, C. Nobre, A. Loredo and R. Rodríguez, *et al.*, Edible mushrooms as a novel protein source for functional foods, *Food Funct.*, 2020, **11**(9), 7400–7414.
2. H. Lu, H. Lou, J. Hu, Z. Liu and Q. Chen, Macrofungi: A review of cultivation strategies, bioactivity, and application of mushrooms, *Compr. Rev. Food Sci. Food Saf.*, 2020, **19**(5), 2333–2356.
3. S. A. Heleno, A. Martins, M. J. R. P. Queiroz and I. C. F. R. Ferreira, Bioactivity of phenolic acids: Metabolites versus parent compounds: A review, *Food Chem.*, 2015, **173**, 501–513.
4. O. Taofiq, A. M. González-Paramás, A. Martins, M. F. Barreiro and I. C. F. R. Ferreira, Mushrooms extracts and compounds in cosmetics, cosmeceuticals and nutricosmetics-A review, *Ind. Crops Prod.*, 2016, **90**, 38–48.
5. X. Luo, N. Li, J. Zhong, Z. Tan, Y. Liu and X. Dong, *et al.*, Grifolin inhibits tumor cells adhesion and migration *via* suppressing interplay between PGC1α and Fra-1/LSF-MMP2/CD44 axes, *Oncotarget*, 2016, **7**(42), 68708–68720.
6. F. S. Reis, A. Martins, M. H. Vasconcelos, P. Morales and I. C. F. R. Ferreira, Functional foods based on extracts or compounds derived from mushrooms, *Trends Food Sci. Technol.*, 2017, **66**, 48–62.
7. F. Souilem, â. Fernandes, R. C. Calhelha, J. C. M. Barreira, L. Barros and F. Skhiri, *et al.*, Wild mushrooms and their mycelia as sources of bioactive compounds: Antioxidant, anti-inflammatory and cytotoxic properties, *Food Chem.*, 2017, **230**, 40–48.
8. A. Gil-Ramírez, C. Pavo-Caballero, E. Baeza, N. Baenas, C. Garcia-Viguera and F. R. Marín, *et al.*, Mushrooms do not contain flavonoids, *J. Funct. Foods*, 2016, **25**, 1–13.
9. C. Evans and J. Harborne, Methods in plant biochemistry: 1. Plant phenolics, *Phytochem. Anal.*, 1991, 48.
10. D. Tsimogiannis and V. Oreopoulou, Classification of Phenolic Compounds in Plants, in *Polyphenols in Plants*, 2019, pp. 263–284.
11. J. Zhuang, Q. Xiao, T. Feng, Q. Huang, C. T. Ho and S. Song, Comparative flavor profile analysis of four different varieties of Boletus mushrooms by instrumental and sensory techniques, *Food Res Int*, 2020, **136**, 109485.

12. M. M. Vuolo, V. S. Lima and M. R. Maróstica Junior, Phenolic Compounds: Structure, Classification, and Antioxidant Power, in *Bioactive Compounds: Health Benefits and Potential Applications*, 2018, pp. 33–50.
13. G. Williamson, The role of polyphenols in modern nutrition, *Nutr Bull*, 2017, **42**(3), 226–235.
14. T. K. Mohanta, Fungi contain genes associated with flavonoid biosynthesis pathway, *J. Funct. Foods*, 2020, **68**, 103910.
15. J. Kalichman, P. M. Kirk and P. B. Matheny, A compendium of generic names of agarics and Agaricales, *Taxon*, 2020, **69**(3), 425–447.
16. K. Tešanović, B. Pejin, F. Šibul, M. Matavulj, M. Rašeta and L. Janjušević, *et al.*, A comparative overview of antioxidative properties and phenolic profiles of different fungal origins: fruiting bodies and submerged cultures of Coprinus comatus and Coprinellus truncorum, *J. Food Sci. Technol.*, 2017, **54**(2), 430–438.
17. N. Stilinović, I. Čapo, S. Vukmirović, A. Rašković, A. Tomas and M. Popović, *et al.*, Chemical composition, nutritional profile and in vivo antioxidant properties of the cultivated mushroom Coprinus comatus: Comprimus comatus properties, *R. Soc. Open Sci.*, 2020, **7**(9), 200900.
18. M. Karaman, K. Atlagić, A. Novaković, F. Šibul, M. Živić and K. Stevanović, *et al.*, Fatty acids predominantly affect anti-hydroxyl radical activity and FRAP value: The case study of two edible mushrooms, *Antioxidants*, 2019, **8**(10), 1–12.
19. M. Gąsecka, Z. Magdziak, M. Siwulski and M. Mleczek, Profile of phenolic and organic acids, antioxidant properties and ergosterol content in cultivated and wild growing species of Agaricus, *Eur. Food Res. Technol.*, 2018, **244**(2), 259–268.
20. K. Kała, A. Krakowska, A. Szewczyk, B. Ostachowicz, K. Szczurek and A. Fijałkowska, *et al.*, Determining the amount of potentially bioavailable phenolic compounds and bioelements in edible mushroom mycelia of Agaricus bisporus, Cantharellus cibarius, and Lentinula edodes, *Food Chem.*, 2021, **352**, 129456.
21. B. Muszyńska, J. Piotrowska, A. Krakowska, A. Gruba, K. Kała and K. Sułkowska-Ziaja, *et al.*, Study of physiologically active components in different parts of fruiting bodies of varieties of Agaricus bisporus (white mushroom), *Eur. Food Res. Technol.*, 2017, **243**(12), 2135–2145.
22. M. Kakoti, D. J. Hazarika, A. Parveen, S. Dullah, A. Ghosh and D. Saha, *et al.*, Nutritional Properties, Antioxidant and Antihaemolytic Activities of the Dry Fruiting Bodies of Wild Edible Mushrooms Consumed by Ethnic Communities of Northeast India, *Pol. J. Food Nutr. Sci.*, 2021, **71**(4), 463–480.
23. E. A. Erdoğan Eliuz, Antibacterial activity and antibacterial mechanism of ethanol extracts of Lentinula edodes (Shiitake) and Agaricus bisporus (button mushroom), *Int. J. Environ. Health Res.*, 2021, 1–14.
24. S. M. Elhusseiny, T. S. El-Mahdy, M. F. Awad, N. S. Elleboudy, M. M. S. Farag and K. M. Aboshanab, *et al.*, Antiviral, cytotoxic, and antioxidant activities of three edible agaricomycetes mushrooms: Pleurotus columbinus, pleurotus sajor-caju, and agaricus bisporus, *J. Fungi*, 2021, 7(8), 645.
25. J. Glamočlija, D. Stojković, M. Nikolić, A. Ćirić, F. S. Reis and L. Barros, *et al.*, A comparative study on edible Agaricus mushrooms as functional foods, *Food Funct.*, 2015, **6**(6), 1900–1910.
26. F. A. de Souza Campos Junior, M. H. Petrarca, A. D. Meinhart, M. de Jesus Filho and H. T. Godoy, Multivariate optimization of extraction and validation of phenolic acids in edible mushrooms by capillary electrophoresis, *Food Res. Int.*, 2019, **126**, 108685.
27. O. Taofiq, R. C. Calhelha, S. Heleno, L. Barros, A. Martins and C. Santos-Buelga, *et al.*, The contribution of phenolic acids to the anti-inflammatory activity of mushrooms: Screening in phenolic extracts, individual parent molecules and synthesized glucuronated and methylated derivatives, *Food Res. Int.*, 2015, **76**, 821–827.

28. M. Alkin, E. Söğüt and A. C. Seydim, Determination of bioactive properties of different edible mushrooms from Turkey, *J. Food Meas. Charact.*, 2021, **15**(4), 3608–3617.
29. K. Kała, A. Krakowska, K. Sułkowska-Ziaja, A. Szewczyk, W. Reczyński and W. Opoka, *et al.*, Kinetics of extracted bioactive components from mushrooms in artificial digestive juices, *Int. J. Food Prop.*, 2017, **20**(8), 1796–1817.
30. M. Fogarasi, S. A. Socaci, F. V. Dulf, Z. M. Diaconeasa, A. C. Fărcas and M. Tofană, *et al.*, Bioactive compounds and volatile profiles of five transylvanian wild edible mushrooms, *Molecules*, 2018, **23**(12), 3272.
31. S. Lee, J. Y. Park, D. Lee, S. Seok, Y. J. Kwon and T. S. Jang, *et al.*, Chemical constituents from the rare mushroom Calvatia nipponica inhibit the promotion of angiogenesis in HUVECs, *Bioorg. Med. Chem. Lett.*, 2017, **27**(17), 4122–4127.
32. F. Çayan, G. Tel-Çayan, E. Deveci and M. E. Duru, A comprehensive study on phenolic compounds and bioactive properties of five mushroom species *via* chemometric approach, *J. Food Process. Preserv.*, 2021, **45**(9), e15695.
33. L. Barros, M. Dueñas, I. C. F. R. Ferreira, P. Baptista and C. Santos-Buelga, Phenolic acids determination by HPLC–DAD–ESI/MS in sixteen different Portuguese wild mushrooms species, *Food Chem. Toxicol.*, 2009, **47**(6), 1076–1079.
34. S. Sezgin, A. Dalar and Y. Uzun, Determination of antioxidant activities and chemical composition of sequential fractions of five edible mushrooms from Turkey, *J. Food Sci. Technol.*, 2020, **57**(5), 1866–1876.
35. E. M. Yahia, F. Gutiérrez-Orozco and M. A. Moreno-Pérez, Identification of phenolic compounds by liquid chromatography-mass spectrometry in seventeen species of wild mushrooms in Central Mexico and determination of their antioxidant activity and bioactive compounds, *Food Chem.*, 2017, **226**, 14–22.
36. A. Ćirić, I. Kruljević, D. Stojković, â. Fernandes, L. Barros and R. C. Calhelha, *et al.*, Comparative investigation on edible mushrooms Macrolepiota mastoidea, M. rhacodes and M. procera: functional foods with diverse biological activities, *Food Funct.*, 2019, **10**(12), 7678–7686.
37. M. Karaman, K. Tesanovic, S. Gorjanovic, F. T. Pastor, M. Simonovic and M. Glumac, *et al.*, Polarography as a technique of choice for the evaluation of total antioxidant activity: The case study of selected Coprinus Comatus extracts and quinic acid, their antidiabetic ingredient, *Nat. Prod. Res.*, 2021, **35**(10), 1711–1716.
38. E. Vamanu, Antioxidant components and antioxidant activities of hydroglyceroalcoholic extract from fresh mushroom mycelium, *Rev. Chim.*, 2017, **68**(3), 534–540.
39. J. Zhuang, Q. Xiao, T. Feng, Q. Huang, C. T. Ho and S. Song, Comparative flavor profile analysis of four different varieties of Boletus mushrooms by instrumental and sensory techniques, *Food Res. Int.*, 2020, **136**, 109485.
40. M. Fogarasi, M. I. Socaciu, C. D. Sălăgean, F. Ranga, A. C. Fărcaş and S. A. Socaci, *et al.*, Comparison of Different Extraction Solvents for Characterization of Antioxidant Potential and Polyphenolic Composition in Boletus edulis and Cantharellus cibarius Mushrooms from Romania, *Molecules*, 2021, **26**(24), 7508.
41. Q. Zheng, X. Li, T. Liu, Y. Zhang, J. Liu and H. Zhang, *et al.*, Effects of air-impingement jet drying on drying kinetics, color, polyphenol compounds, and antioxidant activities of Boletus aereus slices, *J. Food Sci.*, 2021, **86**(5), 2131–2144.
42. M. Khalili, M. A. Ebrahimzadeh, M. Kosaryan, A. Abbasi and M. Azadbakht, Iron chelation and liver disease healing activity of edible mushroom (Cantharellus cibarius), in vitro and in vivo assays, *RSC Adv.*, 2015, **5**(7), 4804–4810.

43. M. Çol Ayvaz, F. Aksu and F. Kır, Phenolic profile of three wild edible mushroom extracts from Ordu, Turkey and their antioxidant properties, enzyme inhibitory activities, *Br. Food J.*, 2019, **121**(6), 1248–1260.
44. M. Kosanić, B. Ranković, T. Stanojković, M. Radović-Jakovljević, A. Ćirić and D. Grujičić, *et al.*, Craterellus cornucopioides Edible Mushroom as Source of Biologically Active Compounds, *Nat. Prod. Commun.*, 2019, **14**(5), 1934578X19843610.
45. E. P. Vasdekis, A. Karkabounas, I. Giannakopoulos, D. Savvas and M. E. Lekka, Screening of mushrooms bioactivity: piceatannol was identified as a bioactive ingredient in the order Cantharellales, *Eur. Food Res. Technol.*, 2018, **244**(5), 861–871.
46. K. C. Duru, E. G. Kovaleva, I. G. Danilova and P. van der Bijl, The pharmacological potential and possible molecular mechanisms of action of Inonotus obliquus from preclinical studies, *Phytother. Res.*, 2019, **33**(8), 1966–1980.
47. P. He, Y. Zhang and N. Li, The phytochemistry and pharmacology of medicinal fungi of the genus: Phellinus: a review, *Food Funct.*, 2021, **12**(5), 1856–1881.
48. Y. M. Wei, L. Yang, W. L. Mei, H. Q. Chen, C. H. Cai and W. Li, *et al.*, Phenolic compounds from the sclerotia of Inonotus obliquus, *Nat. Prod. Res.*, 2020, 2413–2417.
49. R. W. Kou, S. T. Du, B. Xia, Q. Zhang, X. Yin and J. M. Gao, Phenolic and Steroidal Metabolites from the Cultivated Edible Inonotus hispidus Mushroom and Their Bioactivities, *J. Agric. Food Chem.*, 2021, **69**(2), 668–675.
50. Q. Ren, X. Y. Lu, J. X. Han, H. A. Aisa and T. Yuan, Triterpenoids and phenolics from the fruiting bodies of Inonotus hispidus and their activations of melanogenesis and tyrosinase. Chinese, *Chem. Lett.*, 2017, **28**(5), 1052–1056.
51. F. Çayan, G. Tel-Çayan, E. Deveci and M. E. Duru, HPLC–DAD characterization of phenolic profile and in vitro antioxidant, anticholinesterase, and antidiabetic activities of five mushroom species from Turkey, *3 Biotech*, 2021, **11**(6), 1–11.
52. K. Liu, X. Xiao, J. Wang, C. Y. O. Chen and H. Hu, Polyphenolic composition and antioxidant, antiproliferative, and antimicrobial activities of mushroom Inonotus sanghuang, *LWT–Food Sci. Technol.*, 2017, **82**, 154–161.
53. K. Sułkowska-Ziaja, A. Maślanka, A. Szewczyk and B. Muszyńska, Physiologically Active Compounds in Four Species of Phellinus, *Nat. Prod. Commun.*, 2017, **12**(3), 363–366.
54. P. Suabjakyong, R. Saiki, L. J. L. D. Van Griensven, K. Higashi, K. Nishimura and K. Igarashi, *et al.*, Polyphenol Extract from Phellinus igniarius Protects against Acrolein Toxicity In Vitro and Provides Protection in a Mouse Stroke Model, *PLoS One*, 2015, **10**(3), e0122733.
55. S. Datta, J. Dubey, S. Gupta, A. Paul, P. Gupta and A. K. Mitra, Tropical milky white mushroom, calocybe indica (Agaricomycetes): An effective antimicrobial agent working in synergism with standard antibiotics, *Int. J. Med. Mushrooms*, 2020, **22**(4), 335–346.
56. N. Alam, M. M. Sikder, M. A. Karim and S. M. R. Amin, Antioxidant and antityrosinase activities of milky white mushroom, *Bangladesh J. Bot.*, 2019, **48**(4), 1065–1073.
57. S. K. Ghosh and T. Bera, Pal S. Antiproliferative, apoptotic, and antimigration property of ethyl acetate extract of calocybe indica against hela and caski cell lines of cervical cancer, and its antioxidant and mycochemistry analysis, *Middle East J. Cancer*, 2020, **11**(4), 454–468.
58. C. I. Ukaegbu, S. R. Shah, H. A. Hamid, O. R. Alara and M. Z. I. Sarker, Phenolic Compounds of Aqueous and Methanol Extracts of Hypsizygus tessellatus

(brown and white var.) and Flammulina velutipes caps: Antioxidant and Antiproliferative Activities, *Pharm. Chem. J.*, 2020, **54**(2), 170–183.
59. Q. Xu, H. Wang, T. Li, L. Chen, B. Zheng and R. H. Liu, Comparison of phenolics, antioxidant, and antiproliferative activities of two Hypsizygus marmoreus varieties, *J. Food Sci.*, 2020, **85**(7), 2227–2235.
60. T. Islam, X. Yu and B. Xu, Phenolic profiles, antioxidant capacities and metal chelating ability of edible mushrooms commonly consumed in China, *LWT–Food Sci. Technol.*, 2016, **72**, 423–431.
61. H. H. Doğan, M. D. Kars, Ö. Özdemir and U. Gündüz, Fomes fomentarius and tricholoma anatolicum (Agaricomycetes) extracts exhibit significant multiple drug-resistant modulation activity in drug-resistant breast cancer cells, *Int. J. Med. Mushrooms*, 2020, **22**(2), 105–114.
62. N. Nowacka, R. Nowak, M. Drozd, M. Olech, R. Los and A. Malm, Antibacterial, antiradical potential and phenolic compounds of thirty-one polish mushrooms, *PLoS One*, 2015, **10**(10), 1–13.
63. C. Bal, H. Akgul, M. Sevindik, I. Akata and O. Yumrutas, Determination of the Anti-Oxidative Activities of Six Mushrooms, *Fresenius Environ. Bull.*, 2017, **26**(10), 6246–6252.
64. F. Çayan, E. Deveci, G. Tel-Çayan and M. E. Duru, Identification and quantification of phenolic acid compounds of twenty-six mushrooms by HPLC–DAD, *J. Food Meas. Charact.*, 2020, **14**(3), 1690–1698.
65. D. N. Olennikov, L. M. Tankhaeva and S. V. Agafonova, Antioxidant components of Laetiporus sulphureus (Bull.: Fr.) Murr. fruit bodies, *Appl. Biochem. Microbiol.*, 2011, **47**(4), 419–425.
66. A. Mocan, A. Fernandes, L. Barros, G. Crişan, M. Smiljkovic and M. Sokovic, *et al.*, Chemical composition and bioactive properties of the wild mushroom *Polyporus squamosus* (Huds.) Fr: a study with samples from Romania, *Food Funct.*, 2017, 160–170.
67. â. Fernandes, Polyporus squamosus (Huds.) Fr from different origins: Chemical characterization, screening of the bioactive properties and specific antimicrobial effects against Pseudomonas aeruginosa, *LWT–Food Sci. Technol.*, 2017, 91–97.
68. G. Tel-Çayan, F. Çayan, E. Deveci and M. E. Duru, Phenolic profile, antioxidant and cholinesterase inhibitory activities of four Trametes species: T. bicolor, T. pubescens, T. suaveolens, and T. versicolor, *J. Food Meas. Charact.*, 2021, **15**(5), 4608–4616.
69. R. M. Pop, I. C. Puia, A. Puia, V. S. Chedea, N. Leopold and I. C. Bocsan, *et al.*, Characterization of Trametes versicolor: Medicinal mushroom with important health benefits, *Not. Bot. Horti Agrobot. Cluj-Napoca*, 2018, **46**(2), 343–349.
70. M. Rašeta, M. Popović, P. Knežević, F. Šibul, S. Kaišarević and M. Karaman, Bioactive Phenolic Compounds of Two Medicinal Mushroom Species Trametes versicolor and Stereum subtomentosum as Antioxidant and Antiproliferative Agents, *Chem. Biodiversity*, 2020, **17**(12), e2000683.
71. L. Janjušević, M. Karaman, F. Šibul, G. Tommonaro, C. Iodice and D. Jakovljević, *et al.*, The lignicolous fungus Trametes versicolor (L.) Lloyd (1920): a promising natural source of antiradical and AChE inhibitory agents, *J. Enzyme Inhib. Med. Chem.*, 2017, **32**(1), 355–362.
72. A. Hasan, A. Aslan, I. Akata, S. Gunal, C. Bal and H. Baba, Phenolic Content and Biological Activities of Trametes Hirsuta, *Fresenius Environ. Bull.*, 2021, **30**(4), 4130–4135.
73. G. Zengin, A. Karanfil, M. C. Uren, M. S. Kocak, C. Sarikurkcu and H. Gungor, *et al.*, Phenolic content, antioxidant and enzyme inhibitory capacity of two: Trametes species, *RSC Adv.*, 2016, **6**(77), 73351–73357.

74. G. M. Aranha, A. G. Contato, J. C. Salgado, S. dos, T. B. de Oliveira, K. M. Retamiro and G. G. Ortolan, *et al.*, Biochemical characterization and biological properties of mycelium extracts from Lepista sordida GMA-05 and Trametes hirsuta GMA-01: new mushroom strains isolated in Brazil, *Braz. J. Microbiol.*, 2022, 1–10.
75. K. Wang, L. Bao, Q. Qi, F. Zhao, K. Ma and Y. Pei, *et al.*, Erinacerins c-l, isoindolin-1-ones with α-glucosidase inhibitory activity from cultures of the medicinal mushroom hericium erinaceus, *J. Nat. Prod.*, 2015, **78**(1), 146–154.
76. J. Tubić, D. Grujičić, M. R. Jakovljević, B. Ranković, M. Kosanić and T. Stanojković, *et al.*, Investigation of biological activities and secondary metabolites of hydnum repandum acetone extract, *Farmacia*, 2019, **67**(1), 174–183.
77. K. Sułkowska-Ziaja, B. Muszyńska and A. Szewczyk, Antioxidant components of selected indigenous edible mushrooms of the obsolete order Aphyllophorales, *Rev. Iberoam. Micol.*, 2015, **32**(2), 99–102.
78. F. Cayan, E. Deveci, G. Tel-Cayan and M. E. Duru, Phenolic Acid Profile of Six Wild Mushroom Species by HPLC-DAD, *Chem. Nat. Compd.*, 2018, **54**(5), 985–986.
79. M. Mohammed Ali Al-Harrasi, A. Mohammed Al-Sadi, A. M. Al-Tamimi, J. N. Al-Sabahi and R. Velazhahan, *In vitro* production of antifungal phenolic acids by Hypomyces perniciosus, the causal agent of wet bubble disease of Agaricus bisporus, *All Life*, 2021, **14**(1), 948–953.
80. E. M. Yahia, F. Gutiérrez-Orozco and M. A. Moreno-Pérez, Identification of phenolic compounds by liquid chromatography-mass spectrometry in seventeen species of wild mushrooms in Central Mexico and determination of their antioxidant activity and bioactive compounds, *Food Chem.*, 2017, **226**, 14–22.
81. J. A. Vaz, L. Barros, A. Martins and J. S. Morais, Vasconcelos MH, Ferreira ICFR. Phenolic profile of seventeen Portuguese wild mushrooms, *LWT–Food Sci. Technol.*, 2011, **44**(1), 343–346.
82. D. N. Olennikov, L. M. Tankhaeva and S. V. Agafonova, Antioxidant components of Laetiporus sulphureus (Bull.: Fr.) Murr. fruit bodies, *Appl. Biochem. Microbiol.*, 2011, **47**(4), 419–425.
83. M. Nam, J. Y. Choi and M. S. Kim, Metabolic Profiles, Bioactive Compounds, and Antioxidant Capacity in Lentinula edodes Cultivated on Log versus Sawdust Substrates, *Biomolecules*, 2021, **11**(11), 1654.
84. M. Y. Kim, P. Seguin, J. K. Ahn, J. J. Kim, S. C. Chun and E. H. Kim, *et al.*, Phenolic Compound Concentration and Antioxidant Activities of Edible and Medicinal Mushrooms from Korea, *J. Agric. Food Chem.*, 2008, **56**(16), 7265–7270.
85. G. M. Aranha, A. G. Contato, J. C. dos Santos Salgado, T. B. de Oliveira, K. M. Retamiro and G. G. Ortolan, *et al.*, Biochemical characterization and biological properties of mycelium extracts from Lepista sordida GMA-05 and Trametes hirsuta GMA-01: new mushroom strains isolated in Brazil, *Braz. J. Microbiol.*, 2022, 1–10.
86. V. Vieira, L. Barros and A. Martins, Ferreira ICFR. Nutritional and Biochemical Profiling of Leucopaxillus candidus (Bres.) Singer Wild Mushroom, *Molecules*, 2016, **21**(1), 99.
87. M. Sevindik, Anticancer, antimicrobial, antioxidant and DNA protective potential of mushroom Leucopaxillus gentianeus (Quél.) Kotl, *Indian J. Exp. Biol.*, 2021, **59**(05), 310–315.
88. Z. Tietel and S. Masaphy, True morels (*Morchella*)—nutritional and phytochemical composition, health benefits and flavor: A review, *Crit. Rev. Food Sci. Nutr.*, 2017, **58**(11), 1888–1901.
89. F. Çayan, G. Tel-Çayan, E. Deveci and M. E. Duru, HPLC–DAD characterization of phenolic profile and in vitro antioxidant, anticholinesterase, and antidiabetic activities of five mushroom species from Turkey, *3 Biotech*, 2021, **11**(6), 1–11.

90. T. Islam, X. Yu and B. Xu, Phenolic profiles, antioxidant capacities and metal chelating ability of edible mushrooms commonly consumed in China, *LWT–Food Sci. Technol.*, 2016, **72**, 423–431.
91. S. Khatua, P. Mitra, S. Chandra and K. Acharya, *In Vitro* Protective Ability of Ramaria aurea Against Free Radical and Identification of Main Phenolic Acids by HPLC, *J. Herbs, Spices Med. Plants*, 2015, **21**(4), 380–391.
92. A. C. Aprotosoaie, D. E. Zavastin, C. T. Mihai, G. Voichita, D. Gherghel and M. Silion, *et al.*, Antioxidant and antigenotoxic potential of Ramaria largentii Marr & D. E. Stuntz, a wild edible mushroom collected from Northeast Romania, *Food Chem. Toxicol.*, 2017, **108**, 429–437.
93. N. Shomali, O. Onar, T. Alkan, N. Demirtaş, I. Akata and Ö. Yildirim, Investigation of the polyphenol composition, biological activities, and detoxification properties of some medicinal mushrooms from turkey, *Turk. J. Pharm. Sci.*, 2019, **16**(2), 155–160.
94. K. A. Kouassi, E. J. P. Kouadio, K. M. Djè, A. E. Dué, L. P. Kouamé and K. A. Kouassi, *et al.*, Edible Ectomycorrhizal Mushrooms Russula spp. of Côte d'Ivoire: Total Phenolic Content, HPLC-Profiles of Phenolic Compounds and Organic Acids, Antioxidant Activities, *J. Agric. Chem. Environ.*, 2016, **5**(2), 73–84.
95. K. Sułkowska-Ziaja, A. Szewczyk, J. Gdula-Argasińska, H. Ekiert, J. Jaśkiewicz and B. Muszyńska, Chemical compounds of extracts from Sarcodon imbricatus at optimized growth conditions, *Acta Mycol.*, 2016, **51**(2), 1086.
96. Y. Hui-Min, W. Gan, L. Ya-Ping, R. Ming-Qiang, S. Chuan-Bin and Y. Xiu-Wen, *et al.*, Phenolic acids isolated from the fungus Schizophyllum commune exert analgesic activity by inhibiting voltage-gated sodium channels, *Chin. J. Nat. Med.*, 2016, **14**(9), 661–670.
97. Z. Zhang, W. Cai, T. Song, L. Fan and G. Lv, Targeted identification of antioxidant compounds from Sparassis latifolia extracts and their antioxidant activities, *J. Food Process. Preserv.*, 2021, **45**(12), e16068.
98. F. Souilem, â. Fernandes, R. C. Calhelha, J. C. M. Barreira, L. Barros and F. Skhiri, *et al.*, Wild mushrooms and their mycelia as sources of bioactive compounds: Antioxidant, anti-inflammatory and cytotoxic properties, *Food Chem.*, 2017, **230**, 40–48.
99. M. Deng, J. Wang, Y. L. Li, H. X. Chen, M. Tai and L. Deng, *et al.*, Impact of polyphenols extracted from Tricholoma matsutake on UVB-induced photoaging in mouse skin, *J. Cosmet. Dermatol.*, 2022, **21**(2), 781–793.
100. Z. Selamoglu, M. Sevindik, C. Bal, B. Ozaltun, I. Sen and A. Pasdaran, Antioxidant, antimicrobial and DNA protection activities of phenolic content of Tricholoma virgatum (Fr.) P.Kumm, *Biointerface Res. Appl. Chem.*, 2020, **10**(3), 5500–5506.
101. B. Koc, L. Akyuz, Y. S. Cakmak, I. Sargin, A. M. Salaberria and J. Labidi, *et al.*, Production and characterization of chitosan-fungal extract films, *Food Biosci.*, 2020, **35**, 100545.
102. M. Gąsecka, P. Rzymski, M. Mleczek, M. Siwulski, S. Budzyńska and Z. Magdziak, *et al.*, The relationship between metal composition, phenolic acid and flavonoid content in Imleria badia from non-polluted and polluted areas, *J. Environ. Sci. Health, - Part B Pestic Food Contam., Agri. Wastes*, 2017, **52**(3), 171–177.
103. M. Dimitrijević, V. Stankov Jovanović, J. Cvetković, M. Mitić, G. Petrović and A. Đorđević, *et al.*, Phenolics, Antioxidant Potentials, and Antimicrobial Activities of Six Wild Boletaceae Mushrooms, *Anal. Lett.*, 2017, **50**(10), 1691–1709.

CHAPTER 5

Bioactive Properties of Mushrooms with Potential Health Benefits

ROSSANA V. C. CARDOSO,[a] TAOFIQ OLUDEMI,[a,b] ÂNGELA FERNANDES,*[a] ISABEL C. F. R. FERREIRA[a] AND LILLIAN BARROS*[a]

[a] Centro de Investigação de Montanha (CIMO), Instituto Politécnico de Bragança, Campus de Santa Apolónia, 5300-253 Bragança, Portugal; [b] Nutrition and Bromatology Group, Department of Analytical Chemistry and Food Science, Faculty of Science, Univeridade de Vigo, E-32004 Ourense, Spain
*Email: afeitor@ipb.pt

5.1 Introduction

Macrofungi are a group of organisms from the Fungi kingdom generally grouped in the families Ascomycota and Basidiomycota, which under special conditions develop a fruiting body of sufficient size to be seen with the naked eye and grow under the soil.[1] In 2019, global production of mushrooms and truffles was around 11.9 million tons worldwide. China is by far the largest producer of mushrooms globally, producing around 8.93 million metric tons of mushrooms and truffles annually. For producing regions, Asia is the largest producer, and is responsible for 82.8% of total global production, followed by Europe, the Americas, Oceania, and Africa with 12%, 4.5%, 0.4%, and 0.3%, respectively. Poland is the largest European producer

Food Chemistry, Function and Analysis No. 36
Edible Fungi: Chemical Composition, Nutrition and Health Effects
Edited by Dejan Stojković and Lillian Barros

Published by the Royal Society of Chemistry, www.rsc.org

of mushrooms, with a total production of 362 million metric tons, representing 3.05% of the total global volume.[2]

For a long time, mushrooms have played an important role in several aspects of human activity. Particularly, edible mushrooms are used extensively in cooking and make up part of a low-calorie diet. In different world regions, the demand and consumption of mushrooms are mostly related to different characteristics such as organoleptic properties, attractive taste and flavour, and a very accessible and marketable product.[1] Wild and cultivated mushrooms contain a huge diversity of biomolecules with nutritional (since they are rich in minerals, proteins, fibres, and carbohydrates, and are low-caloric foods due to a low content in fat) and/or medicinal properties. Due to these properties, they have been recognized as functional foods and as a source for developing medicines and nutraceuticals.[3–5] In order to survive, mushrooms possess the capacity to generate and accumulate a variety of functional compounds, such as carbohydrates (polysaccharides, glucose, fructose, maltose, rhamnose, arabinose, sucrose, mannitol, trehalose, and xylose), proteins (lectins), fatty acids (linoleic, oleic and palmitic acids), phenolic compounds (caffeic, gallic, cinnamic, melatonin, *p*-hydroxybenzoic, *p*-coumaric, and protocatechuic acids), indole compounds (L-tryptophan and 5-hydroxy-L-tryptophan), vitamins (vitamins B complex, vitamin C, and tocopherols), terpenoids (carotenoids such as β-carotene and lycopene), enzymes (laccases, cellulases, lignins, peroxidases, ribonucleases), and unique molecules (such as ergothioneine and glutathione). These bioactive molecules have immunoregulatory effects, and cardio- and liver-protective, antifibrotic, anti-inflammatory, antidiabetic, antiviral, anti-diabetes, anti-angiogenic, antioxidant, antitumour, and antimicrobial properties.[1,3,6–8] The molecular structure, molecular weight, size, branching pattern, conformation, solubility, and intra- and intermolecular association of these compounds are important for showing biological responses.[3] Their pharmacological action and therapeutic interest in promoting human health have been known for thousands of years. In the recent past, mushrooms have become increasingly important as potential natural agents to prevent and treat numerous diseases, such as cancer, cardiovascular diseases, diabetes mellitus and neurodegenerative diseases.[1,5]

In this respect, this chapter aims to analyse and summarize current knowledge and trends regarding the bioactive properties of mushrooms. In addition, important biological properties are discussed with their potential health benefits, which could be useful for the full use of mushrooms. A compilation of the published information from 2015 onwards has been explored.

5.2 Mushroom Consumption and Potential Health Benefits

The shift towards natural-based diets has been accelerating at varying rates due to the increased awareness of the health and environmental impacts of excessive and ultra-processed foods.[9] Since ancient times, mushrooms have traditionally been collected in various geographical locations for human consumption; and this practice has increased in recent years, mainly due to the growing awareness of their therapeutic and beneficial effects.[10,11] Edible, medicinal, and wild mushrooms are the major components of the global mushroom industry. Medicinal and edible mushrooms are mostly found in higher basidiomycetes with a saprophytic growth habit, which allows them to grow on various agricultural waste-based substrates.[12] Edible mushrooms are the leading component, accounting for approximately 54% of global mushroom production, while medicinal and wild mushrooms make up 38% and 8% of the total, respectively.[13,14] *Agaricus*, *Lentinula*, *Pleurotus*, *Auricularia*, and *Flammulina* represent the five main genera, constituting over 85% of the world's commercial mushroom production. Other mushroom genera with well-documented nutritional and medicinal properties include *Ganoderma*, *Grifola*, *Trametes*, *Clitocybe*, *Antrodia*, *Cordyceps*, *Xerocomus*, *Calvatia*, *Hericium*, *Volvariella*, *Schizophyllum*, *Inonotus*, *Inocybe*, *Lactarius*, *Albatrellus*, *Russula*, *Boletus*, *Cantharellus*, *Lactarius*, *Morchella*, *Macrolepiota*, and *Fomes*.[15,16] *Agaricus bisporus* (white and brown button mushroom) represents about 40% of worldwide production, followed by *Lentinula edodes*, *Pleurotus ostreatus*, and *Flammulina velutipes*.[17] The global mushroom market share is expected to grow significantly in the coming years due to these nutritional advantages and health-promoting benefits. In addition, mushroom production and processing generate a large volume of by-products in the form of caps, stipes, mushrooms that do not fit with commercial standards, and spent mushroom substrate (SMS), being predominantly discarded with a high environmental impact and treatment costs for the industry.[18,19] These mushroom by-products constitute a real economic loss due to the abundance of bioactive compounds, which can be extracted and utilized as value-added ingredients. Hence, some integral approaches for the sustainable valorization of these by-products in developing innovative food and pharmaceutical formulations have been adopted further to improve the economic performance of the mushroom processing industry and promote a circular economy in line with current consumer preferences towards sustainability. The widespread utilizations of mushroom extracts and their associated metabolites in the development of nutraceutical and functional food formulations are due to the extensive range of nutritional, preventive, therapeutic, and structurally diverse bioactive compounds.[10] These biomolecules include polysaccharides (chitosans, β-glucans, fucogalactan, and lentinan), proteins (lectin and cordymin), fatty acids (linoleic, oleic, and palmitic), terpenoids and steroids (ergosterol, betulin, zhankuic acid, and ganoderic acid), phenolic compounds (caffeic acid, gallic acid, cinnamic acid, inotilone, hispolon, *p*-hydroxybenzoic acid, *p*-coumaric acid, and protocatechuic acid), vitamins (thiamine, riboflavin, biotin, and tocopherols), and other unique molecules (ergothioneine, cordycepin, and hispidin).[16,20] Mushrooms have also been well characterized in their mineral content and have been shown to be rich in elements such as potassium, phosphorus, calcium, magnesium, sodium, iron, zinc, and copper.[21]

Apart from their desirable status in the culinary realm, their health-promoting properties are also significant, namely antioxidant, anti-inflammatory, antimicrobial, immunomodulatory, cytotoxic, antimutagenic, cardioprotective, hepatoprotective, antidiabetic, and anti-ageing properties, validated via *in vitro* and *in vivo* studies (see Figure 5.1).[16,22–24] Studies have indicated that mushrooms exposed to ultraviolet (UV), gamma, and electron-beam irradiation under certain conditions convert ergosterol to vitamin D2.[25,26] These irradiated mushrooms effectively maintain serum 25(OH)D levels and are currently being utilized as food fortification ingredients to reduce vitamin D deficiency.[27] *In vivo* studies in a mouse model of osteoporosis have shown that the consumption of vitamin D2-enhanced mushrooms have also been associated with improved bone health.[28]

The use of mushroom extracts in the development of food formulations is widespread in baked goods (pasta, bread, and cookies), sausages, hamburgers, dairy products (cheese), beverages, and soups.[29,30] Additionally, considering the need to reduce foods from animal origins to enable better global sustainability, there is a growing search for plant-based protein

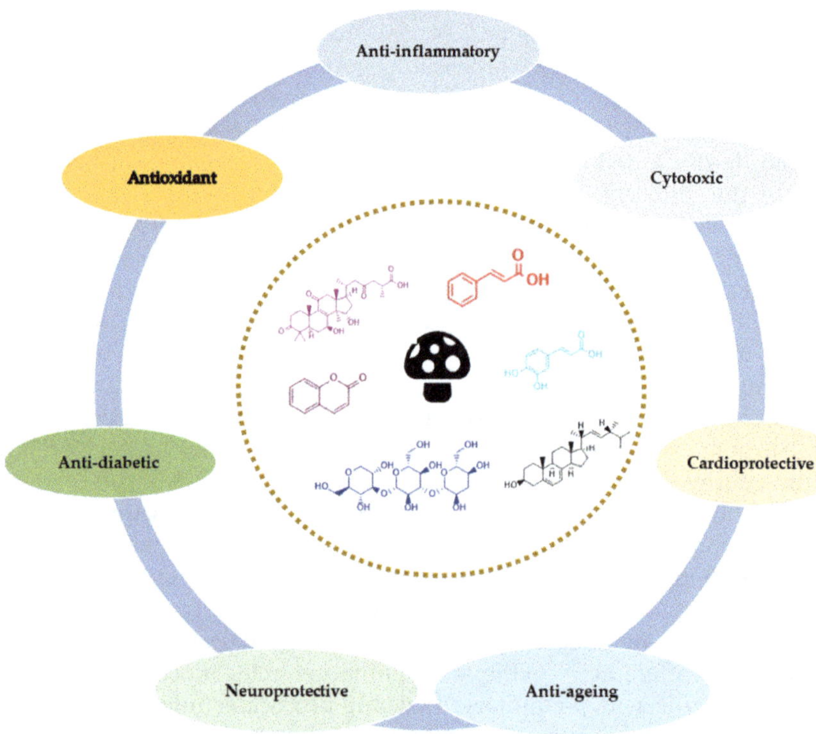

Figure 5.1 Bioactive properties of mushroom extracts.

sources as an alternative to animal protein, and mushrooms represent a viable option for obtaining high-quality protein with a complete profile of essential amino acids, often covering the recommended dietary requirement.[31] Several commercially available products containing blends of mushroom extracts or their metabolites have been developed into patents and sold in the market as daily health supplements.

5.3 Extracts and Individual Compounds from Mushrooms with Bioactive Properties

5.3.1 Antioxidant Activity

Reactive oxygen species (ROS) and reactive nitrogen species (RNS) accompany several cellular metabolic processes. Oxygen is an essential component for the oxidation of organic compounds and energy production,[32] and when a small amount of oxygen is reduced, it produces highly reactive chemical substances called ROS. Thus, ROS represents the most important class of all free radicals generated in living organisms.[33–35] They are related to several physiological processes and are beneficial to cells when produced in small quantities. However, in situations in which the balance between the production of these reactive species and the antioxidant defences is destroyed, oxidative stress begins.[36,37] ROS are highly reactive and when in excess they can cause irreversible damage to the structure of cells. The production of free radicals can be of endogenous or exogenous origin, referring to metabolic interactions or environmental factors, respectively (see Figure 5.2). Free radicals are related to several human diseases and participate as essential elements in many of them, which shows how extensive is the oxidative damage caused by them.[38,39] Natural or synthetic antioxidants are chemical compounds that significantly delay or prevent oxidation even at low concentrations compared to an oxidizable substrate. Organisms have antioxidant defence and repair systems that eliminate free radicals and protect the body from oxidative damage. The harmful effect of free radicals occurs when they are in excessive amounts in the body, exceeding its capacity to neutralize them with its natural defence systems (see Figure 5.2). Antioxidants have the power to intercept free radicals formed by cell metabolism or by exogenous sources, preventing oxidative stress and consequent damage to cells. Thus, antioxidant supplements, or natural products with antioxidant capacity, can help reduce oxidative damage in the body, preventing hundreds of diseases, including many types of cancer, diabetes, obesity, and cardiovascular diseases.[40–42] The appreciation of the relationship between food and health is growing every day, and healthy nutrition focused on natural products, mainly of vegetal origin, can help combat free radicals. Many natural products have bioactive molecules with various beneficial health claims.[38,43]

Mushrooms have been present in the human diet for many years and have long been valued as tasty, nutritious, healthy, and valuable food. They are rich

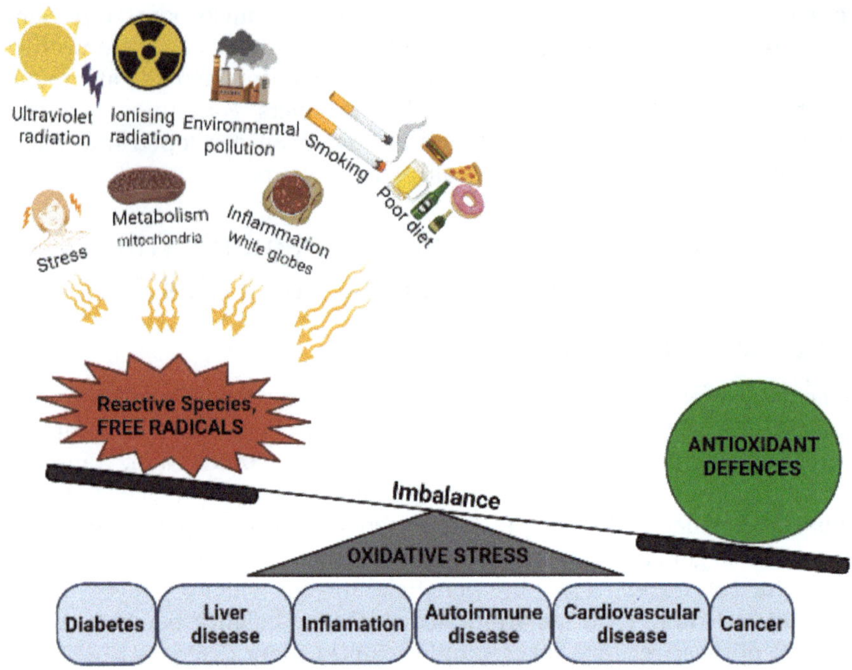

Figure 5.2 The results of the imbalance between reactive species and antioxidant defences.

in bioactive molecules with health benefits, and in recent years the amounts consumed have increased significantly, involving many species.[43,44] Mushrooms are rich in antioxidants, including polysaccharides, phenols, proteins, peptides, carotenoids, ergosterol (pre-vitamin D_2), and vitamins C and E.[32,36,37,45,46] Several studies are available on the antioxidant properties of mushrooms, such as *Agaricus bisporus* samples,[47,48] *Agaricus campestris* L.,[49] *Pleurotus ostreatus*,[47,50–52] *Lentinula edodes*,[53,54] *Macrolepiota procera*,[49] *Pleurotus pulmonarius* (Fr.),[55] among others (see Table 5.1). Therefore, with the growing interest in the beneficial health effects of antioxidants, it is essential to study the antioxidant potential of the most consumed mushroom species worldwide and the varieties of existing methods to determine the antioxidant activity of extracts or substances obtained in *in vitro* and *ex vivo* conditions.[36,44,56] Several methods have been developed to evaluate the antioxidant capacity of edible and medicinal mushroom species (see Table 5.1).

Given the *in vitro* condition, the most commonly applied are the DPPH radical scavenging (2,2-diphenyl-1-picrylhydrazyl), $ABTS^{\bullet +}$ radical scavenging capacity (2,2′-azino-bis-3-ethylbenzthiazoline-6-sulfonic acid), reducing power (RP), and β-carotene bleaching 2,2′-azino-bis-3-ethylbenzthiazoline-6-sulfonic acid. Polysaccharides such as heteroglycan, β-glucan, and lentinan obtained from different edible and medicinal mushrooms have presented high ferrous ion chelating, radical scavenging, and reducing power properties that have

Table 5.1 Antioxidant potential in mushroom extracts.[a]

Mushroom species	Extracts	Methods	Ref.
A. bisporus, *F. velutipes*, *L. edodes*, and *A. brasiliensis*	Ethanolic	DPPH, ABTS and FRAP	44
A. bisporus and *P. ostreatus*	Water, methanolic and ethanolic	ABTS, and FRAP	218
A. campestris and *B. edulis*	Methanolic extract	DPPH and RP	219
Agaricus sp., *A. cylindracea*, *B. loyo*, *C. lebre*, *C. espinosae*, *F. velutipes*, *G. gargal*, *L. deliciosus*, *L. edodes*, *M. procera*, *M. cónica*, *Pleurotus* sp., *Ramaria* sp., *Suillus* sp., *T. terreum*, and *X. chrysenteron*	Dichloromethane and methanolic	DPPH	220
A. silvaticus, *H. rufescens*, and *M. giganteus*	Methanol and ethyl acetate extract	DPPH, ABTS, FRAP, and catalase	221
Amanita sp., *L. volemus*, *Russula* sp., *Termitomyces* sp., *T. crissum*, *V. volvacea*, *A. hygrometricus*, *A. trappei*, *A. auricula*, *C. cibarius*, *C. Craterellus*, and *Lentinus* sp.	Methanolic extract	DPPH and FRAP	222
A. lanipes	Methanolic extract	ABTS and TOS	223
B. edulis	Ethanolic and water extract	DPPH, FRAP, hydroxyl oxide, nitric oxide, and superoxide anion	224
B. edulis, *B. pinophilus*, *B. aureus*, *A. mellea*, *T. aestivum*, *L. piperatus*, *L. deliciosus*, *P. eryngii*, *R. botrytis*, and *R. virescens*	Ethanolic extract	DPPH, chelating, RP, and inhibition of lipid peroxidation	225
B. griseipurpureus	Dichloromethane and methanolic extract	DPPH, ABTS, and ORAC	226
C. cinereus, *C. pistillaris*, *C. nebularis*, and *H. punicea*	Methanol, ethanol, and water extract	DPPH, ABTS, and RP	227
C. cylindracea, *F. velutipes*, *L. duriusculum*, *Pleurotus* sp., *Agaricus* sp., and *L. edodes*	Methanolic	DPPH	48
C. cornucopioides	Acetone	DPPH, superoxide anion, and RP	228
G. lucidum and *A. bisporus*	Water extract	DPPH	229
H. marmoreus, *A. vaginata*, *A. auricular*, *G. lucidum*, *L. edodes*, *L. polychrous*, *L. squarrosulus*, *P. ostreatus*, *P. ostreatus*, *P. pulmonarius*, *P. sajor-caju*, *A. cytindracea*, *T. fusiformis*, *F. velutipes*, and *V. volvacea*	Hexane, ethyl acetate, ethanol, methanol, and water extract	ABTS	230
I. obliquus, *G. frondosa*, *Ganoderma* sp., *L. edodes*, *T. versicolor*, and *H. erinaceus*	Hydroalcoholic and water extract	ORAC, FRAP, DPPH, and NanoCerac	53

Table 5.1 (*Continued*)

Mushroom species	Extracts	Methods	Ref.
L. leucothites	Ethanolic extract	DPPH	231
M. lobayensis	Ethanolic extract	DPPH, ABTS superoxide radical, hydroxyl radical quenching, chelating ability of metal ion, RP, and TAC	232
	Methanolic extract	DPPH, ABTS, RP, and chelating ability of ferrous ion	233
M. giganteus	Methanolic extract	DPPH, RP, β-carotene bleaching, and TBARS	109
M. cognata and *M. stridula*	Ethyl acetate, methanolic, and water	DPPH, ABTS, and RP, phospho-molybdenum, and metal chelating	234
P. eryngii and *S. belinii*	Methanolic extract	DPPH, RP, TBARS, and β-carotene bleaching	235
P. florida	Methanolic, ethanolic, water, ethyl acetate, and hexane	DPPH and RP	52
P. ostreatus	Water	DPPH, ABTS, RP, metal chelating, and hydroxyl radical	38
	Ethanolic	DPPH and hydroxyl radical scavenging	236
	Water	DPPH	50
	Ethyl acetate and *n*-butanol extract	DPPH and ABTS	237
P. ostreatus var. *Florida*	Ethanolic	TBARS and OxHLIA	51
P. ostreatus and *A. bisporus*	Methanolic extract	DPPH, RP, β-carotene, TBARS	47
Pleurotus sp.	Hydroalcoholic extract	DPPH, ORAC, ABTS, and β-carotene bleaching	238
P. pulmonarius	Ethanolic extract	TBARS, Conjugated dienes formation, DPPH, and RP	55
R. subalpina	Methanolic extract	DPPH, ferrous ion chelating, and RP	239
Russula sp.	Methanolic and ethanolic	OxHLIA and TBARS	240
T. indicum	Methanol and ethanol extract	DPPH and ABTS	241
T. equestre	Water and methanol extract	DPPH	242

[a]TBARS, thiobarbituric acid reactive substances; OxHLIA, oxidative haemolysis inhibition assay; DPPH, 2,2′-diphenyl-1-picrylhydrazyl; RP, reducing power; ORAC, oxygen radical absorbance capacity; FRAP, ferric reducing antioxidant power; ABTS, 2,2′-azino-bis-3-ethylbenzthiazoline-6-sulfonic acid; TAC, total antioxidant capacity; TOS, total oxidant status.

contributed to their multifunctional health-promoting function. Some of these polysaccharide-rich mushrooms include *Termitomyces clypeatus*,[57] *Hohenbuehelia serotina*,[58] and *Meripilus giganteus*.[59] An optimized extract of *Pleurotus ostreatus* rich in heteropolysaccharides, β-glucans, α-glucans, and oligosaccharides, besides presenting a solid radical scavenging activity, also showed a promising capacity to protect cells from oxidative damage induced by hydrogen peroxide (H_2O_2).[50] In addition to the well-reported *in vitro* antioxidant effects of these mushrooms and their related metabolites, some *ex vivo* and *in vivo* studies involving animal models have also shown promising results. Among these, Popović *et al.*[60] evaluated antioxidative activity using an aqueous suspension of *Coprinus comatus*, with oxidative stress being induced in rats using alloxane and carbon tetrachloride. Jayakumar *et al.*[61] conducted a review of the *in vivo* antioxidant activity of *Pleurotus ostreatus*. The antioxidant potential was evaluated using an acute state of oxidative stress in animal models induced by carbon tetrachloride (CCl_4) and a chronic state of stress (ageing). In addition to the significant reduction in lipid peroxidation, the levels of enzyme antioxidant defences such as superoxide dismutase (SOD), glutathione peroxidase (GSH-Px), and catalase (CAT), which are involved in the destruction of ROS, were significantly elevated. Recently, the *in vivo* antioxidant effect of *Coprinus comatus* was developed using an animal model of CCl_4-induced hepatotoxicity.[62] The authors reported that oral ingestion of its methanolic extract suppressed CCL_4-induced liver damage, reduced lipid peroxidation, and increased levels of GSH-Px. The beneficial antioxidant and renoprotective effects of an acidic hydrolytic polysaccharide extract from *Lentinula edodes* extract were also conducted using an *in vivo* model. The extract significantly enhanced the levels of SOD, GSH-Px, and CAT enzymes, while also suppressing the level of malondialdehyde (MDA), a vital indicator of lipid peroxidation, responsible for cell membrane disruption and oxidative induced cell damage.[63] These findings imply that mushrooms extract contains various molecular lead compounds with potent antioxidant effects and they are currently used as additives in nutraceutical formulations to manage diseases associated with oxidative stress.

5.3.2 Anti-inflammatory, Neuroinflammatory, and Neuroprotective Activities

Inflammation is a complex biological process triggered in response to external stimuli such as infections, injuries, irritants, ultraviolet radiations, and toxins.[64] However, persistent inflammation has been associated with diabetes, arthritis, obesity, metabolic syndrome, cancer, and several cardiovascular diseases.[65] Pro-inflammatory cells (mainly macrophages and monocytes) play a key role in natural immunity by releasing specialized inflammatory mediators, which include vasoactive amines, eicosanoids, pro-inflammatory cytokines, and acute-phase proteins, which mediate the inflammatory process by preventing further tissue damage and ultimately resulting in healing and restoration of tissue function. However, oversecretion of these inflammatory mediators causes accumulation of oxidative products and increased cytotoxicity, leading to tumour development, DNA damage, and programmed cell death.[65] External stimuli such as bacterial lipopolysaccharides, viral proteins, toxins, or endogenous proteins activate cascades of various transcription factors, namely activator protein 1 (AP-1), signal transducer and activator of transcription 3 (STAT3), and nuclear factor kappa β (NF-κB). NF-κB comprises five structurally related proteins, NF-κB1 (p50), NF-κB2 (p52), RelA (p65), RelB, and c-Rel, responsible for the expression of a vast array of inflammatory mediators.[64] In the absence of stimuli, NF-κB complexes (usually in the forms of p60 and p65 subunits), are confined in the cytoplasm and attached to inhibitory proteins, including from the IκB family. Once the toll-like receptor (TLR) ligands are stimulated, IκB proteins are rapidly phosphorylated, which leads to the ubiquitination and degradation of IκBα (see Figure 5.3). This frees the NF-κβ complex (mainly p65 and p50) to translocate from the cytosol into the nucleus to instantaneously provoke the release of pro-inflammatory mediators, including inducible nitric oxide synthase (iNOS), cyclooxygenase-2 (COX-2), prostaglandin E2 (PGE2), and cytokines like interferon-gamma (IFN-γ), interleukins (IL-1β and IL-6), and tumour necrosis factor alpha (TNF-α).[66]

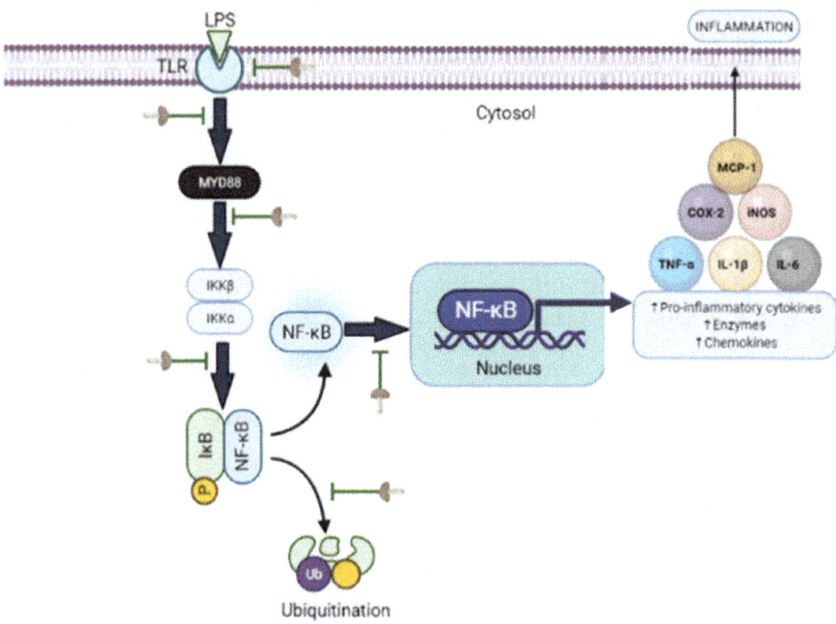

Figure 5.3 The nuclear factor-kappa beta (NF-κβ) pathway. Abbreviations: toll-like receptors (TLR), lipopolysaccharides (LPS), myeloid differentiation protein 88 (MyD88), IL (interleukins), cyclooxygenase-2 (COX-2), inducible nitric oxide synthase (iNOS), tumour necrosis factor (TNF-α), monocyte chemoattractant protein-1 (MCP-1), IkappaB kinase (IKK).

The mechanism behind the inflammatory process is mainly controlled by the NF-κβ pathway and trying to inhibit some of these specific steps in the pathway leading to NF-kβ release is crucial to suppressing the inflammation process (see Figure 5.3).

In vitro models such as lipopolysaccharide (LPS)-induced RAW 264.7 macrophage cells, mouse splenocytes, NIH/3T3 cell, HaCaT, RBL-2H3, and IFN-γ activated murine macrophages have been utilized as effective models to evaluate the anti-inflammatory activities of mushroom extracts and their related metabolites (see Table 5.2).[67–70] Additionally, numerous animal models such as a dextran sulfate sodium (DSS)-induced colitis model, carrageenan-induced mice models, formalin-induced mice models, and an acetic acid-induced mice model have been utilized to determine the anti-inflammatory properties of bioactive ingredients derived from mushroom (see Table 5.2).[71] Polysaccharide and sulfated polysaccharide-rich extracts obtained from *Antrodia cinnamomea* significantly reduced levels of TNF-α, IL-1β, and IL-6 in LPS-induced RAW 264.7 macrophage cells.[72,73] At the same time, a more distinct anti-inflammatory effect via downregulation of NF-κβ expression was reported for its oligosaccharide and ergostatrien-3β-ol rich extracts.[74,75] Polysaccharide rich fractions from *Grifola frondosa*,[76] *Hericium erinaceus*,[77] *Lentinula edodes* (acidic polysaccharide and β-D-glucan),[78,79] *Phellinus baumii* (heteropolysaccharide),[80] and *Schizophyllum commune* (exopolysaccharide)[71,81] presented interesting anti-inflammatory effects in RAW 264.7 macrophage cells, human monocyte THP-1 cell line, carrageenan-induced paw oedema, and DSS-induced colitis models. Sterol-rich extract from *G. lucidum* shows anti-inflammatory effects via attenuation of the NF-κβ pathway by restraining degradation of IκB-α and phosphorylation of NF-κβ p65.[82] Its phenolic-rich extract significantly suppressed gene expression of COX-2, TNF-α, iNOS, IL-1β, IL-6, IL-10, and NF-κB in RAW 264.7 macrophage cells and a DSS-induced colitis model.[83] The mechanism behind the anti-inflammatory effect of mushroom extracts and their bioactive metabolites has been associated with the downregulation and dephosphorylation of NF-κB, inhibition of mRNA expression of TLR4, inhibition of NF-κB translocation into the nucleus, and degradation of IκB-α.[82] Due to the current pharmaceutical practices involving highly purified compounds, the utilization of complex mixtures of bioactive compounds with unknown concentrations is difficult to reconcile. Moreover, the recovery of complex mushroom-derived compounds such as polysaccharides requires energy-consuming and costly extraction processes. Hence, low-molecular-weight compounds, such as *N*6-(2-hydroxyethyl) adenosine (HEA), cordycepin, methyl 9-oxo-(10*E*,12*E*)-octadecadienoate, polyozellin, and poricoic acid C, also have been studied (see Table 5.2) and known to inhibit inflammatory mediator production, block steps leading to NF-κβ activation, and suppress *in vivo* induced inflammation. As shown in Table 5.2, different research studies have highlighted mushrooms' very promising anti-inflammatory effects, mainly based on their extracts and bioactive compounds. Nevertheless, intensive preclinical and clinical research into these bioactive molecules is still lacking.

Table 5.2 Anti-inflammatory and neuroinflammatory effects of mushroom extracts and their individual compounds.

Species	Bioactive form	Bioactive property	Model	Key findings and suggested mechanism of action	Ref.
A. blazei	Polypeptide	Anti-inflammatory effect	NIH/3T3 cell	↓ downregulate the expression of NF-κBp65 and TLR4 ↓ levels of TNF-α, IL-1β and IL-6	68
A. rugosum	Ethanolic extract	Anti-inflammatory effect	RAW 264.7 macrophage cells	↓ production of TNF-α and NO	69
A. cinnamomea	Triterpenoid-enriched fractions	Neuroprotection	LPS induced BV2 microglia cells	↓ production of NO, ↓ mRNA expression of iNOS and COX-2	86
	Sulfated polysaccharide	Anti-inflammatory effect	RAW 264.7 macrophage cells	↓ levels of TNF-α, IL-1β and IL-6	73
	Polysaccharides and sulfated polysaccharides	Anti-inflammatory effect	RAW 264.7 macrophage cells	↓ levels of TNF-α and IL-6	72
	Oligosaccharides	Anti-inflammatory effect	RAW 264.7 macrophage cells and LPS-induced mouse model	↓ mRNA expression of IL-6, IL-8, IL-1, TNF-α, and MCP-1 ↓ activation of MAPK and Akt signalling pathways	74
	Ergostatrien-3β-ol	Anti-inflammatory effect	Mouse skin ischaemia model induced by skin flap surgery	↓ gene expression of iNOS, IL-6, TNF-α, IκB, and NF-κB	75
A. mellea	Xylosyl 1,3-galactofucan	Anti-inflammatory effect	RAW 264.7 macrophage cells	↓ levels of TNF-α and cytokine monocyte chemotactic protein-1 (MCP-1)	243
	Ethyl acetate extract	Neuroprotection	LPS induced BV2 microglia cells	↓ production of NO, IL-6, TNF-α, and IL-1β ↓ phosphorylation of IκB-α and NF-κB p65	84
C. cicadae	*N*6-(2-Hydroxyethyl) adenosine (HEA)	Anti-inflammatory effect	RAW 264.7 macrophage cells	↓ gene expression of TLR4, IκB, p-IκB, and COX-2 ↓ production of TNF-α and PGE2	244

C. militaris	Cordycepin	Neuroprotection	LPS induced BV2 microglia cells and hippocampal neurons from the brain of prenatal mouse	↓ levels of TNF-α and IL-1β ↓ mRNA expression of iNOS and COX-2 ↓ mRNA expression of p65, IκBα, and phosphorylated IκBα (p-IκBα)	66
C. africanus	Polyoxygenated cyathane diterpenoids	Neuroprotection	LPS induced BV2 microglia cells	↓ production of NO	87
D. indusiata	Polysaccharide	Anti-inflammatory effect	DSS-induced colitis model	↓ levels of TNF-α, IFN-γ, IL-1β, IL-6, IL-12, IL-17, IL-4, and IL-10 ↓ dephosphorylation of NF-κB and MAPK, ↓ mRNA expression of iNOS, COX-2, TNF-α, and IL-6	245
F. fomentarius	Methyl 9-oxo-(10*E*,12*E*)-octadecadienoate	Anti-inflammatory effect	Murine macrophage cells	↓ secretion of NO and PGE2, ↓ downregulation of iNOS and COX-2, ↓ levels of TNF-α and IL-6 ↓ activation of STAT3	70
G. lucidum	Polysaccharide	Neuroprotection	LPS- and Aβ-induced BV2 microglia and primary mouse microglia cells	↓ mRNA expressions of IL-1β, IL-6, iNOS and MCP-1 ↓ mRNA expressions of TGFβ	85
	Ganodermanontriol	Anti-inflammatory effect	Ana-1 macrophage cell line	↓ levels of IL-1β, IL-6, TNF-α, and PEG2 ↓ translocation of NF-κB ↓ mRNA expression of TLR4	246
	Phenolic rich extract	Anti-inflammatory effect	RAW 264.7 macrophage cells and dextran sulfate sodium (DSS)-induced colitis model	↓ gene expression of COX-2, TNF-α, iNOS, IL-1β, IL-6, IL-10, NF-κB	83

Table 5.2 (*Continued*)

Species	Bioactive form	Bioactive property	Model	Key findings and suggested mechanism of action	Ref.
G. frondosa	Polysaccharide-rich extract	Anti-inflammatory effect	RAW 264.7 macrophage cells	↓ production of NO, PGE2, IL-6, TNF-α and IL-1β ↓ activation of NF-κB	76
H. erinaceus	Polysaccharide	Anti-inflammatory effect	DSS-induced colitis model	↓ Serum levels of NO, IL-6, IL-1β, and TNF-α ↓ mRNA expression of COX-2 and iNOS ↓ phosphorylation of NF-κB p65, IκB-α, MAPK and Akt	77
	Erinacine C	Neuroprotection	BV-2 microglial cells	↓ production of NO, IL-6, and TNF-α ↓ mRNA expression of iNOS ↓ mRNA expression of NF-κB and IκB-α	88
I. obliquus	Lanostane-type triterpenoids	Neuroprotection	BV-2 microglial cells	↓ production of NO ↓ mRNA expression of iNOS	89
L. edodes	Acidic polysaccharides	Anti-inflammatory effect	LPS-induced mouse model	↓ serum levels of TNF-α, IL-6, and IL-1β	78
	Polysaccharides	Anti-inflammatory effect	LPS-induced mouse model	↓ serum levels of TNF-α, IL-6, and IL-1β	247
	α- and β-D-Glucans	Anti-inflammatory effect	Human monocyte THP-1 cell line	↓ secretion of TNF-α, IL-6, and IL-1β	79
L. rhinocerus	Extract	Neuroprotection	HT22 Hippocampal neuronal cell model	↓ number of apoptotic cells	248

P. baumii	Heteropolysaccharide	Anti-inflammatory effect	RAW 264.7 macrophage cells and DSS-induced colitis model	↓ phosphorylation of STAT-1 ↓ mRNA expression of iNOS and TNF-α	80
P. linteus	Hispidin	Anti-inflammatory effect	RAW 264.7 macrophage cells	↓ translocation of NF-κB and IκB degradation	249
P. eryngii	Polysaccharide	Anti-inflammatory effect	RAW 264.7 macrophage cells	↓ production of IL-1β, IL-6, IL-10, and TNF-α	250
P. multiplex	Polyozellin	Anti-inflammatory effect	*In vitro* cells (HaCaT and RBL-2H3) and DFE/DNCB-induced AD animal model	↓ gene expression levels of TNF-α, IFN-γ, IL-4, IL-13, IL-31, IL-17A, STAT1, and NF-κB	67
P. cocos	Poricoic acid C	Anti-inflammatory effect	RAW 264.7 macrophage cells	↓ production of NO ↓ gene expression of iNOS and COX-2 ↓ downregulation of NF-κB	251
	Lanostane triterpenoids	Anti-inflammatory effect	RAW 264.7 macrophage cells	↓ production of NO and PGE2 ↓ mRNA expression of iNOS and COX-2	252
P. cubensis	Psilocybin-rich extract	Anti-inflammatory effect	Human U937 macrophage cells	↓ level of TNF-α, IL-1β, IL-6, IL-10, and COX-2	253
S. commune	Exopolysaccharide	Anti-inflammatory effect	DSS-induced colitis model	↓ serum levels of IFN-γ, IL-4, IL-10, and IL-17	254
X. nigripes	Aqueous and hydroethanolic mycelial extract	Anti-inflammatory effect	RAW 264.7 macrophage cells	↓ production of NO, IL-6, TNF-α and COX-2	255

Neuroinflammation is a critical defence mechanism against infectious agents and neuronal injuries in the central nervous system (CNS). When microglia cells are activated in response to brain injury or immunological stimuli, such as LPS, amyloid beta (Aβ), and lipoteichoic acid, pro-inflammatory cytokines and neurotoxic mediators are released.[44] However, prolonged release of these inflammatory mediators may lead to aberrant phagocytosis, causing neuronal damage, thereby contributing to the pathogenesis of neurodegenerative disorders, including Parkinson's disease, Alzheimer's disease (AD), Huntington's disease, motor neuron disease, prion disease, and multiple sclerosis.[84,85] Presently, the use of non-steroidal anti-inflammatory drugs against neurodegenerative disorders only slows down the disease progression transiently, and as such, there is increasing attention to finding potent and safe neuroprotective agents from natural sources that are able to prevent or treat these disorders. Mushroom extracts have been reported to display neuroprotective effects mainly via *in vitro* studies conducted in LPS and Aβ induced BV2 microglial cells, whose mechanisms of action are largely associated with downregulation of the gene expression of different inflammatory mediators.[84] Bioactive compounds such as cordycepin, polysaccharides, and phenolic acids have been reported to present a neuroprotective effect in different *in vitro* and *in vivo* models.[66,85] Mushroom-derived triterpenes (see Table 5.2) such as triterpenoid-enriched fractions from *Antrodia cinnamomea*,[86] polyoxygenated cyathane diterpenoids from *Cyathus africanus*,[87] erinacine C from *Hericium erinaceus*,[88] and lanostane-type triterpenoids from *Inonotus obliquus*[89] have been reported to display neuroprotective effects in BV2 microglial cells via downregulation of mRNA expression of IκB-α, NF-κβ, p65, iNOS, and COX-2. Studies have also evidenced the potential of medicinal and edible mushrooms as "brain food" being utilized to manage neurodegenerative disorders. The reported neuroprotective effects of these mushrooms and their bioactive secondary metabolites have been associated with their potential to reduce beta amyloid-induced neurotoxicity, and to promote nerve growth factor (NGF) synthesis, anti-acetylcholinesterase, and anti-neuroinflammatory activity.[90] Inhibitors of acetylcholinesterase improve cholinergic function by preventing the breakdown of acetylcholine into choline and acetate. Extracts or compounds with strong inhibitory capacity against acetylcholinesterase have been widely utilized in the management of AD.[91] The most common inhibitors of acetylcholinesterase utilized as treatment options for the management of AD include galantamine, rivastigmine, and donepezil that are ineffective in the long term with several associated side effects, including hepatotoxicity, nausea, diarrhoea, lack of appetite, gastrointestinal toxicity, and other side effects.[92,93] Hence, several mushroom extracts have been utilized as complementary and alternative medicines in the form of nutraceuticals and functional foods loaded with bioactive molecules that can potentially suppress the severity of AD. Submerged polysaccharide-rich extracts from two different *Schizophyllum commune* strains were examined for their acetylcholinesterase inhibitory capacity.[94] The extracts presented promising inhibitory capacity with IC_{90} values of $79.73 \pm 26.34\ \mu g\,mL^{-1}$ in comparison to donepezil (87.92%), the positive control. Water and hydro-ethanolic extract from *Pholiota adiposa*, among other studied mushrooms, also presented up to 30.9 and 35% acetylcholinesterase (AChE) inhibitory activity, respectively.[95] Butyrylcholinesterase (BChE) inhibitors have also been reported to be viable therapeutic options in managing AD by promoting normal cholinergic function and preventing the breakdown of acetylcholine. Six macrofungi, namely *Coprinus comatus*, *Macrolepiota mastoidea*, *Agaricus campestris*, *Lycoperdon utriforme*, *Macrolepiota procera*, and *Leucoagaricus leucothites* were reported to prevent both AChE and BChE inhibition, and the promising results were expressed as galantamine equivalent.[96] A cholinesterase inhibitory effect of methanol extract rich in phenolic compounds from fruiting bodies of *Pleurotus pulmonarias* was reported to present a strong inhibitory effect against AChE and BChE, with the latter being more effective than galantamine.[97] Bio-guided fractionation performed on the fruiting bodies of mushroom has presented very promising AChE inhibitory activity. *Pleurotus florida* yielded a resveratrol-rich extract that inhibits acetylcholinesterase activity and combat oxidative stress in an *in vivo* animal model,[98] while a gallic acid-rich fraction obtained from *Ganoderma mediosinense* hydromethanolic extract inhibited AChE (IC_{50}, 0.10 ± 0.02 mg mL^{-1}).[99] Regarding the contribution of terpenoids and related compounds from mushrooms on AChE inhibitory activity, five novel meroterpenoids, possessing a γ-lactone motif obtained from *Ganoderma lucidum* presented AChE inhibitory activities with IC_{50} values of 7.37–59.86 μM. Several lanostanoids obtained from the ethyl acetate extract of *Haddowia longipes* showed very low inhibitory capacity against AChE (<40%) at 100 μM.[100] Based on the beta-amyloid cascade hypothesis, several bioactive compounds obtained from mushrooms have shown promising inhibitory effects against Aβ aggregation, thereby reducing Aβ-induced neurotoxicity.[101] Ten monoterpenoids were isolated from the chloroform extracts of *Albatrellus yasudae*, and all the tested compounds presented potential Aβ40-aggregation inhibitory capacity with IC_{50} between 12.3–221 μM, with the carboxyl and farnesyl groups exhibiting a more significant contribution to the bioactivity.[101] A growing body of scientific evidence has shown that dietary interventions involving mushroom extracts and their associated metabolites have shown promising potential to suppress the severity of neurodegenerative disorders due to their neuroprotective, antioxidant, and anti-inflammatory effects. However, studies involving clinical trials are significantly lacking and should be conducted to strengthen their utilization further in combatting these diseases.

5.3.3 Antimicrobial Properties and Impact of Bioactives on Gut Microbiota

Multiple-drug resistance in human pathogenic microorganisms has developed due to the indiscriminate use of commercial antimicrobial drugs commonly used to treat infectious diseases. This situation has forced scientists to search for new antimicrobial substances from various sources to be used as novel antimicrobial chemotherapeutic agents. The scientific community, while searching for new therapeutic alternatives, has studied many kinds of mushrooms and has found several therapeutic activities.[5,102,103]

Although many natural/synthetic antimicrobial agents have been isolated/developed to kill pathogenic microorganisms effectively, global antimicrobial resistance is an increasing public health problem.[104] Mushrooms need antibacterial and antifungal compounds to survive in their natural environment. Therefore, antimicrobial compounds could be isolated from many mushroom species and could be a benefit on humans.[3,8]

Despite the huge diversity of antibacterial compounds, bacterial resistance to first-choice substances has been drastically increasing. Moreover, the association between multiresistant microorganisms and infections highlights the problem and the urgent need for solutions. Natural resources have been exploited in recent years, and among them mushrooms could be an alternative source of new antimicrobials.[5,103,104] The research into mushrooms is extensive, and hundreds of species have demonstrated a broad spectrum of pharmacological activities, including antimicrobial activity. The two most used methodologies to assess the antimicrobial activity of mushroom extracts include the microdilution method and the disk diffusion method. The microdilution method comprises microdilutions of the extract in a liquid medium using microplates to determine the minimum inhibitory concentration (MIC) values. In the disk diffusion method, the extract is incorporated in disks at different concentrations, and the halo of growth inhibition is determined and represented by internal zone diameter (IZD) values.[5] Therefore, the results for antimicrobial activity may be expressed in different units.

Several mushroom extracts have been reported to have antimicrobial activity against Gram-positive and Gram-negative bacteria (see Table 5.3). It is presented as a mean value and, in parentheses, the range of variability of the literature data. *A. blazei* commercial sample from Brazil should be highlighted, as its ethanolic extract revealed MIC mean $= 0.144$ mg mL^{-1} and minimum bactericidal concentration (MBC) mean $= 0.225$ mg mL^{-1} against Gram-positive bacteria, like *S. aureus*, *B. cereus*, *M. flavus*, and *L. monocytogenes*, even lower than the standard ampicillin, and MIC mean $= 0.138$ mg mL^{-1} and MBC mean $= 0.225$ mg mL^{-1} against Gram-negative bacteria, such as *E. coli*, *S. enteritides*, *S. typhimurium*, and *E. cloacae*.[105] *A. bisporus* ethanolic extracts from a commercial sample from Portugal showed an inhibitory effect upon all the tested Gram-positive bacteria (MIC $= 10$ mg mL^{-1}).[106] *A. mellea* methanolic extract also revealed antimicrobial properties against Gram-positive and Gram-negative bacteria strains (MIC mean $= 7.19$ and 9.49 mg mL^{-1}, respectively).[107] The methanolic extract of *H. tessulatus* showed activity against two Gram-positive bacteria (*S. aureus* and *B. subtilis*, MIC mean $= 7.5$ mg mL^{-1}) and Gram-negative bacteria (*E. coli*, *P. aeruginosa*, *S. typhi*, and *K. pneumonia*, MIC

Table 5.3 Mushroom extracts with reported antibacterial activity against Gram-positive and Gram-negative bacteria.[a]

Mushroom	Origin	Samples	Extracts	Activity against Gram (+) and (−) bacteria	Method	Ref.
A. bisporus	Brazil	Commercial	Hydroethanol	Gram (+) bacteria: *S. aureus* and *B. cereus*	MIC = 100 (−)	44
				Gram (−) bacteria: *E. coli* and *S. enteritidis*	MIC = 150 (100–200)	
	Romania	Wild	Methanol	Gram (+) bacteria: *S. aureus*, *B. cereus*	MIC = 59.52 (−) MBC = 59.52 (−)	256
				Gram (−) bacteria: *S. typhimurium*, *P. aeruginosa*, *E. coli*	MIC = 59.52 (−) MBC = 59.52 (−)	
	Brazil	Commercial	Hydroethanol	Gram (+) bacteria: *S. aureus* and *B. cereus*	MIC = 125 (50–200)	44
				Gram (−) bacteria: *E. coli* and *S. enteritidis*	MIC = 200 (−)	
	Portugal	Commercial	Ethanol	Gram (+) bacteria: Enterococcus faecalis, MSSA and MRSA	MIC = 10 (−)	106
				Gram (−) bacteria: *E. coli* and *P. aeruginosa*	MIC ≥20	
A. brasiliensis	Brazil	Commercial	Hydroethanol	Gram (+) bacteria: *S. aureus* and *B. cereus*	MIC = 75 (50–100)	44
				Gram (−) bacteria: *E. coli* and *S. enteritidis*	MIC = 200 (−)	
A. blazei	Brazil	Commercial	Ethanol	Gram (+) bacteria: *S. aureus*, *B. cereus*, *M. flavus* and *L. monocytogenes*	MIC = 0.575 (0.075–0.20) MBC = 0.225 (0.10–0.40)	105
				Gram (−) bacteria: *E. coli*, *S. enteritides*, *S. typhimurium*, *E. cloacae*	MIC = 0.138 (0.025–0.40) MBC = 0.225 (0.05–0.60)	
A. mellea	Serbia	Wild	Methanol	Gram (+) bacteria: *S. aureus*, *B. cereus*, *L. monocytogenes*, and *M. flavus*	MIC = 7.19 (3.75–12.50) MBC = 13.61 (6.25–21.7)	107
				Gram (−) bacteria: *P. aeruginosa*, *E. coli*, *S. typhimurium*, and *E. cloacae*	MIC = 9.49 (3.12–15) MBC = 16.95 (6.25–25)	
B. edulis	Romania	Wild	Methanol	Gram (+) bacteria: *S. aureus*, *B. cereus*	MIC = 20.92 (13.49–28.34) MBC = 28.34	256
				Gram (−) bacteria: *S. typhimurium*, *P. aeruginosa*, *E. coli*	MIC = 20.92 (13.49–28.34) MBC = 28.34 (−)	
C. cibarius	Romania	Wild	Methanol	Gram (+) bacteria: *S. aureus*, *B. cereus*	MIC = 59.52 (−) MBC = 59.52 (−)	256

Table 5.3 (*Continued*)

Mushroom	Origin	Samples	Extracts	Activity against Gram (+) and (−) bacteria	Method	Ref.
				Gram (−) bacteria: *S. typhimurium*, *P. aeruginosa*, *E. coli*	MIC = 59.52 (−) MBC = 59.52 (−)	
F. velutipes	Brazil	Commercial	Hydroethanol	Gram (+) bacteria: *S. aureus* and *B. cereus*	MIC = 37.5 (25–50)	44
				Gram (−) bacteria: *E. coli* and *S. enteritidis*	MIC = (−)	
G. lucidum	Portugal	Commercial	Ethanol	Gram (+) bacteria: *E. faecalis*, *L. monocytogenes*, MSSA, and MRSA	MIC = 10 (5–20)	257
				Gram (−) bacteria: *E. coli*, *E. coli* ESBL, *K. pneumoniae*, *K. pneumoniae* ESBL, *M. morganii*, *P. aeruginosa*, and *P. mirabilis*	MIC = 16 (1–20)	
H. tessulatus	Bangladesh	Commercial	Methanol	Gram (+) bacteria: *S. aureus* and *B. subtilis*	MIC = 7.5 (7–8)	108
				Gram (−) bacteria: *E. coli*, *P. aeruginosa*, *S. typhi*, and *K. pneumonia*	MIC = 7.25 (6–9)	
L. piperatus	Romania	Wild	Methanol	Gram (+) bacteria: *S. aureus*, *B. cereus*	MIC = 41.83 (26.99–56.68) MBC = 56.68 (−)	256
				Gram (−) bacteria: *S. typhimurium*, *P. aeruginosa*, *E. coli*	MIC = 41.83 (26.99–56.68) MBC = 56.68 (−)	
L. edodes	Bangladesh	Commercial	Methanol	Gram (+) bacteria: *S. aureus* and *B. subtilis*	MIC = 14.06 (1.56–12.50)	108
				Gram (−) bacteria: *E. coli*, *P. aeruginosa*, *S. typhi* and *K. pneumonia*	MIC = 3 (−)	
	Portugal	Commercial	Ethanol	Gram (+) bacteria: Enterococcus faecalis, MSSA and MRSA	MIC = 3.33 (2.5–5)	106
				Gram (−) bacteria: *E. coli* and *P. aeruginosa*	MIC ≥20	
	Brazil	Commercial	Hydroethanol	Gram (+) bacteria: *S. aureus* and *B. cereus*	MIC = 100 (−)	44
				Gram (−) bacteria: *E. coli* and *S. enteritidis*	MIC = 3 (2–4)	
M. giganteus	Serbia	Wild	Methanol	Gram (+) bacteria: *S. aureus*, *B. cereus*, *M. flavus*, and *L. monocytogenes*	MIC = 0.23 (0.0125–0.60) MBC = 0.48 (0.035–0.60)	109
				Gram (−) bacteria: *E. coli*, *P. aeruginosa*, *S. typhimurium*, and *E. cloacae*	MIC = 0.819 (0.025–2.50) MBC = 1.72 (0.035–5.00)	

M. conica	Portugal	Wild	Methanol	Gram (+) bacteria: *S. aureus*, *B. cereus*, *M. flavus*, and *L. monocytogenes*	MIC = 2.05 (0.70-3.75) MBC = 3.99 (0.95-7.5)	110
				Gram (−) bacteria: *E. coli*, *P. aeruginosa*, *S. typhimurium*, and *E. cloacae*	MIC = 3.16 (0.70–5.62) MBC = 4.23 (0.95–7.50)	
	Serbia	Wild	Methanol	Gram (+) bacteria: *S. aureus*, *B. cereus*, *M. flavus*, and *L. monocytogenes*	MIC = 3.99 (0.95–3.75) MBC = 7.97 (1.87–15.00)	
				Gram (−) bacteria: *E. coli*, *P. aeruginosa*, *S. typhimurium*, and *E. cloacae*	MIC = 6.56 (3.75–7.5) MBC = 13.13 (7.50–15.00)	
M. esculenta	India	Wild	Butanol	Gram (+) bacteria: *S. aureus* and *B. subtilis*	IZD = 11.5 (8–15)	111
				Gram (−) bacteria: *E. coli*, *P. aeruginosa*, and *P. vulgaris*	IZD = 14 (11–19)	
			Ethyl acetate	Gram (+) bacteria: *S. aureus* and *B. subtilis*	IZD = 11 (8–14)	
				Gram (−) bacteria: *E. coli*, *P. aeruginosa*, and *P. vulgaris*	IZD = 8 (−)	
			Butanol	Gram (+) bacteria: *S. aureus*	MIC = 0.25 (−) MBC = 0.75 (−)	
				Gram (−) bacteria: *E. coli* and *P. aeruginosa*	MIC = 0.5 (0.25–0.75) MBC = 0.75 (0.5–1)	
			Ethyl acetate	Gram (+) bacteria: *S. aureus*	—	
				Gram (−) bacteria: *E. coli* and *P. aeruginosa*	MIC = 0.75 (−) MBC = 1.25 (−)	
P. ostreatus	Bangladesh	Commercial	Methanol	Gram (+) bacteria: *S. aureus* and *B. subtilis*	MIC = 6.5 (6–7)	108
				Gram (−) bacteria: *E. coli*, *P. aeruginosa*, *S. typhi*, and *K. pneumonia*	MIC = 6.5 (5–8)	
	Portugal	Commercial	Ethanol	Gram (+) bacteria: *Enterococcus faecalis*, MSSA, and MRSA	MIC = 5 (2.5–10)	106
				Gram (−) bacteria: *E. coli* and *P. aeruginosa*	MIC ≥20	
	Portugal	Commercial	Hydroethanol	Gram (+) bacteria: *S. aureus*, *B. cereus*, and *L. monocytogenes*	MIC = 0.32 (0.25–0.35) MBC = 0.58 (0.35–0.70)	51
				Gram (−) bacteria: *E. coli*, *E. cloacae*, and *S. typhimurium*	MIC = 0.28 (0.25–0.35) MBC = 0.47 (0.35–0.70)	

Table 5.3 (*Continued*)

Mushroom	Origin	Samples	Extracts	Activity against Gram (+) and (−) bacteria	Method	Ref.
	Romania	Wild	Methanol	Gram (+) bacteria: *S. aureus* and *B. cereus*	MIC = 59.52 (−) MBC = 59.52 (−)	256
				Gram (−) bacteria: *S. typhimurium*, *P. aeruginosa*, and *E. coli*	MIC = 59.52 (−) MBC = 59.52 (−)	
P. sajor-caju	Brazil	Commercial	Hydroethanol	Gram (+) bacteria: *E. faecalis*, *L. monocytogenes*, MSSA, and MRSA	MIC = 10 (−)	258
				Gram (−) bacteria: *P. aeruginosa*, *E. coli*, ESBL, *K. pneumoniae*, *K. pneumoniae* ESBL, and *M. morganii*	MIC = (−)	
P. squamosus	Portugal	Wild	Methanol	Gram (+) bacteria: *S. aureus*, *B. cereus*, *M. flavus*, and *L. monocytogenes*	MIC = 2.00 (0.20–3.13) MBC = 4.00 (0.39–6.25)	259
				Gram (−) bacteria: *E. coli*, *P. aeruginosa*, *S. typhimurium*, and *E. cloacae*	MIC = 2.05 (0.39–3.13) MBC = 4.10 (0.78–6.25)	
	Serbia	Wild	Methanol	Gram (+) bacteria: *S. aureus*, *B. cereus*, *M. flavus*, and *L. monocytogenes*	MIC = 1.23 (0.40–1.50) MBC = 2.44 (0.75–3.00)	
				Gram (−) bacteria: *E. coli*, *P. aeruginosa*, *S. typhimurium*, and *E. cloacae*	MIC = 0.80 (0.40–1.50) MBC = 1.79 (1.14–3.00)	
	Romania	Wild	Methanol	Gram (+) bacteria: *B. cereus*, *M. flavus*, and *S. aureus*	MIC = 10.6 (1.2–20.4) MBC = 21.2 (2.4–40.8)	260
				Gram (−) bacteria: *E. coli*, *P. aeruginosa*, *S. typhimurium*, *L. monocytogenes*, and *E. cloacae*	MIC = 16.44 (0.61–20.4) MBC = 32.8 (1.2–40.8)	
R. integra	Serbia	Wild	Ethanol	Gram (+) bacteria: *M. luteus*, *R. mucilaginosa*, *S. agalactiae*, *S. angiosus*, *S. conselatus*, *S. dysgalactiae*, *S. oralis*, *S. parasanquinis*, *S. pseudopneumoniae*, *S. pyogenes*, *S. salivarius*, *S. aureus*, *S. hominis*, and *S. warneri*	MIC = 4.61 (0.78–12.50) MBC = 7.42 (1.56–12.50)	240
				Gram (−) bacteria: *E. cloacae* and *S. maltophilia*	MIC = 7.81 (3.12–12.50) MBC = 9.38 (6.25–12.50)	

			Methanol	Gram (+) bacteria: *M. luteus*, *R. mucilaginosa*, *S. agalactiae*, *S. angiosus*, *S. conselatus*, *S. dysgalactiae*, *S. oralis*, *S. parasanquinis*, *S. pseudopneumoniae*, *S. pyogenes*, *S. salivarius*, *S. aureus*, *S. hominis*, and *S. warneri*	MIC = 3.74 (0.39–12.50) MBC = 5.40 (0.78–12.50)
				Gram (−) bacteria: *E. cloacae* and *S. maltophilia*	MIC = (−) MBC = (−)
R. nigricans	Serbia	Wild	Ethanol	Gram (+) bacteria: *M. luteus*, *R. mucilaginosa*, *S. agalactiae*, *S. angiosus*, *S. conselatus*, *S. dysgalactiae*, *S. oralis*, *S. parasanquinis*, *S. pseudopneumoniae*, *S. pyogenes*, *S. salivarius*, *S. aureus*, *S. hominis*, and *S. warneri*	MIC = 3.66 (1.56–7.50) MBC = 6.73 (3.12–12.50)
				Gram (−) bacteria: *E. cloacae* and *S. maltophilia*	MIC = 3.91 (1.56–6.25) MBC = 7.81 (3.12–12.50)
			Methanol	Gram (+) bacteria: *M. luteus*, *R. mucilaginosa*, *S. agalactiae*, *S. angiosus*, *S. conselatus*, *S. dysgalactiae*, *S. oralis*, *S. parasanquinis*, *S. pseudopneumoniae*, *S. pyogenes*, *S. salivarius*, *S. aureus*, *S. hominis*, and *S. warneri*	MIC = 3.27 (0.39–12.50) MBC = 5.01 (0.78–12.50)
				Gram (−) bacteria: *E. cloacae* and *S. maltophilia*	MIC = 3.91 (3.12–12.50) MBC = 6.25 (−)
R. rosea	Serbia	Wild	Ethanol	Gram (+) bacteria: *M. luteus*, *R. mucilaginosa*, *S. agalactiae*, *S. angiosus*, *S. conselatus*, *S. dysgalactiae*, *S. oralis*, *S. parasanquinis*, *S. pseudopneumoniae*, *S. pyogenes*, *S. salivarius*, *S. aureus*, *S. hominis*, and *S. warneri*	MIC = 2.93 (0.20–12.50) MBC = 4.26 (0.39–12.50)
				Gram (−) bacteria: *E. cloacae* and *S. maltophilia*	MIC = 9.38 (6.25–12.50) MBC = 12.50 (−)

Table 5.3 (*Continued*)

Mushroom	Origin	Samples	Extracts	Activity against Gram (+) and (−) bacteria	Method	Ref.
			Methanol	Gram (+) bacteria: *M. luteus*, *R. mucilaginosa*, *S. agalactiae*, *S. angiosus*, *S. conselatus*, *S. dysgalactiae*, *S. oralis*, *S. parasanquinis*, *S. pseudopneumoniae*, *S. pyogenes*, *S. salivarius*, *S. aureus*, *S. hominis*, and *S. warneri*	MIC = 3.41 (0.20–12.50) MBC = 5.30 (0.39–12.50)	
				Gram (−) bacteria: *E. cloacae* and *S. maltophilia*	MIC = 6.25 (−) MBC = 12.50 (−)	
T. aestivum	Spain		Ethyl acetate	Gram (+) bacteria: *L. monocytogenes*	IZD = 8.5	
T. gennadii	Spain	Wild	Methanol	Gram (−) bacteria: *S. flexneri*	IZD = 8.1	261
T. magnusii	Italy		Water	Gram (+) bacteria: *S. aureus* and *M. luteus*	IZD = 13.9 (8.1–19.7)	
T. melanosporum	Spain		Water	Gram (+) bacteria: *S. aureus*	IZD = 8.7	
V. bohemica	India	Wild	Butanol	Gram (+) bacteria: *S. aureus* and *B. subtilis*	IZD = 10.5 (9–12)	111
				Gram (−) bacteria: *E. coli*, *P. aeruginosa* and *P. vulgaris*	IZD = 11.7 (9–15)	
			Ethyl acetate	Gram (+) bacteria: *S. aureus* and *B. subtilis*	IZD = 13.5 (12–15)	
				Gram (−) bacteria: *E. coli*, *P. aeruginosa* and *P. vulgaris*	IZD = 11 (9–13)	
			Butanol	Gram (+) bacteria: *S. aureus*	MIC = 0.25 (−) MBC = 0.5 (−)	
				Gram (−) bacteria: *E. coli* and *P. aeruginosa*	MIC = 0.25 (−) MBC = 0.75 (−)	
			Ethyl acetate	Gram (+) bacteria: *S. aureus*	—	
				Gram (−) bacteria: *E. coli* and *P. aeruginosa*	MIC = 0.75 (−) MBC = 1 (−)	

[a](−) MIC and MBC not found for the concentrations tested; MSSA, meticillin-sensitive *Staphylococcus aureus*; MRSA, meticillin-resistant *Staphylococcus aureus*; ESBL, extended spectrum β-lactamases; MIC/MBC, $mg\,mL^{-1}$; IZD, inhibition zone diameters (mm).

mean = 7.25 mg mL^{-1}).[108] *L. edodes* hydroethanolic extract demonstrated good activity against two Gram-negative bacteria, *E. coli* and *S. enteritidis* (MIC mean = 3 mg mL^{-1}).[44] The methanolic extracts of *M. giganteus* wild samples from Serbia showed an antibacterial effect against Gram-positive and Gram-negative bacteria with MIC means of 0.23 and 0.819 mg mL^{-1}, respectively.[109] *M. conica* methanolic extracts from wild samples from Serbia and Portugal also presented an inhibitory effect upon all the tested Gram-positive and Gram-negative bacteria.[110] *P. ostreatus* hydroethanolic extracts from commercial samples from Portugal showed an antibacterial effect against Gram-positive and Gram-negative bacteria with MIC means of 0.32 and 0.28 mg mL^{-1}, respectively.[51]

Shameem *et al.*[111] tested the antimicrobial activity of butanol extract from *M. esculenta* and revealed a significant antibacterial effect against *S. aureus* with MIC and MBC means of 0.25 and 0.75 mg mL^{-1}, respectively and, *E. coli* and *P. aeruginosa* presented MIC and MBC means of 0.75 and 1.25 mg mL^{-1}, respectively.

Data available from the literature indicate that mushroom species showed antimicrobial activity against Gram-positive and Gram-negative bacteria, particularly, against pathogenic microorganisms, including bacteria associated with nosocomial infections (*P. aeruginosa*, *L. monocytogenes*, *S. aureus*, *K. pneumoniae*, *M. morganii*, *P. mirabilis*, *etc.*) and multiresistance (MRSA).

As Table 5.4 shows, different extracts obtained from several mushroom species have been described in the literature as possessing antifungal activity. Stojković *et al.*[109] reported antifungal activity of *M. giganteus* methanolic extracts against *A. fumigatus*, *A. versicolor*, *A. ochraceus*, *A. niger*, *T. viride*, *P. funiculosum*, *P. ochrochloron*, and *P. aurantiogriseum* with MIC and MFC means of 0.138 and 0.31 mg mL^{-1}, respectively. Ethanolic extract of *A. blazei* showed activity against *A. fumigatus*, *A. versicolor*, *A. ochraceus*, *A. niger*, *C. crusei*, *P. funiculosum*, and *P. verrucosum* var. *cyclopium* with MIC and MFC means of 0.75 and 0.94 mg mL^{-1}.[105] Hydroethanolic extract of *P. ostreatus* seems promising against some fungal mentioned before with MIC and MFC means of 0.30 and 0.49 mg mL^{-1}, respectively.[51]

Most studies on mushrooms with antibacterial and antifungal activities describe the action of the extracts, focused on screening without identifying the compounds responsible for this activity.[5] Prebiotics are substances that induce the growth or action of microorganisms that contribute to the well-being of their host. Particularly, mushrooms are considered a potential source of prebiotics as they contain phenolic compounds and several active polysaccharides, such as chitin, hemicellulose, β- and α-glucans, mannans, xylans, and galactans, which make them the right choice for prebiotics. Different bioactive compounds from mushrooms have been shown to alter gut microbiota and improve health status. The gut microbiota comprises trillions of bacteria that contribute to nutrient acquisition and energy regulation. Current research indicates that diet is the most important factor as it significantly affects the composition, diversity, and richness of gut microbiota. The microorganisms present in the gut play an important role in

Table 5.4 Mushroom extracts with reported antifungal activity.[a]

Mushroom	Origin	Samples	Extracts	Activity against	Method	Ref.
A. mellea	Serbia	Wild	Methanol	*A. niger, A. ochraceus, A. versicolor, A. fumigatus, T. viride, P. funiculosum, P. ochrochloron, P. verrucosum* var. *cyclopium*, and *C. albicans*	MIC = 7.28 (6.25–15) MFC = 12.81 (6.25–25)	107
A. blazei	Brazil	Commercial	Ethanol	*A. fumigatus, A. versicolor, A. ochraceus, A. niger, C. crusei, P. funiculosum*, and *P. verrucosum* var. *cyclopium*	MIC = 0.57 (0.20–0.80) MFC = 0.94 (0.40–1.60)	105
G. lucidum	Portugal	Commercial	Ethanol	*C. albicans*	MIC ≥20	257
H. tessulatus	Bangladesh	Commercial	Methanol	*C. albicans* and *S. cerevisiae*	MIC = 5.5. (5–6)	108
L. edodes	Bangladesh	Commercial	Methanol	*C. albicans* and *S. cerevisiae*	MIC = 1 (–)	
M. giganteus	Serbia	Wild	Methanol	*A. fumigatus, A. versicolor, A. ochraceus, A. niger, T. viride, P. funiculosum, P. ochrochloron*, and *P. aurantiogriseum*	MIC = 0.138 (0.025–0.30) MFC = 0.31 (0.050–0.40)	109
M. conica	Portugal	Wild	Methanol	*P. aurantiogriseum, A. versicolor, A. niger*, and *T. viride*	MIC = 3.59 (0.78–6.25) MFC = 5.94 (1.56–12.5)	110
	Serbia				MIC = 9.06 (1.56–12.5) MFC = 13.12 (3.12–25.00)	
M. esculenta	India	Wild	Butanol	*A. niger, C. albicans, C. kruesie, C. paraloposis*, and *A. fumigates*	IZD = 17.25 (11–18)	111
			Ethyl acetate	*A. niger, C. albicans, C. kruesie, C. paraloposis* and *A. fumigates*	IZD = 18 (–)	

P. squamosus	Portugal	Wild	Methanol	*A. fumigatus, A. ochraceus, A. versicolor, A. niger, P. funiculosum, P. ochrochloron, P. aurantiogriseum*, and *T. viride*	MIC = 1.46 (0.78–3.13) MFC = 2.93 (1.56–6.25)	259
	Serbia	Wild	Methanol	*A. fumigatus, A. ochraceus, A. versicolor, A. niger, P. funiculosum, P. ochrochloron, P. aurantiogriseum*, and *T. viride*	MIC = 0.71 (0.40–1.50) MFC = 1.41 (0.75–3.00)	
P. ostreatus	Bangladesh	Commercial	Methanol	*C. albicans* and *S. cerevisiae*	MIC = 4 (−)	108
	Portugal	Commercial	Hydroethanol	*A. fumigatus, A. ochraceus, A. niger, P. funiculosum, P. ochrochloron*, and *P.v. cyclopium*	MIC = 0.30 (0.08–0.75) MFC = 0.49 (0.35–0.70)	51
V. bohemica	India	Wild	Butanol	*A. niger, C. albicans, C. kruesie, C. paraloposis*, and *A. fumigates*	IZD = 14 (9–22)	111
			Ethyl acetate	*A. niger, C. albicans, C. kruesie, C. paraloposis*, and *A. fumigates*	IZD = 10 (7–14)	

[a]MIC/MFC, mg mL^{-1}; IZD, inhibition zone diameters (mm).

the health of the digestive system and influence the immune system. The medicinal metabolites from mushrooms can act as immunomodulatory agents to activate gut microbiota.[112–114] These bioactive ingredients can modulate the gut microbiota and improve body health through various regulatory mechanisms, including enriching microbiota diversity, promoting beneficial bacteria, increasing the cacteroides/firmicutes ratio, decreasing harmful bacteria, and increasing short-chain fatty acid-producing bacteria.[114]

5.3.4 Immunomodulatory and Cytotoxic Properties

The immune system is responsible for recognizing, drowning, and fighting against diseases and toxic microorganisms. Through what is called immunity of two types, acquired or innate immune systems, the immune system acts in multifaceted immune responses to control and eliminate harmful organisms or substances (see Figure 5.4).[115,116] Immunomodulators are substances that work efficiently with the immunologic system to increase or decrease the host response against certain microorganisms.[117] In the immune system, macrophages are the first cells to recognize pathogens. When macrophages are activated, they can fight infectious agents directly by phagocytosis or indirectly by synthesizing and secreting nitric oxide (NO), reactive oxygen species (ROS), interleukins (IL), and tumour necrosis factor-α (TNF-α), which act as regulators of the immune system.

Lately, edible mushrooms have significantly been investigated, as there has been a considerable increase in their production and consumption due to their taste, aroma, and nutritional values, and they possess several bioactive molecules, with stimulant, therapeutic, and medicinal properties for health, which make them potential sources of pharmaceutical drugs and functional foods.[118,119] Edible mushrooms are more aptly considered nutraceutical foods, and nowadays, there is great interest in the medicinal use

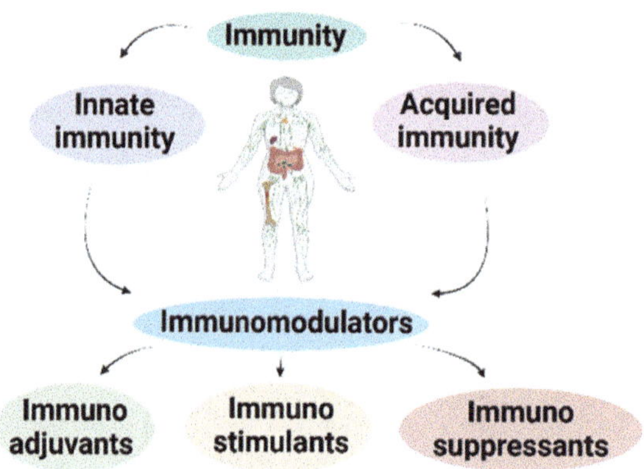

Figure 5.4 Immunity and immunomodulator types.

of those mushrooms which have bioactive compounds with anti-inflammatory, antidiabetic, anticancer, immunomodulatory properties, among others.[120–122] Mushrooms belonging to the Basidiomycota and Ascomycota phylum stand out for their anticancer and immunomodulatory properties. These include *Pleurotus ostreatus*, *Agaricus blazei* Murrill, *Polyporus umbellatus*, *Dictyophora indusiata*, *Lentinus edodes*, *Ganoderma lucidum*, *Inonotus obliquus*, *Boletus edulis*, *Grifola frondosa*, *Calocybe indica*, *Hericium erinaceus*, *Phellinus linteus*, *Ganoderma formosanum*, *Sparassis crispa*, and *Cordyceps militaris*, among others (see Table 5.5).[118,123] The anticancer activities of mushroom polysaccharides (see Table 5.5) have been extensively studied in recent years. The activation mechanisms of these polysaccharides against cancer cells involve a complex sequence of responses, inducing both innate and adaptive immune systems.[124] Polysaccharides obtained from natural matrices possess immense beneficial properties, such as antitumour and immunomodulatory activities. They help the body resist diverse biological tensions and increase immunity against the progression of cancer cells, thus boosting immunity by stimulating the central biological systems.[125] Immunomodulatory polysaccharides have been reported to present beneficial effects on the regulation of macrophage immune function.[126,127] When activated by mushroom polysaccharides, the innate immune system stimulates the body's defence mechanisms by activating macrophages, T-lymphocytes, B-lymphocytes, cytotoxic T-lymphocytes (CTL), and natural killer cells to produce cytokines such as tumour necrosis factor-alpha (TNF-α), interferon-gamma (IFN-γ), and interleukin 1 beta (IL-1β), and they have invariable antiproliferative activity.[124,128] Edible mushrooms have been much highlighted because of their immense benefits and bioactive compounds, and principally their immunomodulatory activity. Additionally, some species have been studied recently for their immunomodulatory and cytotoxic properties. Polysaccharides of mushrooms are found principally in the cell walls, in the fruiting bodies where they concentrate the most significant quantity,[129,130] and glycogen, cellulose, and starch are the most commonly occurring polysaccharides.[131] β-Glucans are glucose polymers linked up together through 1,3 linear β-glycosidic chains and are essential active constituents derived from mushrooms. Furthermore, depending on the source of β-glucans, variations in chain size occur, and many of these variations will determine their bioactive properties. Therefore, because of their health benefits, there has been growing interest in their use in treating disease.[127,132] Lately, edible mushrooms have significantly been investigated, as there has been a considerable increase in their production and consumption due to their taste, aroma, and nutritional values, and they possess several bioactive molecules, with stimulant, therapeutic, and medicinal properties for health, which make them potential sources of pharmaceutical drugs and functional foods.[118,119] Edible mushrooms are more aptly considered nutraceutical foods, and nowadays, there is great interest in the medicinal use of these mushrooms that have bioactive compounds with anti-inflammatory, antidiabetic, anticancer,

Table 5.5 Immunomodulatory and cytotoxic effect of mushroom extracts and their compounds.

Species	Compounds	Experimental models	Immunomodulatory/cytotoxic effects	Ref.
A. auricula-judae	Polysaccharide	RAW 264.7 cells	Promote the release of NO and secretion of TNF-α, IL-6, and IL-10	262
A. bitorquis	Polysaccharide	RAW 264.7 cells	Promoted viability and phagocytic ability, increasing levels of NO, ROS, TNF-α	122
A. bisporus	Polysaccharide	SW 620 cells	Cell growth inhibition	263
A. bisporus and *L. edodes*	—	PC3 cells	Decreased nuclear and total NF-κB activity/ decreased the proliferation of cells	264
A. blazei	—	MIAPaCa-2, PCI-35, and PK-8 cells	Inhibited cell proliferation and upregulated of pro-apoptotic	265
A. blazei, *H. erinaceus*, and *G. frondosa*	—	KG1a, HL 60, Meg 01, INA-6, RPMI-8226, and U226 cells	Caused a cytotoxic and immunomodulatory effect in cells	266
A. camphorata	—	Mice, PLC/PRF/5, HepG2, SMMC-7721, Huh7, MHCC 97, MHCC 97L, HCCLM3, and MIHA cells	Downregulated protein levels of phosphorylated and total STAT3 and JAK2 in cells/reduced cell viability, induced apoptosis, and retarded migration and invasion in cultured cells	267
	—	SW620 and SW480 cells	Inhibitory effect on emigration and invasion of cancer cell	268
	—	B16-F0 cells	Exhibited cytotoxic effect and inhibited the migration ability of cells.	269
A. cinnamomea	—	Mice, LLC, and CL1-5 cells	Inhibited tumour growth and metastasis	270
A. cinnamomea	—	Mice and T47D cells	Inhibited cells proliferation, induced autophagy, decreased tumour volume, and inhibited tumour growth	271
A. mellea	Polysaccharide	RAW264.7 cells	Increases phagocytosis of macrophages and induces secretion of NO, ROS, TNF-α, IL-1β, and IL-6	129
B. edulis	Phenolic and flavonoid	MCF-7 cells	Showed antiproliferative effects	224
	—	MCF-7 cells	Antiproliferative effects against the cell line	224
C. cibarius, *C. comatus*, *L. perlatum*, and *L. deliciousus*	—	SVGp12, U87MG, and LN-18 cells	Inhibition of proliferation and induction of apoptosis	121
C. militaris	Polysaccharide	RAW 264.7 cells	Release of NO, TNF-α, IL-6, and IL-10, and mainly induced M1 polarization of macrophages	272

	Polysaccharide	Mice, splenic lymphocytes, RAW264.7 cells, YAC-1 lymphoma cell, and natural killer (NK) cells	Promoted the proliferation of lymphocytes and promoted lymphocyte secretion of the NO, TNF-α, and IL-2, strengthened the phagocytosis of macrophages and induced M1 polarization/enhanced cytotoxicity of NK cells	273
C. versicolor	Polysaccharide	RAW 264.7 cells	Induction of NO production and induced nitric oxide synthase (iNOS) and TNF-αmRNA expression level	274
C. radicata	Polysaccharide	RAW 264.7 cells	Improved the proliferation and phagocytosis and induced the secretion of NO, iNOS, TNF-α, IFN-γ, IL-6, IL-1β and IL-10	275
C. versicolor and *L. edodes*	Polysaccharide	Peripheral blood mononuclear cells	Reduction of Th2 cytokines and IL-10 in cells cultures	128
F. velutipes	Polysaccharide	RAW 264.7 cells	Regulate the expression of NO, TNF-α, IL-6	276
G. frondosa	Polysaccharide	Mice and H22 hepatoma cells	Increasing the percentage of $CD4^+CD8^+$ T cells, B cells, and Treg cells, and the secretion of IL-2, IL-12p70, TNF-α, and IFN-γ/inhibit liver tumours growth	134
G. lucidum	Polysaccharide	Mice, MDA-MB-231, and 4T1 cells	Significantly suppressed the growth of tumours	277
	Polysaccharide	Mice	Produced better protection of the spleen and thymus, promoting haematopoiesis, and improving IgA levels in serum	278
	Polysaccharide	GBM8901, U87, MDA-MB-231, and 4T1 cells	Inhibited the growth of cancer cells, inducing apoptosis, inhibited cell migration and cell cycle arrest	277 and 279
	Polysaccharide	Mice	Protect spleen and thymus and promoting haematopoiesis, and improve IgA levels	278
G. neo-japonicum	Terpenoid and sterol	CCD-18Co, HCT 116, and HT 29 cells	Exerted a potent cytotoxic effect on colonic carcinoma cells	280
G. applanatum	—	SEC and Caco-2 cell	Decrease in the volume of the tumour mass and tumour growth	281
G. atrum	Polysaccharide	T lymphocytes	Increased the secretion of IL-2, IFN-γ, and IL-12	282
G. leucocontextum	Polysaccharide	RAW 264.7 cells	Activating mitogen-activated protein kinases (MAPKs), phosphatidylinositol-3-kinase (PI3K)/Akt, and nuclear factor-kappa B (NF-κB) signalling pathways	283

Table 5.5 *(Continued)*

Species	Compounds	Experimental models	Immunomodulatory/cytotoxic effects	Ref.
G. tsugae	—	AN3 CA, HEC-1-A and KLE cells	Inhibited cell proliferation, induced cell cycle arrest, and inhibited the Akt signalling pathway	284
G. frondosa	Polysaccharide	Mice	Improve or reverse the CTX-induced immunosuppression, enhance the spleen and thymus indices, spleen lymphocyte proliferation and cytokine production	285
	—	SMMC-7721 cells	Reduction of cell viability and induction of cell apoptosis	286
	Polysaccharide	RAW 264.7 cells	Increased TNF-α, IL-6, IFN-γ, MIP-1β, MIP-1α, and MIP-2 levels	287
G. lucidum	Polysaccharides and triterpenes	Mice, SNU719, and MKN1-EBV cells	Induced lytic reactivation of the Epstein–Barr virus – risk factor of gastric cancer/induced cancer cell apoptosis	288
	Polysaccharide	Rats/ RG2 glioma cells and human myelogenous leukaemia cells K562	Increase the concentration of serum IL-2, TNF-α, and INF-γ/enhance the cytotoxic activity of NK cells and T cells in glioma-bearing rats.	289
	—	Mice, HEK293T, and HEK293FT Cells, MDA-MB-231 cell, and 4T1 cell	Inhibited Wnt-induced hyperproliferation of cancer cells	290
	Triterpenoids	Mice and HCT16 cells	Inhibited cell migration, proliferation, and xenograft tumour growth	291
H. erinaceus	Polysaccharide	Human monocytic cell lines THP-1	Enhance the levels of TNF-α, IL-1β, and IL-6 and induce the proliferation of lymphocytes	292
I. baumii	—	SMMC-7721 cells	Induce apoptosis and autophagy in tumours cells, and inhibit the growth of tumours	293
I. obliquus and *P. linteus*	—	Mice, yeast, RAW264.7, A2780, HeLa, AML1, S18, and CL-81 cells	Enhanced the phagocytosis of macrophages/ inhibitory effect on tumour development and progression	294
L. edodes	Polysaccharide	Mice and MCF-7 cells	Induction of apoptosis	295
	—	Hep-2 and MRC-5 cells	Inhibition of cell proliferation and induced depolarization of mitochondria	296
L. volemus	Polysaccharide	RAW 264.7, H1299, and MCF-7 cells	Induced the secretion of NO, IL-6, and TNF-α/inhibit the proliferation of cells	136

L. tigris	—	Mice and MCF-7, A549, PC3, and 184B5 cells	Show antiproliferative effect and inhibited the growth of cells	297
M. importuna	Polysaccharide	RAW 264.7 and 4T1 cells	Promote the secretion of TNF-α, IL-6, and NO and enhance the phagocytosis of macrophages/ enhance the antitumour activity of doxorubicin (DOX) and inhibit the growth of tumours	298 and 299
M. oreades	—	MCF-7, MDA-MB-231, and HT-29 cells	Inhibited all human cancer cell lines	300
M. giganteus	—	Jurkat and HL-60 cells	Induced apoptosis in Jurkat and HL-60 cell and induced cell-cycle arrest	301
M. esculenta	Fatty acids and sterols	A549, H1264, H1299, and Calu-6 cells	Induction of apoptosis	302
P. highking	—	MCF-7 cells	Reduced the proliferation and viability of cells and reduced size and number of the tumour spheres	303
P. nameko	Polysaccharide	Mice and MCF-7 cells	Activated the death receptor pathway and mitochondrial apoptosis pathway/inhibited malignant proliferation of tumours	304
P. eryngii	Polysaccharide	THP-1 cells	Increased the secretion of IL-1β, IL-10, and NO	305
	Protein	Mice, CCD-18Co, HCT116, and MC38 cells	Suppressed tumour development and proliferation-induced cell cycle arrest and led to cellular apoptosis	306
	Polysaccharide	Mice	Increased concentrations of TNF-α, IFN-γ, IL-1, IL-2, and IL-6	307
P. highking	—	MDA-MB-231 and HCC-1937 cells	Reduced the number and size of the tumour-spheres, suppressed the migratory ability of the cells, and reduced the mRNA expression of Ki-67, MMP-9, and vimentin in the treated tumour-sphere cells	308
P. ostreatus	—	KG-1 and Jurkat cells	Inhibited cell proliferation, migration, expression, and promoted apoptosis	309
P. ostreatus	—	KG-1 cells	Induced apoptosis and increased anticancer effects	310
P. ferulae	—	Mice, H22 and HepG2 cells	Inhibited the growth of cells, increased ROS generation, and suppressed migration	311
P. sajor-caju	—	HCT116 and MRC-5 cells	Induced apoptosis and cell cycle arrest	312

Table 5.5 (*Continued*)

Species	Compounds	Experimental models	Immunomodulatory/cytotoxic effects	Ref.
P. grammocephalus	Polysaccharide	Mice and RAW 264.7 cells	Activation of macrophage and splenocyte and thymocyte stimulatory properties	120
P. rhinocerus	Polysaccharide	Murine immature BMDCs	Induced BMDC maturation, upregulated the expression of membrane phenotypic marker, stimulated the release of MIP-1α, MIP-2, and IL-2, and upregulation of the expression of IL-2, IL-6, MIP-1α, MIP-2, RANTES, IL-12p40p70, IL-12p70, TIMP-1, IFN-γ, KC, MCP-1, and GCSF	313
S. imbricatus	—	Mice and MDA-MB-231, MDA-MB-468, MCF-7, MT-1, and 4T1 cells	Inhibited the proliferation and invasive properties of breast cancer cells and decreased the tumour volume and weight in mice	314
S. commune	Polysaccharide	RAW264.7 cells	Produced NO and cytokines by upregulating mRNA expression levels	315
T. lobayense	Polysaccharide	RAW264.7 cells	Enhanced the phagocytic uptake capacity, increased NO production, increased secretion of IL-6 and TNF-α	316
T. versicolor	Polysaccharide	Hela cells	Induction of cell death by apoptosis	317
T. heimii	Polysaccharide	Mice, Vero, and HCT cells	Diminished hyperplasia in 1,2-dimethylhydrazine (DMH)-induced colon cancer	318

immunomodulatory properties, among others.[120–122] The polysaccharide from *Pleurotus citrinopileatus* revealed antitumor activity on H22 tumour-bearing mice.[133] *Coprinus comatus* and *Lactarius deliciosus* showed an inhibitory effect on U87MG and LN-18 glioblastoma cells viability.[121] *Grifola frondosa* contains promising immunomodulatory agents and effective antitumour polysaccharides.[134] The polysaccharide of *Agaricus bitorquis* shows an immunomodulatory effect on RAW264.7.[122] *Pholiota nameko* induces apoptosis of human breast adenocarcinoma.[135] Polysaccharides from *Lactarius volemus* were investigated for antiproliferative and immunomodulatory activities on RAW264.7 murine macrophages, human lung cancer cells H1299, and human breast cancer cells MCF-7.[136] More studies involving mushrooms and their immunomodulatory and cytotoxic properties are presented in Table 5.5.

5.3.5 Cardioprotective Properties

Cardiovascular diseases (CVD) are among the leading cause of death in the world, afflicting the heart and blood vessels. They include high blood pressure, coronary heart disease, cerebrovascular disease, atherosclerosis, heart failure, and peripheral vascular disease.[137] Diets rich in functional ingredients such as antioxidants, dietary fibre, omega-3 polyunsaturated fatty acids (PUFAs), vitamins, minerals, and phenolic compounds have been associated with a lower risk of cardiovascular disease in numerous studies.[138,139] Cholesterol has long been considered a significant risk factor for heart diseases. Plant sterols, also known as phytosterols, are found in several plant species and have been shown to compete with dietary cholesterol by inhibiting absorption and enabling its elimination from the body.[137] Mycosterols including ergosterol (ergosta-5,7,22-trien-3*b*-ol), ergosta-5,8,22-trien-3-ol, ergosta-7,22-dien-3-ol, ergosta-5,7-dien-3-olm and fungisterol (ergosta-7-en-3-ol) either occurring in their free or esterified forms have gained prominence in recent years due to their well-reported beneficial effects on human health.[140,141] Because of their structural similarities to the dietary mixed micelles (DMM) during *in vitro* digestion, these biomolecules have a cholesterol-lowering impact by competing with cholesterol. In simulated gastrointestinal digestion utilizing Caco-2 cells, ergosterol-rich extracts produced by supercritical fluid extraction (SFE) from *Agaricus bisporus* were found to successfully displace cholesterol (up to 67%) from the DMM.[142] The extract mentioned above's precise mechanism of action was linked to a significant reduction in mRNA expression of sterol regulatory element-binding protein 2 (SREBF2), a protein-coding gene that regulates the expression of numerous genes involved in cholesterol homeostasis, absorption, and biosynthesis.[139] A diet supplemented with an antioxidant-rich extract from *A. brasiliensis* was reported to significantly reduce the total cholesterol level *in vivo* and downregulate mRNA expression of cholesterol metabolism genes, including SREBP-2, HMG-CoA reductase, LDLR, and PPAR-α.[143] The effect of ergosterol peroxide derived from *Ganoderma lucidum* on SREBP-1 mRNA expression in 3T3-L1 preadipocyte cells has also been described.[144] The cholesterol-lowering effect of lard functionalized with extracts from *P. ostreatus* and *L. edodes* rich in β-glucans, water-soluble polysaccharides, and ergosterol was analysed.[138] A significant reduction (up to 22 – 42%) in plasma levels of total cholesterol was observed in the studied *in vivo* model. However, mRNA expressions of cholesterol-related genes (NPC1L1 and ABCG5) in the jejunum, cecum, and liver were not suppressed. Although the precise molecular mechanism is still not completely defined, the cardioprotective effect was associated with hindering the bioavailability of cholesterol and inhibiting the expression of several genes responsible for cholesterol absorption and metabolism in intestinal models.[143,144]

Atherosclerosis is a multifaceted disease that contributes to the pathogenesis of CVDs, characterized by endothelial dysfunction, lipid deposition, inflammatory cell infiltration, smooth muscle cell proliferation, intravascular wall neoangiogenesis, and plaque development.[145] Although the exact cause of atherosclerosis is unknown, persistent inflammation of the artery walls triggered by procoagulant proteins, proinflammatory cytokines, metalloproteinases, and growth factors have all been identified as major contributions to the disease's aetiology.[145,146] Commonly consumed mushrooms, such as *A. bisporus*, *L. edodes*, and *P. ostreatus*, abundant in nutrients, soluble dietary fibres, and several health-promoting bioactive compounds, are beneficial in suppressing inflammation, improving lipid profiles, and ameliorating heart disease and atherosclerosis.[146] The potential of *A. bisporus* (portobello) and *L. edodes* extracts rich in ergothioneine in ameliorating atherosclerosis was conducted *in vivo*. The results showed a significant decrease in plasma lipid concentrations (cholesterol and triglycerides), mRNA expression of vascular cell adhesion molecule 1 (VCAM-1), and mean aortic lesion area.[146] *In vivo* studies involving the infusion of angiotensin II in a mouse model and subsequent supplementation with a diet rich in *Grifola gargal* extract presented a significant decrease in the weight of the heart, vessels, and atheromatous lesions. In addition, the extract also induces the secretion of transforming growth factor beta (TGF-β) and vascular endothelial growth factor (VEGF), both vital signalling molecules responsible for promoting angiogenesis.[147] Several studies have shown that mushroom extracts and their associated bioactive molecules possess antioxidant, anti-inflammatory, and lipid-lowering effects that can potentially slow down the progression of atherosclerosis, thereby potentiating a cardioprotective effect.[148] The mechanism behind these cardioprotective properties of mushroom extracts remains obscure but it has been attributed to suppression of mRNA expression of several genes related to cholesterol absorption and metabolism, decreased oxidative damage due to their potent radical scavenging and metal chelating effect, and downregulation of the expression of inflammatory mediators.[143,146] To the best of our knowledge, no studies have been conducted in clinical settings to evidence the high therapeutic potential of these mushroom metabolites against any cardiovascular diseases. Hence, more work still needs to be performed to determine these bioactive ingredients' clinical effectiveness and potential toxicity effect.

5.3.6 Antidiabetic Properties

Diabetes mellitus (DM) is a chronic metabolic disease, and millions of people worldwide suffer from this disease and its complications, with its predominance constantly increasing. DM is one of the most common life-threatening endocrine diseases caused by defects in insulin secretion, insulin resistance, or both.[149,150] There are different types of DM: type 1, earlier recognized as insulin-dependent DM, where hormonal insulin is not produced due to damage to pancreatic β-cells, and people with this type of DM need daily insulin administration;[151,152] type 2 diabetes, previously known as non-insulin-dependent DM, is more predominant and is recognized by a progressive reduction in insulin secretion by pancreatic β-cells and a relative diminution in the sensitivity of target tissues to the action of this hormone. Type 2 DM can be prevented or delayed by lifestyle intervention.[153] Other forms of DM include gestational DM, which occurs due to any degree of glucose intolerance with it beginning or first being recognized during pregnancy;[154] other types of DM can also be induced by genetic defects, endocrine disorders, other types of diseases, therapy, and malnutrition.[155]

DM is an irreversible disease with no cure, however glycaemic control is fundamental for both the prevention of diabetic complications and the reduction of mortality. In uncontrolled DM, hyperglycaemia is a common occurrence and eventually leads to infections, serious injuries, dysfunction, and multiple organ failure, particularly in the heart, kidneys, eye, blood vessels, and nerves.[156,157] Depending on the nature of DM, insulin and other synthetic medicines are widely used in its treatment. However, with prolonged medicine administration, some insulin resistance and side effects cases may develop, which leads to the search for more natural, safe, and efficient alternatives.[158,159]

The treatment of DM and its chronic and vascular complications have a significant economic impact on people's lives and the wider economy. However, the best weapon is the prevention of DM, focusing on avoiding or delaying its onset and complications with dietary control and exercise.[151] Thus, with concern to treating this disease, the key is in identifying the individually targeted factors that can lead to better glycaemic control. In addition to regular oral and injectable medications, natural product-based therapy is highly indicated because of its efficiency, few adverse effects, and low cost. Thus, due to the benefits of current antidiabetic medicines, the acceptance and use of these natural products are increasing dramatically.[149,160]

Natural products, including mushrooms, are of great help in managing DM[152,161] (see Figure 5.5). They have been reported to potentially control DM by reducing oxidative stress, mimicking insulin action, increasing insulin secretion and activity, regulating glucose transport type 4, glucose uptake,

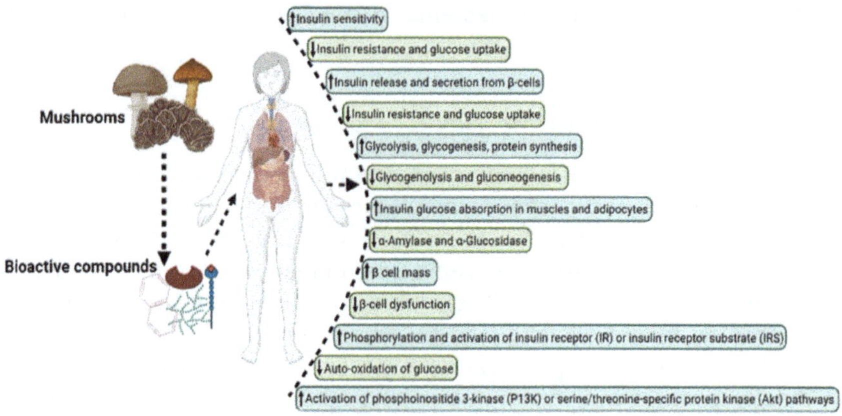

Figure 5.5 Favourable effects of mushroom bioactive compounds on DM.

inhibiting α-amylase and α-glucosidase, and protecting pancreatic β-cells.[153,162] Over the years, several bioactive compounds have been isolated from mushrooms (see Table 5.6), such as polysaccharides, dietary fibre, and other compounds of therapeutic interest such as lectins, lactones, terpenoids, alkaloids, metal chelating agents, and also several enzymes such as laccase, superoxide dismutase, glucose oxidase, and peroxidase, with favourable effects on glucose homeostasis[153,159,163] (see Figure 5.5). In addition, mushrooms have been prestigious as a source of these compounds and have been highly valued as potent sources of antidiabetic and hypoglycaemic compounds, promising in treating diabetes as biological antihyperglycaemic agents (see Table 5.6).[162]

Several species of mushrooms are known for their hypoglycaemic or antidiabetic activities (see Table 5.6), including *Agaricus bisporus*,[164] *Agaricus subrufescens*,[165] *Auricularia auricular*,[166,167] *Antrodia cinnanomea*,[168] *Coprinus comatus*,[169] *Cordyceps militaris*,[170] *Ganoderma lucidum*,[171] *Grifola frondosa*,[172] *Inonotus obliquus*,[173] *Lentinus edodes*,[174] *Pleurotus ostreatus*,[175] *Pleurotus eryngii*,[176] *Pleurotus citrinopileatus*,[177] *Pleurotus pulmonarius*,[178] and *Wolfiporia extensa*.[179] The bioactive compounds in mushrooms have significantly contributed to regulating and managing various diseases, including DM. Thus, the various mushrooms presenting health benefits provide an opportunity to plan and develop new formulations and therapies for DM.[153] However, extensive studies are needed to investigate the untapped mushrooms and extract and purify antidiabetic biomolecules that can be used to treat and manage DM.

5.3.7 Anti-ageing Properties

Skin ageing is a complex biological process that is induced through two independent intrinsic and extrinsic routes, responsible for the progressive loss of structural integrity, elasticity, and physiological functions of the

Table 5.6 Antidiabetic effect of mushroom extracts on rodents.

Samples	Parts/components	Antidiabetic effect	Ref.
A. auricula and *A. polytricha*	Polysaccharides	Reduction of fasting blood glucose, promotion of insulin synthesis, and reduction of the level of plasma glucose	167
A. auricular	Polysaccharides	Reduced HbA1 serum levels	319
	Dried fruiting body/ polysaccharides	Reduction of blood glucose level	166
A. bisponus and *P. ostreatus*	Powder	Decrease in serum glucose level	150
A. bisporus	Ethanol and methanol extract	Decrease in the levels of glucose	164
A. cinnamomea	Mycelium powder	Decreased plasma glucose levels and improved insulin resistance	168
C. militaris	Acidic-extractable polysaccharides	Decreased blood glucose, improved glucose and insulin resistance, enhanced antioxidant enzyme activities, and attenuated injuries of the liver, kidney, and pancreas	320
C. comatus	Ethanol extract	Reduced the blood glucose and glycosylated haemoglobin (HbA1c) levels and increased the plasma insulin level	169
C. gigantea	Methanol extract	Reduced the blood glucose level	321
C. ventricosum	Mycelial polysaccharides	Decrease blood glucose levels	322
C. rutilus	Ethanol extract	Decrease blood glucose concentration	323
G. frondosa	Polysaccharide-chromium(III)	Regulated the mRNA expression related to glucose	172
	Polysaccharides	Regulated mRNA expression levels of the genes responsible for hepatic glucose metabolism	324
	Polysaccharides	Decrease in fasting blood glucose levels and reduction of the inflammatory factor content, and prevented renal fibrosis	325
	Polysaccharides	Decreased fasting blood glucose levels and glucose tolerance	326
	Mycelium/polysaccharides	Reduced fasting blood glucose levels	327

Table 5.6 (*Continued*)

Samples	Parts/components	Antidiabetic effect	Ref.
	Extracts	Improved glucose intolerance	328
	Heteropolysaccharide	Improved blood glucose levels and glucose intolerance	329
	Polysaccharides	Decreased the fasting blood glucose level, improved oral glucose tolerance, alleviated insulin resistance, and protected against liver and kidney injury	330
G. lucidum	Polysaccharides	Decreased the levels of fasting blood glucose and insulin	331
	Spores	Reduced the blood glucose and oxidative stress	171
	Beta-glucans	Decreased blood glucose levels	332
	Ethanol extract	Decreased blood glucose level, increased insulin level, and decreased HbA1c level	333
G. pfeifferi and *G. resinaceum*	Aqueous and ethanol extracts/ polysaccharides	Prevented significant increase in glycaemia values	334
I. obliquus	Polysaccharides	Decreased fasting glucose level, an improvement in the liver, kidney, pancreas, glucose tolerance capacity, and increased hepatic and muscle glycogen	173
I. obliquus	Polysaccharides	Reduced fasting blood glucose levels, improved glucose tolerance ability, increased hepatic glycogen level, and ameliorated insulin resistance	335
L. edodes	Powder	Increased maternal insulin levels	336
L. edodes	β-Glucans extract	Reduced blood glucose levels	174
L. edodes	Freeze-dried powder	Improved the glucose tolerance, potentiating insulin secretion, and reduced insulin resistance	337
L. rhinocerotis	Freeze-dried powder	Reduced elevated blood glucose concentrations	338

M. conica	Methanol extract	Reduced elevated blood glucose and improved liver and kidney damages	339
P. ostreatus	Ergosterol	Improved insulin resistance, reduced fasting blood glucose levels, and protected pancreas and liver	340
	Powder	Reduction of the serum glucose level	341
P. ostreatus and *L. subnudus*	Dried mushrooms	Reduction of fasting blood glucose, the activity of α-amylase, α-glucosidase, and angiotensin-1 converting enzyme	175
P. cocos	Gold nanoparticles	Reduced blood glucose status	179
P. linteus	Extract/polysaccharides	Improves insulin resistance	342
P. pulmonarius	Mycelia	Reduced blood glucose levels, decreased injuries in the pancreas, liver, and kidney	178
P. eryngii	Polysaccharides	Showed renoprotective effects, decreased glucose levels, and improved antioxidant status	343
P. citrinopileatus	Aqueous extract	Inhibitory activities on α-amylase, α-glucosidase	344
P. tuberregium	Aqueous extract	Reduced plasma sodium and raised plasma potassium levels and had no deleterious effects on the liver and kidney	345
S. luridus	Polysaccharides	Decreased blood glucose levels and attenuated kidney and liver injures	346
	Ethanol extract	Decreased blood glucose levels	347
T. melanosporum	Aqueous extract	Reduced glucose levels and attenuated oxidative stress	348

skin.[180] Ultraviolet radiation is the main environmental mutagen responsible for the oxidative damage to cellular components such as cell walls, lipid membranes, mitochondria, and skin DNA by generating free radicals. These free radicals (reactive oxygen species) activate many skin disease-related enzymes, such as collagenase, elastase, hyaluronidase, tyrosinase, and xanthine oxidase, and increase the expression of matrix metalloproteinases (MMPs) responsible for degrading the different components of the extracellular matrix (collagen, elastin, fibronectin, and hyaluronic acid), thus resulting in skin damage and premature skin ageing.[181] These ECM components are responsible for accelerating wound healing, contributing to mechanical resilience and maintaining the structural integrity of the skin.[182] Biomolecules with strong free radical scavenging and key enzyme inhibitory effects are vital ingredients in developing nutraceutical or cosmeceutical formulations to prevent skin ageing.[183] Numerous extracts and compounds obtained from edible and medicinal mushrooms, such as *Antrodia camphorata*, *Auricularia auricula-judae*, *Ganoderma lucidum*, *Inonotus obliquus*, *Lentinula edodes*, *Phellinus vaninii*, *Pholiota nameko*, *Piptoporus betulinus*, *Pleurotus citrinopileatus*, *Russula capensis*, *Schizophyllum commune*, *Tricholoma magnivelare*, and *Tricholoma matsutake*, have been widely investigated and found to possess strong antioxidant, anticollagenase, antihyaluronidase, antielastase, wound healing, anti-inflammatory, antimicrobial, antityrosinase, and photoprotective effects against UV-B and UV-A radiation as shown in Table 5.7. These biological activities are associated with major cell signal transduction pathways related to ageing, inflammation, hyperpigmentation, and extracellular matrix degradation (see Table 5.7).[184,185]

Inhibition of enzymes such as collagenase, hyaluronidase, and elastase has been associated with maintaining the skin's structural integrity. Fractions prepared from *P. citrinopileatus* (*n*-butanol, aqueous, and methanol extract) showed up to 9.7%, 10.8%, and 25.4% hyaluronidase inhibition activity, respectively.[186] In comparison, an aqueous and acetonic extract prepared from the mycelial suspension of *Trametes lactinea* presented 88% antihyaluronidase activity at 100 $\mu g\,mL^{-1}$.[187] Aqueous extracts obtained from *P. ostreatus*, *G. lucidum*, and *Auricularia polytricha* presented over 50% inhibition of hyaluronidase enzyme.[188] To the authors' best knowledge, only these findings have been reported so far on the hyaluronidase inhibition activity of mushroom extracts or their individual compounds. Polysaccharide-rich extracts obtained from *Auricularia auricula-judae*, *G. lucidum*, and *Pholiota nameko* have been reported to significantly inhibit collagenase enzyme activity and promote collagen synthesis in a fibroblast cell model.[189–191] Extracts obtained from different mushrooms have also been reported to present a potent elastase inhibitory effect, as shown in Table 5.7. MMPs are specialized enzymes produced by epithelial cells, fibroblasts, and inflammatory cells and are associated with degrading the ECM, collagen proteolysis, and contribute to other destructive processes such as tumour invasion and inflammation.[192] The studies presented in Table 5.7 showed

Table 5.7 Anti-ageing effect of mushroom extracts and their individual compounds.

Species	Bioactive form	Bioactive property	Key findings and suggested mechanism of action	Ref.
A. camphorata	Extract	Wound healing	– Cell viability of fibroblast cells was maintained at 100 μg mL^{-1} – Closure of 60% was observed in the wound healing scratch assay	349
A. auricula-judae	Water-soluble polysaccharide-rich extract	Wound healing and collagen synthesis	– Fibroblast and keratinocyte cell migration and proliferation were promoted at 20 μg mL^{-1} – Collagen synthesis increased up to 40% at 25 μg mL^{-1} – Wound closure was accelerated in mice wound-healing model	189
G. lucidum	Crude polysaccharide	Antityrosinase, antielastase, and anticollagenase	– Extract effectively inhibited tyrosinase, elastase, and collagenase up to 50% at 0.37, 1.01, and 0.49 mg mL^{-1}, respectively	190
	Ethanolic extract	Wound healing	– Cell viability of keratinocyte cells was maintained at 640 μg mL^{-1} – At 10 μg mL^{-1}, significant improvement in the cell migration capacity was observed in the scratch assay	205
	Cold water and hot water extract	Antityrosinase and antihyaluronidase	– Tyrosinase inhibition activity reached 45% – Up to 72% inhibition of hyaluronidase was observed	188
I. obliquus	Triterpenoid-rich fractions	Antityrosinase and melanin inhibition	– Petroleum ether fraction (3.81 μg mL^{-1}) showed a stronger tyrosinase inhibitory effect than kojic acid (5.23 μg mL^{-1}) – Fractions presented little or no cytotoxicity on B16 melanoma cells – Cellular tyrosinase activity and melanin content were inhibited in α-MSH-stimulated B16 melanoma cells	195

Table 5.7 (*Continued*)

Species	Bioactive form	Bioactive property	Key findings and suggested mechanism of action	Ref.
L. edodes	Hydroethanolic extract	MMP inhibition assay	– Cell viability was maintained in HaCaT keratinocyte cells at 50 μg mL^{-1} – The expressions of MMP-1 and MMP-9 were significantly inhibited – The expression of type I procollagen in UVA and UVB-irradiated HaCaT keratinocytes was increased	193
P. vaninii	Methanolic extract	Antityrosinase, antielastase, and anticollagenase	– The cell viability of B16-F10 murine melanoma cells was maintained up to 85% at 750 μg mL^{-1} – The tyrosinase inhibitory activity ranged from 55.83% to 96.16% at 0.125–2.0 mg mL^{-1} – Melanin synthesis was suppressed (71.18% to 27.61%) in B16-F10 melanoma cells at 25–500 μg mL^{-1} – Elastase and collagenase were inhibited up to 71.24 and 63.02% at 2.0 mg mL^{-1}, respectively	200
P. nameko	Polysaccharides	Anticollagenase and wound healing	– Collagenase was inhibited up to 61% at 500 μg mL^{-1} – Cell migration and proliferation in L929 mouse fibroblast cells were maintained – A significant increase in closure rate (54.75%) was observed in the wound-healing scratch assay	191
P. betulinus	Water-soluble β-D-glucan	Wound healing	– Cell viability was maintained at 1000 μg mL^{-1} – Cell migration was enhanced presenting a closure of 55% in the wound-healing scratch assay	207
P. citrinopileatus	Methanol extract	Antityrosinase, antihyaluronidase,	– Cell viability of B16 melanoma cells was maintained at 100 μg mL^{-1}	186

		and melanin inhibition	– The tyrosinase inhibitory activity ranged from 2.5% to 10.5% at 0.125–2.0 mg mL^{-1} – Melanin content was suppressed in B16 melanoma cells – Hyaluronidase enzyme was inhibited up to 25% at 100 μg mL^{-1}	
R. capensis	Aqueous and methanolic extract	Anticollagenase and wound healing	– Extract (100 μg mL^{-1}) mildly inhibits dermal fibroblast proliferation and significantly improved cell migration – Extract inhibited collagenase up to 20% at 200 μg mL^{-1} – Fibroblast migration in MRHF cells was enhanced and up to 40% wound closure rate was observed	204
S. commune	Cold water extract	Antityrosinase and antihyaluronidase	– Tyrosinase inhibition activity reached 98% comparable to kojic acid (94.4%) – Up to 40% inhibition of hyaluronidase was observed	188
T. fuciformis	Polysaccharides	Collagen synthesis	Collagen levels were elevated in UVA- and UVB-exposed animal models	350
T. magnivelare	Methyl cinnamate	Antityrosinase and melanin inhibition	– Cell viability of B16-F10 melanoma cells was maintained – Total melanin production was significantly suppressed in a concentration-dependent manner	201
T. matsutake	Mycelial extract	Antielastase	– Extract presented potent elastase inhibitory effect ($81.4 \pm 3.92\%$) similar to phophoramidon ($89.6 \pm 7.74\%$) used as positive control – Cell viability of human fibroblasts was maintained at 100 μg mL^{-1} – mRNA expressions of MMP-1 and MMP-3 in fibroblast treated cells were decreased in a dose-independent manner	192

that mushroom extracts and their associated metabolites present potential anti-ageing effects by inhibiting collagenase, elastase, and hyaluronidase. The anti-elastase activity has been attributed to downregulation of mRNA expression of MMPs (MMP-1, MMP-3, and MMP-9) and increased expression of type 1 procollagen in UVA- and UVB-irradiated keratinocyte cells.[192,193] Hence, more work is still needed to determine the mechanism behind the anticollagenase potential of these mushroom extracts or their individual compounds.

Tyrosinase is the key enzyme involved in the synthesis of melanin. Melanin production is crucial for preventing UV-induced skin damage, and the overproduction of this pigment has resulted in hyperpigmentation.[194] Bioactive ingredients with the capacity to bind to the active site of the tyrosinase enzyme and cause enzyme inactivation have become the most prominent and successful targets for melanogenesis inhibitors. Besides the competitive and non-competitive inhibition potential against tyrosinase, mushroom extracts have also been reported to decrease melanocyte metabolism, resulting in reduced melanin synthesis. Fractions from *Inonotus obliquus* containing betulin, trametenolic acid, inotodiol, and lanoserol are reported to present very little or no cytotoxic effect in B16 melanoma cells.[195] The extract reduced cellular tyrosinase activity and significantly lowered melanin content in α-MSH-stimulated B16 melanoma cells. Several extracts obtained from medicinal and edible mushrooms, as shown in Table 5.7, are potent inhibitors of tyrosinase enzyme, being also effective in suppressing melanin production in cultured melanocytes and other *in vitro* models. These findings make these extracts potential ingredients to develop formulations against diseases resulting from aberrant melanin production. Most tyrosinase inhibition activities currently reported have also presented incomparable/irreproducible IC_{50} values due to varied assay conditions, including using different substrates (L-dihydroxyphenylalanine and tyrosine) and substrate concentrations, various incubation times, and the use of other assay ingredients. Cosmetics or skin-lightening formulations commercially available contain hydroquinone, kojic acid, arbutin, azelaic acid, ellagic acid, and resveratrol.[196] Despite being widely used, these compounds have been reported to alter melanosome formation, promote their destruction, and present unsatisfactory clinical efficacy.[197] The lack of efficacy of the available tyrosinase-targeting compounds might be associated with most studies being conducted exclusively using tyrosinase isolated from the mushroom *Agaricus bisporus* (mushTYR).[198] In the past, mushTYR was the only commercially available enzyme, but recent studies have shown that its catalytic activities and substrate specificities are significantly different from human tyrosinase (humTYR). Very few studies have been published on humTYR inhibitors, with thiamidol being one of the most promising candidates.[199] To the best of our knowledge, most tyrosinase inhibition properties of mushrooms or their associated metabolites heavily relied on enzymatic assays performed using mushroom tyrosinase (mushTYR),[190,200,201] mainly because there were no available humTYR-based assays and mushTYR assays are relatively inexpensive. Hence, with the recent advances in humTYR structural information,[199,202] more work needs to be carried out to find bioactive ingredients from mushrooms with the best humTYR and melanin inhibition potential to develop effective and safe cosmeceutical and nutricosmetic formulations. All signalling pathways associated with melanogenesis are controlled by the master regulator, microphthalmia-associated transcription factor (MITF), a melanocyte-specific transcription factor that is responsible for differentiation of melanocyte cells that modulates the expression of tyrosinase and tyrosinase-related proteins (TRP-1 and TRP-2).[203] Hence, most tyrosinase-inhibiting bioactives tend to target the downregulation of MITF expression to inhibit the whole process of melanogenesis, and as such more targeting of the expression of these melanogenesis-related proteins, including MITF, TRP-1, and TRP-2 should be conducted.

The wound-healing effect of mushrooms has been one of the most studied beneficial effects, and is well described in different scientific findings.[204,205] The wound-healing process is characterized by a complex sequence of cellular and molecular events involving inflammation, fibroplasia, neovascularization, collagen deposition, epithelialization, and wound contraction.[189] Most mushroom extracts and their compounds have been investigated (see Table 5.7), targeting relevant wound repair mechanisms such as cells (fibroblasts and keratinocyte) proliferation and migration, inhibition of reactive oxygen species (ROS) production, modulation of inflammatory mediators secretion, and deposition of ECM components (collagen and hyaluronic acid).[189,204–207] The wound-healing potential of most mushroom species has been associated with their rich content in polysaccharides such as water-soluble β-D-glucan from *P. betulinus*, a low-molecular-weight polysaccharide from *P. nameko*, and water-soluble polysaccharide from *A. auricula-judae*.[189,191,207] However, their related mechanism of action has not been fully elucidated. Nonetheless, mushrooms' anti-ageing and wound-healing potential have increasingly been reported, linked to their richness in triterpenes, polysaccharides, and phenolic compounds. These multi-functional bioactive properties have intensified the use of mushrooms and their related metabolites as components in cosmetic formulation development, a niche that has received increased attention in recent years.

5.3.8 Hepatoprotective Properties

The liver is a vital and complex organ of the human body, maintaining homeostasis by regulating various physiological functions such as carbohydrate, protein, and fat metabolism, detoxification, secretion of bile, and storage of vitamins. The liver is also involved in most biochemical pathways related to growth, combatting diseases, nutrient supply, energy provision, and reproduction.[208]

Liver damage is a widespread disease that, in most cases, involves oxidative stress. Normally, host cells are protected from oxygen-derived radical injury by naturally occurring free radical scavengers and antioxidants. When these natural defences are overwhelmed by excessive generation of pro-oxidants, a situation of oxidative stress evolves, and cellular macromolecules suffer oxidative damage.[208,209]

Both reactive oxygen and nitrogen species are the major initiators of oxidative stress, and they appear due to various pathogenic diseases, exposure to radiation, tissue injury, *etc.* Most of the hepatotoxins induce tissue injury after having been metabolized to free radicals and cause subsequent cell damage through a mechanism of covalent binding and lipid peroxidation. Oxidative stress has been implicated in the pathogenesis of acute and chronic liver injury in a variety of pathophysiological conditions such as hepatotoxin exposure, intrahepatic cholestasis, alcoholic liver injury, hepatocellular carcinoma, liver ischaemia and viral hepatitis, fibrosis, cirrhosis, steatohepatitis and biliary disease, inflammation, and necrosis of the liver, among others. Thus, maintaining a healthy liver is crucial for overall health and human well-being.[208–210]

Clinically, serum enzymes [aspartate aminotransferase (AST) and alanine aminotransferase (ALT), hepatic lipids (total cholesterol and triglycerides (TG)], and enzymes involved in alcohol metabolism [alcohol dehydrogenase (ADH) and aldehyde dehydrogenase (ALDH)] are commonly used as biochemical markers for early diagnosis of hepatic injury. Serum ALT and AST activities are elevated when a hepatic injury occurs since ALT and AST could leach out of hepatocytes into the blood circulation.[211]

Carbon tetrachloride (CCl_4), a selective hepatotoxic chemical agent, is one of the most widely used toxins for the experimental induction of liver fibrosis in laboratory animals. The principal causes of CCl_4-induced hepatic damage are lipid peroxidation, decreased activities of antioxidant enzymes, and the generation of free radicals leading to liver fibrosis and cirrhosis.[209,212]

As mentioned earlier, the particular characteristics of growth and development of mushrooms in nature result in the accumulation of a great variety of secondary metabolites. All these related activities give mushrooms great potential as generators of bioactive compounds helpful in promoting human health. This includes their consumption as foods and their role as producers of biomolecules with specific pharmacological properties in laboratory cultures under strictly controlled conditions.[208]

The hepatoprotective properties and the greatest contributions described by different authors in several mushroom species are summarized in Table 5.8.

The protective effects of *L. deliciosus* and *A. cylindracea* supplementation against CCl_4-induced oxidative stress were investigated by measuring the adenosine deaminase (ADA) and myeloperoxidase (MPO) levels and histopathological changes in liver and kidney tissues in an *in vivo* model.[213] The CCl_4-induced histopathology was increased significantly in the *A. cylindracea* and $CCl_4 + $ *A. cylindracea* groups due to the toxic effects of *A. cylindracea*. On the other hand, *L. deliciosus*-treated rats exhibited significantly fewer

Table 5.8 Main mushrooms with reported hepatoprotective properties and the main contribution.

Mushroom species	Liver injury	Main role	Ref.
A. cylindracea	CCl_4	*A. cylindracea* supplementation causes severe pathology in liver and kidney tissues in rats	213
C. comatus	Alcohol	A modified polysaccharide from *C. comatus* significantly attenuated the hepatic and serum lipid levels, enhanced antioxidant enzyme activities, markedly improved alcohol metabolism system and inflammatory response, and mitigated alcohol-induced liver injury histopathologically	217
G. lucidum	CCl_4	*G. lucidum* extract can significantly prevent the CCl_4-induced liver and kidney damage, restored MDA, H_2O_2 contents, SOD, CAT, and GSH levels	214
H. marmoreus	CCl_4	All extracts prepared from garlic and buna-shimeji with low and high contents of S-allyl-L-cysteine sulfoxide and arginine or ornithine significantly suppressed CCl_4-induced hepatic injury in rats	215
L. deliciosus		Supplementation with *L. deliciosus* exhibits beneficial effects in liver and kidney tissues of CCl_4-treated rats	213
L. sulphureus	Acute alcohol	Both hot-water-extractable polysaccharides and enzymatic-extractable polysaccharides confer effective hepatoprotection against acute alcohol-induced alcoholic liver disease	211
P. ostreatus	CCl_4	The crude phenolic-rich extract treatments showed significant increases in SOD and GPx hepatic activities	212
		Daily supplementation of powdered oyster mushroom reduced dyslipidaemia, hepatotoxicity, and haematotoxicity	216
P. tuber-regium	CCl_4	The activity of aspartate aminotransferase, alanine transaminase, and alkaline phosphatase, and the concentration of bilirubin significantly increased in *P. tuber-regium* extract-treated hepatotoxic rats	210

degenerative and necrotic changes in liver and kidney in the $CCl_4 + $ *L. deliciosus* group compared to rats in the CCl_4-treated group. Purified polysaccharide isolated from *L. deliciosus* is a potential source of antioxidants and natural immunostimulants, which could explain the protective effect of this mushroom.[213]

In another study, CCl_4-induced oxidative stress was tested in an adult male Sprague–Dawley rat. The authors found that CCl_4 toxicity increased blood urea, uric acid, and creatinine levels compared to the control samples. Pretreatment of *G. lucidum* extracts significantly decreased urea, uric acid, and creatinine levels, proposing that it would counteract nephrotoxicity caused by CCl_4. Otherwise, treatment with *G. lucidum* extract restored malondialdehyde (MDA), hydrogen peroxide (H_2O_2) contents, superoxide dismutase (SOD), catalase (CAT), and glutathione (GSH) levels.[214] The restoration activities of oxidative stress parameters by *G. lucidum* can protect the enzymes; this species contains several antioxidant compounds and exhibits significant strong scavenging free radical activity.[214]

The combination of the garlic-derived amino acid, *S*-allyl-L-cysteine sulfoxide (ACSO), and ornithine or arginine from *H. marmoreus* on CCl_4-induced hepatic injury was examined by Yamaguchi *et al.*[215] Some mushrooms have highly active arginase that catalyses the hydrolysis of arginine to ornithine and urea, accumulating a considerable amount of urea in the fruit bodies to regulate osmotic pressure. All extracts prepared from garlic and *H. marmoreus* with low and high contents of ACSO and arginine or ornithine significantly suppressed CCl_4-induced hepatic injury in rats. The co-administration of ornithine/arginine with ACSO suppressed CCl_4-induced hepatic injury in rats more potently than a single administration of ACSO, probably owing to synergistic effects.

In another study, the hepatoprotective activity of the crude phenolic-rich extract (CPRE) isolated from *P. ostreatus* on CCl_4-induced oxidative stress was investigated in albino rats and several biochemical parameters were evaluated. Administration of CCl_4 for 28 days exhibited a significant increase in serum markers of liver damage [*i.e.*, ALT, AST, alkaline phosphatase (ALP), lactate dehydrogenase (LDH), urea, creatinine, total lipids and TG]. CCl_4 exposure significantly decreased the hepatic antioxidant enzyme activities such as SOD and glutathione peroxidase (GSH-Px). Contrarily, the CPRE treatments showed a significant increase in these hepatic activities.[212] The potential antioxidant activity of CPRE isolated from *P. ostreatus* against several oxidation systems *in vivo* contributed to its hepatoprotective effects in CCl_4-induced liver injury in male albino rats.[212]

Ubhenin *et al.*[216] evaluate the hepatoprotective, haematoprotective, and hypolipidaemic effects of *P. ostreatus* in CCl_4-induced liver injury in Wistar rats. The results revealed that CCl_4 caused a significant increase in lipid peroxidation due to the significantly elevated level of MDA in the hepatic tissues, whereas the levels or activities of reduced GSH, CAT, SOD, GPx, and glutathione *S*-transferase (GST) in the liver tissues were significantly reduced. Dyslipidaemia and haematotoxicity were manifested by a significant increase in the serum levels of TG, total cholesterol, very low-density lipoprotein cholesterol (VLDL-C), low-density lipoprotein cholesterol (LDL-C), and white blood cells count. The ability of the powdered *P. ostreatus* to mitigate against CCl_4-induced hepatotoxicity is probably due to its antioxidant and enzyme modulatory effects.

The hepatoprotective effects of modified polysaccharide from *C. comatus* on alcohol-induced liver injury were investigated. The mushroom extract significantly increased the SOD, GSHPx, and CAT activities, which may be because the extract activated nuclear factor erythroid 2-related factor-2 in mouse livers. These results indicated that the activities of antioxidant enzymes might be one of the main mechanisms for contributing to the protective effects of modified polysaccharides from *C. comatus* against alcohol-induced liver injury.[217]

Chemical screening and analysis of the hepatoprotective efficacy of *P. tuber-regium* extracts in CCl_4 hepatotoxic rat was evaluated by the authors.[210] When *P. tuber-regium* extract was administered to hepatotoxic rats, AST, ALT, and ALP activity significantly decreased, and the concentration of serum albumin and protein significantly increased compared to the hepatotoxic group. *P. tuber-regium* extract possessed antioxidant activity in a dose-dependent manner, possibly due to a relatively high concentration of bioactive compounds like flavonoids, tannins, alkaloids, and phenolics.[210]

The hepatoprotective effects and antioxidant activities of hot-water-extractable polysaccharides and enzymatic-extractable polysaccharides from *L. sulphureus* in acute alcohol-induced alcoholic liver disease mice were investigated by the authors.[211] The elevation of AST and ALT activities could be attenuated by supplementation of both hot-water-extractable polysaccharides and enzymatic-extractable polysaccharides. Polysaccharides confer effective hepatoprotection against acute alcohol-induced alcoholic liver disease, possibly by reducing oxidative stress. Enzymatic-extractable polysaccharides showed superior effects, indicating that enzymatic hydrolysis has a potential effect on enhancing bioactivities.[211]

5.4 Conclusion and Future Trends

This chapter focuses on the beneficial biological effects of mushroom extracts and their related metabolites. Mushrooms have received significant interest in developing functional foods due to their diverse nutritional profile and a variety of bioactive molecules with effective therapeutic and medicinal properties. Some of these functional ingredients containing mushroom extracts or their metabolites have been utilized to prevent and treat a variety of diseases, and they continue to be patented and sold as a daily health supplement. A considerable body of literature has demonstrated the potential health beneficial effect of consuming mushroom extracts and their associated metabolites. Mushroom consumption has demonstrable antioxidant, anti-inflammatory, antimicrobial, immunomodulatory, anti-diabetic, anti-ageing, neuroprotective, hepatoprotective, and cardioprotective effects through a variety of signalling pathways that have the potential to contribute to the suppression or delay of disease severity. Despite this, there is a significant gap in validating some of these beneficial effects in pre-clinical or clinical settings to evidence their therapeutic effectiveness. In addition, the safety and toxicity profile of mushroom extracts used for different applications, although widely conducted, still needs more convincing research involving improved models. Also, there remains the need to standardize extract preparation and consumption method to avoid inconsistencies in the desired therapeutic effect.

Abbreviations

AChE	Acetylcholinesterase
AP-1	Activator protein 1
ADA	Adenosine deaminase
ALT	Alanine aminotransferase
ALP	Alkaline phosphatase
ADH	Alcohol dehydrogenase
ALDH	Aldehyde dehydrogenase
α-MSH	Alpha-melanocyte-stimulating hormone
AD	Alzheimer's disease
Aβ	Amyloid beta
AST	Aspartate aminotransferase
ABTS	2,2′-Azino-bis-3-ethylbenzthiazoline-6-sulfonic acid
BMDCs	Bone-marrow-derived dendritic cells
BChE	Butyrylcholinesterase
CCl_4	Carbon tetrachloride
CVD	Cardiovascular diseases
CAT	Catalase
CNS	Central nervous system
COX-2	Cyclooxygenase-2
CTL	Cytotoxic T-lymphocytes
DM	Diabetes mellitus
DMM	Dietary mixed micelles
DPPH	2,2′-Diphenyl-1-picrylhydrazyl
DMH	1,2-Dimethylhydrazine
DSS	Dextran sulfate sodium
DOX	Doxorubicin
ESBL	Extended-spectrum β-lactamases
ECM	Extracellular matrix
FSG	Fasting serum glucose
FSI	Fasting serum insulin
FRAP	Ferric reducing antioxidant power
GSH	Glutathione
GSH-Px	Glutathione peroxidase
GST	Glutathione *S*-transferase
HbA1c	Glycosylated haemoglobin
GCSF	Granulocyte colony stimulating factor
HbA1	Haemoglobin subunit alpha 1
H22	Hepatoma cells
HOMA-IR	Homeostasis model assessment of insulin resistance
HEA	Hydroxyethyl adenosine
H_2O_2	Hydrogen peroxide
Humtyr	Human tyrosinase
iNOS	Inducible nitric oxide synthase
IFN-γ	Interferon gamma

IL	Interleukins
IL-1β	Interleukin 1 beta
IL-10	Interleukin 10
IZD	Internal zone diameter
LDH	Lactate dehydrogenase
LPS	Lipopolysaccharides
LDL-C	Low-density lipoprotein cholesterol
KC	Keratinocyte chemoattractant
MIP-1α	Macrophage inflammatory protein-1 alpha
MIP-1β	Macrophage inflammatory protein-1 beta
MIP-2	Macrophage inflammatory protein-2
MDA	Malondialdehyde
MMP	Matrix metalloproteinases
MMP-9	Matrix metallopeptidase 9
MITF	Microphthalmia-associated transcription factor
MSSA	Meticillin-sensitive *Staphylococcus aureus*
MRSA	Meticillin-resistant *Staphylococcus aureus*
mRNA	Messenger ribonucleic acid
MBC	Minimum bactericidal concentration
MIC	Minimum inhibitory concentration
MAPKs	Mitogen-activated protein kinases
MCP-1	Monocyte chemoattractant protein 1
Mushtyr	Mushroom tyrosinase
MYD88	Myeloid differentiation 88
MPO	Myeloperoxidase
NGF	Nerve growth factor
NO	Nitric oxide
NF-κβ	Nuclear factor kappa B
OxHLIA	Oxidative haemolysis inhibition assay
ORAC	Oxygen radical absorbance capacity
PI3K	Phosphatidylinositol-3-kinase
PIH	Post-inflammatory hyperpigmentation
PUFA	Polyunsaturated fatty acids
PGE2	Prostaglandin E2
RNS	Reactive nitrogen species
ROS	Reactive oxygen species
RP	Reducing power
RANTES	Regulated on activation, normal T-cell expressed and secreted
ACSO	*S*-Allyl-L-cysteine sulfoxide
SFE	Supercritical fluid extraction
SOD	Superoxide dismutase
SREBF2	Sterol regulatory element-binding protein 2
STAT3	Signal transducer and activator of transcription 3
TBARS	Thiobarbituric acid reactive substances
TRIF	TIR-domain-containing adapter-inducing interferon-β

TIMP-1	Tissue inhibitor of metalloproteinase 1
TGF-β	Transforming growth factor beta
TG	Triglycerides
TLR	Toll-like receptor
TLR4	Toll-like receptor 4
TAC	Total antioxidant capacity
TOS	Total oxidant status
TNF-α	Tumour necrosis factor alpha
TRP	Tyrosinase-related proteins
UV	Ultraviolet
VCAM-1	Vascular cell adhesion molecule 1
VEGF	Vascular endothelial growth factor
VLDL-C	Very low-density lipoprotein cholesterol

Acknowledgements

The authors are grateful to the Foundation for Science and Technology (FCT, Portugal) for financial support through national funds FCT/MCTES to CIMO (UIDB/00690/2020). L. Barros and A. Fernandes are thankful for national funding by FCT, P. I., through the institutional scientific employment program for their contract. This work is also supported by MICINN supporting the Juan de la Cierva Formation contract for T. Oludemi (FJC2019-042549-I). R. V. C. Cardoso thanks FCT for her PhD grant (SFRH/BD/137436/2018).

References

1. G. A. Martinez-Medina, M. L. Chávez-González, D. K. Verma, L. A. Prado-Barragán, J. L. Martínez-Hernández, A. C. Flores-Gallegos, M. Thakur, P. P. Srivastav and C. N. Aguilar, *J. Funct. Foods*, 2021, 77, 104326.
2. Market Insight Reports, 2020 Industry Report: Mushroom, 2021.
3. P. Maity, I. K. Sen, I. Chakraborty, S. Mondal, H. Bar, S. K. Bhanja, S. Mandal and G. N. Maity, *Int. J. Biol. Macromol.*, 2021, **172**, 408–417.
4. G. Gebreyohannes, A. Nyerere, C. Bii and D. Berhe Sbhatu, *Evidence-Based Complementary Altern. Med.*, 2019, **2019**, 1–7.
5. M. Alves, I. F. R. Ferreira, J. Dias, V. Teixeira, A. Martins and M. Pintado, *Planta Med.*, 2012, **78**, 1707–1718.
6. A. Cruz, L. Pimentel, L. M. Rodríguez-Alcalá, T. Fernandes and M. Pintado, *J. Food Nutr. Res.*, 2016, **4**, 773–781.
7. G. Gebreyohannes, A. Nyerere, C. Bii and D. Berhe Sbhatu, *Evidence-Based Complementary Altern. Med.*, 2019, **2019**, 7.
8. J. Garcia, A. Afonso, C. Fernandes, F. M. Nunes, G. Marques and M. J. Saavedra, *S. Afr. J. Chem. Eng.*, 2021, **35**, 98–106.
9. A. Fardet and E. Rock, *Sustain*, 2020, **12**, 6280.
10. F. S. Reis, A. Martins, M. H. Vasconcelos, P. Morales and I. C. F. R. Ferreira, *Trends Food Sci. Technol.*, 2017, **66**, 48–62.

11. M. Mleczek, M. Siwulski, A. Budka, P. Mleczek, S. Budzyńska, M. Szostek, N. Kuczyńska-Kippen, P. Kalač, P. Niedzielski, M. Gąsecka, P. Goliński, Z. Magdziak and P. Rzymski, *Chemosphere*, 2021, **263**, 128095.
12. M. B. Bellettini, F. A. Fiorda, H. A. Maieves, G. L. Teixeira, S. Ávila, P. S. Hornung, A. M. Júnior and R. H. Ribani, *Saudi J. Biol. Sci.*, 2019, **26**, 633–646.
13. D. J. Royse, J. Baars and Q. Tan, *Edible and Medicinal Mushrooms: Technology and Applications*, 2017, ch. 2.
14. S. Marçal, A. S. Sousa, O. Taofiq, F. Antunes, A. M. M. B. Morais, A. C. Freitas, L. Barros, I. C. F. R. Ferreira and M. Pintado, *Trends Food Sci. Technol.*, 2021, **110**, 418–431.
15. O. Taofiq, R. C. Calhelha, S. Heleno, L. Barros, A. Martins, C. Santos-Buelga, M. J. R. P. Queiroz and I. C. F. R. Ferreira, *Food Res. Int.*, 2015, **76**, 821–827.
16. O. Taofiq, S. A. Heleno, R. C. Calhelha, M. J. Alves, L. Barros, A. M. González-Paramás, M. F. Barreiro and I. C. F. R. Ferreira, *Food Chem. Toxicol.*, 2017, **108**, 139–147.
17. D. Yadav and P. S. Negi, *Food Res. Int.*, 2021, **148**, 110599.
18. F. Antunes, S. Marçal, O. Taofiq, A. M. M. B. Morais, A. C. Freitas, I. C. F. R. Ferreira and M. Pintado, *Molecules*, 2020, **25**, 2672.
19. A. R. Silva, T. Oludemi, C. Costa, J. Barros, I. Ferreira, J. Nunes, M. A. Prieto, J. Simal-Gandara, L. Barros and I. C. F. R. Ferreira, *Food Bioprod. Process.*, 2020, **122**, 183–192.
20. G. A. Martinez-Medina, M. L. Chávez-González, D. K. Verma, L. A. Prado-Barragán, J. L. Martínez-Hernández, A. C. Flores-Gallegos, M. Thakur, P. P. Srivastav and C. N. Aguilar, *J. Funct. Foods*, 2021, **77**, 104326.
21. A. Haro, A. Trescastro, L. Lara, I. Fernández-Fígares, R. Nieto and I. Seiquer, *J. Food Compos. Anal.*, 2020, **91**, 103504.
22. I. C. F. R. Ferreira, S. A. Heleno, F. S. Reis, D. Stojkovic, M. J. R. P. Queiroz, M. H. Vasconcelos and M. Sokovic, *Phytochemistry*, 2015, **114**, 38–55.
23. F. S. Reis, R. T. Lima, P. Morales, I. C. F. R. Ferreira and M. H. Vasconcelos, *Molecules*, 2015, **20**, 17872–17882.
24. Â. Fernandes, J. C. M. Barreira, A. L. Antonio, M. B. P. P. Oliveira, A. Martins and I. C. F. R. Ferreira, *LWT–Food Sci. Technol.*, 2016, **67**, 99–105.
25. R. V. C. Cardoso, Â. Fernandes, J. C. M. Barreira, S. C. Verde, A. L. Antonio, A. M. Gonzaléz-Paramás, L. Barros and I. C. F. R. Ferreira, *Food Chem.*, 2019, **278**, 760–766.
26. D. Morales, A. Gil-Ramirez, F. R. Smiderle, A. J. Piris, A. Ruiz-Rodriguez and C. Soler-Rivas, *Innovative Food Sci. Emerging Technol.*, 2017, **41**, 330–336.
27. K. D. Cashman, *J. Steroid Biochem. Mol. Biol.*, 2015, **148**, 19–26.
28. S. Y. Chen, H. T. Yu, J. P. Kao, C. C. Yang, S. S. Chiang, D. O. Mishchuk, J. L. Mau and C. M. Slupsky, *J. Nutr. Biochem.*, 2015, **26**, 696–703.

29. C. R. L. Francisco, S. A. Heleno, I. P. M. Fernandes, J. C. M. Barreira, R. C. Calhelha, L. Barros, O. H. Gonçalves, I. C. F. R. Ferreira and M. F. Barreiro, *Food Chem.*, 2018, **245**, 845–853.
30. I. Patinho, E. Saldaña, M. M. Selani, A. C. B. Teixeira, B. S. Menegali, T. C. Merlo, J. D. Rios-Mera, M. D. B. Dargelio, H. Rodrigues and C. J. Contreras-Castillo, *Food Res. Int.*, 2021, **147**, 110551.
31. A. González, M. Cruz, C. Losoya, C. Nobre, A. Loredo, R. Rodríguez, J. Contreras and R. Belmares, *Food Funct.*, 2020, **11**, 7400–7414.
32. D. Yadav and P. S. Negi, *Food Res. Int.*, 2021, **148**, 110599.
33. H. Chou, L. Liu, C. Chen, I. Lin, D. Ali, A. Y. Lee and H. D. Wang, *Food Chem. Toxicol.*, 2015, **159**, 113–121.
34. A. Pessoa, C. F. Miranda, M. Batista, M. Bosio, G. Marques, F. Nunes, R. M. Quinta-Ferreiraa and M. E. Quinta-Ferreira, *Energy Rep.*, 2020, **6**, 891–896.
35. A. M. Pisoschi, A. Pop, F. Iordache, L. Stanca, G. Predoi and A. I. Serban, *Eur. J. Med. Chem.*, 2021, **209**, 112891.
36. M. Kozarski, A. Klaus, D. Jakovljevic, N. Todorovic, J. Vunduk, P. Petrović, M. Niksic, M. M. Vrvic and L. Van Griensven, *Molecules*, 2015, **20**, 19489–19525.
37. C. Sánchez, *Synth. Syst. Biotechnol.*, 2017, **2**, 13–22.
38. B. Goswami, S. Majumdar, A. Das, A. Barui and J. Bhowal, *LWT–Food Sci. Technol.*, 2021, **149**, 111768.
39. Y. Sakemi, M. Hagiwara, A. Oikawa, M. Sato, S. Sato, N. Sawa, H. Nishizawa and K. Shindo, *Food Chem.*, 2021, **361**, 130028.
40. M. Carocho, P. Morales and I. C. F. R. Ferreira, *Trends Food Sci. Technol.*, 2018, **71**, 107–120.
41. S. Costa, M. Reina Couto, A. Albino-Teixeira and T. Sousa, *Rev. Port. Cardiol.*, 2016, **35**, 41–57.
42. E. Sharpe, A. P. Farragher-Gnadt, M. Igbanugo, T. Huber, J. C. Michelotti, A. Milenkowic, S. Ludlam, M. Walker, D. Hanes, R. Bradley and F. Bou-Abdallah, *J. Agric. Food Res.*, 2021, **4**, 100130.
43. L. Živković, V. Bajić, M. Bruić, S. Borozan, K. Popić, D. Topalović, J. Santibanez and B. Spremo-Potparević, *Mutat. Res., Genet. Toxicol. Environ. Mutagen.*, 2021, **15**, 433–462.
44. F. Bach, A. A. F. Zielinski, C. V. Helm, G. M. Maciel, A. C. Pedro, A. P. Stafussa, S. Ávila and C. W. I. Haminiuk, *Lwt*, 2019, **107**, 214–220.
45. T. Islam, K. Ganesan and B. Xu, *Int. J. Med. Mushrooms*, 2019, **21**, 237–251.
46. F. S. Reis, A. Martins, M. H. Vasconcelos, P. Morales and I. C. F. R. Ferreira, *Trends Food Sci. Technol.*, 2017, **66**, 48–62.
47. R. V. C. Cardoso, Â. Fernandes, M. B. P. P. Oliveira, R. C. Calhelha, L. Barros, A. Martins and I. C. F. R. Ferreira, *Food Funct.*, 2017, **8**, 2155–2164.
48. D. Krüzselyi, Á. M. Móricz and J. Vetter, *LWT–Food Sci. Technol.*, 2020, **127**, 109436.

49. I. Akata, G. Zengin, C. M. N. Picot and M. F. Mahomoodally, *S. Afr. J. Bot.*, 2019, **120**, 95–99.
50. J. R. Barbosa, M. M. S. Freitas, L. C. Oliveira, L. H. S. Martins, A. O. Almada-Vilhena, R. M. Oliveira, J. C. Pieczarka, D. do, S. B. Brasil and R. N. Carvalho Junior, *Food Chem.*, 2020, **330**, 127173.
51. R. V. C. Cardoso, M. Carocho, Â. Fernandes, J. Pinela, D. Stojković, M. Soković, D. C. Zied, J. D. V. Cobos, A. M. González-Paramás, I. C. F. R. Ferreira and L. Barros, *Sustain*, 2021, **13**, 5019.
52. D. Menaga, P. K. S. M. Rahman, S. Rajakumar and P. M. Ayyasamy, *J. Bioresour. Bioprod.*, 2021, **6**, 338–349.
53. E. Sharpe, A. P. Farragher-Gnadt, M. Igbanugo, T. Huber, J. C. Michelotti, A. Milenkowic, S. Ludlam, M. Walker, D. Hanes, R. Bradley and F. Bou-Abdallah, *J. Agric. Food Res.*, 2021, **4**, 100130.
54. S. Boonsong, W. Klaypradit and P. Wilaipun, *Agric. Nat. Resour*, 2016, **50**, 89–97.
55. N. A. Amirullah, N. Z. Abidin, N. Abdullah and S. Manickam, *Biocatal. Agric. Biotechnol.*, 2020, **31**, 101881.
56. N. Nowacka-Jechalke, M. Olech and R. Nowak, in *Polyphenols: Prevention and Treatment of Human Disease*, ed. R. R. Watson, V. R. Preedy and S. Zibadi, Academic Press, Elsevier Inc., 2nd edn, 2018, pp. 137–150.
57. M. Pattanayak, S. Samanta, P. Maity, I. K. Sen, A. K. Nandi, D. K. Manna, P. Mitra, K. Acharya and S. S. Islam, *Carbohydr. Res.*, 2015, **413**, 30–36.
58. X. Li and L. Wang, *Int. J. Biol. Macromol.*, 2016, **83**, 270–276.
59. P. Maity, A. K. Nandi, D. K. Manna, M. Pattanayak, I. K. Sen, S. K. Bhanja, S. Samanta, B. C. Panda, S. Paloi, K. Acharya and S. S. Islam, *Carbohydr. Polym.*, 2017, **157**, 1237–1245.
60. M. Popović, S. Vukmirović, N. Stilinović, I. Čapo and V. Jakovljević, *Molecules*, 2010, **15**, 4564–4571.
61. T. Jayakumar, P. A. Thomas, J. R. Sheu and P. Geraldine, *Food Res. Int.*, 2011, **44**, 851–861.
62. N. Stilinović, I. Čapo, S. Vukmirović, A. Rašković, A. Tomas, M. Popović and A. Sabo, *R. Soc. Open Sci.*, 2020, 7, 200900.
63. X. Song, Z. Ren, X. Wang, L. Jia and C. Zhang, *Int. J. Biol. Macromol.*, 2020, **151**, 1267–1276.
64. O. Taofiq, A. Martins, M. F. Barreiro and I. C. F. R. Ferreira, *Trends Food Sci. Technol.*, 2016, **50**, 193–210.
65. B. Muszyńska, A. Grzywacz-Kisielewska, K. Kała and J. Gdula-Argasińska, *Food Chem.*, 2018, **243**, 373–381.
66. J. Peng, P. Wang, H. Ge, X. Qu and X. Jin, *PLoS One*, 2015, **10**, 1–18.
67. N. H. Jeong, S. Lee, J. K. Choi, Y. A. Choi, M. J. Kim, H. S. Lee, T. Y. Shin, Y. H. Jang, K. S. Song and S. H. Kim, *Biomed. Pharmacother.*, 2020, **122**, 109743.
68. Q. Feng, X. Lu, G. Yuan, Q. Zhang and L. An, *J. Funct. Foods*, 2020, **72**, 104037.

69. P. M. Chan, Y. S. Tan, K. H. Chua, V. Sabaratnam and U. R. Kuppusamy, *PLoS One*, 2015, **10**, 1–21.
70. J. H. Choe, Y. J. Yi, M. S. Lee, D. W. Seo, B. S. Yun and S. M. Lee, *Mycobiology*, 2015, **43**, 319–326.
71. A. T. M. Lima, M. N. Santos, L. A. R. De Souza, T. S. Pinheiro, A. A. O. Paiva, C. M. P. G. Dore, M. S. S. P. Costa, N. D. Santos, Y. G. Baseia, R. M. Araújo and E. L. Leite, *Carbohydr. Polym.*, 2016, **144**, 400–409.
72. J. J. Cheng, C. H. Chao, P. C. Chang and M. K. Lu, *Food Hydrocoll.*, 2016, **53**, 37–45.
73. J. J. Cheng, C. H. Chao and M. K. Lu, *Int. J. Biol. Macromol.*, 2018, **113**, 1198–1205.
74. J. Zheng, S. Jiao, Q. Li, P. Jia, H. Yin, X. Zhao, Y. Du and H. Liu, *Molecules*, 2018, **23**, 1–16.
75. T. C. Tsai, Y. T. Tung, Y. H. Kuo, J. W. Liao, H. C. Tsai, K. Y. Chong, H. L. Chen and C. M. Chen, *J. Ethnopharmacol.*, 2015, **159**, 113–121.
76. S.-J. Wu, Y.-W. Chen, C.-Y. Wang and Y.-T. Shyu, *Int. J. Food Sci. Technol.*, 2017, **52**, 671–678.
77. Y. Ren, Y. Geng, Y. Du, W. Li, Z. M. Lu, H. Y. Xu, G. H. Xu, J. S. Shi and Z. H. Xu, *J. Nutr. Biochem.*, 2018, **57**, 67–76.
78. X. Song, Z. Ren, X. Wang, L. Jia and C. Zhang, *Int. J. Biol. Macromol.*, 2020, **151**, 1267–1276.
79. D. Morales, R. Rutckeviski, M. Villalva, H. Abreu, C. Soler-Rivas, S. Santoyo, M. Iacomini and F. R. Smiderle, *Carbohydr. Polym.*, 2020, **229**, 115521.
80. Y. Sun, J. Huo, S. Zhong, J. Zhu, Y. Li and X. Li, *Carbohydr. Polym.*, 2021, **268**, 118214.
81. B. Du, Y. Yang, Z. Bian and B. Xu, *Front. Pharmacol.*, 2017, **8**, 252.
82. J. Xu, C. M. Xiao, H. S. Xu, S. X. Yang, Z. M. Chen, H. Z. Wang, B. S. Zheng, B. Z. Mao and X. Q. Wu, *Food Chem. Toxicol.*, 2021, **150**, 112073.
83. M. A. Hasnat, M. Pervin, K. M. Cha, S. K. Kim and B. O. Lim, *Phytochemistry*, 2015, **114**, 125–136.
84. Y. Geng, S. Zhu, P. Cheng, Z. M. Lu, H. Y. Xu, J. S. Shi and Z. H. Xu, *Phytomedicine*, 2017, **26**, 55–61.
85. Q. Cai, Y. Li and G. Pei, *J. Neuroinflammation*, 2017, **14**, 1–13.
86. Y. T. Chen, Y. C. Shen, M. C. Chang and M. K. Lu, *Process Biochem.*, 2016, **51**, 941–949.
87. J. Wei, W. H. Guo, C. Y. Cao, R. W. Kou, Y. Z. Xu, M. Górecki, L. Di Bari, G. Pescitelli and J. M. Gao, *Sci. Rep.*, 2018, **8**, 1–15.
88. L. Wang and C. Chuang, *Molecules*, 2019, **24**, 1–12.
89. R. W. Kou, R. Han, Y. Q. Gao, D. Li, X. Yin and J. M. Gao, *Phytochemistry*, 2021, **184**, 112647.
90. C. W. Phan, P. David, M. Naidu, K. H. Wong and V. Sabaratnam, *Crit. Rev. Biotechnol.*, 2015, **35**, 355–368.
91. C. N. Pope and S. Brimijoin, *Biochem. Pharmacol.*, 2018, **153**, 205–216.

92. M. A. Rahman, N. Abdullah and N. Aminudin, *Crit. Rev. Biotechnol.*, 2016, **36**, 1131–1142.
93. J. A. Takahashi, D. Sande, G. da Silva Lima, M. A. F. E. Moura and M. T. N. S. Lima, *Fungal Metabolites as Promising New Drug Leads for the Treatment of Alzheimer's Disease*, 2019, vol. 62.
94. J. Mišković, M. Karaman, M. Rašeta, N. Krsmanović, S. Berežni, D. Jakovljević, F. Piattoni, A. Zambonelli, M. L. Gargano and G. Venturella, *J. Fungi*, 2021, 7, 1–17.
95. D. Kim, S. Bae, S. Han and J. Lee, *Korean J. Mycol.*, 2016, **44**, 314–317.
96. I. Akata, G. Zengin, C. M. N. Picot and M. F. Mahomoodally, *S. Afr. J. Bot.*, 2019, **120**, 95–99.
97. T. K. Nguyen, K. H. Im, J. Choi, P. G. Shin and T. S. Lee, *Mycobiology*, 2016, **44**, 291–301.
98. K. Randhawa, V. Singh, S. Kaur, R. Kaur, S. Kumar and R. Shri, *Food Sci. Hum. Wellness*, 2021, **10**, 490–496.
99. A. Kaur, K. Randhawa, V. Singh and R. Shria, *Int. J. Med. Mushrooms*, 2019, **21**, 755–763.
100. S. S. Zhang, Q. Y. Ma, S. Z. Huang, H. F. Dai, Z. K. Guo, Z. F. Yu and Y. X. Zhao, *Phytochemistry*, 2015, **110**, 133–139.
101. M. Akiba, K. Kinoshita, Y. Kino, J. Ichi Sato and K. Koyama, *Bioorg. Med. Chem. Lett.*, 2020, **30**, 126808.
102. M. J. Alves, I. C. F. R. Ferreira, J. Dias, V. Teixeira, A. Martins and M. Pintado, *Curr. Top. Med. Chem.*, 2013, **13**, 2648.
103. H. S. Shen, S. Shao, J. C. Chen and T. Zhou, *Compr. Rev. Food Sci. Food Saf.*, 2017, **16**, 316–329.
104. D. Jakubczyk and F. Dussart, *Molecules*, 2020, **25**, 1–18.
105. R. C. G. Corrêa, L. Barros, Â. Fernandes, M. Sokovic, A. Bracht, R. M. Peralta and I. C. F. R. Ferreira, *Food Funct.*, 2018, **9**, 1465–1474.
106. O. Taofiq, S. A. Heleno, R. C. Calhelha, M. J. Alves, L. Barros, M. F. Barreiro, A. M. González-Paramás and I. C. F. R. Ferreira, *Molecules*, 2016, **21**, 1372.
107. M. Kostić, M. Smiljković, J. Petrović, J. Glamočlija, L. Barros, I. C. F. R. Ferreira, A. Ćirić and M. Soković, *Food Funct.*, 2017, **8**, 3239–3249.
108. H. M. H. Chowdhury, K. Kubra and R. R. Ahmed, *Ann. Clin. Microbiol. Antimicrob.*, 2015, **14**, 1–6.
109. D. S. Stojković, N. Kovačević-Grujičić, F. S. Reis, S. Davidović, L. Barros, J. Popović, I. Petrović, A. Pavić, J. Glamočlija, A. Ćirić, M. Stevanović, I. C. F. R. Ferreira and M. Soković, *LWT–Food Sci. Technol.*, 2017, **79**, 454–462.
110. V. Vieira, Â. Fernandes, L. Barros, J. Glamočlija, A. Ćirić, D. Stojković, A. Martins, M. Soković and I. C. F. R. Ferreira, *J. Sci. Food Agric.*, 2016, **96**, 90–98.
111. N. Shameem, A. N. Kamili, M. Ahmad, F. A. Masoodi and J. A. Parray, *Microb. Pathog.*, 2017, **105**, 356–360.
112. M. Jayachandran, J. Xiao and B. Xu, *Int. J. Mol. Sci.*, 2017, **18**, 18.

113. C. Yin, G. D. Noratto, X. Fan, Z. Chen, F. Yao, D. Shi and H. Gao, *Carbohydr. Polym.*, 2020, **250**, 116942.
114. M. Li, L. Yu, J. Zhao, H. Zhang, W. Chen, Q. Zhai and F. Tian, *J. Funct. Foods*, 2021, **83**, 104538.
115. R. K. Sindhu, A. Goyal, J. Das, S. Choden and P. Kumar, *Carbohydr. Polym. Technol. Appl.*, 2021, **2**, 100044.
116. Z. Yin, Z. Liang, C. Li and J. Wang, *Food Sci. Hum. Wellness*, 2021, **10**, 393–400.
117. R. Nicoletti, *Encyclopedia of Mycology*, Elsevier Ltd., 2021, vol. 2, pp. 683–709.
118. S. M. Hyder and D. S. Dutta, *Biocatal. Agric. Biotechnol.*, 2021, **35**, 102085.
119. N. Landi, S. Ragucci, R. Culurciello, R. Russo, M. Valletta, P. V Pedone, E. Pizzo and A. Di, *Food Chem.*, 2021, **359**, 129931.
120. S. Patra, P. Maity, I. Chakraborty, I. Kumar and D. Ghosh, *Int. J. Biol. Macromol.*, 2021, **168**, 649–655.
121. P. Nowakowski, R. Markiewicz-Żukowska, K. Gromkowska-Kępka, S. K. Naliwajko, J. Moskwa, J. Bielecka, M. Grabia, M. Borawska and K. Socha, *Biomed. Pharmacother.*, 2021, **133**, 111090.
122. X. Lin, W. Li, H. Yuen, M. Yuen and Q. Peng, *Int. J. Biol. Macromol.*, 2021, **183**, 79–89.
123. X. Meng, H. Liang and L. Luo, *Carbohydr. Res.*, 2016, **424**, 30–41.
124. D. Kothari, S. Patel and S. Kim, *Biomed. Pharmacother.*, 2018, **105**, 377–394.
125. A. Dwi, T. Cong, V. Chandra, J. Park, A. Teferra, G. Kim and B. Chun, *J. Supercrit. Fluids*, 2021, **179**, 105412.
126. W. Wang, Y. Zou, Q. Li, R. Mao, X. Shao, D. Jin, D. Zheng, T. Zhao, H. Zhu, L. Zhang, L. Yang and X. Wu, *Process Biochem.*, 2016, **51**, 542–553.
127. S. M. Wani, A. Gani, S. A. Mir, F. A. Masoodi and F. A. Khanday, *Int. J. Biol. Macromol.*, 2021, **182**, 1229–1237.
128. A. Sknepnek, S. Tomić, D. Miletić, S. Lević, M. Čolić, V. Nedovića and M. Nikšić, *Food Chem.*, 2021, **342**, 128344.
129. J. Yan, Z. Han, Y. Qu, C. Yao, D. Shen, G. Tai, H. Cheng and Y. Zhou, *Food Chem.*, 2018, **240**, 534–543.
130. S. Zhao, Q. Gao, C. Rong, S. Wang, Z. Zhao, Y. Liu and J. Xu, *Fungi*, 2020, **6**, 269.
131. D. M. D. L. Navarro, J. J. Abelilla and H. H. Stein, *J. Anim. Sci. Biotechnol.*, 2019, **10**, 39.
132. R. Kaur, M. Sharma, D. Ji, M. Xu and D. Agyei, *Fibers*, 2020, **8**, 1.
133. Q. Wang, L. Niu, H. Liu, Y. Wu, M. Li and Q. Jia, *Int. J. Biol. Macromol.*, 2021, **168**, 251–260.
134. F. Zhao, Z. Guo, Y. Zhang, L. Song, L. Ma and J. Zhao, *J. Funct. Foods*, 2021, **87**, 104797.
135. J. A. Takahashi, D. Sande, G. da Silva Lima, M. A. F. E. Moura and M. T. N. S. Lima, *Studies in Natural Products Chemistry*, 2019, vol. 62, pp. 1–39.

136. R. Zhong, J. Yang, J. Geng and J. Chen, *Int. J. Biol. Macromol.*, 2021, **192**, 967–977.
137. R. C. G. Corrêa, R. M. Peralta, A. Bracht and I. C. F. R. Ferreira, *Trends Food Sci. Technol.*, 2017, **67**, 19–35.
138. V. Caz, A. Gil-Ramírez, M. Santamaría, M. Tabernero, C. Soler-Rivas, R. Martín-Hernández, F. R. Marín, G. Reglero and C. Largo, *J. Agric. Food Chem.*, 2016, **64**, 1686–1694.
139. A. Gil-Ramírez, V. Caz, R. Martin-Hernandez, F. R. Marín, C. Largo, A. Rodríguez-Casado, M. Tabernero, A. Ruiz-Rodríguez, G. Reglero and C. Soler-Rivas, *Eur. J. Nutr.*, 2016, **55**, 1041–1057.
140. S. A. Heleno, A. R. Rudke, R. C. Calhelha, M. Carocho, L. Barros, O. H. Gonçalves, M. F. Barreiro and I. C. F. R. Ferreira, *Food Funct.*, 2017, **8**, 103–110.
141. O. Taofiq, S. A. Heleno, R. C. Calhelha, I. P. Fernandes, M. J. Alves, L. Barros, A. M. González-Paramás, I. C. F. R. Ferreira and M. F. Barreiro, *Microchem. J.*, 2019, **147**, 469–477.
142. A. Gil-Ramírez, L. Aldars-García, M. Palanisamy, R. M. Jiverdeanu, A. Ruiz-Rodríguez, F. R. Marín, G. Reglero and C. Soler-Rivas, *Innovative Food Sci. Emerging Technol.*, 2013, **18**, 101–107.
143. A. M. de Miranda, J. V. Rossoni Júnior, L. Souza e Silva, R. C. dos Santos, M. E. Silva and M. L. Pedrosa, *Eur. J. Nutr.*, 2017, **56**, 1707–1717.
144. Y. U. Jeong and Y. J. Park, *Int. J. Mol. Sci.*, 2020, **21**, 1–13.
145. P. L. Hsu, Y. C. Lin, H. Ni and F. E. Mo, *Oxid. Med. Cell. Longevity*, 2018, **2018**, 1–11.
146. S. H. Kim, M. J. Thomas, D. Wu, C. V. Carman, J. M. Ordovás and M. Meydani, *J. Nutr.*, 2019, **149**, 1377–1384.
147. E. Harada, C. N. D'Alessandro-Gabazza, M. Toda, T. Morizono, A. L. Chelakkot-Govindalayathil, Z. Roeen, M. Urawa, T. Yasuma, Y. Yano, T. Sumiya and E. C. Gabazza, *J. Med. Food*, 2015, **18**, 872–881.
148. C. Krittanawong, A. Isath, J. Hahn, Z. Wang, S. E. Fogg, D. Bandyopadhyay, H. Jneid, S. S. Virani and W. H. W. Tang, *Am. J. Med.*, 2021, **134**, 637–642.
149. B. L. Furman, M. Candasamy, S. K. Bhattamisra and S. K. Veettil, *J. Ethnopharmacol.*, 2020, **147**, 112264.
150. C. C. Nweze, N. O. Rasaq and B. I. Istifanus, *Sci. Afr.*, 2020, 7, 00209.
151. S. K. Bharti, S. Krishnan, A. Kumar and A. Kumar, *Ther. Adv. Endocrinol. Metab.*, 2018, **9**, 81–100.
152. I. Ismail Iid, S. Kumar, S. Shukla, V. Kumar and R. Sharma, *Trends Food Sci. Technol.*, 2020, **97**, 317–340.
153. R. Khursheed, S. K. Singh, S. Wadhwa, M. Gulati and A. Awasthi, *Int. J. Biol. Macromol.*, 2020, **164**, 1194–1205.
154. V. Gulati, M. Dass Singh and P. Gulati, *AIMS Med. Sci.*, 2019, **6**, 49–66.
155. American Diabetes Association, *Diabetes Care*, 2021, **44**, 13–33.
156. M. Asrafuzzaman, Y. Cao, R. Afroz, D. Kamato, S. Gray and P. J. Little, *Biomed. Pharmacother.*, 2017, **89**, 1242–1251.

157. O. O. Ogunlana, B. O. Adetuyi, M. Rotimi, L. Esalomi, A. Adeyemi, J. Akinyele, O. E. Ogunlana, O. A. Adetuyi, O. A. Adebisi, E. K. Opata, R. S. Baty and G. E.-S. Batiha, *Clin. Phytosci.*, 2021, 7, 1–9.
158. M. I. Kazeem and T. C. Davies, *J. Funct. Foods*, 2016, **20**, 122–138.
159. P. K. Prabhakar, *New and Future Developments in Microbial Biotechnology and Bioengineering: Recent Advances in Application of Fungi and Fungal Metabolites: Applications in Healthcare*, Elsevier B.V., 2020, pp. 197–208.
160. F. Alam, A. Islam, M. A. Kamal and S. H. Gan, *Curr. Med. Chem.*, 2016, **25**, 5395–5431.
161. C. Necyk and L. Zubach-Cassano, *Can. J. Diabetes*, 2017, **41**, 642–647.
162. D. Stojkovic, M. Smiljkovic, A. Ciric, J. Glamoclija, L. Van Griensven, I. C. F. R. Ferreira and M. Sokovic, *S. Afr. J. Bot.*, 2019, **120**, 100–103.
163. E. Deveci, F. Çayan, G. Tel-Çayan and M. E. Duru, *S. Afr. J. Bot.*, 2021, **137**, 19–23.
164. P. Balakrishnan and C. T. Loganayagi, *Int. J. Pharma Res. Heal. Sci.*, 2018, **6**, 2475–2479.
165. G. S. Henriques, C. V. Helm, A. P. Busato and M. L. F. Simeone, *Acta Sci., Health Sci.*, 2016, **38**, 71–79.
166. A. Lu, M. Shen, Z. Fang, Y. Xu, M. Yu, S. Wang, Y. Zhang and W. Wang, *Nat. Prod. Commun.*, 2018, **13**, 195–200.
167. H. Xiang, D. Sun-Waterhouse and C. Cui, *Food Sci. Hum. Wellness*, 2021, **10**, 87–93.
168. Y.-C. Chung, C.-Y. Tzeng, Y.-I. Chen, S.-W. Chang, T.-H. Hsu, W.-J. Ho, Y.-H. Kuo, P.-H. Hung and S.-L. Chang, *Mol. Med. Rep.*, 2018, **17**, 3260–3266.
169. N. I. Ratnaningtyas, H. Hernayanti, N. Ekowati, D. Sukmawati and H. Widianti, *Biocatal. Agric. Biotechnol.*, 2019, **19**, 101050.
170. D. J. Kim, Y. H. Kang, K. K. Kim, T. W. Kim, J. B. Park and M. Choe, *Nutr. Res. Pract.*, 2017, **11**, 180–189.
171. F. Shaher, S. Wang, H. Qiu, Y. Hu, Y. Zhang, W. Wang, H. AL-Ward, M. A. Abdulghani, S. Baldi and S. Zhou, *Diabetes, Metab. Syndr. Obes.: Targets Ther.*, 2020, **13**, 4809–4822.
172. W. L. Guo, M. Chen, W. L. Pan, Q. Zhang, J. X. Xu, Y. C. Lin, L. Li, B. Liu, W. D. Bai, Y. Y. Zhang, L. Ni, P. F. Rao and X. C. Lv, *Int. J. Biol. Macromol.*, 2020, **145**, 1208–1218.
173. C. Wang, Z. Chen, Y. Pan, X. Gao and H. Chen, *Food Chem. Toxicol.*, 2017, **108**, 498–509.
174. F. Afiati, S. F. Firza, K. Kusmiati and L. S. Aliya, *AIP Conf. Proc.*, 2019, **2120**, 070006.
175. O. M. Agunloye and G. Oboh, *Food Front.*, 2021, **1**, 1–11.
176. Y. Zhao, X. Chen, Y. Zhao, W. Jia, X. Chang, H. Liu and N. Liu, *R. Soc. Chem. Adv.*, 2020, **10**, 11918–11928.
177. Y. Sheng, C. Zhao, S. Zheng, X. Mei, K. Huang, G. Wang and X. He, *Food Sci. Nutr.*, 2019, 7, 1295–1301.

178. P. Balaji, R. Madhanraj, K. Rameshkumar, V. Veeramanikandan, M. Eyini, A. Arun, B. Thulasinathan, D. A. Al Farraj, M. S. Elshikh, A. M. Alokda, A. H. Mahmoud, J. C. Tack and H. J. Kim, *Saudi J. Biol. Sci.*, 2020, **27**, 913–924.
179. W. Li, H. Wan, S. Yan, Z. Yan, Y. Chen, P. Guo, T. Ramesh, Y. Cui and L. Ning, *Arabian J. Chem.*, 2020, **13**, 5966–5977.
180. A. Kammeyer and R. M. Luiten, *Ageing Res. Rev.*, 2015, **21**, 16–29.
181. O. Taofiq, M. F. Barreiro and I. C. F. R. Ferreira, *Curr. Med. Chem.*, 2020, **27**, 6926–6965.
182. T. Yadav, S. Mishra, S. Das, S. Aggarwal and V. Rani, *Environ. Toxicol. Pharmacol.*, 2015, **39**, 384–391.
183. S. Li, H. Liu, W. Wang, X. Wang, C. Zhang, J. Zhang, H. Jing, Z. Ren, Z. Gao, X. Song and L. Jia, *Int. J. Biol. Macromol.*, 2018, **106**, 1297–1306.
184. O. Taofiq, A. M. González-Paramás, A. Martins, M. F. Barreiro and I. C. F. R. Ferreira, *Ind. Crops Prod.*, 2016, **90**, 38–48.
185. K. M. Park, K. M. Kwon and S. H. Lee, *Evidence-Based Complementary Altern. Med.*, 2015, **2015**, 1–7.
186. T. X. Meng, S. Furuta, S. Fukamizu, R. Yamamoto, H. Ishikawa, E. T. Arung, K. Shimizu, S. Ohga and R. Kondo, *J. Wood Sci.*, 2011, **57**, 452–458.
187. Y. A. Yahaya and M. M. Don, *J. Phys. Sci.*, 2012, **23**, 1–15.
188. D. L. Abd Razak, A. Jamaluddin, N. Y. Abd Rashid, N. A. Sani and M. Abdul Manan, *J: Multidiscip. Sci. J.*, 2020, **3**, 329–342.
189. S. Mapoung, S. Umsumarng, W. Semmarath, P. Arjsri, P. Thippraphan, S. Yodkeeree and L. Pornngarm, *J. Fungi*, 2021, 7, 247.
190. M. Kozarski, A. Klaus, D. Jakovljević, N. Todorović, W. A. A. Q. Imad Wan-Mohtar and M. Nikšić, *Arch. Biol. Sci.*, 2019, **71**, 253–264.
191. T. J. Sung, Y. Y. Wang, K. L. Liu, C. H. Chou, P. S. Lai and C. W. Hsieh, *Antioxidants*, 2020, **9**, 65.
192. S. Y. Kim, K. C. Go, Y. S. Song, Y. S. Jeong, E. J. Kim and B. J. Kim, *Int. J. Mol. Med.*, 2014, **34**, 1613–1621.
193. J. I. Lee, J. H. Oh, F. Karadeniz, S. Y. Park, H. R. Kim, H. J. Jo, K. I. Jung, B.-J. Jeon and C.-S. Kong, *Asian Pac. J. Trop. Biomed.*, 2021, **11**, 164.
194. P. K. Mukherjee, R. Biswas, A. Sharma, S. Banerjee, S. Biswas and C. K. Katiyar, *J. Herb. Med.*, 2018, **14**, 1–16.
195. Z. F. Yan, Y. Yang, F. H. Tian, X. X. Mao, Y. Li and C. T. Li, *Evidence-Based Complementary Altern. Med.*, 2014, **2014**, 1–11.
196. M. Kashif, N. Akhtar and R. Mustafa, *Braz. J. Pharmacogn.*, 2017, **27**, 650–662.
197. C. Arrowitz, A. M. Schoelermann, T. Mann, L. I. Jiang, T. Weber and L. Kolbe, *J. Invest. Dermatol.*, 2019, **139**, 1691–1698.
198. M. Ashooriha, M. Khoshneviszadeh, M. Khoshneviszadeh, A. Rafiei, M. Kardan, R. Yazdian-Robati and S. Emami, *Eur. J. Med. Chem.*, 2020, **201**, 112480.

199. T. Mann, W. Gerwat, J. Batzer, K. Eggers, C. Scherner, H. Wenck, F. Stäb, V. J. Hearing, K. H. Röhm and L. Kolbe, *J. Invest. Dermatol.*, 2018, **138**, 1601–1608.
200. K. H. Im, S. A. Baek, J. Choi and T. S. Lee, *Mycobiology*, 2019, **47**, 494–505.
201. H. Satooka, P. Cerda, H. J. Kim, W. F. Wood and I. Kubo, *Biochem. Biophys. Res. Commun.*, 2017, **487**, 840–846.
202. I. Choi, Y. Park, I. Y. Ryu, H. J. Jung, S. Ullah, H. Choi, C. Park, D. Kang, S. Lee, P. Chun, H. Young Chung and H. R. Moon, *Comput. Struct. Biotechnol. J.*, 2021, **19**, 37–50.
203. Y. M. Chen, W. C. Su, C. Li, Y. Shi, Q. X. Chen, J. Zheng, D. L. Tang, S. M. Chen and Q. Wang, *Int. J. Biol. Macromol.*, 2019, **123**, 723–731.
204. N. A. Pringle, M. van de Venter, G. J. Boukes and T. C. Koekemoer, *S. Afr. J. Bot.*, 2021, **138**, 337–347.
205. M. Abate, G. Pepe, R. Randino, S. Pisanti, M. G. Basilicata, V. Covelli, M. Bifulco, W. Cabri, A. M. D'ursi, P. Campiglia and M. Rodriquez, *Pharmaceuticals*, 2020, **13**, 1–16.
206. Z. A. Amin, H. M. Ali, M. A. Alshawsh, P. H. Darvish and M. A. Abdulla, *Evidence-Based Complementary Altern. Med.*, 2015, **2015**, 1–14.
207. L. I. de Jesus, F. R. Smiderle, A. C. Ruthes, F. Vilaplana, F. T. Dal'Lin, D. Maria-Ferreira, M. F. Werner, L. J. L. D. Van Griensven and M. Iacomini, *Int. J. Biol. Macromol.*, 2018, **117**, 1361–1366.
208. A. A. Soares, A. B. De Sá-Nakanishi, A. Bracht, S. M. G. Da Costa, E. A. Koehnlein, C. G. M. De Souza and R. M. Peralta, *Molecules*, 2013, **18**, 7609–7630.
209. B. Nitha, P. V. Fijesh and K. K. Janardhanan, *Exp. Toxicol. Pathol.*, 2013, **65**, 105–112.
210. S. Dandapat, M. P. Sinha, M. Kumar and Y. Jaggi, *Environ. Exp. Biol.*, 2015, **13**, 103–108.
211. H. Zhao, Y. Lan, H. Liu, Y. Zhu, W. Liu, J. Zhang and L. Jia, *Oxid. Med. Cell. Longevity*, 2017, **2017**, 12.
212. A. Osman and A. Toliba, *Egypt. J. Food Sci.*, 2019, **47**, 157–164.
213. A. Dogan, A. Uyar, S. Hasar and O. F. Keles, *Biotech. Histochem.*, 2021, 1–9.
214. B. R. Dabdoub, R. H. Mohammed and H. L. Abdulhadi, *Syst. Rev. Pharm.*, 2020, **11**, 1704–1709.
215. Y. Yamaguchi, Y. Hirata, T. Saito and H. Kumagai, *Foods*, 2021, **10**, 1491.
216. A. E. Ubhenin, F. A. Adamude, C. C. Nweze and E. J. Dingwoke, *J. Med. Plants Res.*, 2019, **13**, 104–111.
217. H. Zhao, H. Li, Q. Lai, Q. Yang, Y. Dong, X. Liu, W. Wang, J. Zhang and L. Jia, *Int. J. Biol. Macromol.*, 2019, **127**, 476–485.
218. L. Wang, M. A. Brennan, W. Guan, J. Liu, H. Zhao and C. S. Brennan, *Food Chem.*, 2021, **351**, 129320.
219. M. Kosanić, B. Ranković, A. Rančić and T. Stanojković, *Emirates J. Food Agric.*, 2017, **29**, 98–103.

220. B. Jacinto-Azevedo, N. Valderrama, K. Henríquez, M. Aranda and P. Aqueveque, *Food Chem.*, 2021, **356**, 129651.
221. M. Garrab, H. Edziri, R. El Mokni, M. Mastouri, H. Mabrouk and W. Douki, *S. Afr. J. Bot.*, 2019, **124**, 359–363.
222. L. Butkhup, W. Samappito and S. Jorjong, *Food Sci. Biotechnol.*, 2018, **27**, 193–202.
223. O. Kaygusuz, M. Kaygusuz, Y. Dodurga, M. Seçme, E. N. Herken and K. Gezer, *Cytotechnology*, 2017, **69**, 135–144.
224. A. Novakovic, M. Karaman, S. Kaisarevic, T. Radusin and N. Llic, *Int. J. Med. Mushrooms*, 2017, **19**, 27–34.
225. E. Vamanu, *Nat. Prod. Res.*, 2018, **32**, 440–443.
226. Y. Sudjaroen and K. Thongkao, *Int. J. Green Pharm.*, 2017, **11**, 174–181.
227. M. V. Dimitrijevic, V. D. Mitic, J. S. Nikolic, A. S. Djordjevic, J. J. Mutic, V. P. Stankov Jovanovic and G. S. Stojanovic, *Chem. Biodiversity*, 2019, **16**, 1800492.
228. M. Kosanić, B. Ranković, T. Stanojković, M. Radović-Jakovljević, A. Ćirić, D. Grujičić and O. Milošević-Djordjević, *Nat. Prod. Commun.*, 2019, **14**, 1–6.
229. M. Sriramulu and S. Sumathi, *Adv. Nat. Sci.: Nanosci. Nanotechnol.*, 2017, **8**, 045012.
230. M. Chaiharn, W. S. Phutdhawong, D. Amornlerdpison and W. Phutdhawong, *Chiang Mai J. Sci.*, 2018, **45**, 1713–1727.
231. M. Sevindik, A. Rasul, G. Hussain, H. Anwar, M. K. Zahoor, I. Sarfraz, K. S. Kamran, H. Akgul, I. Akata and Z. Selamoglu, *Pak. J. Pharm. Sci.*, 2018, **31**, 2163–2168.
232. S. Khatua, S. Ghosh and K. Acharya, *Int. J. Res. Pharm. Sci.*, 2019, **10**, 564–571.
233. S. Khatua, S. Ghosh and K. Acharya, *J. Appl. Pharm. Sci.*, 2017, **7**, 144–151.
234. M. B. Bahadori, C. Sarikurkcu, O. U. Yalcin, M. Cengiz and H. Gungor, *Microchem. J.*, 2019, **150**, 104172.
235. F. Souilem, A. Fernandes, R. C. Calhelha, J. C. M. Barreira, L. Barros, F. Skhiri, A. Martins and I. C. F. R. Ferreira, *Food Chem.*, 2017, **230**, 40–48.
236. Z. Zhang, Y. Zhang, H. Liu, J. Wang, D. Wang, Z. Deng, T. Li, Y. He, Y. Yang and S. Zhong, *Int. J. Biol. Macromol.*, 2021, **168**, 356–370.
237. X. Lu, B. Feng, S. Chen, D. Zhao, G. Chen, H.-F. Wang and Y. Pei, *J. Asian Nat. Prod. Res.*, 2017, **19**, 1160–1171.
238. E. A. Adebayo, D. Martíınez-Carrera, P. Morales, M. Sobal, H. Escudero, M. E. Meneses, A. Avila-Nava, I. Castillo and M. Bonilla, *Int. J. Food Sci. Technol.*, 2018, **53**, 1316–1330.
239. K. Acharya, K. Das, S. Paloi, A. K. Dutta, M. E. Hembrom, S. Khatua and A. Parihar, *Pharmacogn. J.*, 2017, **9**, 30–34.
240. M. Kostić, M. Ivanov, Â. Fernandes, J. Pinela, R. C. Calhelha, J. Glamočlija, L. Barros, I. C. F. R. Ferreira, M. Soković and A. Ćirić, *Molecules*, 2020, **25**, 1–14.

241. J. M. Li, H. Q. Liang, P. Qiao, K. M. Su, P. G. Liu, S. X. Guo and J. Chen, *Chem. Biodivers.*, 2019, **16**, 1800609.
242. B. Muszyńska, K. Kała, J. Radović, K. Sułkowska-Ziaja, A. Krakowska, J. Gdula-Argasińska, W. Opoka and T. Kundaković, *Eur. Food Res. Technol.*, 2018, **244**, 2255–2264.
243. C. C. Chang, J. J. Cheng, I. J. Lee and M. K. Lu, *Int. J. Biol. Macromol.*, 2018, **114**, 584–591.
244. M. Y. Lu, C. C. Chen, L. Y. Lee, T. W. Lin and C. F. Kuo, *J. Nat. Prod.*, 2015, **78**, 2452–2460.
245. S. Kanwal, T. P. Joseph, S. Aliya, S. Song, M. Z. Saleem, M. A. Nisar, Y. Wang, A. Meyiah, Y. Ma and Y. Xin, *J. Funct. Foods*, 2020, **64**, 103641.
246. Z. Hu, R. Du, L. Xiu, Z. Bian, C. Ma, N. Sato, M. Hattori, H. Zhang, Y. Liang, S. Yu and X. Wang, *Cytokine*, 2020, **127**, 154917.
247. Z. Ren, J. Li, X. Song, J. Zhang, W. Wang, X. Wang, Z. Gao, H. Jing, S. Li and L. Jia, *Int. J. Biol. Macromol.*, 2018, **106**, 185–192.
248. P. Kittimongkolsuk, N. Pattarachotanant, S. Chuchawankul, M. Wink and T. Tencomnao, *Biology*, 2021, **10**, 1–16.
249. H. J. Shao, J. B. Jeong, K. J. Kim and S. H. Lee, *J. Sci. Food Agric.*, 2015, **95**, 2482–2486.
250. S. Li and N. P. Shah, *J. Funct. Foods*, 2016, **23**, 80–86.
251. S. Lee, D. Lee, S. O. Lee, J. Y. Ryu, S. Z. Choi, K. S. Kang and K. H. Kim, *J. Funct. Foods*, 2017, **32**, 27–36.
252. S. R. Lee, S. Lee, E. Moon, H. J. Park, H. B. Park and K. H. Kim, *Bioorg. Chem.*, 2017, **70**, 94–99.
253. S. M. Nkadimeng, C. M. Steinmann and J. N. Eloff, *J. Inflammation Res.*, 2021, **14**, 3729–3738.
254. B. Du, Y. Yang, Z. Bian and B. Xu, *Carbohydr. Polym.*, 2017, **172**, 68–77.
255. R. D. Divate and Y. C. Chung, *J. Funct. Foods*, 2017, **35**, 81–89.
256. M. Fogarasi, Z. M. Diaconeasa, C. R. Pop, S. Fogarasi, C. A. Semeniuc, A. C. Fărcaş, D. Ţibulcă, C. D. Sălăgean, M. Tofană and S. A. Socaci, *Agronomy*, 2020, **10**, 1972.
257. O. Taofiq, S. A. Heleno, R. C. Calhelha, M. J. Alves, L. Barros, A. M. González-Paramás, M. F. Barreiro and I. C. F. R. Ferreira, *Food Chem. Toxicol.*, 2017, **108**, 139–147.
258. T. C. Finimundy, L. Barros, R. C. Calhelha, M. J. Alves, M. A. Prieto, R. M. V. Abreu, A. J. P. Dillon, J. A. P. Henriques, M. Roesch-Ely and I. C. F. R. Ferreira, *Food Chem.*, 2018, **245**, 150–158.
259. Â. Fernandes, J. Petrović, D. Stojković, L. Barros, J. Glamočlija, M. Soković, A. Martins and I. C. F. R. Ferreira, *LWT–Food Sci. Technol.*, 2016, **69**, 91–97.
260. A. Mocan, Â. Fernandes, L. Barros, G. Crişan, M. Smiljković, M. Soković and I. C. F. R. Ferreira, *Food Funct.*, 2018, **9**, 160–170.
261. E. Tejedor-Calvo, K. Amara, F. S. Reis, L. Barros, A. Martins, R. C. Calhelha, M. E. Venturini, D. Blanco, D. Redondo, P. Marco and I. C. F. R. Ferreira, *Food Res. Int.*, 2021, **140**, 110071.

262. Y. Zhang, Y. Zeng, Y. Men, J. Zhang, H. Liu and Y. Sun, *Int. J. Biol. Macromol.*, 2018, **115**, 978–984.
263. M. Sriramulu, S. Shanmugam and V. Kumar, *Colloid Interface Sci. Commun.*, 2020, **35**, 100254.
264. C. French, L. C. Chandra, S. Clarke, E. A. Lucas, B. J. Smith, S. Lightfoot and S. R. Kuvibidila, *Int. J. Med. Mushrooms*, 2019, **21**, 645–656.
265. Y. Matsushita, Y. Furutani, R. Matsuoka and T. Furukawa, *BMC Complement. Altern. Med.*, 2018, **18**, 319.
266. J. Tangen, T. Holien, M. R. Mirlashari, K. Misund and G. Hetland, *Biomed Res. Int.*, 2017, **2017**, 1–7.
267. P. Zhu, X. Fu, J. Li, A. K. Tse, H. Guo and J. Chen, *Front. Pharmacol.*, 2018, **9**, 1449.
268. Y.-C. Hseu, Y.-H. Chao, K.-Y. Lin, T.-D. Way, H.-Y. Lin, V. Thiyagarajan and H.-L. Yang, *J. Ethnopharmacol.*, 2017, **208**, 72–83.
269. J. Wang, C. Wu, C. Lee, S. Hsieh, J. Chen and C. Lee, *PLoS One*, 2017, **12**, 0170924.
270. T. Huang, Y.-W. Lan, C.-M. Chen, Y.-F. Ko, D. M. Ojcius, J. Martel, J. D. Young and K.-Y. Chong, *Sci. Rep.*, 2019, **9**, 5145.
271. Y.-C. Chen, Y. Liu, M. El-shazly, T. Wu, J.-G. Chang and Y.-C. Wu, *Int. J. Mol. Sci.*, 2019, **20**, 833.
272. Y. Zhang, Y. Zeng, Y. Cui, H. Liu, C. Dong and Y. Sun, *Carbohydr. Polym.*, 2020, **235**, 115969.
273. S. Bi, Y. Jing, Q. Zhou, X. Hu and J. Zhu, *Food Funct.*, 2018, **9**, 279–293.
274. X. Zhang, Z. Cai, H. Mao, P. Hu and X. Li, *Int. J. Biol. Macromol.*, 2021, **166**, 1387–1395.
275. Y. Wang, Y. Tian, J. Shao, X. Shu, J. Jia, X. Ren and Y. Guan, *Int. J. Biol. Macromol.*, 2018, **108**, 300–306.
276. J. Ye, X. Wang, K. Wang, Y. Deng, Y. Yang, R. Ali, F. Chen, Z. Wu, W. Liao and L. Mao, *J. Anim. Physiol. Anim. Nutr.*, 2020, **104**, 735–748.
277. C. Jiao, W. Chen, X. Tan, H. Liang, J. Li, H. Yun, C. He, J. Chen, X. Ma, Y. Xie and B. B. Yang, *J. Ethnopharmacol.*, 2020, **247**, 112256.
278. J. Li, F. Gu, C. Cai, M. Hu, L. Fan, J. Hao and G. Yu, *Int. J. Biol. Macromol.*, 2020, **143**, 806–813.
279. A. Cheng, Y. Chien, H. Lee, Y. Hsieh and Y.-L. Yu, *Molecules*, 2020, **25**, 3585.
280. M. F. Lau, K. H. Chua, V. Sabaratnam and U. R. Kuppusamy, *Biotechnol. Appl. Biochem.*, 2020, **68**, 902–917.
281. W. A. Elkhateeb, G. M. Zaghlol, I. M. El-Garawani, E. F. Ahmed, M. E. Rateb and A. E. Abdel Moneim, *Biomed. Pharmacother.*, 2018, **101**, 264–277.
282. Q.-D. Xiang, Q. Yu, H. Wang, M.-M. Zhao, S.-Y. Liu, S.-P. Nie and M.-Y. Xie, *J. Agric. Food Chem.*, 2017, **65**, 5306–5315.
283. X. Gao, J. Qi, C. Ho, B. Li, J. Mu, Y. Zhang, H. Hu, W. Mo, Z. Chen and Y. Xie, *Carbohydr. Polym.*, 2020, **249**, 116874.
284. Y.-T. Tsai, P.-H. Kuo, H.-P. Kuo, C.-Y. Hsu, Y.-J. Lee, C.-L. Kuo, J.-Y. Liu, S.-L. Lee and M.-C. Kao, *Environ. Toxicol.*, 2021, **36**, 320–327.

285. Q. Lia, F. Zhang, G. Chen, Y. Chen, W. Zhang, G. Mao, T. Zhao, M. Zhang, L. Yange and X. Wu, *Int. J. Biol. Macromol.*, 2018, **111**, 1293–1303.
286. F. Zhao, Y. F. Wang, L. Song, J. X. Jin, Y. Q. Zhang, H. Y. Gan and K. H. Yang, *Integr. Cancer Ther.*, 2017, **16**, 205–214.
287. M. Meng, D. Cheng, L. Han, Y. Chen and C. Wang, *Carbohydr. Polym.*, 2017, **157**, 1134–1143.
288. S. Huh, S. Lee, S. J. Choi, Z. Wu, J. Cho, L. Kim, Y. S. Shin, B. W. Kang, J. G. Kim, K. Liu, H. Cho and H. Kang, *Molecules*, 2019, **24**, 3834.
289. C. Wang, S. Shi, Q. Chen, S. Lin, R. Wang, S. Wang and C. Chen, *Integr. Cancer Ther.*, 2018, **17**, 674–683.
290. Y. Zhang, *Biochem. Biophys. Res. Commun.*, 2017, **488**, 679–684.
291. K. Li, K. Na, T. Sang, K. Wu, Y. Wang and X. Wang, *Oncol. Rep.*, 2017, **38**, 2803–2813.
292. D. Wu, C. Tang, Y. Liu, Q. Li, W. Wang, S. Zhou, Z. Zhang, F. Cui and Y. Yang, *Carbohydr. Polym.*, 2019, **222**, 114996.
293. Y. Yang, P. He and N. Li, *Evidence-Based Complementary Altern. Med.*, 2019, **2019**, 12.
294. J. Fang, S. Gao, R. Islam, Y. Teramoto and H. Maeda, *Nutrient*, 2020, **12**, 2279.
295. W. Li, J. Wang, H. Hu, Q. Li, Y. Liu and K. Wang, *J. Funct. Foods*, 2018, **45**, 75–85.
296. T. C. Finimundy, G. Scola, F. J. Scariot, A. J. P. Dillon, S. Moura, S. Echeverrigaray, J. P. Henriques and M. Roesch-Ely, *Int. J. Med. Mushrooms*, 2018, **20**, 31–46.
297. B. H. Kong, K. H. Teoh, N. H. Tan, C. S. Tan, S. T. Ng and S. Y. Fung, *PeerJ*, 2020, **8**, 9650.
298. D. Peng, Y. Wen, S. Bi, C. Huang, J. Yang, Z. Guo, W. Huang, J. Zhu, R. Yu and L. Song, *Int. J. Biol. Macromol.*, 2021, **192**, 1134–1149.
299. Y. Wen, S. Bi, X. Hu, J. Yang, C. Li, H. Li, D. Bo, J. Zhu, L. Song and R. Yu, *Int. J. Biol. Macromol.*, 2021, **183**, 145–157.
300. N. Shomali, O. Onar, B. Karaca, N. Demirtas, A. C. Cihan, I. Akata and O. Yildirim, *Int. J. Med. Mushrooms*, 2019, **21**, 571–582.
301. M. Lenzi, V. Cocchi, A. Novaković, M. Karaman, M. Sakač, A. Mandić, M. Pojić, M. C. Barbalace, C. Angeloni, P. Hrelia, M. Malaguti and S. Hrelia, *BMC Complementary Altern. Med.*, 2018, **18**, 300.
302. S. R. Lee, H. S. Roh, S. Lee, H. B. Park, T. S. Jang, Y. J. Ko, K. H. Baek and K. H. Kim, *J. Funct. Foods*, 2018, **40**, 249–260.
303. A. Haque and A. U. Islam, *Medicina*, 2019, **55**, 716.
304. Y. Zhang, Y. Zhang, W. Gao, R. Zhou, F. Liu and T. Bun, *Int. J. Biol. Macromol.*, 2020, **164**, 3171–3178.
305. H. Abreu, M. Zavadinack, F. R. Smiderle, T. R. Cipriani, L. M. C. Cordeiro and M. Iacomini, *Carbohydr. Polym.*, 2021, **252**, 117177.
306. B. Yuan, N. Ma, L. Zhao, E. Zhao, Z. Gao, W. Wang, M. Song, G. Zhang, Q. Hu and H. Xiao, *Food Funct.*, 2017, **8**, 3553–3562.

307. G. Ma, B. M. Kimatu, L. Zhao, W. Yang, F. Peib and Q. Hu, *Food Funct.*, 2017, **8**, 1810–1821.
308. A. Haque, A. S. M. A. Reza, S. Nasrin and A. Rahman, *Integr. Cancer Ther.*, 2020, **19**, 1–9.
309. A. Ebrahimi, A. Atashi, M. Soleimani, M. Mashhadikhan, A. Barahimi and A. Maghari, *J. Basic Clin. Physiol. Pharmacol.*, 2018, **29**, 95–102.
310. A. Ebrahimi, A. Atashi, M. Soleimani, M. Mashhadikhan, A. Barahimi and S. Kaviani, *J. Complementary Integr. Med.*, 2018, **15**, 20160136.
311. Y. Yang, P. Yuan, X. Wei, C. Fu, J. Li, W. Wang, X. Wang, Y. Li and J. Li, *Sci. Rep.*, 2018, **8**, 13984.
312. T. C. Finimundy, R. M. V Abreu, N. Bonetto, F. J. Scariot, A. J. P. Dillon, S. Echeverrigaray, L. Barros, I. C. F. R. Ferreira, J. A. P. Henriques and M. Roesch-ely, *Food Chem. Toxicol.*, 2018, **112**, 383–392.
313. B. Marrow, D. Cells, C. Liu, M. W. Choi, X. Xue and P. C. K. Cheung, *J. Agric. Food Chem.*, 2019, **67**, 12137–12143.
314. X. Tan, W. Chen, C. Jiao, H. Liang, H. Yun, C. He, J. Chen, X. Ma and Y. Xie, *Food Funct.*, 2020, **11**, 1110–1121.
315. K. Yelithao, U. Surayot, C. Lee, S. Palanisamy, N. M. Prabhud, J. Leee and S. You, *Carbohydr. Polym.*, 2019, **218**, 37–45.
316. M. Zhang, X. Tian, Y. Wang, D. Wang, W. Li, L. Chen, W. Pan, S. Mehmood and Y. Chen, *Int. J. Biol. Macromol.*, 2018, **107**, 2679–2685.
317. K. F. Wang, K. Y. Sui, C. Guo and C. Z. Liu, *Int. J. Biol. Macromol.*, 2017, **104**, 377–383.
318. K. Singha, P. K. Hor, J. P. Soren, J. Mondal, K. C. Mondal, B. R. Pati and P. K. Das, *Bioact. Carbohydr. Diet. Fibre*, 2021, **25**, 100255.
319. X. Hu, C. Liu, X. Wang, D. Jia, W. Lu, X. Sun, Y. Liu and L. Yuan, *Exp. Ther. Med.*, 2017, **13**, 352–358.
320. H. Zhao, Q. Lai, J. Zhang, C. Huang and L. Jia, *Oxid. Med. Cell. Longevity*, 2018, **2018**, 1–15.
321. O. O. Ogbole, A. O. Nkumah, A. U. Linus and M. O. Falade, *Mycology*, 2019, **10**, 166–173.
322. Y. Liu, Y. You, Y. Li, L. Zhang, L. Yin, Y. Shen, C. Li, H. Chen, S. Chen, B. Hu and D. Chen, *Carbohydr. Polym.*, 2017, **174**, 72–81.
323. J. Zhang, X. Zhao, L. Q. Zhao, J. Zhao, Z. Qi and L. Wang, *Int. J. Med. Mushrooms*, 2017, **19**, 905–913.
324. W. L. Guo, J.-C. Deng, Y.-Y. Pan, J.-X. Xu, J.-L. Hong, F. F. Shi, G.-L. Liu, M. Qian, W.-D. Bai, W. Zhang, B. Liu, Y.-Y. Zhang, P.-J. Luo, L. Ni, P.-F. Rao and X.-C. Lv, *Int. J. Biol. Macromol.*, 2020, **153**, 1231–1240.
325. T. Jiang, L. Wang, A. Ma, Y. Wu, Q. Wu, Q. Wu, J. Lu and T. Zhong, *J. Food Biochem.*, 2020, **44**, e13515.
326. W. L. Guo, F. F. Shi, L. Li, J. X. Xu, M. Chen, L. Wu, J. L. Hong, M. Qian, W. D. Bai, B. Liu, Y. Y. Zhang, L. Ni, P. F. Rao and X. C. Lv, *Int. J. Biol. Macromol.*, 2019, **131**, 81–88.
327. L. Kou, M. Du, P. Liu, B. Zhang, Y. Zhang, P. Yang, M. Shang and X. Wang, *Appl. Biochem. Biotechnol.*, 2019, **187**, 310–322.

328. H. Aoki, M. Hanayama, K. Mori and R. Sato, *Biosci., Biotechnol., Biochem.*, 2018, **82**, 1550–1559.
329. Y. Chen, Y. Liu, M. M. R. Sarker, X. Yan, C. Yang, L. Zhao, X. Lv, B. Liu and C. Zhao, *Carbohydr. Polym.*, 2018, **198**, 452–461.
330. Y. Chen, D. Liu, D. Wang, S. Lai, R. Zhong, Y. Liu, C. Yang, B. Liu, M. R. Sarker and C. Zhao, *Food Chem. Toxicol.*, 2019, **126**, 295–302.
331. M. Chen, D. Xiao, W. Liu, Y. Song, B. Zou, L. Li, P. Li, Y. Cai, D. Liu, Q. Liao and Z. Xie, *Int. J. Biol. Macromol.*, 2020, **155**, 890–902.
332. E. E. Bach, E. M. B. Hi, A. M. C. Martins, P. A. M. Nascimento and N. S. Y. Wadt, *Medicines*, 2018, **5**, 78.
333. N. I. Ratnaningtyas, H. Hernayanti, S. Andarwanti, N. Ekowati, E. S. Purwanti and D. Sukmawati, *Biosaintifika*, 2018, **10**, 642–647.
334. M. Rašeta, M. Popović, I. Čapo, N. Stilinović, S. Vukmirović, B. Milošević and M. Karaman, *RSC Adv.*, 2020, **10**, 10382–10393.
335. J. Wang, C. Wang, S. Li, W. Li, G. Yuan, Y. Pan and H. Chen, *Biomed. Pharmacother.*, 2017, **95**, 1669–1677.
336. L. F. Laurino, F. J. M. Viroel, E. Caetano, S. Spim, T. B. Pickler, R. M. Rosa-Castro, E. A. Vasconcelos, A. F. Jozala, A. Hataka, D. Grotto and M. Gerenutti, *Nutrients*, 2019, **11**, 2720.
337. H. J. Yang, M. J. Kim, D. Y. Kwon, D. S. Kim, T. Zhang, C. Ha and S. Park, *Nutrients*, 2018, **10**, 948.
338. K. L. Nyam, C. F. Chow, C. S. Tan and S. T. Ng, *Int. J. Med. Mushrooms*, 2017, **19**, 607–617.
339. N. Begum, A. Nasir, Z. Parveen, T. Muhammad, A. Ahmed, S. Farman, N. Jamila, M. Shah, N. S. Bibi, A. Khurshid, Z. Huma, A. A. K. Khalil, A. Albrakati and G. E. S. Batiha, *Front. Pharmacol.*, 2021, **12**, 661803.
340. M. Xiong, Y. Huang, Y. Liu, M. Huang, G. Song, Q. Ming, X. Ma, J. Yang, S. Deng, Y. Wen, J. Shen, Q.-H. Liu, P. Zhao and X. Yang, *Mol. Nutr. Food Res.*, 2017, **62**, 1700444.
341. M. Asrafuzzaman, M. M. Rahman, M. Mandal, M. Marjuque, A. Bhowmik, B. Rokeya, Z. Hassan and M. O. Faruque, *J. Taibah Univ. Med. Sci.*, 2018, **13**, 465–471.
342. Y. Liu, C. Wang, J. Li, T. Li, Y. Zhang, Y. Liang and Y. Mei, *J. Fed. Am. Soc. Exp. Biol.*, 2020, **34**, 1065–1078.
343. C. Zhang, L. Zhang, H. Liu, J. Zhang, C. Hu and L. Jia, *Int. J. Biol. Macromol.*, 2018, **111**, 219–228.
344. P. Chen, Y. Weng, S. Lin, Z. Yu and B. Wang, *J. Food Sci.*, 2017, **82**, 2456–2461.
345. C. C. Ikewuchi, J. C. Ikewuchi and M. O. Ifeanacho, *Biomed. Pharmacother.*, 2017, **95**, 1809–1814.
346. L. Zhang, Y. Liu, Y. Ke, Y. L. A, X. Luo, C. Li, Z. Zhang, A. Liu, L. Shen, H. Chen, B. Hu, H. Wu, W. Wu, D. Lin and S. Li, *Int. J. Biol. Macromol.*, 2018, **119**, 134–140.
347. N. Ekowati, N. I. Yuniati, H. Hernayanti and N. I. Ratnaningtyas, *Biosaintifika*, 2018, **10**, 655–662.

348. T. Zhang, M. Jayachandran, K. Ganesan and B. Xu, *Front. Pharmacol.*, 2018, **9**, 1257.
349. Z. A. Amin, H. M. Ali, M. A. Alshawsh, P. H. Darvish and M. A. Abdulla, *Evidence-Based Complementary Altern. Med.*, 2015, **2015**, 14.
350. L. Wen, Q. Gao, C. Wah Ma, Y. Ge, L. You, R. H. Liu, X. Fu and D. Liu, *J. Funct. Foods*, 2016, **20**, 400–410.

CHAPTER 6

Macromolecules in Fungi with Pharmaceutical Potential

A. HELENA ARAÚJO-RODRIGUES,[†] B. ANA SOFIA SOUSA[†] AND C. MANUELA E. PINTADO*

Universidade Católica Portuguesa, CBQF – Centro de Biotecnologia e Química Fina – Laboratório Associado, Escola Superior de Biotecnologia, Rua Diogo Botelho 1327, 4169-005 Porto, Portugal
*Email: mpintado@ucp.pt

6.1 Introduction

Since ancient times, based on empirical knowledge, natural products have been applied as pharmaceuticals[1,2] in the form of concentrates, extracts, and others.[3] In traditional Asian medicine, particularly in China and some other Asian countries, mushrooms have acquired a medicinal status over the years, a concept that has been extended to folk medicine in Western countries.[1,4–7] Over the last few decades, the scientific community has started to study the function and biological mechanisms of the different macromolecules present in diverse mushroom species.[1] Examples of these explored mushrooms include medicinal (*e.g.*, *Coriolus versicolor*, *Antrodia cinnamomea*, and *Cordyceps sinensis*) and edible species, including those most extensively cultivated and consumed worldwide (*e.g.*, *Pleurotus* spp., *Agaricus bisporus*, and *Lentinula edodes*).[8,9]

[†]Both authors contributed equally to this work.

Food Chemistry, Function and Analysis No. 36
Edible Fungi: Chemical Composition, Nutrition and Health Effects
Edited by Dejan Stojković and Lillian Barros

Published by the Royal Society of Chemistry, www.rsc.org

Beyond their rich nutritional composition, mushrooms contain several bioactive compounds that play crucial roles in human health and well-being.[3,5,6,9–11] Polysaccharides, proteins, phenolic compounds, terpenes, unsaturated fatty acids, and phenolic compounds are examples of bioactive macromolecules present in mushrooms.[3,8,9,11–13] Specific metabolites belonging to these groups include β-glucans, lectins, ganoderic, phenolic, and linoleic acids.[3,8,9,14] Consequently, mushrooms have been extensively used to prevent or treat several clinical conditions,[9,15,16] and have been classified as a next-generation food.[13] Worldwide, numerous scientists have investigated these groups of macromolecules due to their medicinal potential and for the development of functional foods, nutraceuticals, and even to explore new potential drugs.[3,6,8,13] Unfortunately, there is no internationally recognized standard definition for "functional foods" or "nutraceuticals".[208] Indeed, these two terms are often confused and interchanged.[206,208]

The Functional Food Center (FFC), in 2018, proposed a definition, describing functional foods as "natural or processed foods that contain biologically active compounds; which, in defined, effective, and non-toxic amounts, provide a clinically proven and documented health benefit utilizing specific biomarkers for the prevention, management, or treatment of a chronic disease or its symptom".[209] On the other hand, the concept of "nutraceuticals" can be defined as an isolated product, obtained from one or more food ingredients "that provides medical or health benefits, including the prevention and treatment of a disease".[207,210] These are usually referred to as pharma-foods, and are generally sold in pharmaceutical form (*e.g.*, tablets, pills, capsules, *etc.*) to be used beyond the diet and before drugs, to prevent and treat certain chronic diseases.[206,211]

Most research studies that focus on 'mushrooms' functional and biological potential use extracts of isolated macromolecules (*e.g.*, lentinan or pleuran) or specific groups of macromolecules (*e.g.*, polysaccharide or protein fraction). However, several studies have proposed a beneficial synergistic effect between the different 'mushrooms' bioactive macromolecules.[8,17] Consequently, their health-promoting effects can be correlated with entire mushrooms or specific macromolecules or fractions.[9] Several parameters, such as the mushroom species, growing environment, and development stage, as well as storage and processing conditions, influence the concentration and bioactive levels of mushrooms.[8,18] In parallel, fruiting bodies or mycelium in the same mushroom species can also possess different bioactive profiles.[3,6,18]

Several beneficial effects on human health and well-being have been associated with whole mushrooms and extracts consumption, or with the presence of specific macromolecules such as prebiotic, antitumour and antiproliferative, immunomodulating, anti-inflammatory, and antiviral effects.[3,5,7–9,12,14,19] These bioactive potentials have been supported by *in vitro* and *in vivo* studies coupled with clinical trials.[3,6–8] More than 700 mushroom species have been associated with human health benefits.[14,20] However, of these species, only 270 specific macromolecules have been associated with particular bioactive properties.[9] Therefore, it has become evident that macrofungi are still underexploited,[9] with greater involvement of the scientific community in studying the correlation between mushrooms macromolecules or species to specific bioactivities and health benefits being essential.

The numerous bioactive compounds present in edible mushrooms are summarized in Figure 6.1 and discussed throughout this chapter, emphasizing their medicinal properties and pharmaceutical potential.

6.2 Polysaccharides

The carbohydrates represent between 50% and 65% of the total mushroom constitution on a dry basis.[2,9,20] High concentrations of digestible (glucose, glycogen, mannitol, and trehalose) and non-digestible (peptide/protein-polysaccharides, β-glucan, α-glucan, chitin, galactans, hemicellulose, xylans, and mannans) carbohydrates are found in macrofungi.[9,21] The mushroom carbohydrate profile differs according to the variety, growing conditions (*e.g.*, environment and type of substratum), development phase, storage conditions, and other processing and extraction methodologies.[8,9,13] This group of macromolecules includes mono-, di- and polysaccharides,[2,13] with the latter being the most important carbohydrate group, considered as "biological response modifiers" due to their numerous biological and physiological properties.[3,8,20] Some biologically active properties include immunomodulatory, anti-inflammatory, antiviral, antimicrobial, anticancer, antioxidant, and antidiabetic effects.[3,8,9,12,13]

Bioactive polysaccharides are typically characterized by their high molecular weights, holding at least 1 monosaccharide linked by glycosidic bonds[11] and a β-linked glucose backbone.[13,22] Also, a mannose and galactose backbone is found in mushroom polysaccharides.[13] These macromolecules may have one type of monosaccharides (homopolysaccharides) or, alternatively, two or more types of monosaccharides (heteropolysaccharides), which can be associated with protein or peptides, forming protein– or peptide–polysaccharide complexes.[8,22,23]

The polysaccharide groups can differ from each other in their basic structure (chemical composition), glycosidic bonds (type and number), side chains (nature and number), and three-dimensional configuration.[8,11,21,24] A structure–function relationship is usually reported in polysaccharides. In other words, according to their structural properties, distinct biological properties are described.[3,7,11,20] For example, there is a positive correlation between molecular weight and bioactive potential, resulting from their higher affinity to the immune cell receptors connection.[11,24]

α-, β-, or mixed glucans are the major polysaccharide group found in edible mushrooms.[11] β-Glucan and chitin are the major macrofungal cell wall components.[2,4,8,9] However, β-glucans are the most crucial polysaccharide group since they are the most highly prevalent and studied

Figure 6.1 Biological properties of mushroom macromolecules in human health and well-being.

macromolecule, holding numerous biological and physiological activities in human health and well-being.[3,4,8] The main pharmaceutical potential of polysaccharides is discussed throughout this section. The biological properties of protein–polysaccharide complexes, namely lectins, are described in Section 6.3.

6.2.1 β-Glucan

β-Glucans are D-glucose monomers linked by 1,3 and 1,6 β-glycosidic bonds,[4,8,22] whose molecular weight ranges from ten to thousands of kDa.[11] In the case of mushrooms, β-glucan content ranges between 3.1 and 46.5%.[3,4,8,21,22] The β-glucan structure and, consequently, physical and biological properties depend on the chain lengths, molecular weight, branching degree, positions of glycosidic bonds, and three-dimensional arrangement (*e.g.*, single, double, or triple helix).[3,4,21,22] Similarly to carbohydrates, specific β-glucan structures vary according to the mushroom species, growing conditions, isolation approaches, and other parameters.[3,4]

Well-characterized β-glucans are, for example, lentinan (isolated from *L. edodes*), pleuran (isolated from *Pleurotus ostreatus*), grifolan (isolated from *Grifola frondosa*), and ganoderan (isolated from *Ganoderma lucidum*),[3,22] with these currently being studied towards their medicinal market application.[8] The main biological properties of these and other mushroom β-glucans are discussed in the following sections, and some examples of corresponding scientific studies are also summarized in Table 6.1.

Table 6.1 Examples of studies reported different health-promoting properties of β-glucans.

Bioactivity	Ref.
Prebiotic	4, 8, 12 and 25–33
Immunomodulatory	21, 27 and 34–40
Anti-inflammatory	21, 25, 41 and 42
Antitumour and antiproliferative	3, 4, 8, 9, 11, 21, 24 and 43–46
Antidiabetic	7, 9, 11, 15 and 21
Hypocholesterolaemic and antilipidaemic	4, 7, 9, 25 and 47–50
Antimicrobial	7, 9, 17, 21 and 50–57
Antiviral	3 and 58–63
Antioxidant	11, 35, 50 and 64–69
Protective effect on DNA damage	70 and 71
Radioprotective effect	72
Analgesic	73
Wound healing	74
Bone health	75 and 76
Improved cognitive capacity	77
Antiallergic	78–80
Immune system booster	81 and 82

6.2.1.1 Prebiotic Effect

Probiotics, prebiotics, and symbiotics impact the gut microbial community, resulting in beneficial microorganism growth and the production of numerous metabolites involved in human health and well-being.[8] According to the current definition, prebiotics are "a substrate that is selectively utilized by host microorganisms, conferring a health benefit".[83] Numerous factors influence the gut microbiota composition, where diet is one of the most impactful parameters.[8,12,84] Due to their β-glycosidic bonds, mushroom polysaccharides, such as β-glucans, α-glucans, and chitin, are indigestible by the acidity and enzymes of the gastric and upper gut tract,[3,4] reaching the colon without any gastrointestinal tract digestion and resisting gastrointestinal absorption.[11] Nevertheless, gut microbiota can metabolize mushroom polysaccharides,[8,9,28] enhancing the modulation of microbiota dynamics by increasing the probiotic strains.[8,9,11]

The prebiotic properties of polysaccharides, polysaccharide fractions, or entire edible mushroom species, such as *Pleurotus* spp., *A. bisporus*, *L. edodes*, *Tremella fuciformis*, and *G. lucidum* have been extensively reported.[27–29] These investigations have suggested that they were able to promote the growth of numerous intestinal fermenting genera (*e.g.*, *Lactobacillus* and *Enterococcus*).[30] For example, lentinan promotes the growth and primary metabolism of short-chain fatty acids (SCFAs)-related bacteria (*e.g.*, *Lactobacillus*, *Bifidobacterium*, *Oligella*, and *Allobaculum*) and, at the same time, decreases *Enterococcus* genera grown (young rats supplementation).[31] In a human faecal fermentation study, the β-glucans isolated from *Schizophylum commune* Fr and *Auricularia auricula-judae* also promoted bifidobacterial and lactobacilli growth.[32]

SCFAs are produced during the fermentation of β-glucans and other polysaccharides (*e.g.*, α-glucan, hemicelluloses, and chitin).[3,8,9,12] Other gut microbial-derived metabolites are serotonin, gamma-aminobutyric acid (GABA), acetylcholine, and histamine.[8] These molecules are key metabolites for gut homeostasis and, at the same time, decrease the risk of several disorders of the immune, neuroendocrine, and neuronal systems.[84] Gut microbial-derived metabolites influence, for example, the blood pressure, lipid and glucose metabolism, immune function, cancer development[9,11,84] and, through the gut–brain axis, neuronal and mental function.[8] Consequently, dysregulation of the intestinal ecosystem has been associated with numerous human diseases. Table 6.2 shows some examples of gut microbiota modulation-mediated effects through mushroom β-glucan consumption on the overall human organism.

6.2.1.2 Immunomodulatory, Anti-inflammatory, Antitumour, and Antiproliferative Effects

Several mushroom-derived polysaccharides exhibit immunomodulating properties, namely, β-glucans, hetero-β-glucans, as well as glycans and heteroglycans.[85] Some examples include the β-glucans of *Agaricus* spp., *Ganoderma* spp., *Pleurotus* spp.,[27,34–36] and *Poria cocos*,[37] and lentinans from *L. edodes*[38,39]

Table 6.2 Some gut microbiota modulation-mediated impacts through mushroom β-glucans consumption.[a]

Main impacts	Ref.
Provides energy for intestinal epithelium cells	12 and 33
Decreases luminal pH (pathogenic 'microorganisms' inhibition and essential nutrient and mineral absorption)	12
Promotes intestinal integrity and permeability	8, 12 and 33
Positive influence on gene expression and cell signalling pathways (immune responses and metabolic homeostasis)	12 and 33
Blood pressure regulation (cardiovascular diseases risk reduction)	33
Cholesterol-lowering (inhibition of cholesterol synthesis and promotion of LDL cholesterol catabolism)	4, 25 and 33
Antidiabetic effects (alleviate type-2 diabetes)	26 and 33
Tumour growth inhibition	12
Gut–brain axis crosstalk (normal brain function)	8 and 30

[a]LDL, low-density lipoprotein.

and *Schizophyllan commune*.[21,40] In the intestinal epithelium, β-glucans can be internalized and interact with the membrane receptors of immune cells,[4,8] recognizing (1→3) β-D-glucan patterns that possess branched side chains.[22] For example, β-glucans can interact with receptors of neutrophils, macrophages, monocytes, dendritic, and natural killer cells.[22] After receptor interaction and recognition, intracellular signal transduction and activation occur.[4,7,8] Consequently, different interleukins, tumour necrosis factor-α (TNF-α), and cytokines are produced.[7,8,11,85]

Mushroom β-glucans may exhibit antitumour effects by different direct and indirect mechanisms,[8,11] which may prevent or delay the cancer onset and, at the same time, the migration of tumour cells and metastasis.[11] Indirect mechanisms may result from their immunomodulatory potential through the activation of the immune system and response, mainly through the production of several crucial cell messengers.[7,8,11,85] Nowadays, this approach has been merged with traditional treatments of chemotherapy.[11] Direct mechanisms include the expression of chemical messengers[8,11,85] directly involved in inhibiting the growth of cancer cells or apoptosis.[11]

Beyond the immunomodulatory properties, lentinan has been shown to possess antimetastatic and antitumour effects in clinical trials.[4,8,11,44] In this context, this macromolecule has been extensively used after its approval for adjuvant cancer treatments,[4,8,11,24] with demonstrated efficacy and impacts on response rates. This macromolecule has also been associated with the alleviation of adverse chemotherapy symptoms.[44] Lentinan has been extensively used in lung,[45] colorectal,[46] ovarian[43] as well as cardiac, non-Hodgkin's lymphoma, gastric, and other types of cancer, combined with other traditional therapies.[3,8,9,11,44] Although lentinan is the most reported and studied anticancer macromolecule, other β-glucans, such as grifolan, have been reported to hold antitumour potential.[9,21]

Additionally, anti-inflammatory properties have been associated with mushroom β-glucans,[21,41] with a similar potential to glucocorticoids and non-steroidal anti-inflammatory drugs.[21] The mechanisms involve the inhibition of the nitric oxide synthase (NOS), cyclooxygenase-2 (COX2), and pro-inflammatory cytokines,[21,41] which could be very promising for the treatment of inflammatory disorders[21]. The anti-inflammatory potential of *Pleurotus* spp. β-glucans (*e.g.*, *P. ostreatus*) has been widely suggested, for instance, in arthritis in a mouse model,[25] while lentinan also revealed anti-inflammatory potential in mice colitis.[42]

6.2.1.3 Antidiabetic, Hypocholesterolaemic, and Antilipidaemic Effects

Diabetes is a metabolic and chronic disease characterized by high blood glucose levels caused by insulin imbalances within the organism.[86] In this context, several works have indicated that mushroom polysaccharides may perform hypoglycaemic and antihyperglycaemic activities.[7,9,11,15] Mushroom polysaccharides may promote increased glucose uptake through stimulation of insulin signalling.[11] Recent studies have tested new hypoglycaemic agents, suggesting good competitiveness of mushroom polysaccharides, which could be part of a promising approach for the prevention of diabetes and the development of new hypoglycaemic drugs.[11] For example, grifolan has been reported to be one of the most effective β-glucans in the treatment of diabetes.[21]

Through diabetes progression, other conditions such as hyperlipidaemia, hypercholesterolaemia, and hypertriglyceridaemia are also often reported.[87] Beyond the antidiabetic effect, several studies have also demonstrated the hypolipidaemic potential of mushroom polysaccharides.[7,9] Some examples include β-glucans isolated from *A. blazei*[47] and polysaccharides extracted from *Pleurotus eryngii*[48] tested in diabetic mice. Moreover, in hypercholesterolaemic patients, β-glucan supplementation promoted a reduction in low-density lipoprotein (LDL) and cholesterol levels.[25]

One mechanism that may explain the antihyperlipidaemic potential of β-glucans and other polysaccharides is its binding to cholesterol and lipids in the intestines, decreasing their absorption and promoting their excretion.[4] Alternatively, Caz *et al.*[49] reported modulation of cholesterol-related genes mediated by a glucan fraction extracted from *P. ostreatus* in a mice model, analogous to hypocholesterolaemic drugs. In parallel, Zhu *et al.*[50] also reported a reduction of the lipid concentration by promoting the activity of antioxidant enzymes through consumption of *L. edodes* polysaccharides.

6.2.1.4 Antimicrobial and Antiviral Activities

In numerous studies, the antimicrobial and antiviral potentials of mushroom β-glucans have been described against Gram-positive and Gram-negative bacteria, yeasts, mycelia, fungi, and viruses.[7,9,50] The antimicrobial and antiviral mechanisms of β-glucans include indirect stimulation of the immune system, triggering a host immune response.[21] As previously mentioned, these mechanisms involve, for instance, the promotion of phagocytosis and macrophage and neutrophil production.[7,17,21,53] On the other hand, mushrooms may also produce and secrete biomolecules, which play essential roles in defence, communication, and pathogenecity.[54] The oxidation of schizophyllan, an exopolysaccharide secreted by *S. commune*, results in biomolecules (*e.g.*, aldehyde groups) with Gram-positive and Gram-negative bacteria-inhibiting potential.[55]

Several studies have reported the antimicrobial and antiviral potentials of β-glucans of *L. edodes*.[56] For example, the antibacterial activity of lentinan against *Mycobacterium tuberculosis*[51,57] and *Listeria monocytogenes*[52] was described. Regarding its antiviral potential, lentinan has been tested in the treatment of human immunodeficiency virus (HIV). These studies showed an improvement of the host immune response coupled with adjuvant antiretroviral didanosine treatment,[58] constituting a target for HIV treatment and other impactful viral diseases (*e.g.*, malaria).[3] Furthermore, the scientific community has proposed the use of lentinan in flu viral infections and coronavirus disease 2019 (COVID-19), due to its health-stimulating potential combined with a reported decrease in lung epithelial inflammation.[59]

Other β-glucans isolated from distinct mushroom species have also been associated with antimicrobial and antiviral properties. Pleuran[60] has revealed the potential to inhibit the herpes simplex virus (HSV) and to regenerate the epithelium during respiratory infections.[61] In this context, the supplementation of children's diet with β-glucans seems to reduce the incidence of respiratory infections associated with flu viral infections.[62] Moreover, *Aspergillus brasiliensis* β-glucans showed antiviral potential against bovine herpesvirus 1, while *A. auricula-judae* β-glucans had potential against Newcastle disease virus.[63]

6.2.1.5 Antioxidant Potential

Concerning the antioxidant and free radical scavenging potential, the polysaccharide fraction of *P. ostreatus*, *Pleurotus sajor-caju*, *P. eryngii*, *Pleurotus abalonus*, and *Pleurotus djamor*,[35,50,65,66] lentinan,[44,67,88,89] as well as *G. lucidum* polysaccharide fraction,[64] showed strong antioxidant activity. Also, β-glucans extracted from *G. frondosa*,[69] *A. bisporus*, *P. ostreatus*, and *Coprinus comatus* evidenced higher antioxidant potential.[68] The scientific community has proposed that the β-glucan molecular length and type of binding in the side chains impact the antioxidant capacity.[11]

6.2.1.6 Other Bioactivities

Other biological and functional properties have been described for mushroom β-glucans, such as antiaging,[11,64] analgesic,[73] DNA-protective,[70,71] antiallergic,[78,79] healing,[74] and radioprotective.[72] Some studies have suggested a protective effect against DNA damage of β-glucans from *Agaricus* spp.,[70,71] while a radioprotective effect on radiation-induced damage was verified.[72]

Yamada *et al.*[78] associated an antiallergic effect to pleuran, while Jesenak *et al.*[79] ascribed it to lentinan. These macromolecules have also been related to bone health in arthritis,[75,76] as well as mental health,[77] by improving cognitive capacity. Abreu *et al.*[73] also reported an analgesic effect on the injury mice model and Kwon *et al.*[74] reported wound regeneration and healing potential.

6.2.2 Other Polysaccharides

The numerous bioactivities and studies reported in Section 6.2.1. corroborate the high value of mushroom β-glucans in several human diseases and disorders in the pharmaceutical, nutraceutical, and cosmetic fields, which is expected to increase their market value.[90] However, other less studied polysaccharides, such as α-glucan and chitin, also play important roles in human health.

α-Glucan is a less frequent macromolecule, with α type in contrast to β type linkages of β-glucan.[4,21] *G. frondose* α-glucans showed antidiabetic effects due to their capacity to increase insulin sensitivity and improve insulin resistance.[91] This macromolecule has also been associated with immunomodulatory properties.[59,92] Some examples were reported by Volman *et al.*[92] that showed the potential of *A. bisporus* α-glucans and Murphy *et al.*[59] of *L. edodes* α-glucans in immune system stimulation. Although α-glucans may hold an immunomodulatory potential, the results of a comparative study of the bioactive properties of α- and β-glucans extracted from *L. edodes* suggested that only β-glucans exhibit antitumour and antioxidant potential.[93] This study corroborated the relationship between the tridimensional conformation and the physical and biological potential previously mentioned.[3,7,11,20]

Chitin is a polysaccharide with a complex biopolymer network, a β(1–4)-linked *N*-acetyl-glucosamine polymer, which is covalently linked to a β-glucan and forms a protein–glucan–chitin complex.[94,95] As previously mentioned, the mushroom cell wall is rich in both β-glucans and chitin.[4,8,9] The prebiotic potential has also been associated with chitin.[9,94] Chitosan may be produced through the deacetylation reaction of chitin.[94,95] Some of the 'chitins' and 'chitosans' main bioactive properties are their antimicrobial potential, coupled with a biocompatible and biodegradable profile. These properties result in numerous possible applications in the pharmaceutical, nutraceutical, and cosmetic areas. However, the scientific studies in this field are scarce and limited.[95] Some recent clinical trials also suggest their wound-healing properties,[95] increasing their interest and raising the importance of expanding the research in this field.

6.3 Proteins

In addition to polysaccharides, mushrooms also produce many proteins and peptides with health benefits due to their immunomodulatory, antitumour, antiproliferative, anti-inflammatory, antioxidant, antimicrobial, antiviral, antihypertensive, antihypercholesterolaemic, antihyperlipidaemic, and antidiabetic properties.[3,9,16,94,96]

Generally, mushrooms are made up of 19–35% protein based on the dry weight. This amount is relatively high compared to most vegetable protein suppliers, making mushrooms an excellent dietary protein source.[9] Nevertheless, mushroom protein content (as well as peptides and amino acids) differs according to the species, substrate composition, pile size, and harvest time.[9,10] In addition, many mushroom proteins are yet to be identified and characterized.[96] Even though mushrooms are not extremophilic fungi, most proteins have substantial pH and thermal stability. These and other characteristics make mushroom proteins unique and often different from vegetable, animal, and even microbial proteins.[96]

Apart from proteins, mushrooms are a good source of all essential amino acids required for the normal health and growth of humans.[9] Moreover, mushrooms also have substantial amounts of two amino acids with important physiological activities in the human organism, namely GABA and ornithine. The first is a neurotransmitter that reduces the neuron activity in the central nervous system, while the latter is a precursor in arginine synthesis.[9,13,97] Also, ergothioneine is another interesting amino acid synthesized by mushrooms once it acts as an antioxidant to protect against mitochondrial oxidative stress.[94]

According to recent investigations, mushrooms hold many nutritional and pharmacologically useful proteins, with the most relevant structures including lectins (protein–polysaccharide complexes), fungal immunomodulatory proteins (FIPs), and proteins with enzymatic activity.[3,94,98,99] Moreover, another group of proteins isolated from fungi is ubiquitin and ubiquitin-like proteins (Ubls). These are signalling messengers that monitor cellular functions such as cell proliferation, apoptosis, cell cycle and DNA repair, immune response, among others. In this way, Ubls can exhibit antitumour and anti-inflammatory effects. An example of these proteins is a Ubl from *Agrocybe cylindracea* that showed anticancer effects on a human liver cancer cell line (HepG2) and mouse leukaemia cell line (M1), as well as immunomodulatory effects in murine peritoneal macrophages.[100] Therefore, further studies of these proteins are needed to develop a promising class of targets for human therapy.[101,102]

This section summarizes the biochemical and biological properties with their biomedical applications of the most relevant groups of edible mushroom proteins.

6.3.1 Lectins

Lectins are non-immune proteins or carbohydrate-binding proteins with a non-catalytic carbohydrate-binding site (CBS).[98,99] Due to the existence of the CBS, lectins can recognize and reversibly bind to various carbohydrate moieties of the cell surface of the glycoconjugate (glycolipid or glycoprotein) without modifying their covalent structure.[99,103] This activity can lead to the agglutination of different cell types and the precipitation of glycoconjugates.

Indeed, some lectins are described as haemagglutinins as they exhibit haemagglutination on interaction with erythrocytes.[98,104]

Besides, these lectin–glycan interactions play important roles in several biological processes at the cellular level, including signalling, differentiation, morphogenesis and adhesion, targeting, molecular recognition and pathogenesis, scavenging of glycoconjugate from the circulatory system, cell–cell interactions in the immune system, among others.[99,105,106] Additionally, the lectin amino acid sequence and arrangement determine its specificity for CBS, consequently defining its functions and potential biological applications.[103] Thus, based on their glycan specificity, lectins can be used in several therapeutic applications, acting as a diagnostic tool, studying cellular pathology and physiology, and controlling drug release delivery.[3,103] In this manner, a lectin from *A. bisporus* (ABL) is already commercially available in conjugated form with fluorescence markers for detection and bioanalysis purposes.[99]

Among all the mushroom protein groups, lectins are probably the most studied.[98] According to Singh *et al.*,[105] about 144 lectins have been isolated and characterized in edible mushrooms (*e.g.*, *A. bisporus*, *Agrocybe aegerita*, *A cylindracea*, *Boletus edulis*, *P. ostreatus*, and *Volvariella volvacea*). Lectins are widely distributed in mushroom fruiting bodies and mycelia, and their amount can vary depending on the age of the fruiting body and the season.[104,106] The molecular weight of mushroom lectins ranges from 12 to 190 kDa. Generally, lectins are stable at moderate temperatures, resist digestion in the gastrointestinal tract, penetrate the monolayer of intestinal epithelial cells, and enter the blood circulation intact.[3,104,107] As a result, these types of proteins are suitable for oral administration.

Mushroom lectins exhibit antiproliferative, mitogenic, immune-stimulating potential, and antiviral, antimicrobial, antioxidant, and antidiabetic therapeutic effects.[103,104] The main biological properties of mushrooms lectins are summarized in Figure 6.2 and are discussed in the following subsections.

6.3.1.1 *Immunomodulatory, Antiproliferative/Antitumour, and Mitogenic Activities*

Some mushroom lectins possess immunological activity, mainly due to their selectivity and specificity to bind to surface receptors of immune cells. These proteins are capable of activating several downstream signalling pathways, as well as inducing the excretion of interleukins, nitric oxide (NO), NOS, interleukins, TNF-α, interferon-gamma (IFN-γ), and other mediators of innate immunity against pathogens.[103,104,108] Besides, lectins can also stimulate phagocytosis and increase the secretion of toll-like receptors (TLRs), which are crucial in eradicating pathogens.[103] Therefore, mushroom lectins exert immunomodulatory effects through the regulation and proliferation of mediators of innate immunity, such as cytokines, namely by increasing the production of macrophage activating factors and activation of lymphocytes, which are vital immune system cells.[24,104]

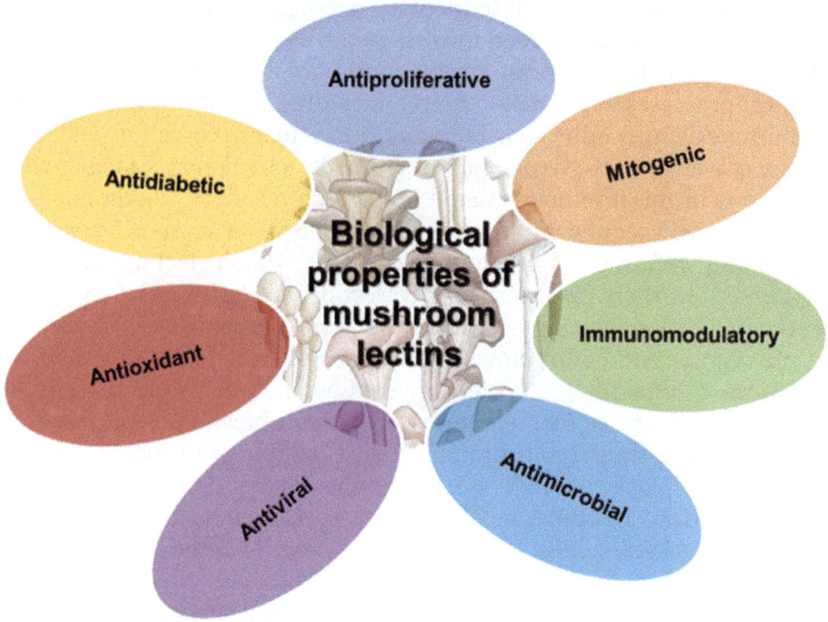

Figure 6.2 Main bio-functional mushrooms of mushroom lectins.

In cancer, tumour cell surfaces vary in the glycoconjugate composition and their terminal carbohydrate units as compared to normal cells,[108] exposing specific disaccharides (hidden in normal cells) on their surface, such as the Thomsen Friedenreich antigen (known as TF-antigen).[103] Thus, due to their carbohydrate specificity, lectins exert an antiproliferative effect by cross-linking the altered glycans on the cancer cell's surface or over immunomodulatory effects. Several cellular mechanisms are induced after this lectin interaction with cancer cells, including apoptosis, leading to cell death.[104] Also, the binding affinity of the lectin to the surface of tumour cells can be used as a prognostic tool and biomarker to detect metastases and diagnose different tumour stages.[103] The antiproliferative/antitumour effects of edible mushroom lectins have been reported against several types of cancers, including breast carcinoma, colon, hepatoma, leukaemia, and sarcoma.[104] Besides, lectins from *Hericium erinaceum*, *Inocybe umbrinella*, and *Russula delica* showed antiproliferative properties against HepG2 and human breast adenocarcinoma cells (MCF-7).

This mitogenic activity can induce mitosis in resting cells into actively growing cells. A few mushroom lectins can stimulate the transformation of small resting cells (such as lymphocytes/splenocytes) to large blast-like cells that undergo mitosis.[108] Typically, this lymphocyte activation and proliferation begins with the lectin binding to T-cell receptors and the consequent activation of the signalling cascade: mobilization of intracellular calcium, expression of the IL-2 gene and, finally, cell proliferation.[109] Some examples of mushroom lectins that exhibit mitogenic effects in murine splenocytes

Table 6.3 Immunomodulatory, antiproliferative/antitumour, and mitogenic activities of selected lectins from edible mushrooms.[a]

Mushroom species	Lectin name	Main therapeutic effects	Ref.
Agrocybe aegerita	AAL	Antitumour activity by apoptosis induction against various human cancer cell lines, *in vitro* and *in vivo*, including BGC-823, HeLa, HL-60, MGC-803, SW-480, SGC-790, and mouse sarcoma S-180	110
Agaricus bisporus	ABL	Immunostimulator of RAW264.7 macrophages by NO and TNF-α production	111
		Anticancer activity against Caco-2, HT-29 and MCF-7 cells by blocking the DNA synthesis during cell proliferation	99
Ganoderma capense	—	Mitogenic activity for mouse splenocytes and antiproliferative activities for HepG2, L1210, and M1 cells	112
Latiporus sulphureus	LSL	Immunopotentiator capable of activating macrophages and inducing cell proliferation	113
Psathyrella asperospora	PAL	Anticancer activity against HT-29 by stopping in the G2/M phase of the cell cycle	114
Tricholoma mongolicum	TML-1 and TML-2	Immunomodulatory and antitumour effects against mouse lymphoblast-like (p815) mastocytoma cells by promoting the excretion of TNF-α, NO	115
Volvariella volvacea	—	Immunomodulatory effect	115
	VVL	Mitogenic activity on T lymphocytes by a calcium-dependent pathway	116

[a]—: not mentioned; BGC-823: human gastric cancer cell line; Caco-2: human colon cancer cell line; HeLa: human cervical cancer cell line; HL-60: human leukaemia cells line; L1210: mouse leukaemia cell line; M1: mousse leukaemia cell line; HT-29: human colon cancer cell line; MCF-7: human breast adenocarcinoma cell line; MGC-803: human gastric adenocarcinoma cell line; NO: nitric oxide; SGC-790: human gastric cancer cell line; SW-480: human colon adenocarcinoma cell line; TNF-α: tumour necrosis factor-α.

can be isolated from *Xerocomus spadiceus*, *Pleurotus citrinopileatus*, *Flammulina velutipes*, *A. cylindracea*, *B. edulis*, *Armillaria luteovirens*, and *Hygrophorus russula*.[106] However, some mushroom lectins exert non-mitogenic or antimitogenic effects.[106] An example of this is ABL which suppresses the production of T and B lymphocytes, inhibiting DNA synthesis.[99]

Table 6.3 lists some examples of lectins isolated from edible mushrooms that exhibit immunomodulatory, anti-inflammatory, antiproliferative, and mitogenic effects.

6.3.1.2 *Antiviral and Antimicrobial Activities*

The antiviral activity of fungi lectins has been studied against viruses such as influenza A/B, hepatitis C, HIV, herpes simplex type 1 and 2, severe acute respiratory syndrome coronavirus (SARS), and severe acute respiratory syndrome coronavirus 2 (SARS-CoV-2), the virus responsible for the current

COVID-19 pandemic.[9,99] Lectins inhibit viral activity by specifically adhering to the virus's surface glycoprotein, which blocks the host receptor or can inhibit the viral polymerase enzyme, biding to its active site.[103,117] Besides, these proteins can inhibit the entry of some viruses into cells, such as SARS-CoV, binding to the viral envelope glycoproteins (spike protein) and consequently preventing cell-virus fusion.[117]

Interestingly, HIV has been the most investigated virus concerning the antiviral activity of mushroom glycoproteins, with mannose-binding lectins (MBLs) being the most studied in this context. MBLs exert their antiretroviral activity by binding to the surface of the glycoprotein gp120 (rich in mannose) and gp41 exposed on the virus's surface, making the pathogen capable of being more easily neutralized by antibodies and the cells of the host's immune system.[118] Additionally, lectins can also inhibit HIV-RT, the crucial enzyme in the HIV life cycle, thus exerting its anti-HIV activity.[119] Therefore, this last group of lectins is of significant interest in developing anti-HIV drugs because of their specific binding ability, particularly HIV-RT. In this regard, lectins from *C. militaris*, *A. bisporus*, and *P. citrinopileatus* are examples of lectins that exhibit potential inhibition of HIV-RT effects.[108]

In the future, further *in vivo* clinical research studies would be interesting to optimize the dose, bioavailability, mechanism of action, and antiviral activity of different types of mushroom lectins. Also, it would be promising to investigate as a new therapeutic approach the combination of lectins, with antiviral or immunomodulatory activity, with available drugs to control viral infection diseases.[103]

As antimicrobial agents, lectins can interfere with the growth, multiplication, and dissemination of microorganisms by agglutination.[120] However, studies of mushroom lectins with antimicrobial activity, mainly against bacteria and fungi, remain scarce.[103] Lectins exert antibacterial activity due to their ability to adhere to carbohydrates existing on the surface of bacterial cell walls (such as *N*-acetylmuramic or teichoic acids in Gram-positive bacteria and lipopolysaccharides in Gram-negative bacteria), causing bacterial motility and preventing its adhesion to the surface.[103,120,121] As a result, this interaction causes bacterial agglutination or cell death by the existence of pores and bubbling in Gram-positive and Gram-negative bacteria cells, respectively.[120,122] Regarding the antifungal effect, lectins can bind to chitin or other glycans present on the fungal surface, interfering with the absorption of nutrients and the growth of fungus, as well as the formation of spores.[120,121] Also, small-sized lectins can penetrate the surface of the fungal cell and inhibit several essential enzymes, altering the morphogenesis of the cell wall.[120]

Mushroom lectins that possess antiviral and antimicrobial activity are presented in Table 6.4.

6.3.1.3 *Antidiabetic Activity*

Lectins with antidiabetic properties may be interesting functional ingredients to be included in different food matrices or to function as nutraceuticals with

Table 6.4 Antiviral and antimicrobial activities of selected edible mushroom lectins.[a]

Mushroom species	Lectin name	Main therapeutic effects	Ref.
Aleuria aurantia	AAL	Antifungal activity against *Mucor racemosus*	123
Agrocybe aegerita	AAL	Antiviral activity against H9N2 viruses in mice because of its interaction with the viral surface glycosylated proteins	124
Boletus speciosus	—	Antiviral activity against HIV by inhibiting HIV-TR	119
Gymnopilus spectabilis	GSL	Antibacterial activity against *Staphylococcus aureus* and antifungal activity against *Aspergillus niger*	121
Pleurotus ostreatus	POL	Immunogenic and antiviral activities against hepatitis B virus by activating the innate immune response	125
Sparassis latifolia	—	Antibacterial activity against *Bacillus subtilis*, *Escherichia coli*, *Listeria monocytogenes*, and *Pseudomonas aeruginosa.* Antifungal activity against *Candida albicans*, *Candida catenulate*, and *Candida glabrata*	126

[a]—: not mentioned; H9N2: avian influenza virus; HIV: human immunodeficiency virus; HIV-RT: human immunodeficiency virus reverse transcriptase activity.

health benefits for diabetes. An example of an excellent lectin to prevent or treat diabetes is ABL, which showed increased *in vivo* insulin secretion when added to the islet of Langerhans of diabetic mice and helped the proliferation of pancreatic β cells through the regulation of cell cycle proteins.[127] Also, the specific interaction between glycoproteins and their cellular oligosaccharide receptors causes conformational changes in the membranes of the islets of Langerhans and β cells, leading to insulin exocytosis.[128]

6.3.1.4 Antioxidant Activity

In addition to the bioactivities described above, lectins from mushrooms can also be used as an antioxidant to decrease free radical-mediated oxidative stress.[104] Actually, mushroom lectins can maintain membrane integrity despite oxidative stress. These macromolecules can scavenge free radicals by interacting with the oxidative cascade, by oxygen quenching and chelation of metal ions, which prevents lipid peroxidation of the membrane.[104] In this context, *Pleurotus florida* lectin demonstrates a potent antioxidant effect against arsenic-induced nephrotoxicity in rats by upregulation of mRNA expression of the superoxide dismutase gene.[129,130]

6.3.2 Fungal Immunomodulatory Proteins

The fungal immunomodulatory proteins (FIPs) are unique bioactive proteins isolated from certain medicinal and edible mushrooms, such as *C. versicolor*[131] and *G. lucidum*,[132] among others, as described by Liu *et al.*[133]

In general, the molecular weight of FIPs is about 13 kDa, and these proteins have 110–114 amino acids, with acetylated-extreme valines and high asparagine concentrations.[94,134] Regarding their structure, FIPs are dimers in a dumbbell-shape, with a design very similar to the variable region of human immunoglobulin heavy chains.[115] As a result of this similarity, they exert remarkable immunomodulatory effects and have been widely investigated for their biochemical and pharmacological applications.[14,133]

Concerning 'FIPs' immunomodulatory activity, they were able to regulate cellular and humoral immunity, which includes increasing the expression of antigen-presenting cells. Additionally, FIPs boost the activation and proliferation of immune cells, such as lymphocytes, T cells, and dendritic cells by inducing the release of cytokines and chemokines.[24,133,135] In parallel, these proteins can also have haemagglutinating, immunomodulatory, anti-inflammatory, antiallergy, and antitumour properties, and have been considered to treat autoimmune diseases, asthma, and cancer conditions.[102,133,134]

Some other examples include a FIP isolated from *F. velutipes* that can exert various biological activities, such as haemagglutinating and activation of T cells *via* cytokine regulation p38/MAPK.[135,136] Similarly, FIP from *V. volvacea* has been shown to perform an immunomodulatory activity by secreting cytokines and enhancing the proliferation of lymphocytes in human peripheral blood.[137] Another FIP with interesting bioactivities is a HEP isolated from *Hericium erinaceus*. HEP shows immunomodulatory activity, auxiliary antitumour effects, anti-inflammatory effects on mice with inflammatory bowel disease, as well as the regulation of the gut microbiota, acting as a potential prebiotic. By regulating the composition and metabolism of the intestinal microbiota, HEP inhibited the growth of certain aerobic and microaerobic bacteria. In parallel, it increased the content of some probiotic bacteria such as *Lactobacillus* and *Bifidobacterium*, among others.[8,14,16]

Genetic engineering techniques make it possible to produce recombinant FIPs (rFIPS) in other organisms, such as *Escherichia coli*. This is important since FIPs are usually present in small concentrations in nature, and consequently, the extraction yield of these proteins is low, which is a significant constraint to their investigation.[24] Besides, rFIPs may have better immunomodulatory activity and more effectiveness in inducing the expression of specific cytokines than those naturally produced in mushrooms.[24,115]

6.3.3 Proteins with Enzymatic Activity

Mushrooms are enzyme producers, which are proteins that catalyse specific phases in metabolism, and some have health-promoting effects.[138] The principal proteins possess enzymatic activity, including ribosome-inactivating proteins (RIPs), ribonucleases, laccases, proteases, and tyrosinases. Table 6.5 lists some examples of these enzymes and their beneficial health effects.

RIPs are enzymes that generally have an RNA *N*-glycosidase domain that irreversibly inactivates ribosomes, preventing protein synthesis.[98] In this context, RIPs can inhibit translation in the cell-free rabbit reticulocyte lysate

Table 6.5 Proteins that possess enzymatic activity with potential health benefits.[a]

Mushroom species	Enzyme	Enzyme name	Main biological functions	Ref.
Agrocybe aegerita	RNase	AA-RNase	Anticancer activity against HepG2, HeLa, and SH-SY5Y	100
		Ageritin	Anticancer activity against HeLa, COLO 320 and Raji cells; nuclease activity in genomic DNA; endonuclease activity in supercoiled plasmid; antiviral activity against the tobacco mosaic virus (RNA virus); antimicrobial activity effects *Micrococcus lysodeikticus* (bacteria) and *Penicillium digitatum* (fungal)	144
Cordyceps militaris	Protease	CMP	Anticancer against MCF-7 and A-549 cells; antifungal effect against the *Fusarium oxysporum*	157
Flammulina velutipes	RIP	Flammulin	Induction of apoptosis in HeLa cells	140
		Flammin and Velin	Translation inhibition in the rabbit reticulocyte lysate system	139
Ganoderma lucidum	RNase	GLR	Anticancer activity against HCT116 and HT-29	146
Hypsizigus marmoreus	RIP	Hypsin	Anticancer activity against L1210, HepG2, and HL60 cells; antifungal activity against *Botrytis cinerea*, *Fusarium oxysporum*, *Mycosphaerella arachidicola*, and *Physalospora piricola*; inhibition of HIV-RT; inhibition of translation in a rabbit reticulocyte lysate system	141
		Marmorin	Anticancer activity against MCF-7 and HepG2 cells, and promotes endoplasmic reticulum stress and the death receptor apoptotic pathway	142
Pleurotus cornucopiae	Laccase	—	Anticancer activity against HepG2 and MCF-7; inhibition of HIV-RT activity	162
Pleurotus eryngii	Laccase	—	Antiviral effects by inhibition of HIV-RT activity	163
	Protease	Pleureryn	Translation inhibition in the rabbit reticulocyte lysate system; inhibition of HIV-RT activity	155
Pleurotus ostreatus	Laccase	—	Antiviral activity against the hepatitis C virus	164
Pleurotus sajor-caju	RNase	PS-RNase	Anticancer activity against HepG2 and L1210; antifungal effects against *Fusarium oxysporum* and *Mycosphaerella arachidicola;* antibacterial activity against *Pseudomonas aeruginosa* and *Staphylococcus aureus*	147
Thelephora ganbajun	RNase	TG-RNase	Antiviral effects by inhibition of HIV-RT activity	148

[a]—: not mentioned; A-549: human lung cancer cell line; COLO 320: human colon cancer cell line; HeLa: human cervical cancer cells line; HepG2: human liver cancer cell line; HIV: human immunodeficiency virus; HIV-RT: human immunodeficiency virus reverse transcriptase activity; HL-60: human leukaemia cell line; L1210: mouse leukaemia cell line; HCT116: human colorectal carcinoma cell line; HT-29: human colon cancer cell line; MCF-7: human breast adenocarcinoma cell line; Raji: human lymphoma cell line; RIPs: ribosome-inactivating proteins; RNases: ribonucleases; SH-SY5Y: human neuroblastoma cell line.

system. Due to their RNA-glycosidase activity, RIPs can precisely remove an adenine from adenosine (A4324) from 28S ribosomal RNA in rats, which stops protein synthesis.[101] So far, RIPs have been identified in several species of edible mushrooms, with some of these fungi being the source of more than one RIP, such as *F. velutipes*[139,140] and *Hypsizygus marmoreus*[141,142] (see Table 6.5).

Other enzymes that have high specificity for RNA are ribonucleases (RNases). These proteins catalyse the hydrolysis of RNA molecules into smaller fragments with distinct degrees of specificity. Therefore, RNAases can have antitumour, angiogenic, immunosuppressive, or antiviral properties, making them potential therapeutic agents.[96,143] However, studies into the mechanism of action of mushroom RNases, namely on antitumour activity, are scarce. Thus, further research into these enzymes could be an area of great interest in the future, exploring their potential therapeutic applications such as their use in treating tumours and the treatment of acquired immunodeficiency syndrome (AIDS) due to their activity antiviral against HIV.[102,143] In this way, various RNases have been isolated from different mushroom species, including *A. aegerita*,[144,145] *G. lucidum*,[146] *Pleurotus sajor-caju*,[147] and *Thelephora ganbajun*,[148] among others, which were able to perform antiproliferative/antitumour, antimicrobial, and/or antiviral activities (see Table 6.5). Additionally, according to Rezvani *et al.*,[102] several RNases that have been isolated from mushrooms still have no biological activity attributed, including those from *A. bisporus*[149] and *V. volvacea*,[150] among others.

Laccases are another ordinary group of enzymes found in mushrooms that can catalyse the oxidation of several substrates, including inorganic and phenolic compounds, and reduce molecular oxygen in the water.[98,101] Beyond their biotechnological and industrial applications, such as lignin degradation, laccases interfere with the morphogenesis, pathogenesis, and immunogenesis of organisms, and the metabolic turnover of complex organic substances, which explains their pharmaceutical and medical applications.[94,98] A variety of laccases with antiproliferative and antiviral activities, as well as immunology effects, have been isolated from several edible mushroom species. Laccases extracted from oyster mushrooms demonstrate health-promoting effects (see Table 6.5).

Proteases (also known as proteinases or peptidases) are proteolytic enzymes that break down proteins through the hydrolysis of peptide bonds.[102] These enzymes are grouped based on their catalytic site and enzymatic mechanisms into asparagine, aspartic, glutamic, cysteine, threonine, serine, and metalloproteases. Proteases are involved in the physiology and developmental processes of all living organisms.[151] In this way, these proteins are known to accomplish biological activities in cells, such as protein turnover, enzymatic modification, nutrition, gene expression regulation, and cell death initiation, among others.[102,151,152]

Indeed, in living organisms, including humans, an equilibrium between proteases and their natural inhibitors is essential, as a disturbance of this balance can lead to many diseases, *e.g.*, cancer. As a result of the substantial implications of proteases in pathological processes, highly selective inhibition of disease-involved proteases would benefit research and medical applications. An example is mycocypine, a cysteine protease inhibitor extracted from basidiomycetes.[96] Besides, proteases can act as antitumour agents at several stages, including tumour initiation, growth, and metastasis, due to their immunomodulatory functions.[102] Fungi, as well as bacteria, are good sources of proteases.[153] These enzymes have been extracted from edible mushrooms such as *A. bisporus*,[154] *P. eryngii*,[155] *P. ostreatus*,[156] *C. militaris*,[157] *F. velutipes*,[158] and *Lyophyllum cinerascens*,[159] among others (see Table 6.5).

The final enzymes under review in this subtopic are tyrosinases. These enzymes, which belong to a group of copper proteins, catalyse successive oxidation steps with several phenolic substrates. For that reason, tyrosinase is also known as polyphenol oxidase (PPO) and it is involved in numerous biological functions.[160]

Tyrosinases are necessary for the first step of melanin biosynthesis since they are able to catalyse the conversion of L-tyrosine to L-3,4-dihydroxyphenylalanine (well-known as L-DOPA), which is then transformed into dopaquinone, a precursor in melanin biosynthesis.[99] Consequently, PPO has been associated with pigment deficiency or hyperpigmentation conditions, such as vitiligo and melanoma, respectively.[96,99] Therefore, there is great interest in researching tyrosinase inhibitors for therapeutic use.[96] Besides, studies have demonstrated that PPO can suppress tumour growth and be applied to produce L-DOPA, one of the compounds used to treat Parkinson's disease. Despite the promising applications of tyrosinases, their direct use in human therapy is limited once they are mutagenic and catalyse reactions that generate harmful radicals such as quinones.[99]

Mushroom tyrosinases may be isolated from *A. bisporus* to study enzymatic browning during development and postharvest storage.[161] Also, this PPO demonstrated health benefits, including anti-inflammatory effects, antioxidant properties, and antibacterial activity against *Staphylococcus aureus*.[10] Interestingly, this mushroom PPO has high homology with mammalian enzymes, making it an excellent model for melanogenesis studies and the search for new tyrosinase inhibitors and substrates.[96] Mushroom tyrosinase is commercially available and has been used on tyrosinase inhibition and as a potent inhibitor for antipigmentation research studies.[99]

6.4 Lipids

The total content of lipids in mushrooms typically ranges between 2% and 6% on a dry basis.[3,9] Depending on the mushroom species and geographic localization, some variations in the total content of lipids are found in the literature data.[165] Generally, their incidence in the fruiting body is low compared with carbohydrates and proteins.[3] Fatty acids consist of carboxylic acids originating from hydrocarbon chains containing between 4 and 36 carbons. Fatty acids result from the condensation of acetate units (two carbons) with an unbranched chain. These can be divided into saturated fatty acids, containing simple bonds between carbon atoms, and unsaturated fatty acids, which possess at least a double bond.[165]

Unsaturated fatty acids, for example, oleic acid (C18:1) and linoleic acid (C18:2), are the most predominant fatty acids (approximately 75% of the total fat), compared with saturated fatty acids.[3,9,13,165] Linoleic acid is an essential ω-6 fatty acid (not synthesized by the human body), and the most incident polyunsaturated fatty acid (PUFA) in mushrooms, followed by oleic acid.[3,165] Thus, mushrooms are a key source of essential fatty acids (ω-6), associated with several human health benefits related to the brain, cardiovascular and kidney functions, and inflammatory and immunological responses.[165]

Numerous studies have been suggesting the antitumour effects of linoleic acid. The reported mechanisms in colon, breast, and prostate cancers in animal models described significant impacts in all stages of tumorigenesis.[13] The lipid fraction of *C. militaris* showed the capacity to modulate the inflammation, and downregulate inflammatory and upregulate anti-inflammatory genes. This fraction contains fatty acids, sterols, and fatty acid esters and may be promising in treating several inflammatory disorders.[166] Phan *et al.*[167] also studied the bioactive properties of the lipid fraction of *Pleurotus giganteus*. The results indicated a strong antifungal potential against *Candida* species infections.

6.5 Terpenes and Terpenoids

Terpenes are a large and diverse group of organic compounds consisting of isoprene five-carbon atom units with a molecular formula $(C_5H_6)_n$ as repeating building blocks.[115,168] Terpenes are classified based on the number of isoprene units from which they are formed. Thus, they may be classified as monoterpenes (two units of isoprene, C_{10}), sesquiterpenes (three units of isoprene, C_{15}), diterpenes (four units of isoprene, C_{20}), sestererpenes (five units of isoprene, C_{25}), triterpenes (six units of isoprene C_{30}), and tetraterpenes (eight units of isoprene C_{40}).

In mushrooms, terpenes are present in their modified form as terpenoids or isoprenoids due to the addiction of a functional group (normally oxygen-containing) to terpenes, producing terpenoids.[24] Different terpenoid classes are present in mushrooms, including sterols (C_{30}), carotenoid pigments (C_{40}), essential oils, monoterpenoids, sesquiterpenoids, diterpenoids, and triterpenoids.[115,138] Triterpenoids are exclusively found in certain macrofungi, mainly Basidiomycetes.[169]

Sterols, important eucaryotic components, are biosynthesized through triterpenoids.[168] In particular, ergosterol is a constituent of the fungal cell membrane, which, naturally under the influence of UV radiation coming from sunlight, is converted to vitamin D2 (ergocalciferol). Also, ergosterol possesses various health beneficial effects, performing antioxidant, anti-inflammatory, antihyperlipidaemic, and antimicrobial activities.[3,138] Agarol is an ergosterol obtained from *A. blazei*, which showed anticancer effects, inducing caspase-independent apoptosis in human carcinoma cell lines (A549, MKN45, HSC-3, and HSC-4) through a mitochondrial pathway.[170] Among the cultivated mushrooms, the *Pleurotus* species and *L. edodes* have been reported to have a high content of ergosterol.[3]

Regarding carotenoids, these compounds play an important role in protecting against disorders related to oxidative stress, including those of the nervous system and circulatory system, and several types of cancer.[171] On the other hand, the main terpenoid essential oils are volatile constituents responsible for many 'mushrooms' characteristic odour and aroma, and they are commercially attractive as a skincare product in cosmetics and flavourings in the food industry.[138] In this sense, D-limonene, α-pinene, β-pinene, β-phellandrene, and β-myrcene are examples of terpenoid essential oils identified in *A. bisporus* and *B. edulis*.[172]

According to Ma *et al.*,[16] the sesquiterpenoids obtained from mushrooms consist of aristolane, bisabolene, cuparene, drimane, fomannosane, lactarane, nordasinane, spiro, and sterpurane. Also, the mushroom diterpenoids are mostly of the cyathane type, while generally, triterpenoid compounds identified are lanostane types, which are reported mainly as anticancer, immunomodulatory, and anti-inflammatory agents, capable of inducing a reduction in NO production and other cytokines.[13,16,115,169]

Additionally, the mushroom terpenoids have been associated with several pharmacological effects, such as antioxidant, anticancer, immunomodulating, anticholinesterase, anti-inflammatory, antiviral, antimalarial, antitubercular, antibacterial, antifungal, and cytotoxic activities.[9,16,24] Some of the terpenoids isolated from different edible mushrooms along with their biological activities are given in Table 6.6.

G. lucidum is known for its high triterpenoid content.[115] Common examples include chemical structures such as methyl ganoderate A acetonide, *n*-butyl ganoderate H, lucidadiol, ganoderol, ganodermanontriol, ganodermadiol, and ganoderic acids.[13,14,169] Most *G. lucidum* triterpenes exhibit comprehensive biological activities, including anticancer, antimetastatic, cytotoxic, antioxidant, hepatoprotective, antimicrobial, antiviral, neurotrophic, anti-inflammatory, and antimicrobial activities.[173] In particular, methyl ganoderate A acetonide and n-butyl ganoderate H exhibited anti-acetylcholinesterase activity. Consequently, using these triterpenoids might effectively prevent and treat Alzheimer's disease and related neurodegenerative diseases.[174] Moreover, ganoderic acid A ameliorated renal damage by reducing serum urea, creatinine, and uric acid levels and alleviated kidney fibrosis.[9] Also, Li *et al.*[14] reported that *G. lucidum* triterpenes might modulate gut microbiota composition and stimulate the relative abundance of SCFA-producing bacteria, acting as potential prebiotics.

6.6 Phenolic Compounds

Phenolic compounds are secondary metabolites present in mushrooms with many health-related properties, namely antioxidant, antitumor, anti-

Table 6.6 Terpenoids isolated from edible mushrooms and their health benefits.[a]

Mushroom species	Type of terpenoid	Health benefits	Ref.
Astraeus odoratus	Triterpenoids	Antibacterial activity against *Mycobacterium tuberculosis*; cytotoxicity against KB, NCI-H187, and MCF-7 cells	175
Boletus edulis	Sesquiterpenoids	Cytotoxicity against five human cancer cell lines	176
Flammulina velutipes	Sesquiterpenoids	Antibacterial activity against *Escherichia coli*, *Bacillus subtilis*, and *Staphylococcus aureus;* antifungal activity against *Aspergillus fumigatus;* cytotoxicity against HepG2, SGC7901, A549, and MCF-7 cells	177 and 178
Ganoderma applanatum	Triterpenoids	Antitumour and immuno-modulating effects in mice	179
Ganoderma lucidum	Triterpenoids	Acetylcholinesterase-inhibitory activity; antiproliferative effects against MCF-7 cells by DNA damage, G1 cell cycle arrest and apoptosis; inhibitory effects against topoisomerases I and II and promote apoptosis; inhibitory activity against Dengue virus NS2B-NS3 protease; inhibitory effects on TNF-α, prostaglandin E_2 and NO production; suppressed interleukin-6 production; antimalarial activity against *Plasmodium falciparum*	174 and 180–184
Macrolepiota procera	Triterpenoids	Cytotoxicity effects against HL-60, SMMC-7721, A-549, SW-480, and MCF-7 cells; antitubercular activity against *Mycobacterium tuberculosis*	185
Pleurotus citrinopileatus	Sesquiterpenoids	Inhibitory effects on protein tyrosine phosphatase 1B, a negative regulator of the insulin-dependent signal cascades	186
Pleurotus cornucopiae	Monoterpenoids	Inhibitory activity on NO production in lipopolysaccharide-activated macrophage	187
	Sesquiterpenoids	Growth inhibition against HeLa and HepG2 cells	
Pleurotus eryngii	Diterpenoids	Inhibit the proliferation of HeLa and HepG2 cells	188 and 189
Thricholoma saponaceum	Diterpenoids	Cytotoxic against HL-60 cells	189

[a]A-549: human lung cancer cell line; HeLa: human cervical cancer cell line; HepG2: human liver cancer cell line; HL-60: human leukaemia cell line; KB: human epidermoid carcinoma cell line; MCF-7: human breast adenocarcinoma cell line; NCI-H187: human lung cancer cell line; NO: nitric oxide; SGC7901: human gastric cancer cell line; SMMC-7721: human hepatocellular carcinoma cell line; SW-480: human colon adenocarcinoma cell line.

inflammatory, cardioprotective, antiallergenic, and antimicrobial activities.[3,5,13] Their basic structure consists of at least one aromatic ring linked to one or more hydroxyl groups, possessing varied complexities from simple macromolecules to complex polymers.[3,5,190] Several phenolic compounds are reported in mushrooms, where phenolic acids were the most relevant, flavonoids, tocopherols and tannins. Some phenolic acids present in mushrooms are caffeic, gallic, *p*-coumaric, cinnamic, ferulic, protocatechuic, and chlorogenic acids.[5,190]

Beyond the several biological properties of phenolic compounds, their antioxidant activity is the most reported and well-established bioactivity. Their hydroxyl groups are responsible for the high antioxidant potential.[5,13] A relationship between the polyphenol content and antioxidant capacity has been reported in mushrooms.[5,190] *L. edodes*, *Pleurotus* spp., *Agaricus* spp., and *G. lucidum* are examples of mushroom species with high antioxidant potential.[13] The antioxidant cascades associated with phenolic compounds (*e.g.*, free radical scavengers and singlet oxygen quenchers) play essential roles in numerous degenerative diseases.[3] Some examples of studies focusing on the antioxidant capacity of mushroom phenolic compounds, as well as other biological potentials, are presented in Table 6.7.

Abdelshafy *et al.*[5] reviewed the high number of studies and increasing interest in mushroom polyphenols. For example, studies focusing on the antitumour potential of these macromolecules have risen significantly in recent years. Generally, these studies use mainly extracts possessing phenolic groups, but also specific isolated phenolic compounds are reported. Some examples of biological properties of phenolic extracts are presented in Table 6.7. Regarding specific phenolic compounds, caffeic acid is present in higher concentrations (approximately 15 $\mu g\, g^{-1}$ on a dry basis) in several mushroom species (*e.g.*, *A. bisporus* and *P. ostreatus*). This macromolecule is one of the most bioactive phenolic compounds, with strong anti-inflammatory potential.[94] Lin *et al.* also reported another example of specific phenolic compounds with bioactive potential. In this study, syringic and chlorogenic acids showed antiangiogenic potential.[191]

Nowadays, macrofungi polyphenols have been extensively used in functional foods and nutraceuticals. In parallel, several studies have suggested the efficacy of phenolic compounds on skincare, being also used in the cosmeceutical industry. However, some challenges limit the industrial applications of phenolics in the medicine, food, and cosmetic fields. The solubility, stability, and bioavailability of these compounds are some of the issues registered during the processing and storage of phenolic-derived products. Thus, complementary studies are needed to circumvent these problems and improve the efficiency of phenolic-based formulations, such as developing novel drug-delivery strategies.[5]

Table 6.7 Some examples of studies related to the bioactive properties of mushroom phenolic extracts.[a]

Mushroom species	Phenolic compounds	Bioactivity	Main results	Ref.
Agaricus bisporus, *Boletus edulis*, *Calocybe gambosa*, *Cantharellus cibarius*, *Craterellus cornucopioides*, *Hygrophorus marzuolus*, *Lactarius deliciosus*, and *Pleurotus ostreatus*	Phenolic extract (*e.g.*, protocatechuic acid, homogentisic acid, gallic acid, *p*-hydroxybenzoic acid)	Antioxidant	Strong antioxidant capacity by the linoleic acid autoxidation assay	192
Melanoleuca cognata and *Melanoleuca stridula*	Phenolic extract (main compounds: syringic and benzoic acid)	Antioxidant	Strong antioxidant capacity by DPPH and ABTS assays	193
Agaricus silvaticus, *Hydnum rufescens*, and *Meripilus giganteus*	Phenolic extract (flavonoids, polyphenols and tannins)	Antioxidant	Strong antioxidant capacity by DPPH, ABTS, CAT, and FRAP assays	194
Pleurotus eryngii	Phenolic extract (*e.g.*, gallic acid derivative, methyl gallate, syringic acid, ellagic acid, and catechin)	Anti-inflammatory antitumour	Inhibition of pro-inflammatory markers overproduction Inhibition of HCT116 cells proliferation	195
Ganoderma lucidum	Phenolic extract (phenolic and triterpenic acids)	Antioxidant protective effect on DNA damage antitumour	The extract exhibit chemopreventive potential (antiproliferative, genotoxic, and proapoptotic activity in a cell model)	196

Inonotus obliquus	Phenolic extract (polyphenolic compounds)	Antitumour	Promote apoptosis in cancer cells without impacting normal cells on human cell lines	197
Agaricus blazei Murrill, *Coprinus comatus* (O. F. Müll.) Pers., *Cordyceps militaris* (L.) Fr., *Inonotus obliquus* (Ach. ex Pers.) Pilát, *Morchella conica* Pers., and *Phellinus linteus* Berk & M. A. Curtis	Phenolic extract	Antihyperglycaemic	*In vitro* inhibition of α-amylase and α-glucosidase, showing promising results for diabetes type 2 treatment	198
Agaricus brasiliensis	Phenolic extract (mainly gallic acid)	Antioxidant Antimicrobial	High antioxidant capacity by DPPH, ABTS and FRAP assays. High antimicrobial potential (MI $\leq$ 200 mg mL^{-1}) against *Bacillus cereus* and *Staphylococcus aureus*	199

[a]DPPH: 1,1-diphenyl-2-picrylhydrazyl; ABTS: 2,2′-azinobis-3-ethylbenzothiazoline-6-sulfonate; CAT: catalase activity assay; FRAP: ferric-reducing antioxidant power; HCT116: human colorectal carcinoma cell line; MIC: minimum inhibitory concentration.

6.7 Other Compounds

Besides the bioactive molecules mentioned above, mushrooms have been identified as a good source of different compounds with nutritional properties, namely vitamins and minerals, necessary for human physiological functions.[9]

6.7.1 Vitamins

Vitamins are vital compounds for good functioning of the human organism, playing an essential role in several functions such as in metabolism, immunity, and digestion. Since mushrooms are valuable sources of vitamins, they can be used as an excellent ingredient for nutraceutical formulations.[3,9] The vitamins of the B complex are abundant, especially thiamine (B1), riboflavin (B2), niacin (B3), pantothenic acid (B5), folic acid/folate (B9), and cobalamin (B12).[3] Besides, they also contain biotin (B8), vitamin C (ascorbic acid), and fat-soluble vitamins such as vitamin A, vitamin E (tocopherol), and vitamin D2 (ergocalciferol).[2,200] *P. ostreatus* has high folic acid, thiamine, and niacin content, while *L. edodes* and *B. edulis* have a high vitamin D content.[3] Moreover, riboflavin and niacin are the most abundant vitamins in *A. bisporus*.[10]

Interestingly, vitamin B12 in mushrooms has high bioavailability and is comparable to that found in animal foods.[9] Furthermore, mushrooms are the only non-animal source of vitamin D. Among others, these characteristics make mushrooms a beneficial food product in vegan/vegetarian diets.[2] In particular, vitamin D impacts bone and muscle health, as it promotes bone formation and mineralization, stimulating osteoblastic cells and inhibiting osteoclasts accountable for bone resorption. Along with this, vitamin D is essential for the regulation of calcium–phosphate homeostasis and in controlling bone turnover.[201] It also helps to maintain a healthy immune system.[3] Besides, vitamins play other relevant physiological roles in cancer and cardiovascular diseases, secondary hyperparathyroidism, multiple sclerosis, rheumatoid arthritis, Crohn disease, systemic lupus erythematosus, atopic dermatitis, liver function, depression, obesity, and diabetes, namely type 1 diabetes.[3,202] Similarly, other mushroom vitamins play important physiological roles, such as vitamin B5 that is essential for building the nervous system and vitamins A, C, and E, which are natural antioxidants involved in the defence of human cells from oxidative damage.[9]

6.7.2 Minerals

The major minerals present in mushrooms are potassium (K), phosphorus (P), or magnesium (Mg), depending on the species.[2] On the other hand, generally, calcium (Ca) and sodium (Na) are those with the lowest content.[200] Indeed, the high potassium and low sodium levels, make mushrooms a promising food product in the prevention and dietary treatment of hypertension.[2] Mushrooms also contain copper (Cu), iron (Fe), manganese (Mn), selenium (Se), and zinc (Zn).[2] For example, *P. ostreatus* is abundant in P, Mg, and K, while *B. edulis* and *L. edodes* are good sources of Se in the diet.[9,203] Also, *A. bisporus* is rich in P, K, Mg, Cu, Fe, and Se.[10]

Among the mushroom minerals, trace elements such as Cu, Fe, Mn, and Zn are cofactors of various enzymes with antioxidant functions, being designated as antioxidant micronutrients.[61] Besides, Se is an important antioxidant that acts in the form of selenoproteins, reducing the cytotoxic effects of ROS.[204] Various Se compounds have been identified in mushrooms, such as selenocysteine, selenomethionine, methylselenocysteine, selenite, and seleno-polysaccharides. Also, Se is a crucial component of several antioxidant enzymes, such as glutathione peroxidase (GPx). Also, GPx and selenoprotein-P are two of the most secreted Se compounds in the blood.[61]

The presence of Cu, Zn, and Se in the fruiting bodies of *L. edodes* has been associated with anti-inflammatory properties, indicating that this mushroom, or part of it, could be used as a potent dietary supplement with anti-inflammatory properties.[203]

Interestingly, Se can be combined with polysaccharides through a chemical process designated as selenylation of polysaccharides. This combination enhances Se and polysaccharide's physiological and pharmacological activities in immunomodulation, anticancer effects, hypoglycaemia, and antibacterial effects.[205] In this context, Ren *et al.*[205] investigated the ability of the combination of selenium with lentinan (seleno-lentinan) to prevent chronic pancreatitis (an inflammatory disorder of the pancreas) and its prebiotic-like effect efficacy on gut microbiota modulation in mice. They concluded that seleno-lentinan effectively hindered chronic pancreatitis progress by improving superoxide dismutase and GPx activities, inhibiting lipid peroxidation and reducing TNF-α and IL-1β, as well as pancreatic hydroxyproline.[205] Regarding the modulation of gut microbiota, seleno-lentinan supplementation had beneficial effects on the proliferation of lactic acid bacteria.[205] This is important as lactic acid bacteria are an SCFA-producing bacteria that can suppress the inflammatory response.[14] Besides, there was a protective effect of seleno-lentinan against the invasion of pathogens and in the upkeep of immunological homeostasis, probably resulting from its ability to inhibit *Alistipes*, *Helicobacter parabacteroides*, and *Ruminococcus*, and, in increasing the proliferation of beneficial batteries such as *Lactobacillus* and *Bacteroides*, among others.[205]

6.8 Conclusions and Future Perspectives

This chapter describes the main bio-functional components that are part of edible mushrooms, as well as their representative human health benefits. As mentioned, macrofungi contain a chemically diverse group of compounds such as polysaccharides, proteins, terpenoids, and polyphenols that have been associated with several medicinal properties, including anticancer, immunomodulatory, anti-inflammatory, antioxidant, antidiabetic, cholesterol-lowering, antibacterial, antiviral, hepatoprotective, and neuroprotective effects, and other bioactivities.[3,9,94]

Nowadays, the concept of food is becoming more complex due to the robust evidence linking diet with life quality and advances in food science and technology.[138,206] New food concepts have been developed in this context, including "functional foods" and "nutraceuticals".[206] In industrialized countries, likely due to rising medical costs, increased interest in health care, and increased life expectancy, the market for functional foods and nutraceuticals is expanding rapidly, with significant consumer preferences for food products that contain functional ingredients associated with health benefits claims.[207,208]

Due to their nutritional and health-promoting properties, mushrooms are considered a premium natural food product, as well as an important functional food and valuable source for developing dietary supplements.[9,138,206] In addition to their pharmacological properties, these macrofungi are hypocaloric, with low carbohydrate and fat contents, but, at the same time, being an excellent source of high-quality dietary fibre. They also have significant digestible proteins and are a good source of essential amino acids and various vitamins and minerals for all age groups, which are often deficient in plants.[13,138]

Therefore, research has been focused on novel functional foods, nutraceuticals, and drugs production by utilizing fruit bodies, mycelia, and the extracts obtained through different preparation methods of mushrooms.[16] Thus, it is expected that this growing interest in functional foods and dietary supplements, especially in Western countries, will cause an increase in mushroom consumption and an increasing demand for product development and innovation, resulting in a remarkable rise in the future market value of nutraceuticals. Furthermore, converging trends and the popularity of Asian herbal medicines, as well as the preference for natural food products, are supporting the growth of the mushroom market.[3]

Indeed, at present, relevant and extensive data about the classification, nutritional composition, structure, biology, medicinal properties, postharvest technologies, and value addition of different species of mushrooms, including edible and medicinal, are available.[13] However, there is a dearth of information on identifying and characterizing the bioactive components of mushrooms, as well as the mechanisms of action of various mushroom-isolated metabolites involved in health promotion (including the treatment or prevention of diseases).[9,13,24,94] More rigorous chemical analysis, standard protocols, and an understanding of individual 'compounds' *in vivo* pharmacokinetics and pharmacodynamics are needed.[8,24]

First, different bioassay guidelines must be used to isolate, determine, and characterize bioactive compounds from mushrooms; as a result, standardized protocols must be investigated and studied.[94] Another promising area of study is identifying the mechanism of action of biologically active components and/or mushroom extracts and their potential molecular targets by genomic, transcriptomic, and proteomic approaches at the cellular/

molecular level or in animal models.[13,94] It is also essential to understand and determine the biological and/or pharmacological activity of mushroom bioactive compounds through further *in vivo* assays. In this way, in recent studies focusing on cell lines and animal models, the results of clinical trials may be favourable for the treatment of patients worldwide with specific pathologies such as cancer in the near future.[44] Besides, more information is required about the effectiveness and safety of these components at the preclinical and clinical trial levels.[8,9] Additionally, potential interactions between bioactive compounds from mushrooms and other matrices such as foods, dietary supplements, and drugs should be analysed to establish guidelines for the safe and effective use of these products.[13,24] For all these reasons, investment in research and development (R&D) to demonstrate health claims, find innovative approaches, and market research are essential to the success of the mushroom nutraceutical industry.[3]

Furthermore, the application of edible mushrooms in combination with therapeutic interventions for certain diseases seems promising, since the combinations of some chemotherapy drugs with edible mushrooms can improve the side effects in the gut microbiota inherent to chemotherapy and promote even more anticancer effects. Still, there aren't many reports in this area, and more research is needed to explore this.[14] In fact, mushroom-derived prebiotics is one of the most recent focuses of researcher attention worldwide, particularly for the relationship between gut microbiota regulation and host health, including gut immune function, gut inflammation, colon cancer, obesity, and neurological diseases.[9]

In the future, to stably and efficiently cultivate edible macrofungi, genetic engineering can be applied to develop stable strains with commercially desirable traits. This will enable mushrooms to be designed and cultivated for large-scale and high-quality production.[8]

To sum up, edible mushrooms can be considered traditional foods with medicinal properties. In this way, the bioactive compounds from mushrooms can be used as an active ingredient in the preparation of functional foods and hold promise for future innovations in supplements and drug development to treat or prevent human diseases. Although the future looks bright, more information and further research focusing on nutraceutical applications of mushroom macromolecules should be conducted to validate their further efficacy and safety.

Abbreviations

ABTS	2,2′-Azinobis-(3-ethylbenzothiazoline-6-sulfonate)
DPPH	1,1-Diphenyl-2-picrylhydrazyl
ABL	*Agaricus bisporus* lectin
H9N2	Avian influenza virus
CBS	Carbohydrate-binding site
CAT	Catalase activity assay

HeLa	Cervical cancer cells
Caco-2	Caco-2 colon cancer
SW-480	Colon adenocarcinoma cells
HCT116	Colorectal cancer cells
COVID-19	Coronavirus disease 2019
DNA	Deoxyribonucleic acid
EPS	Exopolysaccharides
FRAP	Ferric-reducing antioxidant power
FIPs	Fungal immunomodulatory proteins
GABA	Gamma-aminobutyric acid
BGC-823, SGC-790, MKN45	Gastric cancer cells
GPx	Glutathione peroxidase
HBV	Hepatitis B virus
HSV	Herpes simplex virus
MCF-7	Human breast adenocarcinoma cells
Bre-04	Human breast cancer cell
COLO 320, HCT116, HT-29, and SW-480	Human colon cancer
KB	Human epidermoid carcinoma cells
HepG2	Human hepatocarcinoma cells
HIV	Human immunodeficiency virus
HIV-RT	Human immunodeficiency virus reverse transcriptase
HepG2 and SMMC-7721	Human liver cancer cells
A-549 and NCI-H187	Human lung cancer cells
Raji	Human lymphoma cells
HSC-3 and HSC-4	Human oral squamous cells
SH-SY5Y	Human neuroblastoma cells
L-DOPA	L-3,4-Dihydroxyphenylalanine
HL-60, L1210 and M1	Leukaemia cells
LDL	Low-density lipoprotein
MBL	Mannose-binding lectins
MIC	Minimum inhibitory concentration
NO	Nitric oxide
NOS	Nitric oxide synthase
PPO	Polyphenol oxidase
PUFA	Polyunsaturated fatty acids
rFIPS	Recombinant fungal immunomodulatory proteins
ROS	Reactive oxygen species
RNA	Ribonucleic acid
Rnases	Ribonucleases
RIPs	Ribosome inactivating proteins
SARS-CoV-2	Severe acute respiratory syndrome coronavirus 2
SCFAs	Short-chain fatty acids
TLRs	Tool-like receptors

TNF-α	Tumour necrosis factor-α
Ubls	Ubiquitin and ubiquitin-like proteins
UV	Ultraviolet

Acknowledgements

The authors are grateful to the European Regional Development Fund (ERDF) through the Competitiveness and Internationalization Operational Program (POCI), within the scope of project MicoBioExtract [POCI-01-0247-FEDER-033939]. We would also like to thank the National Funds from FCT—Fundação *para* a Ciência e a Tecnologia through project UIDB/50016/2020 and FCT individual PhD grant number 2020.05798.BD.

References

1. M. Zeb and C. H. Lee, *Molecules*, 2021, **26**(2), 251.
2. S. Marçal, A. S. Sousa, O. Taofiq, F. Antunes, A. M. M. B. Morais, A. C. Freitas, L. Barros, I. C. F. R. Ferreira and M. Pintado, *Trends Food Sci. Technol.*, 2021, **110**, 418–431.
3. A. G. Niego, S. Rapior, N. Thongklang, O. Raspé, W. Jaidee, S. Lumyong and K. D. Hyde, *J. Fungi*, 2021, 7, 397.
4. C. Cerletti, S. Esposito and L. Iacoviello, *Nutrients*, 2021, **13**(7), 2195.
5. A. M. Abdelshafy, T. Belwal, Z. Liang, L. Wang, D. Li, Z. Luo and L. Li, *Crit. Rev. Food Sci. Nutr.*, 2022, **62**(22), 6204–6224.
6. G. Venturella, V. Ferraro, F. Cirlincione and M. L. Gargano, *Int. J. Mol. Sci.*, 2021, **22**, 1–31.
7. P. Roupas, J. Keogh, M. Noakes, C. Margetts and P. Taylor, *J. Funct. Foods*, 2012, **4**, 687–709.
8. H. Lu, H. Lou, J. Hu, Z. Liu and Q. Chen, *Compr. Rev. Food Sci. Food Saf.*, 2020, **19**, 2333–2356.
9. D. Yadav and P. S. Negi, *Food Res. Int.*, 2021, **148**, 110599.
10. M. Usman, G. Murtaza and A. Ditta, *Appl. Sci.*, 2021, **11**, 5943.
11. P. Gong, S. Wang, M. Liu, F. Chen, W. Yang, X. Chang, N. Liu, Y. Zhao, J. Wang and X. Chen, *Carbohydr. Res.*, 2020, **494**, 108037.
12. C. Yin, G. D. Noratto, X. Fan, Z. Chen, F. Yao, D. Shi and H. Gao, *Carbohydr. Polym.*, 2020, **250**, 116942.
13. H. Rathore, S. Prasad and S. Sharma, *PharmaNutrition*, 2017, **5**, 35–46.
14. M. Li, L. Yu, J. Zhao, H. Zhang, W. Chen, Q. Zhai and F. Tian, *J. Funct. Foods*, 2021, **83**, 104538.
15. A. Sharma, A. Sharma and A. Tripathi, *J. Food Biochem.*, 2021, **45**, 1–16.
16. G. Ma, W. Yang, L. Zhao, F. Pei, D. Fang and Q. Hu, *Food Sci. Hum. Wellness*, 2018, **7**, 125–133.
17. J. Ferrão, V. Bell, V. Calabrese, L. Pimentel, M. Pintado and T. H. Fernandes, *J. Food Nutr. Res.*, 2017, **5**(4), 226–233.

18. F. Antunes, S. Marçal, O. Taofiq, A. M. M. B. Morais, A. C. Freitas, I. C. F. R. Ferreira and M. Pintado, *Molecules*, 2020, **1**, 1–40.
19. M. Scuto, P. Di Mauro, M. L. Ontario, C. Amato, S. Modafferi, D. Ciavardelli, A. T. Salinaro, L. Maiolino and V. Calabrese, *Int. J. Mol. Sci.*, 2020, **21**, 1–23.
20. X. Huang and S. Nie, *Food Funct.*, 2015, **6**(10), 3205–3217.
21. I. Mironczuk-Chodakowska, K. Kujawowicz and A. M. Witkowska, *Nutrients*, 2021, **13**(11), 3960.
22. N. Chakraborty, A. Banerjee, A. Sarkar, S. Ghosh and K. Acharya, *Biointerface Res. Appl. Chem.*, 2021, **11**, 8915–8930.
23. J. Zhao, S. Gao, Q. Rong, C. Wang, S. Zhao, Z. Liu and Y. Xu, *J. Fungi*, 2020, **6**, 1–37.
24. S. Zhao, Q. Gao, C. Rong, S. Wang, Z. Zhao, Y. Liu and J. Xu, *J. Fungi*, 2020, **6**, 1–37.
25. D. Nidhi, *Int. J. Heal. Sci. Res.*, 2018, **2**, 115–126.
26. L. Zhao, F. Zhang, X. Ding, G. Wu, Y. Y. Lam, X. Wang, H. Fu, X. Xue, C. Lu, J. Ma, L. Yu, C. Xu, Z. Ren, Y. Xu, S. Xu, H. Shen, X. Zhu, Y. Shi, Q. Shen, W. Dong, R. Liu, Y. Ling, Y. Zeng, X. Wang, Q. Zhang, J. Wang, L. Wang, Y. Wu, B. Zeng, H. Wei, M. Zhang, Y. Peng and C. Zhang, *Science*, 2018, **359**, 1151–1156.
27. A. Synytsya, K. Míčková, A. Synytsya, I. Jablonský, J. Spěváček, V. Erban, E. Kovářiková and J. Čopíková, *Carbohydr. Polym.*, 2009, **76**, 548–556.
28. C. Y. Wang, *Food Bioprocess Technol.*, 2020, **13**, 217–228.
29. G. Ma, Q. Xu, H. Du, B. Muinde Kimatu, A. Su, W. Yang, Q. Hu and H. Xiao, *Food Chem.*, 2022, **370**, 131303.
30. B. Dalile, L. Van Oudenhove, B. Vervliet and K. Verbeke, *Nat. Rev. Gastroenterol. Hepatol.*, 2019, **16**, 461–478.
31. M. Wang, Z. Xie, L. Li, Y. Chen, Y. Li, Y. Wang, B. Lu, S. Zhang, F. Ma, C. W. Ma, L. Lin and Q. Liao, *Food Funct.*, 2019, **10**, 2658–2675.
32. C. Chaikliang, S. Wichienchot, W. Youravoug and P. Graidist, *Funct. Foods Heal. Dis.*, 2015, **5**, 395–405.
33. F. Brial, A. Le Lay, M. E. Dumas and D. Gauguier, *Cell. Mol. Life Sci.*, 2018, **75**, 3977–3990.
34. J. E. Ramberg, E. D. Nelson and R. A. Sinnott, *Nutr. J.*, 2010, **9**, 1–22.
35. J. Rodrigues Barbosa, M. M. dos Santos Freitas, L. H. da Silva Martins and R. N. de Carvalho, *Carbohydr. Polym.*, 2020, **229**, 115550.
36. R. M. M. Gern, E. Wisbeck, J. R. Rampinelli, J. L. Ninow and S. A. Furlan, *Bioresour. Technol.*, 2008, **99**, 76–82.
37. G. H. Li, Y. M. Shen and K. Q. Zhang, *J. Microbiol.*, 2005, **43**, 17–20.
38. V. E. C. Ooi and F. Liu, *Curr. Med. Chem.*, 2012, **7**, 715–729.
39. J. A. Bohn and J. N. BeMiller, *Carbohydr. Polym.*, 1995, **28**, 3–14.
40. C. Thongsiri, Y. Nagai-Yoshioka, R. Yamasaki, Y. Adachi, M. Usui, K. Nakashima, T. Nishihara and W. Ariyoshi, *Carbohydr. Polym.*, 2021, **253**(1), 117285.
41. B. Du, C. Lin, Z. Bian and B. Xu, *Trends Food Sci. Technol.*, 2015, **41**, 49–59.

42. Y. Nishitani, L. Zhang, M. Yoshida, T. Azuma, K. Kanazawa, T. Hashimoto and M. Mizuno, *PLoS One*, 2013, 22, 8(4):e62441.
43. K. Fujimoto, M. Tomonaga and S. Goto, *Anticancer Res.*, 2006, **26**, 4015–4018.
44. M. Zhang, Y. Zhang, L. Zhang and Q. Tian, *Mushroom polysaccharide lentinan for treating different types of cancers: A review of 12 years clinical studies in China*, Elsevier Inc., 1st edn, 2019, vol. 163.
45. X. E. Wang, Y. H. Wang, Q. Zhou, M. Peng, J. Zhang, M. Chen, L. J. Ma and G. M. Xie, *Pathol. Oncol. Res.*, 2020, **26**, 499–505.
46. N. H. Segal, P. Gada, N. Senzer, M. A. Gargano, M. L. Patchen and L. B. Saltz, *Clin. Colorectal Cancer*, 2016, **15**, 222–227.
47. Y. W. Kim, K. H. Kim, H. J. Choi and D. S. Lee, *Biotechnol. Lett.*, 2005, **27**, 483–487.
48. K. Le Wang, Z. M. Lu, X. Mao, L. Chen, J. S. Gong, Y. Ren, Y. Geng, H. Li, H. Y. Xu, G. H. Xu, J. S. Shi and Z. H. Xu, *Int. J. Biol. Macromol.*, 2019, **137**, 1102–1111.
49. V. Caz, A. Gil-Ramírez, C. Largo, M. Tabernero, M. Santamaría, R. Martín-Hernández, F. R. Marín, G. Reglero and C. Soler-Rivas, *J. Agric. Food Chem.*, 2015, **63**, 7371–7380.
50. M. Zhu, P. Nie, Y. Liang and B. Wang, *Carbohydr. Polym.*, 2013, **95** 644–648.
51. N. Markova, V. Kussovski, I. Drandarska, S. Nikolaeva, N. Georgieva and T. Radoucheva, *Int. Immunopharmacol.*, 2003, **3**, 1557–1562.
52. C. Kupfahl, G. Geginat and H. Hof, *Int. Immunopharmacol.*, 2006, **6**, 686–696.
53. F. Zhu, B. Du, Z. Bian and B. Xu, *J. Food Compos. Anal.*, 2015, **41**, 165–173.
54. O. O. Osemwegie, C. O. Adetunji, E. A. Ayeni, O. I. Adejobi, R. O. Arise, C. O. Nwonuma and A. O. Oghenekaro, *Heliyon*, 2020, **6**, e04205.
55. G. C. Jayakumar, S. V. Kanth, B. Chandrasekaran, J. Raghava Rao and B. U. Nair, *Carbohydr. Res.*, 2010, **345**, 2213–2219.
56. G. E. Rasmy, W. A. Botros, S. S. A. Kabeil and A. Daba, *Aust. J. basic Appl. Sci.*, 2010, **4**, 5717–5726.
57. N. Markova, L. Michailova, V. Kussovski, M. Jourdanova and T. Radoucheva, *Pharmazie*, 2005, **60**, 42–48.
58. M. Gordon, B. Bihari, E. Goosby, R. Gorter, M. Greco, M. Guralnik, T. Mimura, V. Rudinicki, R. Wong and Y. Kaneko, *J. Med.*, 1998, **29**, 305–330.
59. E. J. Murphy, C. Masterson, E. Rezoagli, D. O. Toole, I. Major, G. D. Stack, M. Lynch, J. G. Laffey and N. J. Rowan, *Sci. Total Environ.*, 2020, **732**, 139330.
60. I. Urbancikova, D. Hudackova, J. Majtan, Z. Rennerova, P. Banovcin and M. Jesenak, *Evidence-based Complement. Altern. Med.*, 2020, **13**, 8562309.
61. M. Kozarski, A. Klaus, D. Jakovljevic, N. Todorovic, J. Vunduk, P. Petrović, M. Niksic, M. M. Vrvic and L. Van Griensven, *Molecules*, 2015, **20**, 19489–19525.

62. M. Jesenak, J. Majtan, Z. Rennerova, J. Kyselovic, P. Banovcin and M. Hrubisko, *Int. Immunopharmacol.*, 2013, **15**, 395–399.
63. T. L. Nguyen, J. Chen, Y. Hu, D. Wang, Y. Fan, J. Wang, S. Abula, J. Zhang, T. Qin, X. Chen, X. Chen, S. K. Khakame and B. K. Dang, *Carbohydr. Polym.*, 2012, **90**, 1254–1258.
64. S. Sudheer, I. Alzorqi, S. Manickam and A. Ali, *Ref. Ser. Phytochem.*, 2019, 1863–1893.
65. B. Muszyńska, A. Grzywacz-Kisielewska, K. Kała and J. Gdula-Argasińska, *Food Chem.*, 2018, **243**, 373–381.
66. H. Xiao, C. Chen, C. Li, Q. Huang and X. Fu, *Int. J. Biol. Macromol.*, 2019, **132**, 308–315.
67. H. Wang, F. Liu, L. Yang, Y. Zu, H. Wang, S. Qu and Y. Zhang, *Food Chem.*, 2011, **128**, 93–99.
68. X. Sheng, J. Yan, Y. Meng, Y. Kang, Z. Han, G. Tai, Y. Zhou and H. Cheng, *Food Funct.*, 2017, **8**, 1020–1027.
69. B. C. Lee, J. T. Bae, H. B. Pyo, T. B. Choe, S. W. Kim, H. J. Hwang and J. W. Yun, *Enzyme Microb. Technol.*, 2003, **32**, 574–581.
70. J. P. F. Angeli, L. R. Ribeiro, M. L. C. Gonzaga, S. D. A. Soares, M. P. S. N. Ricardo, M. S. Tsuboy, R. Stidl, S. Knasmueller, R. E. Linhares and M. S. Mantovani, *Cell Biol. Toxicol.*, 2006, **22**, 285–291.
71. J. P. F. Angeli, L. R. Ribeiro, M. F. Bellini and M. S. Mantovani, *Arch. Toxicol.*, 2009, **83**, 81–86.
72. T. G. Pillai and P. Uma Devi, *Mutat. Res., Genet. Toxicol. Environ. Mutagen.*, 2013, **751**, 109–115.
73. H. Abreu, F. F. Simas, F. R. Smiderle, V. Sovrani, J. L. Dallazen, D. Maria-Ferreira, M. F. Werner, L. M. C. Cordeiro and M. Iacomini, *Int. J. Biol. Macromol.*, 2019, **122**, 1128–1135.
74. A. H. Kwon, Z. Qiu, M. Hashimoto, K. Yamamoto and T. Kimura, *Am. J. Surg.*, 2009, **197**, 503–509.
75. F. R. Smiderle, L. M. Olsen, E. R. Carbonero, C. H. Baggio, C. S. Freitas, R. Marcon, A. R. S. Santos, P. A. J. Gorin and M. Iacomini, *Eur. J. Pharmacol.*, 2008, **597**, 86–91.
76. K. Bauerová, E. Paulovičová, D. Mihalová, K. Švík and S. Poništ, *Toxicol. Ind. Health*, 2009, **25**, 329–335.
77. W. Pan, P. Jiang, J. Zhao, H. Shi, P. Zhang, X. Yang, J. Biazik, M. Hu, H. Hua, X. Ge, X. F. Huang and Y. Yu, *J. Transl. Med.*, 2021, **19**, 1–17.
78. J. Yamada, J. Hamuro, H. Hatanaka, K. Hamabata and S. Kinoshita, *J. Allergy Clin. Immunol.*, 2007, **119**, 1119–1126.
79. M. Jesenak, M. Hrubisko, J. Majtan, Z. Rennerova and P. Banovcin, *Phytother. Res.*, 2014, **28**, 471–474.
80. A. R. Burg, L. Quigley, A. V. Jones, G. M. O'Connor, K. Boelte, D. W. McVicar and S. J. Orr, *SpringerPlus*, 2016, **5**, 1–10.
81. K. Bergendiova, E. Tibenska and J. Majtan, *Eur. J. Appl. Physiol.*, 2011, **111**, 2033–2040.
82. M. Sevindik, *J. Bacteriol. Mycol. Open Access*, 2021, **9**, 73–76.

83. G. R. Gibson, R. Hutkins, M. E. Sanders, S. L. Prescott, R. A. Reimer, S. J. Salminen, K. Scott, C. Stanton, K. S. Swanson, P. D. Cani, K. Verbeke and G. Reid, *Nat. Rev. Gastroenterol. Hepatol.*, 2017, **14**, 491–502.
84. A. Romo-Araiza and A. Ibarra, *Med. Hypotheses*, 2020, **134**, 109410.
85. X. Meng, H. Liang and L. Luo, *Carbohydr. Res.*, 2016, **424**, 30–41.
86. V. K. Chaturvedi, S. Agarwal, K. K. Gupta, P. W. Ramteke and M. P. Singh, *3 Biotech*, 2018, **8**, 1–20.
87. K. Nirosha, M. Divya, S. Vamsi and M. Sadiq, *Int. J. Nov. Trends Pharm. Sci.*, 2013, **3**, 59–70.
88. Y. rong Liu, B. Sun, G. hong Zhu, W. wei Li, Y. xuan Tian, L. meng Wang, S. min Zong, P. zhen Sheng, M. Li, S. Chen, Y. Qin, H. juan Liu, H. gang Zhou, T. Sun and C. Yang, *Toxicol. Appl. Pharmacol.*, 2018, **360**, 1–8.
89. Y. Zi, B. Jiang, C. He and L. Liu, *J. Cosmet. Dermatol.*, 2020, **19**, 502–507.
90. A. G. Niego, S. Rapior, N. Thongklang, O. Raspé, W. Jaidee, S. Lumyong and K. D. Hyde, *J. Fungi*, 2021, 7(5), 397.
91. L. Hong, M. Xun and W. Wutong, *J. Pharm. Pharmacol.*, 2010, **59**, 575–582.
92. J. J. Volman, R. P. Mensink, L. J. L. D. Van Griensven and J. Plat, *Eur. J. Clin. Nutr.*, 2010, **64**, 720–726.
93. D. Morales, R. Rutckeviski, M. Villalva, H. Abreu, C. Soler-Rivas, S. Santoyo, M. Iacomini and F. R. Smiderle, *Carbohydr. Polym.*, 2020, **229**, 115521.
94. G. A. Martinez-Medina, M. L. Chávez-González, D. K. Verma, L. A. Prado-Barragán, J. L. Martínez-Hernández, A. C. Flores-Gallegos, M. Thakur, P. P. Srivastav and C. N. Aguilar, *J. Funct. Foods*, 2021, **77**, 104326.
95. M. Jones, M. Kujundzic, S. John and A. Bismarck, *Mar. Drugs*, 2020, **18**, 64.
96. J. Erjavec, J. Kos, M. Ravnikar, T. Dreo and J. Sabotič, *Trends Biotechnol.*, 2012, **30**, 259–273.
97. P. Manzi, L. Gambelli, S. Marconi, V. Vivanti and L. Pizzoferrato, *Food Chem.*, 1999, **65**, 477–482.
98. X. Xu, H. Yan, J. Chen and X. Zhang, *Biotechnol. Adv.*, 2011, **29**, 667–674.
99. W. T. Ismaya, R. R. Tjandrawinata and H. Rachmawati, *Molecules*, 2020, **25**, 1–16.
100. P. H. K. Ngai, H. X. Wang and T. B. Ng, *Peptides*, 2003, **24**, 639–645.
101. T. B. Ng, *Peptides*, 2004, **25**, 1055–1073.
102. V. Rezvani, H. R. Pourianfar, S. Mohammadnejad, A. Madjid Ansari and L. Farahmand, *Appl. Microbiol. Biotechnol.*, 2020, **104**, 6855–6871.
103. Y. A. El-Maradny, E. M. El-Fakharany, M. M. Abu-Serie, M. H. Hashish and H. S. Selim, *Int. J. Biol. Macromol.*, 2021, **179**, 239–258.
104. R. S. Singh, A. K. Walia and J. F. Kennedy, *Int. J. Biol. Macromol.*, 2020, **151**, 1340–1350.

105. S. S. Singh, H. Wang, Y. S. Chan, W. Pan, X. Dan, C. M. Yin, O. Akkouh and T. B. Ng, *Molecules*, 2015, **20**, 446–469.
106. M. A. A. Hassan, R. Rouf, E. Tiralongo, T. W. May and J. Tiralongo, *Int. J. Mol. Sci.*, 2015, **16**, 7802–7838.
107. Y. Ditamo, L. L. Rupil, V. G. Sendra, G. A. Nores, G. A. Roth and F. J. Irazoqui, *Food Funct.*, 2016, 7, 262–269.
108. R. S. Singh, R. Bhari and H. P. Kaur, *Crit. Rev. Biotechnol.*, 2010, **30**, 99–126.
109. J. C. K. Ho, S. C. W. Sze, W. Z. Shen and W. K. Liu, *Biochim. Biophys. Acta, Gen. Subj.*, 2004, **1671**, 9–17.
110. C. Zha, H. Sun, X. Tong and Y. Qi, *Biochem. J.*, 2003, **374**, 321–327.
111. H. H. Chang, P. J. Chien, M. H. Tong and F. Sheu, *Food Chem.*, 2007, **105**, 597–605.
112. P. H. K. Ngai and T. B. Ng, *Biochem. Biophys. Res. Commun.*, 2004, **314**, 988–993.
113. Y. Wang, Y. Zhang, J. Shao, B. Wu and B. Li, *Int. J. Biol. Macromol.*, 2019, **130**, 399–406.
114. R. Rouf, A. S. Stephens, L. Spaan, N. X. Arndt, C. J. Day, T. W. May, E. Tiralongo and J. Tiralongo, *Glycoconj. J.*, 2014, **31**, 61–70.
115. H. A. El Enshasy and R. Hatti-Kaul, *Trends Biotechnol.*, 2013, **31**, 668–677.
116. S. C. W. Sze, J. C. K. Ho and W. K. Liu, *J. Cell. Biochem.*, 2004, **92**, 1193–1202.
117. E. Keyaerts, L. Vijgen, C. Pannecouque, E. Van Damme, W. Peumans, H. Egberink, J. Balzarini and M. Van Ranst, *Antiviral Res.*, 2007, **75**, 179–187.
118. A. Barre, Y. Bourne, E. J. M. Van Damme and P. Rougé, *Int. J. Mol. Sci.*, 2019, 20.
119. S. Zheng, C. Li, T. B. Ng and H. X. Wang, *Process Biochem.*, 2007, **42**, 1620–1624.
120. I. Lagarda-Diaz, A. M. Guzman-Partida and L. Vazquez-Moreno, *Int. J. Mol. Sci.*, 2017, 18.
121. S. Alborés, P. Mora, M. J. Bustamante, M. P. Cerdeiras and L. Franco Fraguas, *Appl. Biochem. Biotechnol.*, 2014, **172**, 2081–2090.
122. T. Santi-Gadelha, B. A. M. Rocha, C. A. A. Gadelha, H. C. Silva, R. E. R. Castellon, F. J. T. Gonçalves, D. O. Toyama, M. H. Toyama, A. J. F. de Souza, L. O. S. Beriam, J. L. Martins, P. P. Joazeiro and B. S. Cavada, *Pestic. Biochem. Physiol.*, 2012, **103**, 15–22.
123. K. Amano, H. Katayama, A. Saito, A. Ando and Y. Nagata, *Biosci., Biotechnol., Biochem.*, 2012, **76**, 967–970.
124. L. Ma, B. Xu, M. Huang, L. Sun, Q. Yang, Y. Chen, Y. Yin, Q. He and H. Sun, *J. Zhejiang Univ. B*, 2017, **18**, 653–661.
125. M. He, D. Su, Q. Liu, W. Gao and Y. Kang, *Sci. Rep.*, 2017, 7, 5814.
126. G. Chandrasekaran, Y.-C. Lee, H. Park, Y. Wu and H.-J. Shin, *Int. J. Med. Mushrooms*, 2016, **18**(4), 291–299.
127. Y. Wang, Y. Liu, H. Wang, C. Li, P. Qi and J. Bao, *Exp. Biol. Med.*, 2012, **237**, 287–296.

128. R. B. Ewart, S. Kornfeld and D. M. Kipnis, *Diabetes*, 1975, **24**, 705–714.
129. A. K. Bera, T. Rana, S. Das, D. Bhattacharya, D. Pan, S. Bandyopadhyay and S. K. Das, *Hum. Exp. Toxicol.*, 2010, **30**, 940–951.
130. T. Rana, A. K. Bera, S. Das, D. Bhattacharya, D. Pan, S. Bandyopadhyay, D. K. Mondal, S. Samanta, S. Bandyopadhyay and S. K. Das, *Exp. Toxicol. Pathol.*, 2012, **64**, 665–671.
131. F. Li, H. Wen, X. Liu, F. Zhou and G. Chen, *Protein Expr. Purif.*, 2012, **82**, 339–344.
132. S. Tanaka, K. Ko, K. Kino, K. Tsuchiya, A. Yamashita, A. Murasugi, S. Sakuma and H. Tsunoo, *J. Biol. Chem.*, 1989, **264**, 16372–16377.
133. Y. Liu, S. Bastiaan-Net and H. J. Wichers, *Front. Nutr.*, 2020, **7**, 132.
134. Q.-Z. Li, Y.-Z. Zheng and X.-W. Zhou, *Drug Discov. Today*, 2019, **24**, 307–314.
135. P.-H. Wang, C.-I. Hsu, S.-C. Tang, Y.-L. Huang, J.-Y. Lin and J.-L. Ko, *J. Agric. Food Chem.*, 2004, **52**, 2721–2725.
136. Y. Mingyi, T. Belwal, H. P. Devkota, L. Li and Z. Luo, *Trends Food Sci. Technol.*, 2019, **92**, 94–110.
137. H.-C. HSU, C.-I. HSU, R.-H. LIN, C.-L. KAO and J.-Y. LIN, *Biochem. J.*, 1997, **323**, 557–565.
138. F. Cateni, M. L. Gargano, G. Procida, G. Venturella, F. Cirlincione and V. Ferraro, *Mycochemicals in wild and cultivated mushrooms: nutrition and health*, Springer, Netherlands, 2021, vol. 2.
139. T. B. Ng and H. X. Wang, *Peptides*, 2004, **25**, 929–933.
140. H. X. Wang and T. B. Ng, *Biochem. Cell Biol.*, 2000, **78**, 699–702.
141. S. K. Lam and T. B. Ng, *Biochem. Biophys. Res. Commun.*, 2001, **285**, 1071–1075.
142. W. L. Pan, J. H. Wong, E. F. Fang, Y. S. Chan, X. J. Ye and T. B. Ng, *Biochim. Biophys. Acta, Mol. Cell Res.*, 2013, **1833**, 987–996.
143. E. F. Fang and T. B. Ng, *Biochim. Biophys. Acta, Rev. Cancer*, 2011, **1815**, 65–74.
144. L. Citores, S. Ragucci, J. M. Ferreras, A. Di Maro and R. Iglesias, *ACS Chem. Biol.*, 2019, **14**, 1319–1327.
145. Y. Chen, S. Jiang, Y. Jin, Y. Yin, G. Yu, X. Lan, M. Cui, Y. Liang, B. H. C. Wong and L. Guo, *Mol. Nutr. Food Res.*, 2012, **56**, 1729–1738.
146. X. Dan, W. Liu, J. H. Wong and T. B. Ng, *Front. Pharmacol.*, 2016, **7**, 217.
147. P. H. K. Ngai and T. B. Ng, *Peptides*, 2004, **25**, 11–17.
148. H. X. Wang and T. B. Ng, *Biochem. Biophys. Res. Commun.*, 2004, **324**, 855–859.
149. H. X. Wang and T. B. Ng, *Biochem. Cell Biol.*, 2006, **84**, 178–183.
150. H. Wang and T. B. Ng, *Biochem. Biophys. Res. Commun.*, 1999, **264**, 714–718.
151. I. Yike, *Mycopathologia*, 2011, **171**, 299–323.
152. M. Nakamura, A. Iketani and Y. Shioi, *Mycoscience*, 2011, **52**, 234–241.
153. T. Nishiwaki, S. Asano and T. Ohyama, *J. Biosci. Bioeng.*, 2009, **107**, 605–609.
154. K. S. Burton, J. F. Smith, D. A. Wood and C. F. Thurston, *Mycol. Res.*, 1997, **101**, 1341–1347.

155. H. Wang and T. B. Ng, *Biochem. Biophys. Res. Commun.*, 2001, **289**, 750–755.
156. G. Palmieri, C. Bianco, G. Cennamo, P. Giardina, G. Marino, M. Monti and G. Sannia, *Appl. Environ. Microbiol.*, 2001, **67**, 2754–2759.
157. B. T. Park, K. H. Na, E. C. Jung, J. W. Park and H. H. Kim, *Korean J. Physiol. Pharmacol.*, 2009, **13**, 49–54.
158. H.-H. Shin and H.-S. Choi, *J. Microbiol.*, 1998, **36**, 20–25.
159. T. Yoshimoto, A. A. Sattar, W. Hirose and D. Tsuru, *J. Biochem.*, 1988, **104**, 622–627.
160. R. O. De Faria, V. R. Moure, M. A. L. De Almeida Amazonas, N. Krieger and D. A. Mitchell, *Food Technol. Biotechnol.*, 2007, **45**, 287–294.
161. S. Halaouli, M. Asther, J.-C. Sigoillot, M. Hamdi and A. Lomascolo, *J. Appl. Microbiol.*, 2006, **100**, 219–232.
162. X. Wu, C. Huang, Q. Chen, H. Wang and J. Zhang, *Biomed. Chromatogr.*, 2014, **28**, 548–553.
163. H. X. Wang and T. B. Ng, *Appl. Microbiol. Biotechnol.*, 2006, **69**, 521–525.
164. I. Golak-Siwulska, A. Kałużewicz, T. Spiżewski, M. Siwulski and K. Sobieralski, *Folia Hortic.*, 2018, **30**, 191–201.
165. D. Sande, G. P. de Oliveira, M. A. F. E. Moura, B. de, A. Martins, M. T. N. S. Lima and J. A. Takahashi, *Food Res. Int.*, 2019, **125**, 108524.
166. N. Nallathamby, S. N. A. Malek, S. Vidyadaran, C.-W. Phan and V. Sabaratnam, *Int. J. Med. Mushrooms*, 2020, **22**, 1215–1223.
167. C.-W. Phan, G.-S. Lee, I. G. Macreadie, S. N. A. Malek, D. Pamela and V. Sabaratnam, *Nat. Prod. Commun.*, 2013, **8**, 1763–1765.
168. Y. Yaoita, M. Kikuchi and K. Machida, *Nat. Prod. Commun.*, 2014, **9**, 419–426.
169. O. Taofiq, A. Martins, M. F. Barreiro and I. C. F. R. Ferreira, *Trends Food Sci. Technol.*, 2016, **50**, 193–210.
170. T. Shimizu, J. Kawai, K. Ouchi, H. Kikuchi, Y. Osima and R. Hidemi, *Int. J. Oncol.*, 2016, **48**, 1670–1678.
171. A. Podkowa, A. Kryczyk-Poprawa, W. Opoka and B. Muszyńska, *Eur. Food Res. Technol.*, 2021, **247**, 513–533.
172. M. Fogarasi, S. A. Socaci, F. V. Dulf, Z. M. Diaconeasa, A. C. Fărcaş, M. Tofană and C. A. Semeniuc, *Molecules*, 2018, **23**, 3272.
173. C. Liang, D. Tian, Y. Liu, H. Li, J. Zhu, M. Li, M. Xin and J. Xia, *Eur. J. Med. Chem.*, 2019, **174**, 130–141.
174. I. Lee, B. Ahn, J. Choi, M. Hattori, B. Min and K. Bae, *Bioorg. Med. Chem. Lett.*, 2011, **21**, 6603–6607.
175. K. Arpha, C. Phosri, N. Suwannasai, W. Mongkolthanaruk and S. Sodngam, *J. Agric. Food Chem.*, 2012, **60**, 9834–9841.
176. T. Feng, Z.-H. Li, Z.-J. Dong, J. Su, Y. Li and J.-K. Liu, *Nat. Prod. Bioprospect.*, 2011, **1**, 29–32.
177. Y. Wang, L. Bao, X. Yang, H. Dai, H. Guo, X. Yao, L. Zhang and H. Liu, *Helv. Chim. Acta*, 2012, **95**, 261–267.
178. Y. Wang, L. Bao, X. Yang, L. Li, S. Li, H. Gao, X.-S. Yao, H. Wen and H.-W. Liu, *Food Chem.*, 2012, **132**, 1346–1353.

179. Y. Jeong, B. Yang, S. Jeong, S. Kim and C. Song, *Phyther. Res.*, 2008, **22**, 614–619.
180. G.-S. Wu, J.-J. Lu, J.-J. Guo, Y.-B. Li, W. Tan, Y.-Y. Dang, Z.-F. Zhong, Z.-T. Xu, X.-P. Chen and Y.-T. Wang, *Fitoterapia*, 2012, **83**, 408–414.
181. C.-H. Li, P.-Y. Chen, U.-M. Chang, L.-S. Kan, W.-H. Fang, K.-S. Tsai and S.-B. Lin, *Life Sci.*, 2005, 77, 252–265.
182. S. Bharadwaj, K. E. Lee, V. D. Dwivedi, U. Yadava, A. Panwar, S. J. Lucas, A. Pandey and S. G. Kang, *Sci. Rep.*, 2019, **9**, 19059.
183. S. Dudhgaonkar, A. Thyagarajan and D. Sliva, *Int. Immunopharmacol.*, 2009, **9**, 1272–1280.
184. W. Lakornwong, K. Kanokmedhakul, S. Kanokmedhakul, P. Kongsaeree, S. Prabpai, P. Sibounnavong and K. Soytong, *J. Nat. Prod.*, 2014, 77, 1545–1553.
185. H.-P. Chen, Z.-Z. Zhao, Z.-H. Li, Y. Huang, S.-B. Zhang, Y. Tang, J.-N. Yao, L. Chen, M. Isaka, T. Feng and J.-K. Liu, *J. Agric. Food Chem.*, 2018, **66**, 3146–3154.
186. Q.-Q. Tao, K. Ma, L. Bao, K. Wang, J.-J. Han, W.-Z. Wang, J.-X. Zhang, C.-Y. Huang and H.-W. Liu, *Planta Med.*, 2016, **82**, 639–644.
187. S. Wang, L. Bao, F. Zhao, Q. Wang, S. Li, J. Ren, L. Li, H. Wen, L. Guo and H. Liu, *J. Agric. Food Chem.*, 2013, **61**, 5122–5129.
188. S. Wang, Y. Li, L. Bao, J. Han, X. Yang, H. Li, Y. Wang, S. Li and H. Liu, *Org. Lett.*, 2012, **14**, 3672–3675.
189. K. Yoshikawa, M. Kuroboshi, S. Ahagon and S. Arihara, *Chem. Pharm. Bull.*, 2004, **52**, 886–888.
190. S. Khatua, S. Paul and K. Acharya, *Res. J. Pharm. Technol.*, 2013, **6**, 496–505.
191. S. Lin, L. T. Ching, J. Chen and P. C. K. Cheung, *J. Funct. Foods*, 2015, **17**, 802–815.
192. I. Palacios, M. Lozano, C. Moro, M. D'Arrigo, M. A. Rostagno, J. A. Martínez, A. García-Lafuente, E. Guillamón and A. Villares, *Food Chem.*, 2011, **128**, 674–678.
193. M. B. Bahadori, C. Sarikurkcu, O. U. Yalcin, M. Cengiz and H. Gungor, *Microchem. J.*, 2019, **150**, 104172.
194. M. Garrab, H. Edziri, R. El Mokni, M. Mastouri, H. Mabrouk and W. Douki, *S. Afr. J. Bot.*, 2019, **124**, 359–363.
195. Q. Hu, B. Yuan, H. Xiao, L. Zhao, X. Wu, K. Rakariyatham, L. Zhong, Y. Han, B. Muinde Kimatu and W. Yang, *Food Funct.*, 2018, **9**, 1601–1611.
196. R. Saltarelli, F. Palma, A. M. Gioacchini, C. Calcabrini, U. Mancini, R. De Bellis, V. Stocchi and L. Potenza, *J. Ethnopharmacol.*, 2019, **231**, 464–473.
197. Y. Nakajima, H. Nishida, S. Matsugo and T. Konishi, *J. Med. Food*, 2009, **12**, 501–507.
198. D. Stojkovic, M. Smiljkovic, A. Ciric, J. Glamoclija, L. Van Griensven, I. C. F. R. Ferreira and M. Sokovic, *S. Afr. J. Bot.*, 2019, **120**, 100–103.
199. F. Bach, A. A. F. Zielinski, C. V. Helm, G. M. Maciel, A. C. Pedro, A. P. Stafussa, S. Ávila and C. W. I. Haminiuk, *LWT*, 2019, **107**, 214–220.

200. P. Kalač, *J. Sci. Food Agric.*, 2013, **93**, 209–218.
201. A. El-Sharkawy and A. Malki, *Molecules*, 2020, 25.
202. O. Taofiq, â. Fernandes, L. Barros, M. F. Barreiro and I. C. F. R. Ferreira, *Trends Food Sci. Technol.*, 2017, **70**, 82–94.
203. B. Muszyńska, K. Kała, A. Włodarczyk, A. Krakowska, B. Ostachowicz, J. Gdula-Argasińska and P. Suchocki, *Biol. Trace Elem. Res.*, 2020, **194**, 603–613.
204. K. K. Gupta, S. Agarwal, A. Kushwaha, S. Maurya, V. K. Chaturvedi, R. K. Pathak, V. Verma and M. P. Singh, *Incred. World Biotechnol.*, 2017, **1**, 43–57.
205. G. Ren, M. Yu, K. Li, Y. Hu, Y. Wang, X. Xu and J. Qu, *J. Funct. Foods*, 2016, **22**, 177–188.
206. F. S. Reis, A. Martins, M. H. Vasconcelos, P. Morales and I. C. F. R. Ferreira, *Trends Food Sci. Technol.*, 2017, **66**, 48–62.
207. F. Boccia and G. Punzo, *Food Res. Int.*, 2020, **130**, 108888.
208. M. Alongi and M. Anese, *J. Funct. Foods*, 2021, **81**, 104466.
209. J. Gur, M. Mawuntu and D. Martirosyan, *Funct. Foods Heal. Dis.*, 2018, **8**, 385–397.
210. S. A. El Sohaimy, *World Appl. Sci. J.*, 2012, **20**, 691–708.
211. A. Santini, S. M. Cammarata, G. Capone, A. Ianaro, G. C. Tenore, L. Pani and E. Novellino, *Br. J. Clin. Pharmacol.*, 2018, **84**, 659–672.

CHAPTER 7

Terpenes and Steroids in Fungi Used in the Daily Diet

EKATERINA-MICHAELA TOMOU, CHRISTINA BARDA, PANAGIOTIS LYMPERIS, KRYSTALIA LYTRA AND HELEN SKALTSA*

Department of Pharmacognosy & Chemistry of Natural Products, School of Health Sciences, Faculty of Pharmacy, National and Kapodistrian University of Athens, Panepistimiopolis, Zografou, 15771 Athens, Greece
*Email: skaltsa@pharm.uoa.gr

7.1 Introduction

It is estimated that the second largest group of living organisms is fungi, with more than 1.5 million species existing in nature.[1] Among these, over 2000 species of mushrooms are present in biodiversity, while approximately 25 are widely consumed as food, and some of these are commercially cultivated.[2] However, it is noteworthy that there are often reports of poisoning by mushrooms due to the fact that there is no easy distinction between edible and toxic ones.[3] Some of the most common cultivated mushrooms around the world are *Agaricus bisporus*, *Lentinus edodes*, *Pleurotus* spp., and *Flammulina velutipes*.

Over the years, mushroom production has continuously increased, with China being the largest producer worldwide.[4] In Europe, the main country for European mushroom cultivation was the Netherlands, while Germany is the most important country for the imports of fresh and canned mushrooms. Remarkably, Poland has its increased mushroom production during the last 20 years, making it now the largest in Europe.[4]

Food Chemistry, Function and Analysis No. 36
Edible Fungi: Chemical Composition, Nutrition and Health Effects
Edited by Dejan Stojković and Lillian Barros

Published by the Royal Society of Chemistry, www.rsc.org

Mushrooms possess high nutritional and functional value, and they are considered nutraceutical foods because of their important effects on human health, including antidiabetic, antioxidant, antiviral, antibacterial, osteoprotective, nephroprotective, and hepatoprotective activities.[5] Specifically, the high nutritional value of edible mushrooms is attributed to their high contents of proteins, fiber, vitamins, and minerals, and to the low level of fats, as well as their bioactive secondary metabolites such as terpenoids. Many fungi produce a wide range of mycochemicals due to the competitive environment in which they live. The subject of mycochemistry is considered a distinct discipline with an enormous variety of chemical substances, known as "mycochemicals," which originated from mushrooms.[6] The chemical composition of wild and cultivated mushrooms has been greatly investigated around the world. Various scientific studies have reported a plethora of mycochemicals with different chemical backbones, including phenolic compounds, terpenoids, lipids, polysaccharides, and proteins.[7]

7.2 Terpenes and Steroids in Edible Mushrooms

Terpenoids are widespread natural products that are formed from C5 isoprene precursor units, the isopentenyl diphosphate (IPP) and its isomer dimethylallyl diphosphate (DMAPP), leading to their characteristic branched-chain structure. These units are synthesized through two independent metabolic routes, the mevalonate (MVA) pathway and the 2-*C*-methyl-D-erythritol 4-phosphate (MEP) pathway, which are localized in the cytosol and the chloroplast, respectively. It has been generally assumed that the MVA pathway supplies isoprenoid precursors for the synthesis of sesquiterpenes, as well as sterols and triterpenes, while the MEP pathway is responsible for monoterpenes, diterpenes, and carotenoids. Terpenoids are divided into families based on the number of isoprene units from which they are formed. Thus, there are monoterpenoids (C10), sesquiterpenoids (C15), diterpenoids (C20), sesterterpenoids (C25), triterpenoids (C30), and carotenoids (tetraterpenoids, C40). The first step in the biosynthesis of all triterpenes is the cyclization of 2,3-oxidosqualene. At that point the pathway is divided into two modules that lead to: (a) the protosteryl cation, an intermediated moiety, which is the precursor of all phytosterols and (b) the dammarenyl cation, an intermediated moiety, which subsequently forms diverse triterpene skeletons (Figure 7.1).

7.2.1 *Antrodia* Species

The polypore genus *Atrodia* P. Karst includes approximately 50 species worldwide.[8] Species of the genus *Antrodia* play a significant ecological role as wood decay factors as they degrade cellulose and hemicellulose, influencing forest development and carbon elimination.[9] In addition, some are used in pharmaceutical and biotechnological industries due to their important pharmacological properties.[9]

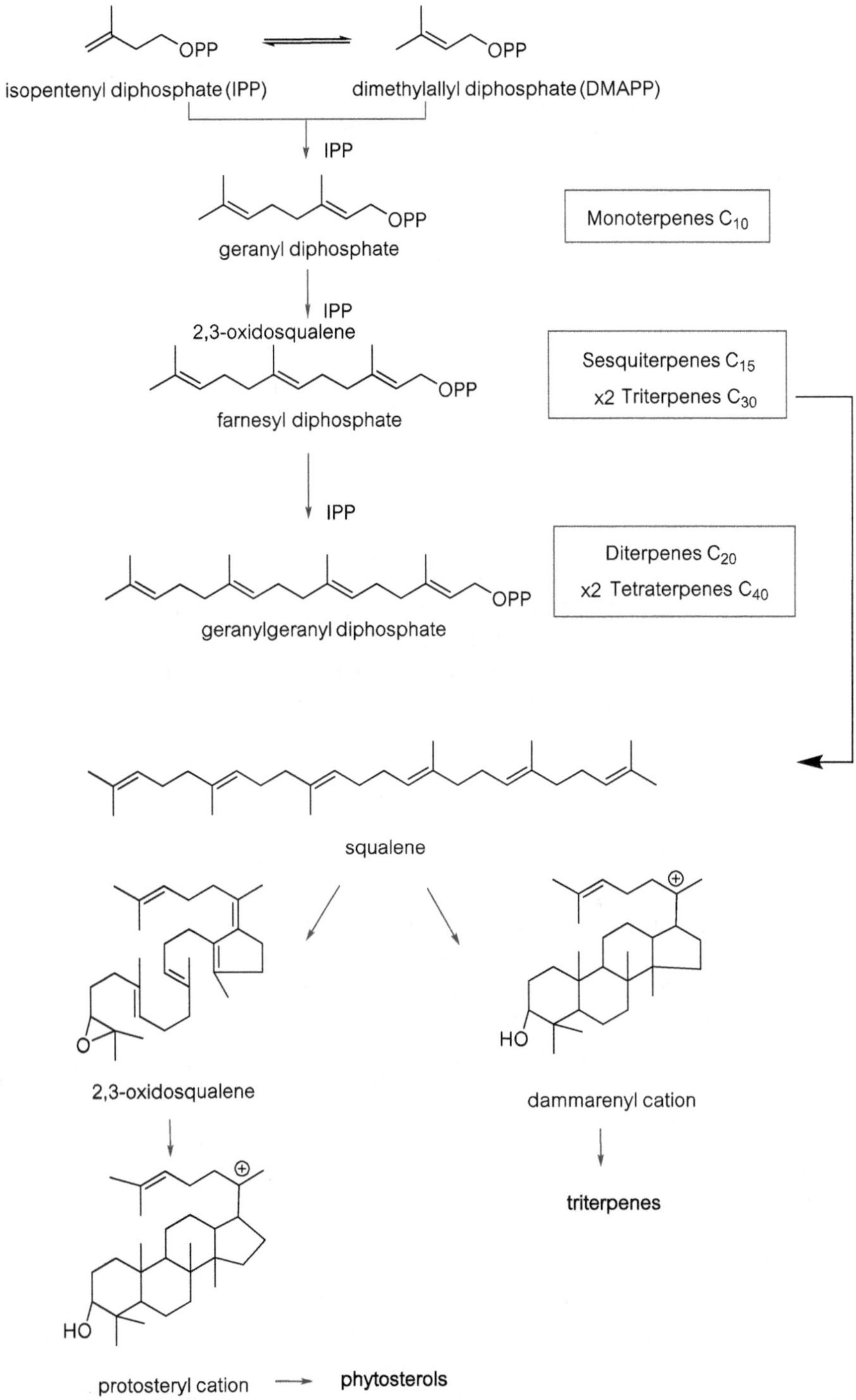

Figure 7.1 Biosynthetic pathway of terpenoids.

7.2.1.1 *A. camphorata*

Synonyms: *A. cinnamomea* Chang & Chou, *Taiwanofungus camphoratus* (Zang & Su) Wu, Yu, Dai & Su, *Ganoderma camphoratum* M. Zang & C. H. Su *A. camphorata* (Zang & Su) Wu, Ryvarden & Chang is a well-known medicinal mushroom in Taiwan with the name "Niu-Chang-Chih".[10] It usually occurs on the decaying empty trunks of the endemic plant of Taiwan, *Cinnamomum kanehirai* (called as Bull camphor tree), which is its only host tree.[1,10,11] The growth of wild *A. camphorata* is extremely slow, since it needs one year for the mushroom to grow to the required size.[1] Some morphological characteristics are diagnostic features of this species such as the red to light cinnamon, resupinate, effused-reflexed to pileate fruit bodies, the very bitter taste, the cylindrical basidiospores, and the weakly amyloid skeletal hyphae.[10] This particular fruit-body morphology and the cultural characteristics apparently indicate that this species belongs to the genus *Antrodia* P. Karst.

Over the years, controversial taxonomical issues have emerged as four different names could be found for this fungus in the international literature: *Ganoderma camphoratum*, *A. cinnamomea*, *A. camphorata*, and *Taiwanofungus camphoratus*.[1,12] In 1990, this species was reported as *G. camphoratum* and then later it was described as *A. cinnamomea*.[13] Based on a phylogenetic study, *A. camphorata* was related to the genus *Taiwanofungus*, namely as *T. camphoratus*.[14] Nowadays, the widely accepted official name for this mushroom is *A. camphorata*, while *A. cinnamomea* and *T. camphoratus* are also used.[12]

Over the years, *A. camphorata* has been widely used in the traditional medicine of Taiwan. The local tribes used this species for the treatment of abdominal pain, metabolic syndrome, cancer, and the improvement of liver function, as well as for alcohol/food intoxication and as an anti-itch agent.[1,11,12,15] It is noteworthy to mention that *A. camphorata* has been used as an adjuvant therapeutic agent in 12% of cancer patients in Taiwan during their treatment.[12] One of the main challenges of these medicinal fungi is their high demand which has far exceeded their supply. It is estimated that the price for the wild fruiting bodies could range from $15 000 to $25 000 per kilo.[1,12] To achieve large-scale commercialization, many cultivation techniques have been developed in recent years, including submerged fermentation, solid support culture, cut wood cultures, and dish cultures.[1]

A. camphorata exhales a characteristic strong odor of sassafras (camphor aroma), resulting from its aromatic components.[1,12] Interestingly, the pure sample of this mushroom does not possess this specific odor. Furthermore, it has a strong bitter taste, which could be attributed to its high terpenoid content.[1]

Regarding the phytochemical profile of this species, *A. camphorata* includes a wide range of chemical constituents such as polysaccharides, terpenoids, maleic and succinic acid derivatives, benzoquinone derivatives, lignans, and benzenoids.[12,16,17] Among these, terpenoids are the largest chemical class of *A. camphorata*. Recently, Kuang *et al.*[12] described in detail the terpenoids isolated or identified from the fruiting bodies and mycelia, and some from the culture broth. Based on their chemical skeletons, they can be categorized into steroids, triterpenoids (ergostane and lanostane-type), meroterpenoids, sesquiterpenoids, and diterpenoids. It should also be mentioned that triterpenoids possess very similar structures, with the only differences being in the number, location, or stereochemistry of hydroxyl groups, which makes their separation difficult.[18] Ergostane-type triterpenoids and meroterpenoids are characteristic compounds of this species.[12]

The ergostane-type triterpenoids, the largest number of ergostanes of *A. camphorata*, bear a specific C29 skeleton with an α-CH_3 group at C-4 and a 24-*exo*-methylene-26-oic acid side chain at C-17b. Furthermore, they are usually found as mixtures of C-25 epimers (25*R*- and 25*S*-structures).[12] Antcins A, B, and C were the first isolated ergostane-type triterpenoids from the dried fruiting bodies of *A. camphorata*[15] (Scheme 7.1).

Lanostane-type triterpenoids are a group of tetracyclic triterpenoids derived from lanosterol, which is formed by the cyclization of squalene-2,3-epoxide in a chair–boat–chair–boat conformation, making protosterol. Then, a consequent backbone rearrangement forms lanosterol, from which different series of triterpenoids are produced, such as dammaranes, tirucallanes, euphanes, cucurbitanes, and lanostanes, which have 30 carbons and a gem-dimethyl at C-4.[19] Most lanostane derivatives in *A. camphorata* present a C31 or C30 skeleton, with a hydroxyl or ketone group at C-3, and a carboxyl group at C-21 or C-26. They usually include Δ^8 and $\Delta^{7,9,(11)}$ double bonds, and appear as a pair.[12] Eburicol (Scheme 7.2) was the

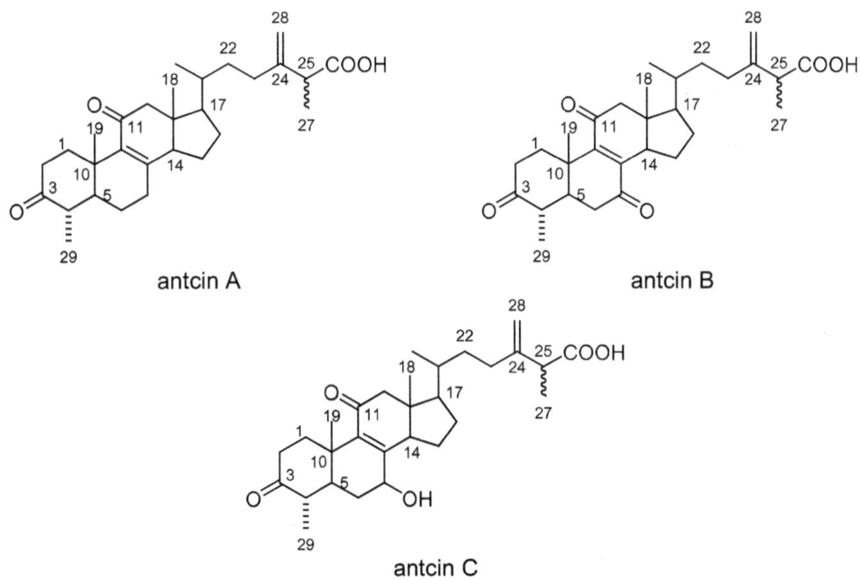

Scheme 7.1 Chemical structures of antcins A, B, and C.

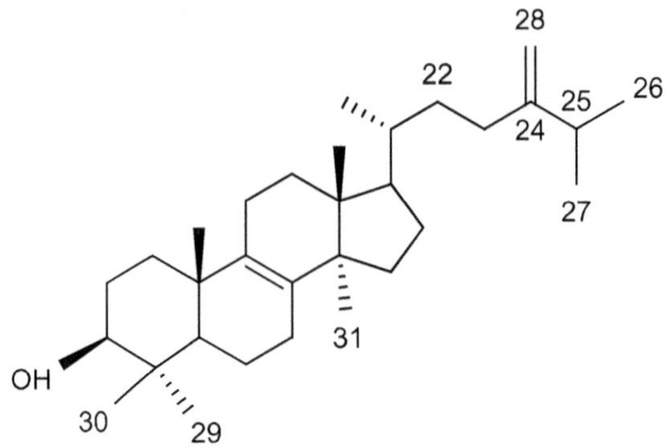

Scheme 7.2 Chemical structure of eburicol.

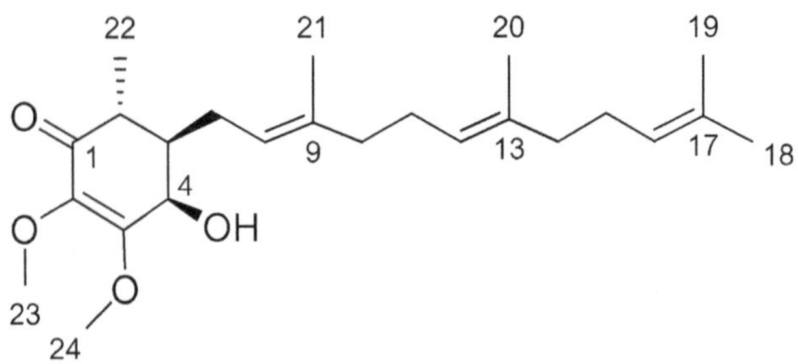

Scheme 7.3 Chemical structure of antroquinonol.

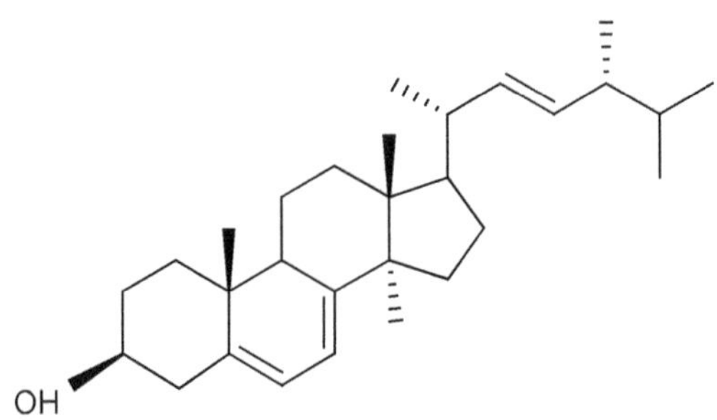

Scheme 7.4 Chemical structure of ergosterol.

first lanostane-type triterpenoid described from this species and some common constituents of this chemical class are eburicoic acid, sulfurenic acid, dehydroeburicoic acid, and dehydrosulfurenic acid.

In addition, antroquinonol (Scheme 7.3) and its derivatives are reported as characteristic compounds of this species.[12] These constituents belong to meroterpenoids and are formed by the combination of an orsellinic acid-derived C7 cyclohexenone and a C15 sesquiterpene unit.[12]

Ergosterol (Scheme 7.4) is the main steroid found in *A. camphorata*, while 22 steroids have been reported based on the review study of Kuang *et al.*[12]

Table 7.1 Chemical names and molecular formulas of triterpenoids of *A. salmonea*.

Chemical name	Molecular formula	Ref.
I. Ergostane-type		
Zhankuic acid A	$C_{29}H_{40}O_5$	22
Zhankuic acid B	$C_{29}H_{42}O_5$	
Antcin C	$C_{29}H_{42}O_5$	
Zhankuic acid C	$C_{29}H_{42}O_6$	
Antcin M	$C_{29}H_{44}O_5$	
Antcin K	$C_{29}H_{44}O_6$	
Methyl antcinate L	$C_{30}H_{46}O_6$	
Methyl antcinate K	$C_{30}H_{46}O_6$	
II. Lanostane-type		
3β,15α-Dihydroxylanosta-7,9(11),24-trien-21-oic acid	$C_{30}H_{46}O_4$	22
3-Ketodehydrosulfurenic acid	$C_{31}H_{46}O_4$	
Dehydrosulfurenic acid	$C_{31}H_{48}O_4$	
Eburicoic acid	$C_{31}H_{50}O_3$	23
Sulfurenic acid	$C_{31}H_{50}O_4$	22
Fomeffinic acid C	$C_{31}H_{50}O_4$	23

7.2.1.2 *A. salmonea*

Synonym: *Taiwanofungus salmoneus* (T. T. Chang & W. N. Chou) Sheng H. Wu, Z. H. Yu, Y. C. Dai & C. H. Su[20] *A. salmonea* T. T. Chang & W. N. Chou was first reported as a new species of genus *Antrodia* in 2004.[21] It grows in the empty rotten trunks of the Taiwan coniferous tree, namely *Cunninghamia konishii*, and causes a brown heart rot.[21] *A. salmonea* is similar to the previously described *A. camphorota* and both species have a bitter taste which could be attributed to their bioactive components.[21] Shen *et al.* isolated three previously undescribed ergostane-type triterpenoids: methyl antcinate L, antcin M, and methyl antcinate K, together with nine known compounds from the ethanol extract of *A. salmonea* (Table 7.1).[22] A year later, the chloroform extract of this species was studied, leading to the isolation of additional bioactive constituents (Table 7.1).[23]

7.2.2 *Coriolus* Species

Coriolus Quél., as well as some other genera, are considered as synonyms of genus *Trametes*.[24]

7.2.2.1 *C. versicolor*

C. versicolor (L.) Quél. is a synonym of *Trametes versicolor* (L.) C. G. Lloyd and belongs to the genus *Trametes* Fr. (see Section 7.2.8.1.)

7.2.2.2 *C. consors*

Homotypic synonym: *Irpex consors* Berk.

C. consors (Berk.) Imazeki belongs to the genus *Irpex* Fr.[25] A sesquiterpene named coriolin is the characteristic secondary metabolite of this species. Coriolin, coriolin B, coriolin C, and dehydrocoriolin C have been identified

Table 7.2 Chemical names and molecular formulas of terpenes of *C. consors*.

Chemical name	Molecular formula	Ref.
Coriolin	$C_{15}H_{20}O_5$	26 and 27
Coriolin B	$C_{23}H_{36}O_6$	27 and 28
Coriolin C	$C_{23}H_{34}O_7$	27
Dehydrocoriolin C	$C_{23}H_{36}O_7$	28
Hirsutene	$C_{15}H_{24}$	27
Hirsutanol D	$C_{15}H_{22}O_2$	28

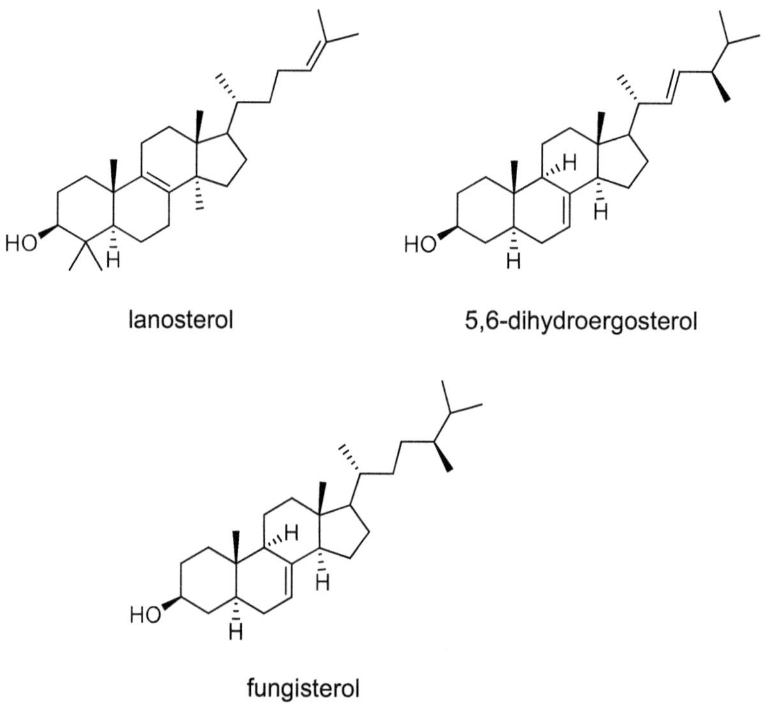

Scheme 7.5 Chemical structures of *C. pargamenus*.

in extracts of *C. consors*.[26–28] Moreover, hirsutene and hirsutanol D were found in the aforementioned species.[27,28] The names and molecular formulas of *C. consors* terpenes are illustrated in Table 7.2.

7.2.2.3 C. pargamenus

Synonym: *Trichaptum biforme* (Fr.) Ryvarden. *C. pargamenus* (Fr.) G. Cunn. belongs to the genus *Trichaptum*.[29] Yokokawa identified the compounds, namely ergosterol, ergost-7-en-3β-ol, lanosterol, and 24-methylene-24,25-dihydrolanosterol in *C. pargamenus*,[30] while Cole *et al.* described the presence of fungisterol, ergosterol, ergosta-5,7-dien-3,β-ol, and 5,6-dihydroergosterol in the same species (Scheme 7.5).[31] The chemical

Table 7.3 Chemical names and molecular formulas of terpenoids and steroids of *C. pargamenus*.

Chemical name	Molecular formula	Ref.
Ergosterol	$C_{28}H_{44}O$	30 and 31
Fungisterol	$C_{28}H_{48}O$	
Ergost-7-en-3β-ol	$C_{28}H_{48}O$	30
Ergosta-5,7-dien-3β-ol	$C_{28}H_{46}O$	31
5,6-Dihydroergosterol	$C_{28}H_{46}O$	
Lanosterol	$C_{30}H_{50}O$	30
24-Methylene-24,25-dihydrolanosterol (=eburicol)	$C_{31}H_{52}O$	

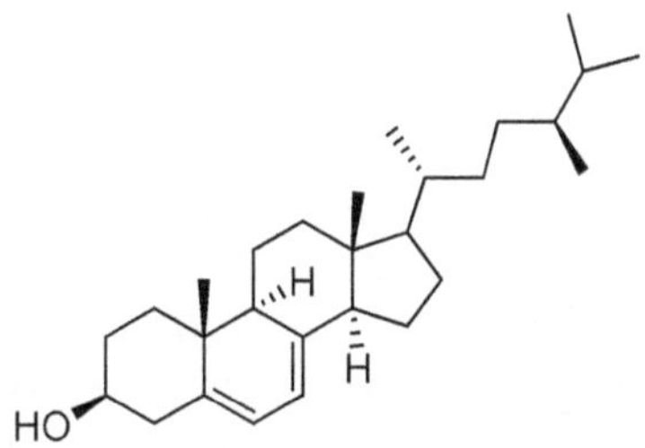

Scheme 7.6 Chemical structure of ergosta-5,7-dien-3β-ol.

names and molecular formulas of terpenoids and steroids of *C. pargamenus* are illustrated in Table 7.3.

7.2.2.4 C. sangamenus

Two sterols, ergosterol (Scheme 7.4) and ergosta-5,7-dien-3β-ol (Scheme 7.6), were described in *C. sangamenus* by Cole *et al.*[31]

7.2.3 *Ganoderma* species

Genus *Ganoderma* (Ganodermataceae family) is a cosmopolitan genus with a wide geographical distribution in many tropical and temperate countries, comprising more than 200 species.[32–35] It is one of the most popular medicinal mushrooms in China, Japan, and Korea.[2] Many species of this genus are pathogenic in nature, as they cause different diseases in plants and cause wood decay.[36] Historically, *Ganoderma* (Lingzhi) has been mentioned in many ancient books, including Shen Nong Ben Cao Jing (Shennong Materia Medica), which was written around 100 B.C. and included experiences gained by the ancient Chinese medical practitioners.[37] Based on the approach of Yin and Yang and the five elements in traditional Chinese medicine (TCM), Shennong Materia Medica classified Lingzhi according to their colors in six kinds:[37]

(a) Cizhi (or Danzhi): bitter taste and it could be used to relieve chest congestion and improve memory

(b) Heizhi (or Xuanzhi): salty taste and it could be used for the treatment of renal problems and increasing consciousness
(c) Qingzhi (or Longzhi): acidic taste and it could be used for the improvement of eyes and liver function
(d) Baizhi (or Yuzhi): pungent taste and it could be used to treat coughs and lung problems
(e) Huangzhi (or Jinzhi): sweet taste and it could be used to treat heart, spleen and stomach ailments and
(f) Zizhi (or Muzhi): sweet taste and it could be used to cure difficulty in hearing and arthritis.

Furthermore, all the above kinds of Lingzhi could be used to promote and improve health, well-being, and longevity. Several myths and poems describing people's love, worship, and beliefs on Lingzhi have been reported in Chinese literature since ancient times. The Pharmacopoeia of the Peoples Republic of China (Part 1) (2000, 2005, 2010, and 2015 editions) reported the fruiting body of *G. lucidum* and *G. sinensis* as legal traditional Chinese medicines.[33] Other *Ganoderma* species, such as *G. capense*, *G. cochlear*, and *G. tsugae* have also been used in folk medicines.[38] In addition, *G. lucidum* exhibited various pharmacological effects for the treatment of current diseases such as cancer, diabetes, and human immunodeficiency virus (HIV).[39] Over the years, China has developed a strong industry for making several *Ganoderma* products such as drugs, supplements, cosmetics, and foods.[37] As a result of the increasing consumption of these products, many cultivation practices have been also developed to meet the demand. It is estimated that the global annual gross output of the *Ganoderma* industry was around US$5.0 billion in 2015, with yields of 160 000 tons of fruiting body and spore, of which China represented 30% of the gross output and 75% in yields.

More than 20 *Ganoderma* species have been investigated for their phytochemical constituents, reporting approximately 600 chemical compounds that mainly belong to the chemical classes of triterpenes, meroterpenoids, steroids, alkaloids, and polysaccharides.[40–42] *G. lucidum* is the species most thoroughly explored for its secondary metabolites and biological effects.[39,41] However, other *Ganoderma* species which have been studied for their phytochemical constituents include *G. applanatum*, *G. colossum*, *G. sinense*, *G. cochlear*, *G. tsugae*, *G. amboinense*, *G. orbiforme*, *G. resinaceum*, *G. hainanense*, *G. concinna*, *G. pfeifferi*, *G. neo-japonicum*, *G. tropicum*, *G. australe*, *G. carnosum*, *G. fornicatum*, *G. lipsiense* (synonym *G. applanatum*), *G. mastoporum*, *G. theaecolum*, *G. boninense*, *G. capense*, and *G. annulare*.[41] Triterpenes are the major chemical category, with over 300 triterpenoids being isolated.[40] *G. lucidum* is one of the most well-studied species producing all lanostane-type triterpenoids, which can be contained in fungi.[19,41] Meroterpenoids also occur in the *Ganoderma* species, and they have a 1,2,4-trisubstituted phenyl group connecting with a polyunsaturated side chain or polycyclic substructure.[38] The first meroterpenoids were ganomycins A and B (Scheme 7.7). Regarding the steroids, ergosterol derivatives are the most common in this genus.[40]

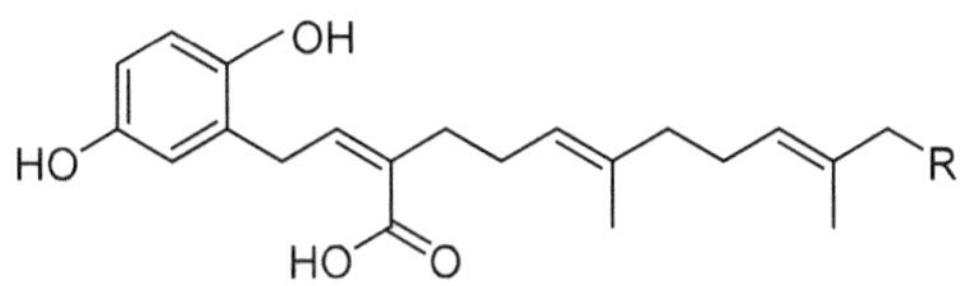

Ganomycin A: R=OH; Ganomycin B: R=H

Scheme 7.7 Chemical structure of ganomycins A and B.

7.2.4 *Laetiporus* species

Laetiporus Murrill is a genus found worldwide, causing brown cubical heart rot on living hardwoods and conifers. It belongs to the order Polyporales and the Fomitopsidaceae family. Some *Laetiporus* species are known as forest pathogens. They grow from cold temperate to tropical zones and are associated with many families such as Betulaceae, Burseraceae, Elaeocarpaceae, Fabaceae, Fagaceae, Meliaceae, Myrtaceae, Oleaceae, Pinaceae, Salicaceae, Sapindaceae, and Taxaceae. *Laetiporus* includes 11 species: *L. ailaoshanensis* B. K. Cui & J. Song, *L. cremeiporus* Y. Ota & T. Hatt., *L. versisporus* and *L. zonatus* B. K. Cui & J. Song from East Asia; *L. cincinnatus* (Morgan) Burds., Banik & T. J. Volk, *L. conifericola* Burds. & Banik and *L. huroniensis* Burds. & Banik from North America; *L. caribensis* Banik & D. L. Lindner from Central America; *L. montanus* Černý ex Tomšovský & Jankovský from East Asia and Europe; *L. sulphureus* from North America, South America, and Europe; *L. gilbertsonii* Burds. and four undescribed taxa found from Hawaii, South America, and Central America. Terpenoids are one of the most abundant secondary metabolites of this genus.[43] The chemical names and molecular formulas of steroids and terpenoids of the genus *Laetiporus* are illustrated in Tables 7.4 and 7.5, respectively.

7.2.4.1 *L. sulphureus*

The edible mushroom *L. sulphureus*, also known as sulfur polypore or sulfur shelf, is a sulfur-yellow-colored polyporous mushroom[43,44] that has been used for centuries in traditional European and Asian ethno-medicines due to its immunostimulative, antioxidant, antidiabetic, antimicrobial, anti-inflammatory, and hepatoprotective effects.[43,44] Moreover, the fruit bodies have been used for the treatment of coughs, gastric cancer, and rheumatism in Europe. This species is also called "chicken polypore" or "chicken of the woods" due to its taste and texture, which resemble those of poultry. In addition, its young fruit bodies are edible. In some parts of Germany and North America, it is considered as a delicacy, and it is often consumed as a substitute for chicken in a vegetarian diet. Furthermore, the burned *L. sulphureus* fruit bodies are thought to repel mosquitoes and midges.[45]

However, gastrointestinal problems have been reported after eating this mushroom, as well as the presence of some negative effects such as allergic

Table 7.4 Chemical names and molecular formulas of steroids of the genus *Laetiporus*.

Chemical name	Molecular formula	Species	Ref.
Sulfurenolides A, B	$C_{22}H_{31}O_6$	*L. sulphureus*	56
Sulfurenolide C	$C_{21}H_{29}O_5$		
Ergost-5,7,9(11),22-tetraen-3-ol (dehydroergosterol)	$C_{28}H_{42}O$		57
Ergosta-5,7,22-trien-3β-ol (ergosterol)	$C_{28}H_{44}O$	*L.* sp., *L. sulphureus*	57 and 59
Ergost-7,22-dien-3-ol	$C_{28}H_{46}O$	*L. sulphureus*	50 and 57
Ergost-7-en-3-ol	$C_{28}H_{48}O$		50 and 57
Ergosterol peroxide	$C_{28}H_{44}O_3$		57 and 58
Ergost-4,6,8(14),22-tetraen-3-one	$C_{28}H_{40}O$		50
Cerevisterol	$C_{28}H_{46}O_3$		50, 52 and 57
24-Ethylcholest-5,7,22-trien-3β-ol	$C_{29}H_{46}O$		50
24-Ethylcholestan-3β-ol	$C_{29}H_{52}O$		50 and 57
4,4-Dimethylergost-24-en-3-ol	$C_{30}H_{52}O$		57
4,4-Dimethylergost-22-en-3β-ol	$C_{30}H_{52}O$		50
3-Acetylergosterol	$C_{30}H_{46}O_2$	*L.* sp.	59

reactions, vomiting, and fever. The consumption of this fungus has also been reported to cause hallucinations. For this reason, it has been thought that it might contain alkaloids like those found in psychoactive plants. Nevertheless, it is very possible that these hallucinogenic effects might rather be attributed to a closely related species.[45]

L. sulphureus has gained a lot of attention as a sustainable food supplement due to its high content of carbohydrates (trehalose > mannitol > fructose), proteins (histidine, isoleucine, leucine, lysine, methionine, threonine), minerals (calcium, phosphorus, magnesium, sodium, potassium, iron, zinc, manganese, copper), vitamins (B, D, E), fatty acids (linoleic acid, oleic acid, palmitic acid), and fiber. Furthermore, it contains a wide range variety of bioactive metabolites. Several steroids and terpenoids with different structures have been identified and isolated from *L. sulphureus*[49–59] (Tables 7.4 and 7.5). It is reported that *L. sulphureus* also contains other compounds such as benzofurans, flavonoids, coumarins, phenolics, and N-containing compounds.[43]

7.2.4.2 *L. versisporus*

Although *L. versisporus* (Lloyd) Imazeki is another species of this genus, its morphological characteristics are entirely different. Aging fruit bodies are semiglobose and have hymenophore in the form of more chlamydospores. Initially, the color of the fruit bodies is lemon yellow, before it turns to white and brown in the mature stage.[60] Lanostane-triterpenoids and lanostanoid glycosides (laetiposides A–G) have been isolated and identified from *L. versisporus*[61,62] (Table 7.5).

Table 7.5 Chemical names and molecular formulas of terpenoids of the genus *Laetiporus*.

Chemical name	Molecular formula	Species	Ref.
Sulfureuine A	$C_{15}H_{22}O_2$	*L. sulphureus*	47
Sulfureuines B, C	$C_{15}H_{28}O_4$		46 and 48
Sulfureuine D	$C_{15}H_{26}O_3$		46
Sulfureuine F	$C_{15}H_{24}O_3$		
Sulfureuine G	$C_{15}H_{28}O_3$		
Sulfureuine E	$C_{15}H_{24}O_4$		
Sulfureuine H	$C_{15}H_{26}O_3$		
3β-Hydroxy-11,12-*O*-isopropyldrimene	$C_{18}H_{30}O_3$		
Agripilol A	$C_{15}H_{28}O_4$		
Laetiporin A	$C_{31}H_{48}O_3$	*L.* sp., *L. sulphureus*	49
Laetiporin B	$C_{34}H_{52}O_7$		
Laetiporin C	$C_{31}H_{50}O_5$	*L. sulphureus*	51
Laetiporin D	$C_{31}H_{48}O_5$		
24-Methylenelanost-8-en-3-ol (eburicol)	$C_{31}H_{52}O$		57
3β-Hydroxy-24-methylenelanost-8-en-21-oic acid (eburicoic acid)	$C_{31}H_{50}O_3$	*L.* sp., *L. sulphureus*	49–51, 54 and 57
Acetyl-eburicoic acid	$C_{33}H_{52}O_4$	*L. sulphureus*	52 and 53
Dehydrosulfurenic acid	$C_{31}H_{48}O_4$	*L.* sp.	49
3β,15α-Dihydroxy-24-methylenelanost-8-en-21-oic acid (sulfurenic acid)	$C_{31}H_{50}O_4$		49 and 50
3-Oxosulfurenic acid	$C_{31}H_{48}O_4$	*L. sulphureus*	52
Acetyl derivative of 3-oxosulfurenic acid	$C_{33}H_{50}O_5$		
Trametenolic acid	$C_{30}H_{48}O_3$	*L.* sp.	49 and 51
Acetyl-tramenolic acid	$C_{32}H_{50}O_4$	*L. sulphureus*	52
15α-Hydroxytrametenolic acid	$C_{30}H_{48}O_4$	*L.* sp., *L. sulphureus*	49, 51, 52 and 55
15α-Hydroxy-3,4-secolanosta-4(28),8,24-triene-3,21-dioic acid	$C_{30}H_{46}O_5$	*L. sulphureus*	55
5α-Hydroxy-3,4-seco-lanosta-4(28),8,24-triene-3,21-dioic acid 3-methyl ester	$C_{31}H_{48}O_5$		
Versisponic acid A	$C_{30}H_{48}O_5$	*L. versisporus*	61
Versisponic acid B	$C_{32}H_{48}O_5$		
Versisponic acid C	$C_{33}H_{50}O_5$		
Versisponic acid D	$C_{33}H_{52}O_5$		55 and 61
Versisponic acid E	$C_{34}H_{52}O_6$		61
Laetiposide A	$C_{37}H_{60}O_9$		
Laetiposide B	$C_{36}H_{58}O_9$		
Laetiposide C	$C_{36}H_{56}O_9$		
Laetiposide D	$C_{37}H_{58}O_9$		
Laetiposide E	$C_{36}H_{58}O_8$		62
Laetiposide F	$C_{37}H_{60}O_{10}$		
Laetiposide G	$C_{37}H_{60}O_9$		

7.2.5 *Meripilus* Species

The genus *Meripilus* (Meripilaceae family) has only six species. *M. sumstinei* is found in the Midwest and temperate North America, while similar species, namely *M. giganteus*, occurs in Europe and Asia, *M. lentifrondosus* and *M. tropicalis* are found in Mexico, as well as the brown-rot species known as *M. persicinus* which was moved to *Laetiporus* spp.[63] The genus name combines the Ancient Greek words "meris" (μερίς) which means part or portion and "pilos" (πῖλος = cap). *M. giganteus* is the only well-studied representative of the genus and is well known to possess a wide range of pharmacological properties. Up to date, its antioxidant and antimicrobial activity, its *in vitro* effect on cancer cell viability and apoptosis, as well as its toxicity on zebrafish have been reported.[64] Multiple studies have revealed several bioactive metabolites from this mushroom, including phenolic compounds, steroids, organic acids, tocopherols, and fatty acids. *M. giganteus* is a very large polypore basidiomycete that appears on stumps and at the base of some living broad-leaf trees (such as beech) being also characterized as saprobic. It is fructifying from summer to autumn. It is known under the common names "giant polypore" or "black-staining polypore".[7] *Meripilus* caps can be half a meter wide, and their fruiting body is edible after cooking, but the fully grown mushroom is considered not edible due to its hard and tough consistency.[64] To date, six ergosterol type compounds have been isolated and identified from *M. giganteus*, *i.e.* ergosterol, 3β-hydroxyergosta-7,22-diene, cerevisterol, 3β-hydroxyergosta-6,8(14),22-triene, 3β-*O*-glucopyranosyl-5,8-epidioxyergosta-6,22-diene, and ergosterol peroxide (Table 7.6).[64]

7.2.6 *Pleurotus* Species

The genus *Pleurotus* (Pleurotaceae family) is a cosmopolitan group of fungi that comprises approximately 30 species and subspecific taxa also known as oyster mushrooms. *Pleurotus* species are members of the phylum Basidiomycota, which is known for producing basidiocarps, *i.e.*, oyster-shaped mushrooms, leading to the use of the term oyster mushrooms. Specifically, the name of the genus *Pleurotus* is derived from two Greek words meaning "side" and "ear", implying its manner of growth. They grow mainly on decayed wood, but they are also capable of parasitizing living trees, infecting them with white rot. Such mushrooms are naturally distributed within almost all latitudes except

Table 7.6 Chemical names and molecular formulas of steroids of *M. giganteus*.

Chemical name	Molecular formula	Ref.
Ergosterol	$C_{28}H_{44}O$	64
3β-Hydroxyergosta-6,8(14),22-triene	$C_{28}H_{44}O$	
3β-Hydroxyergosta-7,22-diene	$C_{28}H_{46}O$	
Ergosterol peroxide	$C_{28}H_{44}O_3$	
Cerevisterol	$C_{28}H_{46}O_3$	
3β-*O*-Glucopyranosyl-5,8-epidioxyergosta-6,22-diene	$C_{34}H_{54}O_8$	

Antarctica, including tropical and subtropical forests.[65,66] Oyster mushrooms are commercially important in the global mushroom market, and many varieties are commercially cultivated on a wide and small scale in several countries.[67] Among the genera, *P. ostreatus* and *P. pulmonarius* are the most consumed. Medicinally, traditional remedies with *Pleurotus* mushrooms have been used to control diabetes, high cholesterol, asthma, arteriosclerosis, and nervous, cardiovascular, and gastrointestinal disorders, among others.[68] However, current data also discuss therapeutic properties like anti-inflammatory, immunomodulatory, anticancer, anti-oxidant, and bacteriostatic effects.[69] To date, five triterpenoids and more than 20 sterols have been reported to this genus. Regarding the triterpenes, oleanolic acid and ursolic acid have been isolated from *P. ostreatus*,[70] while lupeol, 2,3,6,23-tetrahydroxy-urs-12-en-28-oic acid and 2,3,23-trihydroxyurs-12-en-28-oic acid have been isolated from *P. eryngii*.[71] However, the sterol-type structures are the most abundant in the genus and especially compounds with ergostane-type skeletons. Furthermore, campesterol and cholesterol have been also detected in *P. cornucopiae and P. giganteus*, respectively. The chemical names and molecular formulas of steroids and terpenes of the genus *Pleurotus* are illustrated in Table 7.7.

Table 7.7 Chemical names and molecular formulas of terpenes and steroids of genus *Pleurotus*.

Chemical name	Molecular formula	Species	Ref.
Steroids			
Ergosterol	$C_{28}H_{44}O$	*P. cornucopiae*	72–74
Ergosta-5,8,22-trien-3-ol	$C_{28}H_{44}O$		
5,6-Dihydro-ergosterol	$C_{28}H_{46}O$		
Ergosta-7,22-dien-3-ol	$C_{28}H_{46}O$		
Ergosta-14,22-dien-3-ol	$C_{28}H_{46}O$		
Ergosta-7-enol	$C_{28}H_{48}O$		
Campesterol	$C_{28}H_{48}O$		
Ergosta-7,22-diene-3β,5α,6β-triol	$C_{28}H_{46}O_3$		
Ergosta-8(14),22-diene-3β,5α,6β,7α-tetraol	$C_{28}H_{46}O_4$		
5α,9α-Epidioxy-8α,14α-epoxy-(22*E*)-ergosta-6,22-dien-3β-ol	$C_{28}H_{42}O_4$	*P. eryngii*	75
5α,9α-Epidioxy-3β-hydroxy-(22*E*)-ergosta-7,22-dien-6-one	$C_{28}H_{42}O_4$		
Ergosterol	$C_{28}H_{44}O$		
3β,5α-Dihydroxy-(22*E*)-ergosta-7,22-dien-6-one	$C_{28}H_{44}O_3$		
Ergosterol peroxide	$C_{28}H_{44}O_3$		
5α,9α-Epidioxy-(22*E*)-ergosta-7,22-diene-3β,6β-diol	$C_{28}H_{44}O_4$		
3β,5α,9α-Trihydroxy-(22*E*)-ergosta-7,22-dien-6-one	$C_{28}H_{44}O_4$		
3β,5α-Dihydroxyergost-7-en-6-one	$C_{28}H_{46}O_3$		
6β-Acetoxy-(22*E*)-ergosta-7,22-diene-3β,5α-diol	$C_{28}H_{46}O_3$		

Table 7.7 (*Continued*)

Chemical name	Molecular formula	Species	Ref.
Cholesterol	$C_{27}H_{46}O$	*P. giganteus*	76
Ergost-2,5,7,9(11),22-pentaene	$C_{28}H_{40}$	*P. ostreatus*	70 and 77–82
Ergost-5,7,9(11),22-tetraen-3β-ol (dehydroergosterol)	$C_{28}H_{42}O$		
Ergosterol	$C_{28}H_{44}O$		
Ergost-5,7,22-trien-3β-ol	$C_{28}H_{44}O$		
Ergosta-5,7-dienol	$C_{28}H_{46}O$		
Ergosta-7,22-dienol	$C_{28}H_{46}O$		
Ergosta-7-enol (fungisterol)	$C_{28}H_{48}O$		
Ergost-7-en-3β-ol	$C_{28}H_{48}O$		
Ergost-2,5,7,9(11),22-pentaene	$C_{28}H_{40}$	*P. pulmonarius*, *P. sajor-caju*	78
Ergost-5,7,9(11),22-tetraen-3β-ol	$C_{28}H_{42}O$		
Ergost-5,7,22-trien-3β-ol	$C_{28}H_{44}O$		
5α,9α-Epidioxy-(22*E*)-ergosta-7,22-diene-3β,6α-diol	$C_{28}H_{44}O_4$	*P. serotinus*	75
5α,9α-Epidioxy-(22*E*)-ergosta-7,22-diene-3β,6β-diol	$C_{28}H_{44}O_4$		
Triterpenoids			
Lupeol	$C_{30}H_{50}O$	*P. eryngii*	71
2,3,23-Trihydroxy-urs-12-en-28-oic acid	$C_{30}H_{48}O_5$		
2,3,6,23-Tetrahydroxy-urs-12-en-28-oic acid	$C_{30}H_{48}O_6$		
Oleanolic acid	$C_{30}H_{48}O_3$	*P. ostreatus*	70
Ursolic acid	$C_{30}H_{48}O_3$		

7.2.7 *Polyporus* Species

The genus *Polyporus*, commonly known as Agarikon (Agharicon), Ghariqon, and Ghariqun belongs to the phylum Basidiomycota, class Basidiomycetes, order Polyporales, and family Polyporaceae.[83,84] It includes fungi that are characterized by the presence of hymenium, consisting of multiple, small pores.[85] These distinctive pores are the reason why these fungi are called *Polyporus*, since the word has a Greek origin meaning "many pores".[84] The genus is cosmopolitan and includes diverse species. The latest report contained 32 species, with several having been added or resurrected. Most *Polyporus* species grow on dead wood of various tree genera, but several species are found on other substrates. For example, *Polyporus cryptopus* Ellis & Barthol. and *P. rhizophilus* (Pat.) Sacc. occur on grass roots, while *P. phyllostachydis* Sotome, T. Hatt. & Kakish. is found on bamboo roots. *P. squamosus* (Huds.) Fr. frequently grows on living hardwood trees and causes stem rot.

Several species of this genus, such as *P. tuberaster* (Jacq.) Fr. and *P. umbellatus* Fr., are edible and medicinal mushrooms.[85] In general, species of this genus have been of great interest for those looking for novel medicinal compounds due to their rich history of traditional uses by various cultures.[84,86] Furthermore, *Polyporus* species have demonstrated several pharmacological activities such as diuretic, cytotoxic, nephroprotective, hepatoprotective, immuno-enhancing, anti-inflammatory, antioxidative, hair-growing, and antimicrobial activities.[87,88]

Numerous studies have investigated their phytochemical profile, confirming the long history of these species as traditional remedies. The chemical names and molecular formulas of terpenoids and steroids of the genus *Polyporus* are illustrated in Tables 7.8 and 7.9. Sesquiterpenoids have been found in *P. officinalis*[89] and *P. arcularius*.[90] However, *P. officinalis* is a rich source of triterpenoids.[89,91–95] Moreover, two triterpenoids have been isolated from *P. umbellatus*.[96,97] Regarding the sterols, many have been described in *P. umbellatus*[96–101] while two sterols have been isolated from *P. tuberaster*[102] and one sterol has been found in *P. brumalis*.[103]

7.2.7.1 *P. officinalis*

Synonyms: *Fomitopsis officinalis* (Vill.) Bondartsev & Singer (current name),[35] *Fomes officinalis* (Vill.) Bres, *Laricifomes officinalis* (Vill.) Kotl. & Pouzar *P. officinalis* (Vill.) Fr. belongs to the Polyporaceae family. It is a wood-decaying fungus found on living or dead coniferous trees.[89] The fruiting bodies are used as a folk medicine in North America, Western Europe, and Asia, including Mongolia, for the treatment of asthma, cough, gastric cancer, pneumonia, tuberculosis, and rheumatism, as well as for hemostasis.[86,93,104]

7.2.7.2 *P. umbellatus*

P. umbellatus Fries is a saprophytic fungus, growing on withered beech and maple tree roots.[99] It is an edible medicinal mushroom of great interest, mainly used by the food, pharmaceutical, and cosmetic industries. Many products are developed from *P. umbellatus* mycelium, sclerotium, and exopolysaccharides, such as food supplements for the prevention, support, or cure of several diseases. *P. umbellatus* is widely distributed in the temperate regions of the northern hemisphere in Asia, Europe, and North America.[87] *P. umbellatus* is also known as *Grifola umbellata* (Pers.) Pilat and *Dendropolyporus umbellatus* (Pers.) Julich. Its common names include "Zhuling" in Chinese, "Chorei-Maitake" or "Tsuchi-maitake" in Japanese, "Eichhase" in German, "Polypore en ombelle" in French, and "Umbrella polypore" in English.[87]

P. umbellatus has been commonly used in traditional Chinese medicine for over 2500 years.[96] It was referred to in the well-known medical book Shen Nung Pen Tsao Ching between A.D. 25–220 about 1600 years prior to the earliest foreign record (1801). According to Li Shi-chen's Compendium of Materia Medica, *P. umbellatus* "opens up the texture and interspaces of the skin, and muscle, including the sweat pore, cures gonorrheal swelling, beri-beri, leucorrhea, gestational urination, disturbances, fetus swelling and difficulty in urination." Moreover, there are several Chinese traditional herbal formulas that include lignicolous mushrooms as sclerotia of *P. umbellatus*, which can be used to treat conditions such as acute nephritis, systemic dropsy, thirst, difficulty in urination, edema, urination disturbance, heatstroke, watery diarrhea, jaundice, cirrhosis, and ascites. Furthermore, *P. umbellatus* has a diuretic effect on pathogenic dampness and is used in traditional medicine to treat oliguria, edema, diarrhea, strangury, or leucorrhea. The sclerotium of *P. umbellatus* is also a traditional Chinese medicine used for edema and to promote diuresis. Ecdysteroids, which exist in the sclerotia, act as a defensive mechanism and exhibit various biological activities including *in vitro* cytotoxic, *in vivo* antitumor-promoter, and antioxidant effects.[87] Nowadays, the wild sclerotium of *P. umbellatus* is the main source for medicinal uses.[87,99]

7.2.7.3 *P. squamosus*

P. squamosus (Huds.) Fr, also known as dryad's saddle and pheasant's back mushroom, is an edible basidiomycete bracket fungus mushroom that is distributed through Asia, Europe, Australia, and North America. It causes a white rot in both living and dead hardwood trees. It got its name "dryad's saddle" from creatures in Greek mythology known as dryads, which could ride on this mushroom, while the name "pheasant's back" has to do with the pattern of colors found on the bracket and which matches that of a pheasant's back. The fruiting bodies of this fungus occur on living and dead hardwood trees during spring and autumn. Young fruiting bodies are soft but toughen with age. As a result, the young fruiting bodies which have good flavor and taste are preferable for eating. At an early stage, while it is still tender, it is used as a spice.[104]

It has been used as a medicine due to its various biological activities, including diuretic, antibacterial, antifungal, nephroprotective, cytotoxic, immunostimulatory, anti-inflammatory, hepatoprotective, antioxidant, hair-growing, and antimicrobial effects.[88,104]

7.2.7.4 *P. tuberaster*

P. tuberaster Jacquin ex Fr (Polyporaceae) is also known as Canadian tuckahoe and "the stone that grows." Specifically, these names refer to the sclerotium or resting stage of the fungus, a form which preserves the fungus during extreme environmental conditions of dryness and cold. The tuckahoe resembles a large black stone and occurs in the parkland zone of Canada and the northern United States. It appears as a large dark, irregular, slightly flattened spherical mass. The outside of the "stone" is dull dark gray to black in color with numerous ridges, while the internal part of the mass has grains of sand and even small stones embedded in it. Under specific conditions, these "stones" produce fruiting bodies to which the name ("the stone that grows") is referred. The Plains Indians called the tuckahoe a "ground medicine" and used it as a poultice for treating rheumatism or as "Indian bread" although it seems too tough to be consumed. Moreover, the sclerotia are long-lived and persistent in the soil and seem to be remarkably resistant to other fungi and insects.[102]

Table 7.8 Chemical names and molecular formulas of terpenoids of the genus *Polyporus*.

Chemical name	Molecular formula	Species	Ref.
I. Sesquiterpenoids			
3β-Hydroxycinnamolide	$C_{15}H_{22}O_3$	*P. arcularius*	90
3β-Hydroxydihydroconfertifolin	$C_{15}H_{24}O_3$		
Isodrimenediol	$C_{15}H_{26}O_2$		
7-Drimene-3,11,12-triol	$C_{15}H_{26}O_3$		
Laricinolic acid	$C_{15}H_{24}O_3$	*P. officinalis*	89
Fomeffic acid	$C_{15}H_{26}O_4$		
II. Triterpenoids			
3β,6β-Dihydroxy-4,4,14α-trimethyl-Δ^8-5α-pregnene-20-one	$C_{24}H_{38}O_3$	*P. officinalis*	91
Officinalic acid	$C_{30}H_{44}O_6$		105
3-Keto-dehydrosulfurenic acid	$C_{31}H_{46}O_4$		89 and 91
Dehydroeburiconic acid	$C_{31}H_{48}O_3$		89
Dehydrosulfurenic acid	$C_{31}H_{48}O_4$		
Fomefficinol A	$C_{31}H_{48}O_5$		95
Fomefficinol B	$C_{31}H_{48}O_6$		
Fomefficinic acid A	$C_{31}H_{48}O_3$		14
Fomefficinic acid B	$C_{31}H_{48}O_4$		
Fomefficinic acid C	$C_{31}H_{50}O_4$		
Fomefficinic acid D	$C_{31}H_{48}O_4$		
Fomefficinic acid E	$C_{33}H_{48}O_5$		
Fomitopsin C	$C_{34}H_{50}O_7$		89
Fomitopsin D	$C_{34}H_{52}O_8$		95
Fomitopsin E	$C_{34}H_{50}O_7$		
Fomitopsin F	$C_{34}H_{50}O_7$		
Officimalonic acid A	$C_{34}H_{50}O_9$		94
Officimalonic acid B	$C_{34}H_{48}O_8$		
Officimalonic acid C	$C_{34}H_{48}O_9$		
Officimalonic acid D	$C_{35}H_{52}O_9$		
Officimalonic acid E	$C_{35}H_{54}O_8$		
Officimalonic acid F	$C_{35}H_{52}O_8$		
Officimalonic acid G	$C_{35}H_{52}O_8$		
Officimalonic acid H	$C_{35}H_{54}O_7$		
Officimalonic acid I	$C_{34}H_{50}O_8$		93
Officimalonic acid J	$C_{34}H_{50}O_7$		
Officimalonic acid K	$C_{34}H_{48}O_8$		
Officimalonic acid L	$C_{34}H_{50}O_7$		
Officimalonic acid M	$C_{35}H_{50}O_8$		
Officimalonic acid N	$C_{31}H_{48}O_4$		
Officimalonic acid O	$C_{31}H_{48}O_3$		
24-Methylene-lanost-8-ene-3β,21-diol (eburicodiol)	$C_{31}H_{52}O_2$		92
24-Methylene-lanost-8-ene-3β-ol-21-al (eburical)	$C_{31}H_{50}O_2$		
Friedelin	$C_{30}H_{50}O$	*P. umbellatus*	96,97
1β-Hydroxylfriedelin	$C_{30}H_{50}O_2$		
III. Triterpene lactone			
Fomefficinin	$C_{36}H_{54}O_9$	*P. officinalis*	89

Table 7.9 Chemical names and molecular formulas of steroids of the genus *Polyporus*.

Chemical name	Molecular formula	Species	Ref.
Ergosta-5,7,22-trien-3β-ol (ergosterol)	$C_{28}H_{44}O$	*P. brumalis* *P. tuberaster* *P. umbellatus*	96, 97, 100, 102 and 103
Ergosta-4,6,8(14),22-tetraen-3-one (ergone)	$C_{28}H_{40}O$	*P. umbellatus*	96, 100 and 106
Ergost-7,22-dien-3β-ol	$C_{28}H_{46}O$		96 and 106
Ergosta-7,22-dien-3-one	$C_{28}H_{44}O$		96–98
5α,8α-Epidioxy-ergosta-6,22-dien-3-ol (ergosterol peroxide)	$C_{28}H_{44}O_3$	*P. tuberaster* *P. umbellatus*	96, 98, 102 and 106
Ergosta-6,22-dien-3β,5α,6β-triol[a]	$C_{28}H_{46}O_3$	*P. umbellatus*	106
Ergosta-7,22-dien-3β,5α,6β-triol[a]	$C_{28}H_{46}O_3$		96, 97 and 99
19-Norergosta-5,7,9,22-tetraen-3β-ol (neoergosterol)	$C_{27}H_{40}O$		96 and 98
24-Methylcholesta-7,22-dien-3β,5α,6β-triol	$C_{28}H_{46}O_3$		
24-Ethyl-cholesta-7,22-dien-3β,5α,6β-triol	$C_{29}H_{48}O_3$		
Polyporusterone A	$C_{28}H_{46}O_6$		94–97, 99, 101, 104 and 106
Polyporusterone B	$C_{28}H_{44}O_6$		
Polyporusterone C	$C_{28}H_{44}O_6$		96 and 99
Polyporusterone D	$C_{28}H_{44}O_5$		96
Polyporusterone E	$C_{28}H_{44}O_5$		
Polyporusterone F	$C_{28}H_{46}O_5$		
Polyporusterone G	$C_{28}H_{44}O_5$		96 and 99
Polyporusterone I	$C_{30}H_{46}O_7$		96
Polyporusterone II	$C_{29}H_{44}O_6$		
(20*S*,22*R*,24*R*)-16,22-Epoxy-3β,14α,23β,25-tetrahydroxyergost-7-en-6-one	$C_{28}H_{44}O_6$		96 and 101
(23*R*,24*R*,25*R*)-23,26-Epoxy-3β,14α,21α,22α-tetrahydroxyergost-7-en-6-one	$C_{28}H_{44}O_6$		
Polyporoid A	$C_{28}H_{44}O_7$		96
Polyporoid B	$C_{27}H_{44}O_7$		
Polyporoid C	$C_{28}H_{46}O_7$		

[a]Both compounds are identical with different carbon numbering in the cited references.

7.2.8 *Trametes* Species

Genus *Trametes* Fr. is characterized by pileate basidiocarps, poroid hymenophore, di- to tri-mitic hyphal systems, smooth non-dextrinoid spores, and a hymenium usually without true hymenial cystidia.[107–109] It can be found in different types of forest ecosystems (temperate, boreal, tropical areas).[24,107] *Trametes* species are traditionally consumed and are particularly used in Asian countries to enhance longevity and health. The genus is referred to in about 850 records.

There is a difficulty concerning the identification of many *Trametes* species because their spores can develop similar characteristics.[107] They are considered as wood decomposers and are also economically important due to their potential to be used in bioremediation and biodegradation endeavors (including agricultural waste management and wastewater treatment).[109]

The genera *Artolenzites*, *Coriolopsis*, *Coriolus*, *Cubamyces*, *Cyclomycetella*, *Lenzites*, *Poronidulus*, *Pseudotrametes*, and *Pycnoporus* are considered synonyms of *Trametes*.[24]

The chemical names and molecular formulas of terpenes and steroids of the genus *Trametes* are given in Tables 7.10 and 7.11.

7.2.8.1 T. versicolor

Synonyms: Coriolus versicolor (L.) Quél., Polyporus versicolor (L.) Fr., Polystictus versicolor (L.) Fr.

The most widespread representative of this genus is *T. versicolor* (L.) C. G. Lloyd. Some of the common names for this species are "Yun Zhi" (which means cloud-like mushroom) in China, because of its wavy surface covered with fluff, "turkey tail" in European countries because it is fan-shaped, and "kawaratake" (which means mushroom by the riverbank) in Japan.[110] *T. versicolor* is one of the most common species and seems to be associated with many woody plant species. In Asia, it is used as an alternative medicine against diseases and infections[109,111] and it is applied in the treatment of cough, asthma, cancer, hepatitis, bronchitis, and rheumatoid arthritis in China.[112] *T. versicolor* contains dietary fiber and vitamins B1, B2, D2, and C, as well as exhibiting important pharmacological properties attributed to polyphenols, polysaccharides, proteins, and terpenoids. It is reported that sterols and terpenoids are responsible for the antioxidant, antimicrobial, and antiviral activities.[111,113,114] Some studies have mentioned that *T. versicolor* demonstrates health beneficial effects due to its antioxidant, antidiabetic, anti-inflammatory, anticancer, antiviral, immunomodulatory, antiparasitic, and antimicrobial properties, as well as the role of its polysaccharides as prebiotics.[110,111,115] The two best-known bioactive compounds of *T. versicolor* are the polysaccharide krestin (PSK) and the polysaccharopeptide (PSP), which are polysaccharide–protein complexes derived from the mycelia of this fungus and which have exhibited strong biological effects.[116,117]

Among several chemical compounds, terpenoids and steroids have been isolated from different extracts of this species. Chu *et al.* reported the presence of the sesquiterpenes coriolin and deoxycoriolic acid in the mycelia of *Coriolus versicolor* (=*Trametes versicolor*),[110] while Wang *et al.* identified four spiroaxane sesquiterpenes named tramspiroins A–D, two drimane sesquiterpenes named isodrimenediol and funatrol D, as well as a rosenonolactone called 15–16 acetonide E in the ethyl acetate extract of *T. versicolor* fruiting bodies.[118]

Moreover, many triterpenes have been identified in this species. Some examples are betulinic acid, betulin, oleanolic acid, friedelin, glutinol, ursolic acid, corosolic acid, esculentic acid, lupeol, lupenone, and trametenolic acid B.[112,115,119–121] In addition, Gao *et al.* mentioned the triterpenoids, namely ganoderic acid A, lucidumol A, ganoderiol D, ganoderiol F, and ganodermanontriol.[122]

Steroids consist also of a chemical category represented in *T. versicolor*. Some of the sterols identified are ergosterol, ergosterol peroxide (=5α,8α-epidioxy-22*E*-ergosta-6,22-dien-3β-ol, =5α,8α-epidioxyergosta-6,22-dien-3β-ol), stellasterol (=5,6-dihydroergosterol, =5α-ergost-7,22-dien-3β-ol, ergosta-7,22-dien-3β-ol), 5α,8α-epidioxy-22*E*-ergosta-6,9(11)22-trien-3β-ol (=9,11-dehydroergosterol peroxide), 22*E*-7α-methoxy-5α,6α-epoxyergosta-8(14),22-dien-3β-ol, 5α,6α-epoxyergosta-8(14),22-diene-3β,7α-diol, 6-epicerevisterol, cerevisterol (=ergosta-7,22-diene-5α,6β,7β-triol), 8,9-epoxyergosta-5,22-dien-3β,15-diol, ergosta-14,22-dien-3-ol, stigmasta-5α,22-dien-3β-ol, cholest-5-en-3β-ol, ergosta-5,7-dien-3β-ol (=24-methylcholesta-5,7-dien-3β-ol, =22,23-dihydroergosterol), and fungisterol.[30,31,112,115,119,120,123,124]

In addition, volatile compounds have been identified in *T. versicolor*, including linalool, α-cubebene, isoledene, sativene, β-caryophyllene, α-curcumene, β-guaiene, α-muurolene, γ-patchoulene, γ-cadinene, and δ-cadinene.[125]

7.2.8.2 *T. gibbosa*

Synonyms: *Lenzites gibbosa* (Pers.) Hemmi, *Pseudotrametes gibbosa* (Pers.) Bondartsev & Singer *T. gibbosa* (Pers.) Fr. is a saprophytic polypore known as "Lampy Bracket" in English.[126] It is common in Britain and Ireland, and also occurs in temperate regions of mainland Europe and parts of Asia. Its presence in the USA and Canada was described in 2007.[127] *T. gibbosa* has a large, trametoid basidiocarp with thickened base, upper pileus with hirsute zones, and elongated pores.[127] Its dichloromethane extract was found to contain β-bisabolene and α-bisabolol,[128] while its methanol extract revealed the presence of the triterpenes namely lup-20(29)-en-3-one, lupeol, betulin, lanosterol, and the steroids, ergosterol and γ-ergostenol (=fungisterol).[129]

7.2.8.3 *T. orientalis*

Xiao *et al.* investigated the ethyl acetate extract of *T. orientalis* (Yasuda) Imazeki and identified the cytotoxic triterpenes: 24-ene-2′-methyl carboxyacetylquercinicate, 24-ene-carboxyacetyl quercinic acid, 24-ene-dimethylcarboxy-acetylquercinicate, and fomlactone B.[130]

7.2.8.4 *T. suaveolens*

7.2.8.4.1 Synonym: T. bulliardii Fr. *T. suaveolens* (L.) Fr. was found to contain both volatile (β-barbatene, *trans*-nerolidol) and non-volatile secondary metabolites [*e.g.* lupeol, ergosterol peroxide, (22*E*)-ergosta-7,22-dien-3β-ol, (22*E*,24*R*)-5-hydroxy-5α-ergosta-7,22-diene-3,6-dione, 9,11-dehydroergosterol peroxide, 5,8-epidioxy-(5α,8α,22*E*)-ergosta-6,22-dien-3-one, ergosta-7,22-diene-3-one, stigmast-4-en-3-one].[3,131]

Table 7.10 Chemical names and molecular formulas of terpenes of the genus *Trametes*.

Chemical name	Molecular formula	Species	Ref.
I. Volatile sesquiterpenes			
α-Cubebene	$C_{15}H_{24}$	*T. versicolor*	125
Isoledene	$C_{15}H_{24}$		
Sativene	$C_{15}H_{24}$		
β-Caryophyllene	$C_{15}H_{24}$		
β-Guaiene	$C_{15}H_{24}$		
α-Muurolene	$C_{15}H_{24}$		
γ-Patchoulene	$C_{15}H_{24}$		
γ-Cadinene	$C_{15}H_{24}$		
δ-Cadinene	$C_{15}H_{24}$		
α-Curcumene	$C_{15}H_{22}$		
β-Barbatene	$C_{15}H_{24}$	*T. suaveolens*	131
β-Bisabolene	$C_{15}H_{24}$	*T. gibbosa*	128
α-Bisabolol	$C_{15}H_{26}O$		
trans-Nerolidol	$C_{15}H_{26}O$	*T. suaveolens*	131
II. Sesquiterpenoids			
Coriolin	$C_{15}H_{20}O_5$	*T. versicolor*	110
Deoxycoriolic acid	$C_{22}H_{33}O_5$		
Rosenonolactone 15,16-acetonide E	$C_{23}H_{34}O_5$		118
Tramspiroin A	$C_{15}H_{22}O_3$		
Tramspiroin B	$C_{15}H_{22}O_3$		
Tramspiroin C	$C_{15}H_{24}O_2$[a]		
Tramspiroin D	$C_{15}H_{22}O_3$		
Isodrimenediol	$C_{15}H_{26}O_2$		
Funatrol D	$C_{15}H_{26}O_2$		
III. Triterpenoids			
Betulinic acid	$C_{30}H_{48}O_3$	*T. versicolor*	112, 115 and 120
Betulin	$C_{30}H_{50}O_2$		119 and 120
		T. gibbosa	129
Oleanolic acid	$C_{30}H_{48}O_3$	*T. versicolor*	112
Friedelin	$C_{30}H_{50}O$		
Glutinol	$C_{30}H_{50}O$		
Ursolic acid	$C_{30}H_{48}O_3$		
Corosolic acid	$C_{30}H_{48}O_4$		
Esculentic acid	$C_{30}H_{48}O_5$		
Lupeol	$C_{30}H_{50}O$	*T. gibbosa*	129
		T. suaveolens	3
		T. versicolor	112
Lupenone [=lup-20(29)-en-3-one]	$C_{30}H_{48}O$	*T. gibbosa*, *T. versicolor*	112 and 129
Tramenetolic acid B	$C_{30}H_{48}O_3$	*T. cubensis*	3
		T. lactinea	133
		T. versicolor	121
Ganoderic acid A	$C_{30}H_{44}O_7$	*T. versicolor*	122
Ganodermanontriol	$C_{30}H_{48}O_4$		
Ganoderiol D	$C_{30}H_{48}O_5$		
Ganoderiol F	$C_{30}H_{46}O_3$		
Lucidumol A	$C_{30}H_{48}O_4$		

Table 7.10 (*Continued*)

Chemical name	Molecular formula	Species	Ref.
Lanosterol	$C_{30}H_{50}O$	*T. gibbosa*	129
Fomlactone B	$C_{31}H_{48}O_4$	*T. orientalis*	130
24-Ene-2′-methyl carboxy acetylquercinicate	$C_{35}H_{52}O_7$		
24-Ene-carboxyacetylquercinic acid	$C_{34}H_{50}O_7$		
24-Ene-dimethyl-carboxy-acetylquercinicate	$C_{36}H_{54}O_7$		
24(*E*)-3β-Hydroxylanosta-8,24-dien-26-al-21-oic acid	$C_{30}H_{46}O_4$	*T. cubensis*	3
Trametin A	$C_{30}H_{46}O_6$		
Trametin B	$C_{30}H_{46}O_6$		
Hexagonin F	$C_{30}H_{46}O_5$		
Sinensoic acid	$C_{30}H_{48}O_4$		
Eburicoic acid	$C_{31}H_{50}O_3$	*T. versatillis*	134
24-Methylenelanost-8-en-3-ol (=eburicol)	$C_{31}H_{52}O$	*T. cubensis*	3
Eburicodiol	$C_{31}H_{52}O_2$		
Gloeophyllin B	$C_{32}H_{50}O_5$		

[a]Correct formula, herein.

7.2.8.5 *T. cubensis*

7.2.8.5.1 Synonym: Polyporus cubensis Mont. The methanol extract of *T. cubensis* (Mont.) Sacc. has been investigated and two previously undescribed triterpenes, trametin A and trametin B, have been isolated.[3] In the same study, other triterpenes were also isolated from *T. cubensis*, namely hexagonin F, sinensoic acid, 24(*E*)-3β-hydroxylanosta-8,24-dien-26-al-21-oic acid, trametenolic acid, eburicoic acid, 24-methylenelanost-8-en-3-ol, eburicodiol, gloeophyllin B, as well as some steroids, *i.e.*, ergosterol peroxide, ergosterol, and (22*E*,24*R*)-5-hydroxy-5α-ergosta-7,22-diene-3,6-dione.[3]

7.2.8.6 *Other Trametes Species*

Ergosterol and ergosterol peroxide were identified in *T. betulina* (L.) Pilát and *T. hirsuta* (Wulfen) Pilát, while ergosta-14,22-dien-3-ol and ergosta-7,22-dien-3-ol were found in *T. robiniophila* Murr.[124,132] Furthermore, trametenolic acid B, a lanostane-type triterpene, is the main compound of *T. lactinea* (Berk.) Pat.,[133] whereas the triterpene eburicoic acid was found at high levels in *T. versatillis* Berk.[134] Triterpenoids and sterols have also been found in *T. hirsuta*, *T. lactinea*, *T. gibbosa*, *T. orientalis*, *T. elegans*, *T. polyzona*, *T. stwardii*, *T. trogii*, *T. ochracea*, *T. cingulata*, and *T. betulina*.[72,135–144]

Table 7.11 Chemical names and molecular formulas of steroids of the genus *Trametes*.

Chemical name	Molecular formula	Species	Ref.
Cholest-5-en-3β-ol	$C_{27}H_{46}O$	*T. versicolor*	124
Ergosterol	$C_{28}H_{44}O$	*T. betulina*	132
		T. cubensis	3
		T. hirsuta	132
		T. versicolor	30, 31, 120 and 123
Ergosta-7,22-diene-3-one	$C_{28}H_{44}O$	*T. suaveolens*	3
Stellasterol (=5,6-dihydroergosterol, =5α-ergosta-7,22-dien-3β-ol, =ergosta-7,22-dien-3β-ol, =(22*E*)-ergosta-7,22-dien-3β-ol)	$C_{28}H_{46}O$	*T. cubensis*	3
		T. robiniophila	124
		T. suaveolens	3
		T. versicolor	30, 31, 112, 115, 119, 120 and 123
Ergosta-5,7-dien-3β-ol [=24-methylcholesta-5,7-dien-3β-ol, =22,23-dihydroergosterol]	$C_{28}H_{46}O$	*T. versicolor*	31
Ergosta-14,22-dien-3β-ol (3β,5α,22*E*)	$C_{28}H_{46}O$	*T. robiniophila*, *T. versicolor*	124
γ-Ergostenol (=fungisterol)	$C_{28}H_{48}O$	*T. gibbosa*, *T. versicolor*	30, 31 and 129
Stigmast-4-en-3-one	$C_{29}H_{48}O$	*T. suaveolens*	3
Stigmasta-5α,22-dien-3β-ol	$C_{29}H_{48}O$	*T. versicolor*	124
5α,8α-Epidioxy-22*E*-ergosta-6,9(11)22-trien-3β-ol (=9,11- dehydroergosterol peroxide)	$C_{28}H_{42}O_3$	*T. suaveolens*	3
		T. versicolor	112 and 119
(22*E*,24*R*)-5-Hydroxy-5α-ergosta-7,22-diene-3,6-dione	$C_{28}H_{42}O_3$	*T. cubensis*	3
		T. suaveolens	
5,8-Epidioxy-(5α,8α,22*E*)-ergosta-6,22-dien-3-one	$C_{28}H_{42}O_3$		
Ergosterol peroxide (=5α,8α-epidioxy-22*E*-ergosta-6,22-dien-3β-ol, =5α,8α-epidioxyergosta-6,22-dien-3β-ol)	$C_{28}H_{44}O_3$	*T. betulina*	132
		T. cubensis	3
		T. hirsuta	132
		T. suaveolens	3
8,9-Epoxyergosta-5,22-dien-3β,15-diol	$C_{28}H_{44}O_3$	*T. versicolor*	112
5α,6α-Epoxyergosta-8(14),22-diene-3β,7α-diol	$C_{28}H_{44}O_3$		
6-Epicerevisterol	$C_{28}H_{46}O_3$		
Cerevisterol (=ergosta-7,22*E*-diene-3β,5α,6β-triol)	$C_{28}H_{46}O_3$		
22*E*-7α-Methoxy-5α,6α-epoxyergosta-8(14),22-dien-3β-ol	$C_{29}H_{46}O_3$		
Ergosta-7,22-dien-3β-yl palmitate	$C_{44}H_{76}O_2$		115

7.3 Conclusions

Many species of edible mushrooms are important components of daily diets around the world. Over the years, mushrooms have been widely studied for their bioactive compounds along with their important nutritional value.

Among them, terpenoids and steroids have been reported as the main constituents of these fungi, exhibiting a wide range of pharmacological activities. In this chapter the terpenoid and steroid compounds of the genera *Antrodia*, *Coriolus*, *Ganoderma*, *Laetiporus*, *Meripilus*, *Pleurotus*, *Polyporus*, and *Trametes* are thoroughly described and summarized to provide all the available knowledge and highlight the significant chemical profiles of these fungi which have a major role in daily diets.

References

1. M.-C. Lu, M. El-Shazly, T.-Y. Wu, Y.-C. Du, T.-T. Chang and C.-F. Chen, *et al.*, *Pharmacol. Ther.*, 2013, **139**, 124.
2. M. E. Valverde, T. Hernández-Pérez and O. Paredes-López, *Int. J. Microbiol.*, 2015, **2015**, DOI: 10.1155/2015/376387.
3. Y.-C. Li, N. T. Ngan, K.-C. Cheng, T.-L. Hwang, T. D. Thang and N. N. Tuan, *et al.*, *Molecules*, 2021, **26**, DOI: 10.3390/molecules26237311.
4. M. Kozarski, A. Klaus, D. Jakovljevic, N. Todorovic, J. Vunduk and P. Petrović, *et al.*, *Molecules*, 2015, **20**, 19489.
5. G. Anusiya, U. Gowthama Prabu, N. V. Yamini, N. Sivarajasekar, K. Rambabu and G. Bharath, *et al.*, *Bioengineered*, 2021, **12**, 11239.
6. P. M. Dewick, *Medicinal natural products: a biosynthetic approach*, Wiley, Chichester, West Sussex, England, 2nd edn, 2002.
7. F. Cateni, T. Altieri, M. Zacchigna, G. Procida, J. Zilič and D. Žigon, *et al.*, *Nat. Prod. Commun.*, 2015, **10**, 1833.
8. M. Rajchenberg, S. P. Gorjón and M. B. Pildain, *Aust. Syst. Bot.*, 2011, **24**, 111.
9. M. S. Park, Y. Quan, P. E. Jung, S.-Y. Oh, Y. Jang and J.-J. Kim, *et al.*, *Mycobiology*, 2014, **42**, 114.
10. T. T. Chang and W. N. Chou, *Mycol. Res.*, 1995, **99**, 756.
11. M. Geethangili and Y.-M. Tzeng, *Evidence-Based Complementary Altern. Med.*, 2011, **2011**, DOI: 10.1093/ecam/nep108.
12. Y. Kuang, B. Li, Z. Wang, X. Qiao and M. Ye, *Nat. Prod. Rep.*, 2021, **38**, 83.
13. S.-H. Wu, L. Ryvarden and T.-T. Chang, *Bot. Bull. Acad. Sin.*, 1997, **38**, 273.
14. X. Wu, J. Yang, L. Zhou and Y. Dong, *Chem. Pharm. Bull.*, 2004, **52**, 1375.
15. I.-H. Cherng and H.-C. Chiang, *J. Nat. Prod.*, 1995, **58**, 365.
16. T.-H. Lee, C.-C. Chen, J.-J. Chen, H.-F. Liao, H.-S. Chang and P.-J Sung, *et al.*, *Molecules*, 2014, **19**, 21378.
17. N. Ganesan, R. Baskaran, B. K. Velmurugan and N. C. Thanh, *J. Food Biochem.*, 2019, **43**, DOI: 10.1111/jfbc.12936.
18. X. Qiao, W. Song, Q. Wang, K. Liu, Z. Zhang and T. Bo, *et al.*, *RSC Adv.*, 2015, **5**, 47040.
19. J.-L. Ríos, I. Andújar, M.-C. Recio and R.-M. Giner, *J. Nat. Prod.*, 2012, **75**, 2016.
20. CABI databases, http://www.speciesfungorum.org/Index.htm (accessed 2022 Jan 3).
21. T.-T. Chang and W.-N. Chou, *Bot. Bull. Acad. Sin.*, 2004, **45**, 347.

22. C.-C. Shen, Y.-H. Wang, T.-T. Chang, L.-C. Lin, M.-J. Don and Y.-C. Hou, *et al.*, *Planta Med.*, 2007, **73**, 1208.
23. C.-C. Shen, C.-F. Lin, Y.-L. Huang, S.-T. Wan, C.-C. Chen and S.-J. Sheu, *et al.*, *J. Chin. Chem. Soc.*, 2008, **55**, 854.
24. A. Justo and D. S. Hibbett, *Taxon*, 2011, **60**, 1567.
25. Global Biodiversity Information Facility, https://www.gbif.org/species/5245835 (accessed 2022 Jan 3).
26. T. Takeuchi, H. Iinuma, J. Iwanaga, S. Takahashi, T. Takita and H. Umezawa, *J. Antibiot.*, 1969, **22**, 215.
27. W. A. Ayer and L. M. Browne, *Tetrahedron*, 1981, **37**, 2197.
28. G.-Y.-S. Wang, L. M. Abrell, A. Avelar, B. M. Borgeson and P. Crews, *Tetrahedron*, 1998, **54**, 7335.
29. Global Biodiversity Information Facility, https://www.gbif.org/species/5247270 (accessed 2022 Jan 3).
30. H. Yokokawa, *Phytochemistry*, 1980, **19**, 2615.
31. R. J. Cole, B. B. Jarvis and M. A. Schweikert, *Handbook of Secondary Fungal Metabolites*, Academic Press, Cambridge, Massachusetts, USA, 2003, vol. 2.
32. C. Richter, K. Wittstein, P. M. Kirk and M. Stadler, *Fungal Diversity*, 2015, **71**, DOI: 10.1007/s13225-014-0313-6.
33. Z. Du, C.-H. Dong, K. Wang and Y.-J. Yao, Classification, Biological Characteristics and Cultivations of Ganoderma, in *Ganoderma and Health: Biology, Chemistry and Industry*, Springer Singapore, Singapore, 2019.
34. Index Fungorum, http://www.indexfungorum.org/Names/Names.asp (accessed 2022 Jan 3).
35. MYCOBANK Database, https://www.mycobank.org/page/Simple%20names%20search (accessed 2022 Jan 3).
36. T. Luangharn, S. C. Karunarathna, A. K. Dutta, S. Paloi, I. Promputtha and K. D. Hyde, *et al.*, *J. Fungi*, 2021, 7, 819.
37. Z. Lin and B. Yang, *Ganoderma and Health: Biology, Chemistry and Industry*, Springer Singapore, Singapore, 2019.
38. X. Peng and M. Qiu, *Nat. Prod. Bioprospect.*, 2018, **8**, 137.
39. K. S. Bishop, C. H. J. Kao, Y. Xu, M. P. Glucina, R. R. M. Paterson and L. R. Ferguson, *Phytochemistry*, 2015, **114**, 56.
40. T. Gong, R. Yan, J. Kang and R. Chen, Chemical Components of Ganoderma, in *Ganoderma and Health: Biology, Chemistry and Industry*, Springer Singapore, Singapore, 2019.
41. S. Baby, A. J. Johnson and B. Govindan, *Phytochemistry*, 2015, **114**, 66.
42. I. C. F. R. Ferreira, S. A. Heleno, F. S. Reis, D. Stojkovic, M. J. R. P. Queiroz and M. H. Vasconcelos, *et al.*, *Phytochemistry*, 2015, **114**, 38.
43. S. Khatua, S. Ghosh and K. Acharya, *Pharmacogn. J.*, 2017, **9**, 1.
44. J. Petrović, J. Glamočlija, T. Ilić-Tomić, M. Soković, D. Robajac and O. Nedić, *et al.*, *Int. J. Biol. Macromol.*, 2020, **148**, 129.
45. U. Grienke, M. Zöll, U. Peintner and J. M. Rollinger, *J. Ethnopharmacol.*, 2014, **154**, 564.

46. J.-B. He, J. Tao, X.-S. Miao, W. Bu, S. Zhang and Z.-J. Dong, *et al.*, *Fitoterapia*, 2015, **102**, 1.
47. J.-B. He, J. Tao, X.-S. Miao, Y.-P. Feng, W. Bu and Z.-J. Dong, *et al.*, *J. Asian Nat. Prod. Res.*, 2015, **17**, 1054.
48. J.-W. Zhang, G.-L. Wen, L. Zhang, D.-M. Duan and Z.-H. Ren, *Bangladesh J Pharmacol.*, 2015, **10**, 844.
49. C. Chepkirui, J. C. Matasyoh, C. Decock and M. Stadler, *Phytochem. Lett.*, 2017, **20**, 106.
50. E. David, C. Barrera, I. Jeannette and N. Ramírez, *Rev. Fac. Cienc. Básicas*, 2011, 7, 298.
51. K. Hassan, B. Matio Kemkuignou and M. Stadler, *Molecules*, 2021, **26**, 7090.
52. F. León, J. Quintana, A. Rivera, F. Estévez and J. Bermejo, *J. Nat. Prod.*, 2004, **67**, 2008.
53. E. Saba, Y. Son, B. R. Jeon, S.-E. Kim, I.-K. Lee and B.-S. Yun, *et al.*, *Mycobiology*, 2015, **43**, 131.
54. J. Wang, P. Zhang, H. He, X. Se, W. Sun and B. Chen, *et al.*, *Naunyn-Schmiedeberg's Arch. Pharmacol.*, 2017, **390**, 845.
55. X. Yin, Z.-H. Li, Y. Li, T. Feng and J.-K. Liu, *J. Asian Nat. Prod. Res.*, 2015, **17**, 793.
56. Q. Khalilov, L. Li, Y. Liu, W. Liu, S. Numonov and H. A. Aisa, *et al.*, *Steroids*, 2019, **151**, DOI: 10.1016/j.steroids.2019.108468.
57. D. C. B. Ericsson and J. N. R. Ivonne, *Chem. Nat. Compd.*, 2009, **45**, 193.
58. W. Krzyczkowski, E. Malinowska, P. Suchocki, J. Kleps, M. Olejnik and F. Herold, *Food Chem.*, 2009, **113**, 351.
59. M. Martinez, S. T. Alvarez, M. G. Campi, J. A. Bravo and J. L. Vila, *Rev. Boliv. Quim.*, 2015, **32**, 90.
60. S. Gunasekaran, V. Mayakrishnan, S. Al-Ghamdi, M. Alsaidan, A. Geddawy and M. A. Abdelaziz, *et al.*, *J. King Saud Univ., Sci.*, 2021, **33**, DOI: 10.1016/j.jksus.2021.101551.
61. K. Yoshikawa, K. Matsumoto and S. Arihara, *J. Nat. Prod.*, 1999, **62**, 543.
62. K. Yoshikawa, K. Matsumoto, C. Mine, S. Bando and S. Arihara, *Chem. Pharm. Bull.*, 2000, **48**, 1418.
63. M. J. Larsen and F. F. Lombard, *Mycologia*, 1988, **80**, 612.
64. A. Sárközy, Z. Béni, M. Dékány, Z. P. Zomborszki, K. Rudolf and V. Papp, *et al.*, *Molecules*, 2020, **25**, 1395.
65. A. T. Pandey, I. Pandey, P. Kerkar and M. P. Singh, *Vegetos*, 2021, **34**, 619.
66. K. Bhardwaj, A. Sharma, N. Tejwan, S. Bhardwaj, P. Bhardwaj and E. Nepovimova, *et al.*, *J. Fungi*, 2020, 6, 351.
67. E. Adebayo, J. Oloke, A. A. Adegunlola and B. Olukemi, *J. Microbiol. Biotechnol. Res.*, 2012, **2**, 366.
68. G. Venturella, E. Palazzolo, F. Saiano and M. L. Gargano, *Int. J. Med. Mushrooms*, 2015, **17**, 199.
69. M. Grabarczyk, W. Mączka, K. Wińska and C. Uklańska-Pusz, *Biotechnol. Food Sci.*, 2020, 13.

70. G. Koutrotsios, N. Kalogeropoulos, P. Stathopoulos, A. C. Kaliora and G. I. Zervakis, *World J. Microbiol. Biotechnol.*, 2017, **33**, 98.
71. Z. Xue, L. Zhai, W. Yu, H. Wang, X. Kou and L. Peng, *et al.*, *J. Food Biochem.*, 2015, **39**, 19.
72. R. Parmar and D. Kumar, *Biol. Forum - Int. J.*, 2015, **7**, 1057.
73. S. Wang, L. Bao, J. Han, Q. Wang, X. Yang and H. Wen, *et al.*, *J. Nat. Prod.*, 2013, **76**, 45.
74. S. Wang, L. Bao, F. Zhao, Q. Wang, S. Li and J. Ren, *et al.*, *J. Agric. Food Chem.*, 2013, **61**, 5122.
75. Y. Yaoita, Y. Yoshihara, R. Kakuda, K. Machida and M. Kikuchi, *Chem. Pharm. Bull.*, 2002, **50**, 551.
76. C.-W. Phan, W.-L. Wong, P. David, M. Naidu and V. Sabaratnam, *BMC Complementary Altern. Med.*, 2012, **12**, 102.
77. I. Sarma, *Adv. Appl. Sci. Res.*, 2017, **8**, 17.
78. I. J. Nieto and A. C. Chegwin, *J. Chil. Chem. Soc.*, 2013, **58**, 1580.
79. F. Souilem, â. Fernandes, R. C. Calhelha, J. C. M. Barreira, L. Barros and F. Skhiri, *et al.*, *Food Chem.*, 2017, **230**, 40.
80. T. Kikuchi, N. Motoyashiki, T. Yamada, K. Shibatani, K. Ninomiya and T. Morikawa, *et al.*, *Int. J. Mol. Sci.*, 2017, **18**, 2479.
81. T. Kikuchi, K. Kitaura, A. Katsumoto, J. Zhang, T. Yamada and R. Tanaka, *Fitoterapia*, 2018, **129**, 108.
82. T. Kikuchi, M. Isobe, S. Uno, Y. In, J. Zhang and T. Yamada, *Bioorg. Chem.*, 2019, **89**, DOI: 10.1016/j.bioorg.2019.103011.
83. M. M. Muslat and M. N. Owaid, *Int. J. Environ.*, 2015, **4**, 185.
84. S. Olimat, *Prac. Clin. Invest.*, 2020, **4**, DOI: 10.46619/pci.2021.4.1032.
85. K. Sotome, T. Hattori, Y. Ota, C. To-anun, B. Salleh and M. Kakishima, *Mycologia*, 2008, **100**, 603.
86. S. Naranmandakh, T. Murata, B. Odonbayar, K. Suganuma, J. Batkhuu and K. Sasaki, *J. Nat. Med.*, 2018, **72**, 523.
87. A. R. Bandara, S. Rapior, D. J. Bhat, P. Kakumyan, S. Chamyuang and J. Xu, *et al.*, *Cryptogam.: Mycol.*, 2015, **36**, 3.
88. A. Mocan, â. Fernandes, L. Barros, G. Crişan, M. Smiljković and M. Soković, *et al.*, *Food Funct.*, 2018, **9**, 160.
89. W. Feng and J.-S. Yang, *J. Asian Nat. Prod. Res.*, 2015, **17**, 1065.
90. W. F. Fleck, B. Schlegel, P. Hoffmann, M. Ritzau, S. Heinze and U. Gräfe, *J. Nat. Prod.*, 1996, **59**, 780.
91. C. G. Anderson, W. W. Epstein and G. V. Lear, *Phytochemistry*, 1972, **11**, 2847.
92. C. G. Anderson and W. W. Epstein, *Phytochemistry*, 1971, **10**, 2713.
93. J. Han, W. Liu, M. Li, Y. Gu, Y. Zhang and T. Yuan, *Molecules*, 2020, **25**, 4807.
94. J. Han, L. Li, J. Zhong, Z. Tohtaton, Q. Ren and L. Han, *et al.*, *Phytochemistry*, 2016, **130**, 193.
95. B. Muszyńska, A. Fijałkowska, K. Sułkowska-Ziaja, A. Włodarczyk, P. Kaczmarczyk and E. Nogaj, *et al.*, *Chem. Biodiversity*, 2020, **17**, DOI: 10.1002/cbdv.202000213.

96. Y.-Y. Zhao, *J. Ethnopharmacol.*, 2013, **149**, 35.
97. W. Zhou and S. Guo, *Chem. Nat. Compd.*, 2009, **5**, 124.
98. H. S. Lee, I. H. Hwang, J. A. Kim, J. Y. Choi, T.-S. Jang and H. Osada, *et al.*, *ChemInform*, 2011, **42**, 697.
99. Y. Sun and K. Yasukawa, *Bioorg. Med. Chem. Lett.*, 2008, **18**, 3417.
100. Y. Zhao, R. Xie, X. Chao, Y. Zhang, R. Lin and W. Sun, *J. Ethnopharmacol.*, 2009, **126**, 184.
101. W.-W. Zhou, W.-H. Lin and S.-X. Guo, *Chem. Pharm. Bull.*, 2007, **55**, 1148.
102. W. A. Ayer, M. Sun, L. M. Browne, L. S. Brinen and J. Clardy, *J. Nat. Prod.*, 1992, **55**, 649.
103. M. P. Jones, A. C. Lawrie, T. T. Huynh, P. D. Morrison, A. Mautner and A. Bismarck, *et al.*, *Process Biochem.*, 2019, **80**, 95.
104. W. Elkhateeb, G. Daba, M. Elnahas and P. Thomas, *Egypt. Pharm. J.*, 2019, **18**, 285.
105. W. W. Epstein, F. W. Sweat, G. VanLear, F. M. Lovell and E. J. Gabe, *J. Am. Chem. Soc.*, 1979, **101**, 2748.
106. Y.-Y. Zhao, X. Chao, Y. Zhang, R.-C. Lin and W.-J. Sun, *Planta Med.*, 2010, **76**, 1755.
107. L. Ryvarden, *Genera of polypores: Nomenclature and taxonomy*, Fungiflora A/S, Oslo Norway, 1991.
108. K. S. Ko and H. S. Jung, *Antonie van Leeuwenhoek*, 1999, **75**, 191.
109. K. K. Hapuarachchi, S. C. Karunarathna, X.-H. Xu, A. K. Dutta, P. Phengsintham and K. D. Hyde, *et al.*, *Chiang Mai J. Sci.*, 2021, **48**, 674.
110. K. K. W. Chu, S. S. S. Ho and A. H. L. Chow, *J. Clin. Pharmacol.*, 2002, **42**, 976.
111. A. Bains, P. Chawla, S. Kaur, A. Najda, M. Fogarasi and S. Fogarasi, *Materials*, 2021, **14**, 7640.
112. M. Jin, W. Zhou, C. Jin, Z. Jiang, S. Diao and Z. Jin, *et al.*, *Nat. Prod. Res.*, 2019, **33**, 2422.
113. F. S. Reis, A. Martins, M. H. Vasconcelos, P. Morales and I. C. F. R. Ferreira, *Trends Food Sci. Technol.*, 2017, **66**, 48.
114. M. Ramos, N. Burgos, A. Barnard, G. Evans, J. Preece and M. Graz, *et al.*, *Food Chem.*, 2019, **292**, 176.
115. C. Bi, X.-Y. Guo and Q.-M. Che, *J. Chin. Pharm. Sci.*, 2007, **16**, 38.
116. K. Pallav, S. E. Dowd, J. Villafuerte, X. Yang, T. Kabbani and J. Hansen, *et al.*, *Gut Microbes*, 2014, **5**, 458.
117. J. M.-F. Wan, Polysaccaride Krestin (PSK) and Polysaccharopeptide PSP, in *Handbook of Biologically Active Peptides*, Academic Press, Cambridge, Massachusetts, USA, 2nd edn, 2013.
118. S.-R. Wang, L. Zhang, H.-P. Chen, Z.-H. Li, Z.-J. Dong and K. Wei, *et al.*, *Fitoterapia*, 2015, **105**, 127.
119. S. Hybelbauerová, J. Sejbal, M. Dračínský, A. Hahnová and B. Koutek, *Chem. Biodiversity*, 2008, **5**, 743.
120. E. Habibi, S. E. Sadat-Ebrahimi, S. A. Mousazadeh and Y. Amanzadeh, *Int. J. Med. Mushrooms*, 2015, **17**, 255.

121. V. Leliebre-Lara, L. Monzote Fidalgo, E.-M. Pferschy-Wenzig, O. Kunert, C. Nogueiras Lima and R. Bauer, *Molecules*, 2016, **21**, DOI: 10.3390/molecules21081045.
122. J. Gao, N. Sato, M. Hattori and C.-M. Ma, *J. Pharm. Biomed. Anal.*, 2013, **74**, 246.
123. M. S. Borlagdan, M. E. G. De Castro, I. A. van Altena and C. Y. Ragasa, *Res. J. Pharm. Biol. Chem. Sci.*, 2017, **8**, 740.
124. S.-J. Yuan, X.-S. He, X.-H. Yuan and Y. Huang, *South Asian Res. J. Nat. Prod.*, 2019, **2**, 1.
125. K. Drilling and K. Dettner, *Chemoecology*, 2009, **19**, 109.
126. Global Biodiversity Information, https://www.gbif.org/species/9208081 (accessed 2022 Jan 3).
127. J. Kout and J. Vlasák, *Can. J. Bot.*, 2007, **85**, 342.
128. P. Thakeow, S. Angeli, B. Weissbecker and S. Schutz, *Chem. Senses*, 2008, **33**, 379.
129. V. P. Kurchenko, N. V. Sushinskaya, I. S. Kiseleva and A. A. Ermoshin, *AIP Conf. Proc.*, 2022, **2390**, 030045.
130. K.-M. Xiao, J.-M. Wang, L.-Y. Zhou, L. Guo, D.-C. Tan and Q. Yan, *et al.*, *Nat. Prod. Commun.*, 2020, **15**, DOI: 10.1177/1934578X20921621.
131. J. Rösecke, M. Pietsch and W. A. König, *Phytochemistry*, 2000, **54**, 747.
132. R. Nowak, N. Nowacka-Jechalke, W. Pietrzak and U. Gawlik-Dziki, *Food Chem.*, 2022, **369**, DOI: 10.1016/j.foodchem.2021.130927.
133. Q. Zhang, N. Huang, J. Wang, H. Luo, H. He and M. Ding, *et al.*, *Fitoterapia*, 2013, **89**, 210.
134. B.-J. Ma and Y. Ruan, *Z. Naturforsch., C: J. Biosci.*, 2007, **62**, 909.
135. G. Ren, X. Y. Liu, H. K. Zhu, S. Z. Yang and C. X. Fu, *Fitoterapia*, 2006, **77**, 408.
136. T. Mizuno, H. Saito, T. Nishitoba and H. KaWagishi, *Food Rev. Int.*, 1995, **11**, 23.
137. E. E. Essien and S. M. Akpan, *J. Chem. Pharm.*, 2014, **6**, 280.
138. I. A. Sheikh, D. Vyas, R. Lone and V. Singh, *World J. Pharm. Pharm. Sci.*, 2015, **4**, 1695.
139. S. Awala and V. Oyetayo, *J. Adv. Med. Pharm. Sci.*, 2015, **4**, DOI: 10.9734/JAMPS/2015/20059.
140. G. Melappa, A. Roshan, C. Nithi, T. S. Mohummed and C. Poojari CC, *Pharmacogn. J.*, 2015, 7, 136.
141. A. T. Adeyelu, V. Olusegun and S. I. Awala, *J. Altern. Complement. Med.*, 2016, **1**, 4.
142. A. Knežević, M. Stajić, I. Sofrenić, T. Stanojković, I. Milovanović and V. Tešević, *et al.*, *PLoS One*, 2018, **13**, DOI: 10.1371/journal.pone.0203064.
143. O. O. Oduntan, *Int. J. Innovative Pharm. Sci. Res.*, 2020, **5**, 1206.
144. S. Q. Suliaman, S. H. AL-Abbasi, Y. H. Mahmood and H. A. AL-Azzawi, *Cell Arch.*, 2021, **21**, 4533.

CHAPTER 8

Cultivation of Mushrooms Widely Appreciated by Consumers

DIEGO CUNHA ZIED,*[a] ANDRÁS GEÖSEL[b] AND ARTURO PARDO-GIMÉNEZ[c]

[a] Universidade Estadual Paulista, Faculdade de Ciências Agrárias e Tecnológicas, Dracena, Brazil; [b] Hungarian University of Agriculture and Life Sciences, Institute of Horticultural Sciences, Budapest, Hungary; [c] Centro de Investigación, Experimentación y Servicios del Champiñón, Quintanar del Rey, Spain
*Emails: dczied@gmail.com; diego.zied@unesp.br

8.1 Introduction

Mushrooms are a classic example of the transformation of lignocellulolytic materials into food.[1] Other eukaryotes also have this capacity, such as animals (cattle, pig, chicken, and others) which feed on grass/straws and provide the population with meat and other derivatives (milk, eggs, *etc.*).[2,3] However, mushrooms continue to be differentiated, as they are capable of transforming sawdust into food.[4,5] Which animal would be able to feed on sawdust?

They are also fantastic for their nutritional, pharmacological, and medicinal properties, as mentioned in other chapters of this book.[6,7] All this is without the emission of methane, unlike animals. In addition, mushrooms can be produced in small spaces, using residues, obtaining high bioconversion,

Food Chemistry, Function and Analysis No. 36
Edible Fungi: Chemical Composition, Nutrition and Health Effects
Edited by Dejan Stojković and Lillian Barros

Published by the Royal Society of Chemistry, www.rsc.org

without any risk of transmitting zoonoses to humans, with a quick economic return due to their short cultivation cycle.

Currently, the most cultivated mushrooms belong to the genus *Lentinula* (see Figure 8.1A), *Auricularia*, *Pleurotus* (see Figure 8.1B), and *Agaricus*. Although *Lentinula edodes* (shiitake) is the most produced mushroom in the world, *Agaricus bisporus* is the species cultivated in a greater number of countries, which has provided greater technological development due to the great diversity of researchers studying these mushrooms.[8] Currently, there are companies specialized in all stages of production, from the construction of large industries to small sensors that control irrigation.[9]

Table 8.1 presents the 15 most cultivated and consumed mushrooms in the world. Each country adopts a cultivation system with greater or lesser

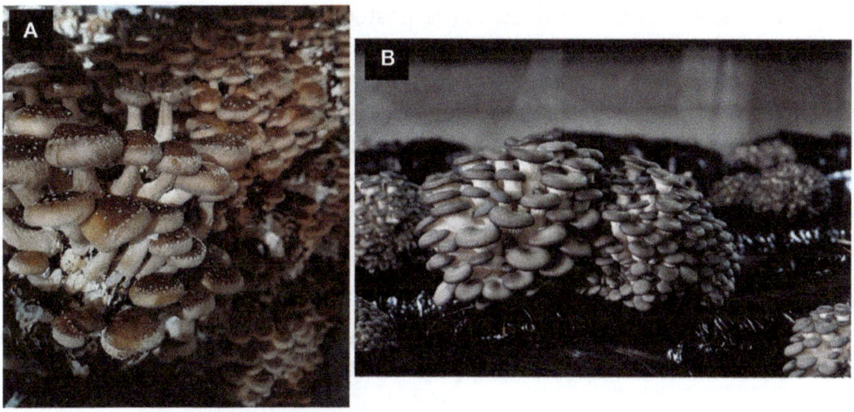

Figure 8.1 (A) *L. edodes* cultivation in sawdust, with 4 production flushes, obtaining 42% yield in 160 days of crop cycle; (B) *P. ostreatus* cultivation on wheat straw, with 3 production flushes, obtaining 36% yield in 45 days of crop cycle.

Table 8.1 Most cultivated and consumed mushrooms in the world.

Number	Species	Popular name
1	*Lentinula edodes*	Shiitake
2	*Auricularia auricula*	Black wood ear
3	*Pleurotus ostreatus*	Oyster mushroom (dark gray-blue pileus)
4	*Agaricus bisporus*	Button mushroom (white color)
5	*Pleurotus pulmonarius*	Indian oyster (pale pileus)
6	*Agaricus bisporus*	Portobello (cream color)
7	*Auricularia polytricha*	Cloud ear fungus
8	*Flammulina velutipes*	Enoki
9	*Volvariella volvacea*	Straw mushroom
10	*Hypsizygus marmoreus*	Shimeji Buna
11	*Pleurotus eryngii*	King oyster
12	*Pleurotus djamor*	Pink oyster mushroom
13	*Pleurotus citrinopiliatus*	Golden oyster mushroom
14	*Pholiota nameko*	Nameko
15	*Grifola frondosa*	Maitake

technology, according to the available financial resources. Mushrooms are very versatile, depending on the cultivation substrate (nutrition) and environmental growth conditions (temperature and relative humidity). Thus, it is possible to cultivate different species of mushrooms in diverse situations, regions, and localities.

Briefly, there are species of mushrooms that demand greater technological investment, as they are grown on sterilized substrates, inoculated in a laminar flow, and cultivated under extremely clean conditions at low temperatures (for example, *F. velutipes*, *P. eryngii*, *P. naneko*, and *H. marmoreus*). On the other hand, there are species that demand less technological investment, which can be cultivated in pasteurized substrate or treated with an alkaline solution and cultivated at higher temperatures (for example, the species of *Pleurotus* and *V. volvacea*). In the middle of this short description, there are the *Agaricaceos* mushrooms, which are cultivated in a pasteurized substrate and, depending on the country of cultivation, extremely advanced technology is adopted.

Several publications describe how to cultivate mushrooms, following the main steps, such as spawn production, substrate preparation, inoculation, and spawn run, the addition of the casing layer (optional), induction of primordial, and harvest (see Figure 8.2).[10–15] In this chapter, we do not discuss how these steps are carried out, but instead the main trends and technologies that are emerging in the market to make mushroom cultivation increasingly competitive and viable from an agronomic, environmental, and economic point of view. The main technologies mentioned in this chapter are applied to the *A. bisporus* strains, popularly known as button mushroom and Portobello, although in general they can be applied to other cultivated mushrooms.

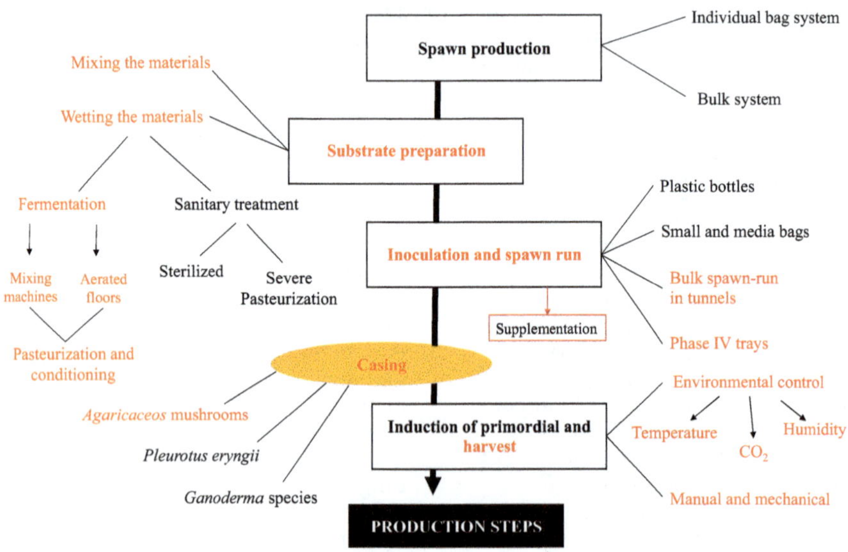

Figure 8.2 Main steps of mushroom production, the letters highlighted in red refer to the main steps and processes that are described in this chapter.

8.2 Main Technological Advances in Commercial *Agaricus* spp. Production

Agaricus is the fifth major genus of cultivated edible mushrooms, contributing about 15% of the global production.[9] Among the different cultivated species, *Agaricus bisporus* is the mushroom whose cultivation is most widespread geographically. Two situations are presented as the main ones responsible for this development and improvements in production aspects.

In the first place, together with the obtaining of new hybrid strains of mycelium, we find the centralized production of composts, basic substrates for cultivation, in large high-capacity composting factories, whose facilities have advanced technologically at an increasing rate, although without substantially altering the chemical, physical, and biological bases of the composting process in two phases, as described in the works of Fermor *et al.* and Gerrits.[16,17] Secondly, the improvement of the growing rooms, equipped with climate control systems, varying degrees of mechanization, and high production capacity, the best example of which is the growth model in shelves developed in the Netherlands and currently used in a large number of producing countries.

8.2.1 Substrate Production

The button mushroom is heterotrophic; the carbon produced by plants serves as a source of nutrients for the mushroom. In cultivation, it requires a special substrate (compost) with physical features and ideal chemical composition for the mycelium. Unlike what occurs with other crops, mushrooms cannot be supplemented with additional nutrients during the cultivation process, the compost and casing must contain and provide all the necessary nutrients and inorganic substances, such as water, carbon, nitrogen, and minerals. Nowadays, most producers use phase III compost in bulk in year-round cultivation to supply the market with fresh mushrooms.[18,19]

The carbon source is essential in mushroom cultivation and originates from agricultural residues of plant production. The straw of cereal contains cellulose, hemicellulose, and lignin. In the compost preparation process, these materials are composted, and the mushroom can easily utilize them.[20] The nitrogen necessary for the production of button mushrooms is supplied by manure, a byproduct of animal husbandry. Due to its accessibility and for logistic reasons, mainly chicken manure is used for this purpose, but because of its composition and structure, horse manure can also serve as a nitrogen source.[21] Aside from the significant cost of using artificial chemical compounds, such as urea for compost preparation, they do not have a positive effect on the structure either. The main reason for their application is that they increase the activity of the compost. To achieve optimal compost structure and pH, water and gypsum are added to the mixture as well.[22]

In phase I, the compost ingredients are thoroughly mixed, and the breakdown of the raw materials begins. In phase II, the compost goes through a pasteurization process, which also means that the breakdown process comes to an end. During phase III, the mycelium colonizes the entire mass of the compost. The aim of composting is to release the nutrients from the raw materials, to change them into a form that is easily accessible for the button mushroom, and to eliminate competing organisms, such as pests and diseases.[23]

Due to the availability of ingredients and machinery, the exact steps and progress of compost preparation can differ from one compost plant to another. In the following sections, we introduce the general process of preparing phase III compost.

8.2.1.1 Phase I in Aerated Bunkers/Indoors

During the first phase of compost preparation, the ingredients are moistened, mixed, and fermented. First, the wheat is soaked in water, which is called pre-wetting.[24] Excess water is continuously collected and re-used for this purpose since it contains a variety of microbes that can facilitate composting. Pre-wetting takes a few days, depending on the quality of the wheat. In the following stage the nitrogen sources, such as chicken and horse manure, are mixed into the material.[25] Manure may need to be ground first if the structure is dense. Gypsum ($CaSO_4$) is added next, either mixed into the manure, or individually (see Figure 8.3A).

This first stage of composting requires 5 to 5.5 tons of water to achieve optimal water content (72–76%) in the mixture before pasteurization. Depending on its nitrogen content, approximately 750 kg of chicken manure is added to 1 ton of dry wheat. This value results in 3.3 to 4 tons of phase I compost. The better the wheat quality, the higher this value becomes. Different machines and equipment can be used for the mixing of raw materials, but the product should always be a homogeneous (both on a micro and macro scale) mixture (see Figure 8.3B).

In the next step of compost preparation, the fermentation process begins. During this stage the nutrients in the raw material continue to break down,

Figure 8.3 (A) Gypsum and manure mixed. (B) Part of a raw material mixing machine, there are 3 lifting points that have the function of improving homogenization and adding some organic or synthetic material.

thus they become accessible and can be utilized by the mycelia of the button mushroom, thereby intensifying the cultivation process.[26] During fermentation, the carbohydrates of the wheat are processed by the microbiota (bacteria, fungi, actinobacteria, *etc.*) of the compost.[27] At the same time, the nitrogen content of the manure turns into ammonia. Biological degradation produces heat, which results in a 65 °C compost temperature. The following chemical processes increase the temperature further, even up to 80 °C. It is not recommended that the temperature exceed 80 °C due to future contamination problems.[25]

Biological degradation requires vast amounts of oxygen, as the microbes need it for their metabolism. For that reason, it is advised using facilities with aeration floors for compost preparation. This type of flooring allows fresh air to be ventilated through gaps in the concrete, which provides an optimal oxygen level in the mixture that contributes to homogeneity and a quicker composting process. To be able to preserve the energy produced in the compost, and to eliminate unfavorable odors, fermentation can be done in bunkers, where spigot floors improve the oxygen level (see Figure 8.4).

In some countries, the entire compost preparation process has to be done indoors due to strict environmental regulations. To achieve optimal

Figure 8.4 (A) Concrete floor, with the spigots coupled to air ducts, where you can see only the surface in each channel (red arrows); (B) overview of bunkers containing approximately 150 tons of compost; (C) mechanical platform responsible for uniform distribution of the compost inside the bunker, avoiding compaction and anaerobic zones in the compost.

Figure 8.5 Indoor compost plant, through closed phase 1, with significantly restricted emission of gases and odor to an outdoor area.

homogeneity and structure, the mixture must be turned multiple times throughout the process (see Figure 8.5).

8.2.1.2 Phase III (Bulk Spawn-run in Tunnels)

The end of phase I is signaled by a decrease in microbe activity, and the changes in the physical and chemical properties of the compost. In phase II, a controlled heat treatment, a pasteurization process sets the final compost structure and quality. Pasteurization takes place in tunnels, where filling machines pile the phase I compost on the special aerated floors of the tunnel for better oxygen supply. During the heat treatment, the remaining pests (*e.g.* mites, flies, nematodes) and pathogens (such as mold spores) are destroyed, and most of the ammonia is eliminated.[28] In the first part of the pasteurization process, which is called leveling, the temperature of the compost mass is evened out by intensive airflow. The warming up of the compost is the next step, when the material in the tunnels is heated up to 56–60 °C.[17] This temperature value is called peak-heat, and is maintained for 6 to 10 hours.

This is followed by the conditioning process when the compost temperature is set at 46–48 °C for a few days, during which time the remaining ammonia gets into proteins to serve as nutrients for the mycelia. The final step of pasteurization is the cooling-down stage, which precedes spawning. At this point, the ammonia level should be below 5 ppm.[25] An the end of the process, the resulting material is a selective, homogeneous compost with optimal physical and chemical properties, ideal for supporting the development of button mushroom mycelia.[29] The big technological advance right now is the emptying of the compost from the tunnel, spawning and filling into a clean tunnel made with a special machine (see Figure 8.6). This equipment makes the spawning process fast and ensures a homogeneous mixture. Another big advantage of this system is the reduction of plastic used to pack the compost.

Modern composters have a fully equipped chemical laboratory to measure the most important compost parameters. The NIR (near-infrared spectroscopy) method makes the measurement easy, cheap, and reliable.[30]

Figure 8.6 (A) Emptying of phase 2 compost from tunnel, spawning, and filling back into another one. (B) Plastic air ducts that distribute clean air throughout the spawning area. (C) Tunnel used for phases 2 and 3 being washed, note the nylon nets that are used to remove the compost.

Without any pre-treatment, the sampled compost is filled into the machine and, within minutes, the results are visualized. By using continuous chemical and physical checks of the developing compost (at bunker filling, before spawning, *etc.*) the necessary treatment (like watering) can be done precisely. In Table 8.2, valid results of compost sampling during the process can be seen applying the NIR method.

Grain spawn is the most commonly used type, while synthetic spawn, which was developed in recent years, is not prevalent. During spawning, the grains colonized by the mycelia of the mushroom are mixed evenly into the compost. The ratio of spawn can be adjusted by how fast the conveyor belt transferring the compost is set. The spawned compost is then moved by a filling cassette into a clean tunnel with the same type of aerated or spigot floor system as the

Table 8.2 Chicken manure and wheat straw-based compost chemical parameters in phase 2 and phase 3 of compost production (measured by NIR method).

	Phase 2 compost				Phase 3 compost			
Parameter (measured by NIR)	Average	Min.	Max.	Relative standard deviation (%)	Average	Min.	Max.	Relative standard deviation (%)
Water content, %	68.62	64.34	74.35	*0.08*	57.07	55.23	58.89	*3.17*
Dry matter, %	31.38	25.65	35.66	*0.16*	42.93	41.11	44.77	*4.22*
Ash (wet), %	8.55	6.77	9.79	*0.18*	12.58	11.79	13.54	*6.62*
Ash (dm), %	27.19	26.40	27.70	*0.03*	29.28	28.68	30.23	*2.46*
Nitrogen (wet), %	0.59	0.43	0.70	*0.24*	1.04	0.98	1.14	*7.11*
Nitrogen (dm), %	1.87	1.66	1.98	*0.10*	2.43	2.37	2.55	*3.47*
pH (NIR), %	7.82	7.73	7.88	*0.01*	5.97	5.80	6.10	*2.36*
Lignin (dm), %	11.67	11.11	12.13	*0.04*	12.04	11.73	12.29	*2.01*
Fiber (dm), %	29.47	28.24	30.58	*0.04*	19.81	18.35	20.62	*5.41*
NDF (dm), %	49.66	48.66	50.93	*0.02*	40.27	38.42	41.28	*3.33*
ADF (dm), %	38.29	36.64	39.81	*0.04*	33.56	32.25	34.42	*2.97*
Cellulose (wet), %	8.32	7.10	9.11	*0.13*	9.23	8.94	9.54	*2.71*
Cellulose (dm), %	26.62	25.53	27.68	*0.04*	21.52	19.96	22.26	*5.04*
Hemicellulose (wet), %	3.58	2.85	4.28	*0.20*	2.88	2.76	3.02	*3.94*
Hemicellulose (dm), %	11.37	10.97	12.02	*0.05*	6.71	6.17	6.98	*5.44*
Protein (wet), %	3.71	2.69	4.38	*0.24*	6.52	6.13	7.13	*7.11*
Protein (dm), %	11.69	10.38	12.38	*0.10*	15.16	14.81	15.94	*3.48*
Biomass, %	19.44	18.85	19.98	*0.03*	23.93	23.78	24.22	*0.83*
Carbon/nitrogen	19.61	18.26	22.17	*0.11*	14.60	13.68	15.05	*4.32*
Ammonia, %	0.02	0.01	0.02	*0.3*	0.07	0.06	0.07	*8.88*

pasteurizing tunnel.[31] To aid further material transport, the compost is placed on nylon nets, which makes it possible to move the entire mass of the compost easily. For intensive mycelial growth, a temperature of 25–28 °C is maintained, accompanied by a constant fresh air supply.[25] Since the development of the mycelium is stimulated by higher CO_2 levels, the air quality is monitored and controlled. Ideally, the spawn run takes 18 to 19 days.

During this period, the mycelium completely colonizes the compost, which is called phase III compost. Although spawn runs can be cut by a few days, it has been proven that an 18 to 19 days period is better for button mushroom mycelia.[18] Phase III compost is removed from the tunnels, usually by using the opening on the opposite side to where it was filled. Perfectly colonized compost can be transported longer distances without a decrease in quality. Refrigerated vehicles can even take the compost thousands of kilometers away for cultivation.

Modern mushroom farms receive the bulk compost transported by special trucks.[29] At the farm, the shelves of the growing houses are filled with 22 to 24 tons of compost by a filling machine. Depending on house types, the growing rooms are equipped with 5 to 7 leveled shelf systems. Each house has 6 (or a multiple thereof) of the same size and capacity rooms, thus a continuous production and supply for the market can be provided. The filling machine not only places the compost on the shelves but also covers it with an even layer and density of casing.[32] One truckload of compost can be processed with the help of 3–4 workers in a few hours while maintaining quality. Depending on the cultivation technology of the given farm, compost can be mixed into the casing (CAC, compost added at casing), or a special casing inoculum can be used to shorten the cultivation cycle.[33]

8.2.2 Supplementation

There are two types of supplements used in mushroom cultivation: plant-based and animal-based products.[34,35] The main ingredient of plant-based supplements is soy flour, and nitrogen-containing keratin for animal-based supplements. Feather meal may also be added to the latter type. Supplements contain high levels of nitrogen and serve as an optimal source of this nutrient to the mycelia (see Figure 8.7A), while only slightly increasing the mineral content of the fruitbody itself. The water content of the compost does not influence the efficacy of the added supplement, but the original nitrogen content of the compost could affect it. Depending on the compost preparation technology, the supplements can be added to the mixture at two different stages.[22] Or it is mixed into the compost together with the spawn (phase II), in which the risk of contamination is higher and the effect on the yield is moderate (see Figure 8.7B). Or the supplement is mixed with the compost at the end of the colonization process (phase III; Figure 8.7C), by this method, a 10–20% increase in yield can be achieved, while the risk of secondary contamination is reduced, as the mycelia are present in such a large mass they can overcome other microorganisms.

Figure 8.7 (A) Machine used for inoculation and supplementation after removal of the compost from the pasteurization tunnel (phase 2), detail of the operator adding the supplement in the distribution box. After inoculation and supplementation, the machine presses the compost blocks and packs them. (B) Installations used in the supplementation process in phase 3, the green machine removes the colonized compost from the tunnel, transfers it to a nylon net, later the supplement was added. In this installation, cultivation can be done using substrate in blocks, trays (phase IV compost) and in bulk Dutch system. (C) Spawn running in the supplemented compost. The golden spots are fragments of soybeans that make up the supplement.

8.2.3 Cultivation in the Dutch System

Dutch-type growing houses were designed and are continuously developed to achieve maximum efficiency.[36] The same type of structure is built on farms all over the world. In the growing rooms, shelves have an average of 5–7 levels, and each room has the same capacity. In each farm 6 or a multiple thereof growing rooms can be found, thus a continuous production and supply for the market can be provided.[14] The growing rooms are designed to receive an entire load of phase III compost from 1–3 trucks.

The compost is placed on the shelves by a special filling machine, which can regulate the thickness and density of the compost layer and lay the casing on top of it at the same time. Depending on its structure and quality, an average of 26–32 cm thick compost layer is used. In case of lower quality less, and better quality more, compost is filled into the shelves.

The casing is usually applied in a 5 cm layer, but this is influenced by the type of product used.[37] Casing facilitates the generative phase of the mycelia and provides optimal water content for the developing fruit bodies.[38] One truckload of compost can be processed with the help of 3–4 workers in a few hours while maintaining quality (see Figure 8.8). Dutch-type houses are equipped with machines for the complete growing process, including filling, watering, or ruffling. High-level production management is aided by air-conditioning systems capable of fine-tuning ambient conditions, such as compost and air temperature, humidity, and CO_2 level. Smart device applications provide the possibility to control and check each growing room remotely.

Depending on the cultivation technology of the given farm, compost can be mixed into the casing (CAC, compost added at casing), or a special casing inoculum can be used to shorten the cultivation cycle. The amount of CAC necessary depends on the casing, but it is usually around 100–400 g

Figure 8.8 Filling machine for phase 3 compost at a Dutch type of farm. The conveyor belt in the back is dedicated to emptying the compost from the truck, while the front conveyor belt delivers the casing.

compost m^{-2}.[33] At the beginning of the growing cycle, the goal is to achieve an optimal water content in the casing and compost, as each kilogram of mushroom requires 2 liters of water to develop. Most of the water is provided by the casing, and partially by the compost.[39] That is why, in the first few days of cultivation, a small amount of water is introduced to the substrate multiple times a day. Generally, 0.5 to 1 L m^{-2} is applied in 24 hours. In the initial 1–4 days of irrigation, an average total of 10 to 30 L m^{-2} water is used.

The growing houses are equipped with watering trees, sprinklers, and recently developed drip irrigation systems (see Figure 8.9). At this stage of growth, the mycelium begins to colonize the casing as well.[40] Low-quality casing may lose structure due to irrigation and become dense, thus an additional step of ruffling (loosening the casing) may become necessary. When the mycelium colonizes the casing to a certain degree, fruitbodies development is initiated by introducing stress factors, such as lower temperature.[41] Moreover, as ventilated fresh air lowers the compost temperature, the CO_2 level decreases as well, which triggers the mycelium to start developing fruit bodies.

In Dutch-type growing houses automatized climate computers run preprogrammed protocols (*e.g.*, recovery, cooling, pinning, fruiting) to regulate and optimize the ambient conditions. Necessarily manual fine-tuning and override are possible. Mushroom farms, shelf structures, and sizes are designed for manual picking, which requires a different climate control setting than that of farms with machine picking. In the former case, the goal is to prolong fruit body development to 5–7 days, to enable continuous picking. This is achieved by slow, gradual cooling. For machines, picking a more intensive cooling is necessary, which results in mass and simultaneous fruitbody development. Cooling causes pinhead formation or pinning when fruitbody initials (primordia) in the size of a few millimeters appear on the

Figure 8.9 (A) First flush of button mushroom produced on a Dutch-type shelf system with fixed sprinklers irrigation system. (B) Mobile watering apparatus in a Dutch system growing room.

surface of the casing.[37] These primordia then begin to grow, doubling their size each day. The first picking can usually be scheduled 16–18 days following filling, after the compost has been placed on the shelves.[25]

Two flushes on average are picked, but with proper maintenance and logistics, a third flush can also be picked.[31] At the end of the growing cycle, to eliminate microbes, the used compost is treated by steam in the cookout process.[18] This process introduces steam into the growing room to achieve a temperature of 60–70 °C in the substrate for 6–8 h. This aims to inactivate all the pathogens and pests that might be present in the growing substrates and growing areas.[42] Following this step, the compost is taken out of the growing room; the surfaces are cleaned and sanitized to be ready for the next load of phase III compost and casing.

8.2.4 Cultivation in Specialized Rooms (Phase IV)

If the phase III compost is cased, the mycelia are allowed to colonize the casing layer under vegetative growth conditions and the crop growth up to the point of pinhead mushrooms before sending to the growing unit or delivering to growers, this is called phase IV compost.[31] Phase IV starts by placing the phase III compost, either supplemented or not, in large trays and topping it with a casing layer (see Figure 8.10). Trays vary considerably in size but are mostly in the range of 0.9–1.2 m by 1.6–3.5 m and are between 20 and 23 cm deep.[18,43] In Europe, trays of size 1.2×2.4 m^2 are commonly used.[32] After filling with compost and casing (CAC optional), the trays are placed in a phase IV growing room (sometimes called a holding room or set back room), stacked 9–11 high, with uniform air movement. After 2 weeks, "pinned" trays are transported to the growing room at the production farm.[32,43]

Figure 8.10 (A) Filling line of phase IV trays. (B) Part of the filling line responsible for pressing and leveling the compost. (C) After adding the casing layer, irrigation and riffling are done. In this image, casing-based soil-peat is used.

Phase III and IV composts have increased in popularity because they allow for an increase in the number of growing cycles growers can expect from their cropping houses. In the Netherlands, in 1985, phase II composting undertaken on shelves produced an average of 4.1 crops per year. In 1999, growers using phase III composts were getting an average of 7.1 crops per year, while an additional gain can be made in the number of crops (10–12 crops per year) when phase IV is used.[31,44,45] In Germany, Dohme farm has 27 growing rooms, and 10–11 rooms are filled per week with phase IV compost.[32] Considering that only two flushes are harvested, this schedule could reach 18 cycles room year.

In addition to requiring less demanding air conditioning equipment, the shorter duration of the growing cycles also implies a lower incidence of pests and diseases. The main drawbacks to this system include less control of fruiting by the grower and high transport costs.

8.2.5 Automated Watering Systems

The application of water on the casing layer during the production of *A. bisporus* was initially carried out manually with the help of a rose face (water-break) spray head.[37] Considering the watering of the casing as an art, the experience and training of watering personnel are a compulsory condition in manual watering. Subsequently, with the implementation of new cultivation systems, fixed and mobile apparatuses have been developed to apply water simultaneously at several levels.[46]

Manually operated watering trees consist of a vertically mounted boom on a wheeled base, with short lateral arms fitted with nozzles, while the spraying lorries, hanging on the picking rail of shelves, are designed to spray from one side across the entire width of the growing bed by spray lances supplied with specially developed nozzles. This hand-propelled water supply equipment has given way to automated systems without the need for manual operation, with a remarkable reduction in labor cost and greater convenience. These automated systems achieve greater uniformity of irrigation, without the drawbacks of the manual model in terms of variability associated with the skill of the operator.

By using fixed sprinklers installed on the beds, the system allows setting the amount of water delivered by selecting the duration of watering, while facilitating the possibility of programming the time of application. Watering has been reduced from 30 to 3 minutes in farms that have substituted the watering tree, manned by an operator, with a system installed on the shelves that allows watering a mushroom house by just pressing a button in a control panel.[47] Mobile devices have also been developed, which consist of nebulizers arranged in a structure suspended from the ceiling of the room that moves through guides along the growing room, simultaneously watering all levels on both sides of the growing room (see Figure 8.9B). The amount of water applied can be regulated automatically with the number of round-trip cycles of the system and its speed.

A special mention is deserved for drip irrigation. An antecedent to the system in terms of providing water without spraying the mushroom beds is found in the works of Noble *et al.*, who developed a simple system for watering the casing layer by capillarity, to maintain the water content of the casing layer between narrow limits, considering it as a possible automatic method of watering commercial crops.[41] In the subsequent work of Lomax different aspects of the automatic irrigation systems are discussed, such as the moisture sensors and the irrigation apparatus, referring to porous tubing for distribution, placed in the casing layer, and providing water without spraying the mushroom caps.[48]

On this same idea, a new system has been developed using drip irrigation.[49–51] This system allows the addition of water to the casing layer during the entire cropping cycle and minimizes the decrease in casing and compost moisture, which can be maintained at optimal levels. Additionally, it enhances water efficiency, decreases the incidence of bacterial blotch, and enhances the quality of mushrooms.[52] Increased yield and quality on one hand, and a reduction in casing soil and energy for drying the mushrooms after superficial irrigation, on the other hand, brings a significant economic advantage.[40] The results of the agronomical assessment of the implementation of a drip irrigation system in a Spanish commercial mushroom farm (see Figure 8.11) have been recently reported.[53]

8.2.6 Harvest

Harvesting has a significant impact on the quality of mushrooms, yield, and production costs.[54] Mushrooms intended for the fresh market are of superior quality and must be manually picked at an optimal stage of development by experienced employees. It is important to avoid applying high

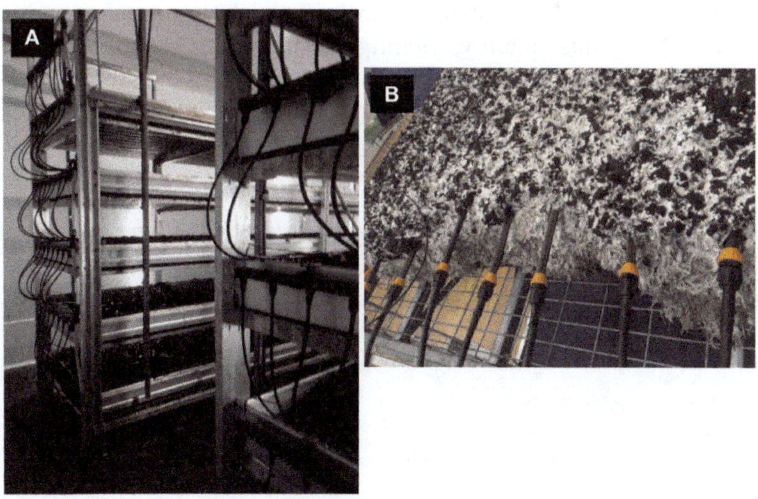

Figure 8.11 (A) Drip irrigation system in a Dutch cultivation. (B) Rip lines accommodated between the compost and the casing layer, 15 cm apart.

physical pressure to the mushrooms during harvest, so that enzymatic browning does not occur during storage. Labor associated with manual harvesting has a high impact on production costs, which has led to the development of the mechanical harvest. In the growing rooms of the Dutch system, this operation usually employs a cutting machine with transverse discharge and all mushrooms over the width of the bed are cut off in one go (see Figure 8.12). In this way, the harvesting time is significantly reduced, although the fact that it is a non-selective method negatively affects quality, so mushrooms harvested in this way are regularly used for canning and processing purposes. In this case, it is possible to carry out optionally a previous manual harvest of first-quality mushrooms for the fresh market.

Technological advances in computer image analysis and robotics have introduced a further advanced harvesting system that remains at the interface between manual and mechanical handling. Bac *et al.* identified four general challenges with R&D directions to obtain a positive trend in performance and to successfully implement harvesting robots in practice: (1) simplifying the task, by modifying the crop environment; (2) enhancing the robot's design and ability; (3) defining requirements and measuring performance indicators; and (4) considering additional requirements for a successful implementation, involving disciplines other than robotics.[55]

In this line of work, numerous attempts have been made in recent years by institutes and companies to address the challenge of achieving a fully automatized harvesting system for mushrooms that can carry out the different tasks of harvesting in a coordinated manner. After positioning the harvester system on the shelves, selection of mushrooms is carried out through a computer-based vision system, capable of identifying mushrooms and pre-determining the picking order, from a vision-processing algorithm like the one developed by Tillet and Batchelor, and the system built and evaluated by Reed and Crook, and their later developments and alternatives.[56–61] Subsequently, picking is carried out through a robot manipulator provided with a control system reproducing the human dexterity, equipped with a picking mechanism provided with a silicone vacuum induced suction cup or a three-fingered gripper.[57,61–63] Recently, Huang *et al.* assessed harvesting motions (bending, twisting, and lifting) for

Figure 8.12 (A) Mechanized harvesting, where the mushrooms are transported to the post-harvest area using a strap. (B) Reception of the mushrooms in the post-harvest area, where electronic sieves separate the mushrooms.

implementation in a robotic end-effector designed to harvest *Agaricus* mushrooms.[64] After detachment, mushrooms are transferred to a suitable conveyor for subsequent trimming and transferring to the packing line.

Now, people who are working on robotic harvesting are achieving picking results of 6 to 12 pieces per minute.[65] The high potential harvesting rates of these automated systems, working 24 hours a day, can generate exceptional productivity and profitability for the grower, but require a high degree of investment.

8.3 Smart Farms

Environmental conditions are a significant factor that needs to be controlled in the mushroom production process and demands a high economic expense. This fact has led to the development of automatic environmental control systems to provide optimum condition to the mushroom production house, acting mainly on the four major microclimate factors: air temperature, air relative humidity, compost temperature, and carbon dioxide concentration.[26,66]

Automation of air conditioning, regulating the climate within several predetermined conditions, provides a continuous and reliable operation in the growing rooms. Measurement of climatic factors at regular intervals makes it possible to immediately establish and correct climatic deviations.[8] Control over the crop allows a substantial improvement in efficiency and productivity, increasing yield and saving costs of operation.

The implementation of modern information and communication technologies, such as the Internet of Things (IoT), provides growers with a real-time monitoring condition of their farm from anywhere and anytime by accessing the Internet. Sensors (see Figure 8.13A) are properly placed at specific regions of the mushroom growing rooms to measure the climatic factors which are transmitted to a remote monitoring station for further

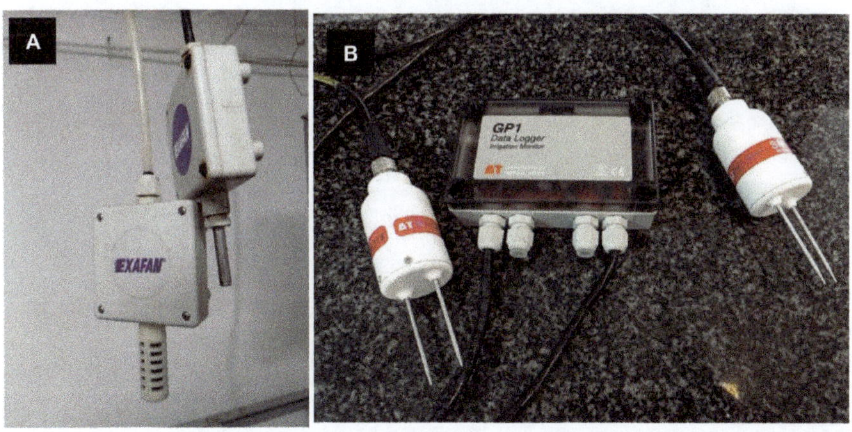

Figure 8.13 (A) Air temperature and carbon dioxide sensors. (B) Casing moisture sensors and data logger irrigation monitor.

actions.[67] In smart irrigation systems, sensors are used in the same way to collect a large amount of real-time data from the casing soil (see Figure 8.13B). A dielectric moisture sensor has been developed to measure the matric potential (water suction pressure). The output from moisture sensors positioned in the casing and compost has been used to control an automated drip irrigation system.[41]

Considering that all these factors are interconnected in the way that one influences the other, there is a need to use a multidimensional control system to maintain the microclimate conditions, through a model that describes all the signal interactions during cultivation.[68] The large amount of data collected from the interactive sensors can be sent to the webserver through an Internet connection using a wireless sensor network (WSN). Finally, the IoT framework analyzes, processes, and assesses the sensed data. The prediction of different parameters will make it possible to schedule technological operations more precisely, such as fruiting induction, harvesting, or irrigation.

This technology has been mainly applied in recent years to *A. bisporus*, although it has also been used for other cultivated species, such as *P. ostreatus*, *L. edodes*, and *F. velutipes*.[67–74]

Finally, mushroom facilities have advanced toward clean energy generation, using wind and solar energy (Figure 8.14). It is known to be a global concern of the 17 objectives of the United Nations (UN) that the achievement

Figure 8.14 (A) Oyster mushroom farm, in the background it is possible to see a group of wind turbines. (B) Button mushroom farm with a smart inflatable solar thermal collector.

of quality food is done through clean energy, without damaging the environment and while maintaining sustainability.

Another crucial point to be considered is the consolidation of the mushroom circular economy, using the companies' residues in the companies themselves.[75] This will reduce raw material purchases, the reuse of spent mushroom substrate in new substrate formulations, and as a casing layer, replacing the use of peat, which is currently a problem in the mushrooms industry. Furthermore, recent studies indicate the use of spent mushroom substrate as fuel, which can be pelleted and used in boilers.[76] To further consolidate the circular economy, researchers are verifying the effect of using the spent mushroom substrate as a gas adsorbent, with the possibility of coupling in the boiler chimney.[77]

References

1. J. Kumla, N. Suwannarach, K. Sujarit, W. Penkhrue, P. Kakumyan, K. Jatuwong and S. Lumyong, *Molecules*, 2020, **25**(12), 2811.
2. R. S. Bhatt, S. Sarkar, A. Sahoo, P. Sharma, L. Soni, V. K. Saxena and A. Soni, *Anim. Feed Sci. Technol.*, 2021, **278**, 114979.
3. T. Wegi, A. Tolera, J. Wamatu, G. Animut and B. Rischkowsky, *Asian-Australas. J. Anim. Sci.*, 2018, **31**(8), 1221.
4. Y. Wang, S. Hu, Z. Yang, Q. Jin, A. Ren, Q. He and M. Zhao, *Cellulose*, 2021, **28**(11), 1.
5. S. Yıldız, A. Yılmaz, Z. Can, C. Kılıç and Ü. C. Yıldız, *J. Food*, 2017, **42**(3), 315.
6. S. T. Chang and S. P. Wasser, *Int. J. Med. Mushrooms*, 2012, **14**(2), 95.
7. V. K. Chaturvedi, S. Agarwal, K. K. Gupta, P. W. Ramteke and M. P. Singh, *3 Biotech*, 2018, **8**(8), 1.
8. J. Carrasco, D. C. Zied, M. J. Navarro, F. J. Gea and A. Pardo-Giménez, in *Advances in Macrofungi: Industrial Avenues and Prospects*, ed. K. R. Sridhar and S. K. Deshmukh, CRC Press, Boca Raton, 2021, ch. 2, pp. 11–40.
9. D. J. Royse, J. Baars and Q. Tan, in *Edible and Medicinal Mushrooms: Technology and Applications*, ed. D. C. Zied and A. Pardo-Giménez, John Wiley & Sons Ltd, Chichester, 2017, ch. 2, pp. 5–13.
10. S. T. Chang and P. G. Miles, *Mushrooms: Cultivation, Nutritional Value, Medicinal Effect, and Environmental Impact*, CRC Press, Boca Raton, 2nd edn, 2004.
11. S. T. Chang and W. A. Hayes, *The Biology and Cultivation of Edible Mushrooms*, Academic Press, London, 1978.
12. P. Oei, *Mushroom Cultivation*, Backhuys Publishers, Leiden, 3rd edn, 2003.
13. J. E. Sánchez, D. J. Royse and H. Leal, *Cultivo, Mercadotecnia e Inocuidad Alimenticia de Agaricus bisporus*, Ecosur, Tapachula, 2007.
14. L. J. L. D. van Griensven, *The Cultivation of Mushrooms*, Mushroom Experimental Station, Horst, 1988.

15. D. C. Zied and A. Pardo-Giménez, *Edible and Medicinal Mushrooms: Technology and Applications*, John Wiley & Sons Ltd, Chichester, 2017.
16. T. R. Fermor, P. E. Randle and J. F. Smith, in *The Biology and Technology of the Cultivated Mushroom*, ed. P. B. Flegg, D. M. Spencer and D. A. Wood, John Wiley & Sons Ltd., Chichester, 1985, ch. 6, pp. 81–109.
17. J. P. G. Gerrits, in *The Cultivation of Mushrooms*, ed. L. J. L. D. van Griensven, Mushroom Experimental Station, Horst, 1988, ch. 2, pp. 29–72.
18. A. Gómez, in *Edible and Medicinal Mushrooms: Technology and Applications*, ed. D. C. Zied and A. Pardo-Giménez, John Wiley & Sons Ltd, Chichester, 2017, ch. 10, pp. 211–220.
19. B. Robinson, K. Winans, A. Kendall, J. Dlott and F. Dlott, *Int. J. Life Cycle Assess.*, 2019, **24**(3), 456.
20. M. A. Kertesz and M. Thai, *Appl. Microbiol. Biotechnol.*, 2018, **102**(4), 1639.
21. Z. Sebaaly, M. A. AlSanad, P. Hayek, L. Kfoury, N. Shaban and Y. N. Sassine, presented in part at XXX International Horticultural Congress IHC2018: International Symposium on Medicinal and Aromatic Plants, Culinary Herbs and Edible Fungi, Istambul, August, 2018.
22. A. Pardo-Giménez, D. C. Zied, M. Álvarez-Ortí, M. Rubio and J. E. Pardo, *J. Sci. Food Agric.*, 2012, **92**(8), 1665.
23. T. Song, Y. Shen, Q. Jin, W. Feng, L. Fan, G. Cao and W. Cai, *PeerJ*, 2021, **9**, e10452.
24. B. L. Dhar, in *Edible and Medicinal Mushrooms: Technology and Applications*, ed. D. C. Zied and A. Pardo-Giménez, John Wiley & Sons Ltd, Chichester, 2017, ch. 14, pp. 271–308.
25. J. Buth, in *Edible and Medicinal Mushrooms: Technology and Applications*, ed. D. C. Zied and A. Pardo-Giménez, John Wiley & Sons Ltd, Chichester, 2017, ch. 6, pp. 129–147.
26. M. Awtoniuk and A. Chochowski, presented in part at 7th International Conference On Engineering, Project, and Production Management (EPPM2016), Białystok, September, 2016.
27. J. Carrasco, C. García-Delgado, R. Lavega, M. L. Tello, M. De Toro, V. Barba-Vicente and G. M. Preston, *Microb. Biotechnol.*, 2020, **13**(6), 1933.
28. C. F. McGee, *Appl. Microbiol. Biotechnol.*, 2018, **102**(3), 1075.
29. M. den Ouden, *Mushroom Signals: A Practical Guide to Optimal Mushroom Growing*, Roodbont Publishers B.V., Zutphen, 2016.
30. H. S. S. Sharma, M. Kilpatrick and L. Burns, *J. Near Infrared Spectrosc.*, 2000, **8**(1), 11.
31. D. J. Royse and R. B. Beelman, *Six Steps to Mushroom Farming*, College of Agricultural Sciences, Agricultural Research and Cooperative Extension, PennState, 2nd edn, 2013.
32. M. den Ouden, *Mushroom Bus.*, 2013, **59**, 18.
33. D. C. Zied, M. T. A. Minhoni, J. E. Pardo-González and A. Pardo-Giménez, *HortScience*, 2010, **45**(11), 1649.

34. J. Carrasco, D. C. Zied, J. E Pardo, G. M. Preston and A. Pardo-Giménez, *AMB Express*, 2018, **8**(1), 1.
35. D. J. Royse and J. E. Sánchez, *World J. Microbiol. Biotechnol.*, 2008, **24**(8), 1333.
36. P. J. C. Vedder, *Cultivo moderno del champiñón*, Mundi-Prensa, Madrid, 1986.
37. A. Pardo-Giménez, J. E. Pardo and D. C. Zied, in *Edible and Medicinal Mushrooms: Technology and Applications*, ed. D. C. Zied and A. Pardo-Giménez, John Wiley & Sons Ltd, Chichester, 2017, ch. 7, pp. 149–174.
38. A. Pardo-Giménez, J. E. Pardo, J. A. de Juan and D. C. Zied, *Arch. Microbiol.*, 2010, **192**(12), 1023.
39. E. S. Dias, D. C. Zied and A. Pardo-Giménez, *Cienc. Agrotecnol.*, 2021, **45**, e0001R21.
40. D. Raz, O. Danay, P. van den Berg, Y. Engel, E. Kobi, I. Barski, D. Levanon and P. van der Meij, *Mushroom Sci.*, 2016, **19**, 31.
41. R. Noble, M. Goodchild, D. Jenkins, K. Burton, D. Walker and E. Moorhouse, *Mushroom News*, 2017, **65**(10), 8.
42. F. J. Gea, M. J. Navarro, M. Santos, F. Diánez and J. Carrasco, *Microorganisms*, 2021, **9**(3), 585.
43. J. T. Fletcher and R. H. Gaze, *Mushroom Pest and Disease Control. A Colour Handbook*, Manson Publishing Ltd, London, 2008, ch. 1, pp. 7–21.
44. S. T. Chang and S. P. Wasser, The Cultivation and Environmental Impact of Mushrooms, in *Environmental Science*, Oxford University Press, Oxford, 2017.
45. G. Lemmers, *Mushroom J.*, 2003, **642**, 17.
46. J. Campbell, *Mushroom News*, 2003, **51**(1), 12.
47. F. M. Arqueros, PhD thesis, National University of Ireland Maynooth, 2011.
48. K. Lomax, *Mushroom News*, 2003, **41**(1), 8.
49. D. Levanon, O. Danai and D. Raz, *US Pat.*, 2015/0096224 A1, 2015.
50. D. Levanon, O. Danai and D. Raz, *European Pat.*, EP 2827699B1, 2016.
51. D. Levanon, O. Danai and D. Raz, *European Pat.*, EP 3106025A1, 2016.
52. O. Danay, R. Kadosh, D. Raz, G. Yasinov, N. Ezoz, M. Miron, Y. Hadar and D. Levanon, *Mushroom Sci.*, 2012, **18**, 863.
53. M. J. Navarro, F. J. Gea, A. Pardo-Giménez, A. Martínez, D. Raz, D. Levanon and O. Dannay, *Sci. Hortic.*, 2020, **265**, 109234.
54. J. Valverde, in *Edible and Medicinal Mushrooms: Technology and Applications*, ed. D. C. Zied and A. Pardo-Giménez, John Wiley & Sons Ltd, Chichester, 2017, ch. 13, pp. 261–270.
55. C. W. Bac, E. J. van Henten, J. Hemming and Y. Edan, *J. Field Rob.*, 2014, **31**(6), 888.
56. R. D. Tillett and B. G. Batchelor, *Comput. Electron. Agric.*, 1991, **6**(3), 191.
57. J. N. Reed and S. Crook, *Mushroom Sci.*, 1995, **14**(1), 385.
58. R. Noble, J. N. Reed, S. Miles, A. F. Jackson and J. Butler, *J. Agric. Eng. Res.*, 1997, **68**, 215.
59. J. N. Reed, S. J. Miles, J. Butler, M. Baldwin and R. Noble, *J. Agric. Eng. Res.*, 2001, **78**(1), 15.

60. C. P. Lu, J. J. Liaw, T. C. Wu and T. F. Hung, *Agronomy*, 2019, **9**, 32.
61. L. He, D. Choi and J. Pecchia, *Mushroom News*, 2021, **69**(1), 4.
62. European Commission, Seventh Framework Program, CORDIS EU Research Reports, https://cordis.europa.eu/project/id/262037 (accessed March 2022).
63. W. Lu, P. Wang, L. Wang and Y. Deng, *Trans. Chin. Soc. Agric. Mach.*, 2020, **51**(11), 28.
64. M. Huang, X. Jiang, L. He, D. Choi and J. Pecchia, presented in part at ASABE 2020 Annual International Meeting, Virtual & On Demand, July, 2020.
65. R. van Doremaele, *Mushroom News*, 2021, **69**(1), 12.
66. M. S. A. Mahmud, S. Buyamin, M. M. Mokji and M. S. Z. Abidin, *Indones. J. Electr. Eng. Comput. Sci.*, 2018, **10**(3), 847.
67. M. A. M. Ariffin, M. I. Ramli, M. N. M. Amin, M. Ismail, Z. Zainol, N. D. Ahmad and N. Jamil, presented in part at 10th International Conference on System Engineering and Technology (ICSET), Shah Alam, November, 2020.
68. M. Awtoniuk, M. Daniun, D. Komarchuk and S. Syrotyuk, *Mater. Sci. Eng.*, 2019, **710**, 012011.
69. S. Pravinthraja, A. P. Roger, S. Nagarani and N. S. Kavitha, *Int. J. Eng. Technol.*, 2018, 7, 1238.
70. A. Subedi, A. Luitel, M. Baskota and T. D. Acharya, *Proceedings*, 2020, **42**(1), 46.
71. M. F. Mohammed, A. Azmi, Z. Zakaria, M. F. N. Tajuddin, Z. M. Isa and S. A. Azmi, *J. Phys.: Conf. Ser.*, 2018, **1019**, 012053.
72. T. H. Nasution, M. Yasir, F. Fahmi, and S. Soeharwinto, presented in part at 3rd International Conference on Electrical Engineering and Computer Science (ICECOS), Batam, October, 2019.
73. M. R. M. Kassim, A. N. Harun, I. M. Yusoff, I. Mat, C. P. Kuen and N. Rahmad, presented in part at 2017 Eleventh International Conference on Sensing Technology (ICST), Sydney, December, 2017.
74. K. W. Lee, J. O. Jeon, K. J. Lee, Y. H. Kim, C. J. Lee and M. J. Jang, *J. Mushrooms*, 2019, **17**(4), 197.
75. D. C. Zied, J. E. Sánchez, R. Noble and A. Pardo-Giménez, *Agronomy*, 2020, **10**(9), 1239.
76. L. S. Alves, B. R. A. Moreira, R. S. Viana, A. Pardo-Giménez, E. S. Dias, R. Noble and D. C. Zied, *J. Cleaner Prod.*, 2021, **313**, 127875.
77. L. S. Alves, B. R. A. Moreira, R. S. Viana, E. S. Dias, D. L. Rinker, A. Pardo-Giménez and D. C. Zied, *Environ. Res.*, 2022, **204**, 111945.

CHAPTER 9

The Cultural Use of Mushrooms

JOVANA VUNDUK*[a,b] AND ALONA YU. BIKETOVA[c,d]

[a] University of Belgrade-Institute of General and Physical Chemistry, Studentski trg 12/V, 11158 Belgrade, Serbia; [b] Ekofungi Ltd., Padinska Skela bb, Belgrade, Serbia; [c] Jodrell Laboratory, Royal Botanic Gardens, Kew, Richmond TW9 3DS, UK; [d] Mycological Society of Israel, P.O. Box 164, Pardesiya 42815, Israel
*Email: vampum00@yahoo.com

9.1 Introduction

When two tribes of ancient people met for the first time, they were inevitably just as amazed, as much as they were interested in each other. Probably, the two groups were curious about many things like what do these other people eat, which language do they speak, where do they live, which spirits occupy their land, what kind of game do they hunt, how do they cure maladies, which gods or forces do they believe in, do they sing, do they have enemies, and are there any laws among these people? In other words, as comprehended by modern society, which culture marks these people? All the listed aspects, which define one culture, were formed, learned, and shared by a group of people over a long period of time.[1,2] Today, cultural values are constantly changing and are strongly influenced by many factors among which political ideologies, economy, and technology are among the most powerful.[3–5] On the other hand, the biophysical environment must have had a major role in culture formation and evolution in early human communities. For example, the colder climate dominating in Siberia suited well *Amanita muscaria* growth and abundance. No wonder that this species found its way into the local native cultures.

Food Chemistry, Function and Analysis No. 36
Edible Fungi: Chemical Composition, Nutrition and Health Effects
Edited by Dejan Stojković and Lillian Barros

Published by the Royal Society of Chemistry, www.rsc.org

These first human communities were heavily dependent on climate and the natural resources around them. Inevitably, the culture expanded around plants, animals, and forces of nature. In this setting, wheat could be assigned several uses, like food, medicine, clothes, or shelter.[6] Another example of plants affecting the ancient culture is alcohol production, which differed across the globe mainly because of the available raw materials.[7,8] In ancient Egypt, beer was used as a payment currency, medicine, a reward for hard work, a class symbol, and even a student's meal.[9] All this speaks of a rich culture evolving around an alcoholic beverage. When it comes to mushrooms, the situation is no different; species were numerous, adapted to different climates and substrates, and thus widely present. Inevitably, early humans started using them although it is still not clear for how many purposes.[10] Nowadays, the science of ethnomycology explores the many aspects of the human–mushroom relationship. Dr R. Gordon Wasson and his wife Valentina Pavlovna Guercken were among the pioneers of this science and, with their personal collection of books and articles dealing with mushrooms, established the first and only ethnomycological library in the world.[11,12] Besides, being convenient as food and medicine there is speculation of the more direct and substantial cultural influence of mushrooms. The theory was proposed by Terrence McKenna. It is known as "The Stoned Ape" and speculates that psychedelic mushrooms enabled changes to the human cortex, which caused the emergence of language and, finally, lead to the development of several cultural aspects.[13] These isolated theoretical cases attracted a lot of attention at the time of their appearance, however, continuous research directed on the cultural use of mushrooms is lacking. The majority of existing data deals with cultures from Mesoamerica and native communities whose knowledge about mushrooms is now heavily deteriorated and lost as a consequence of urbanization and the destruction of forests. The psychedelic mushroom cults developed by ancient cultures of Mesoamerica are good examples of the knowledge and practice constantly changing because of external sociological factors. The initial religious cult went through three phases. The first occurred in the 16th century when Spanish conquistadors tried to eradicate the native mushroom-oriented religion. The cult survived; however it incorporated many aspects of the newcomer's culture, such as language and elements of the Catholic religion. The second phase started in the 1960s right after Wasson's article about a psychedelic session with the native shaman Maria Sabina. The inflow of American hippies interested in the good trip was followed by incidents, military actions, and strict laws forbidding this tradition. Similar occurred in Russia as well, when Soviet scientists discovered Siberian shamans and marked this practice as a dark religion.[14] Finally, globalization, the internet, and social media rediscovered these cults and new interested parties emerged – tourists. Using mushrooms as part of a sacred religion became a lucrative business for the native people, which inevitably affected the ritual and its purpose. The still unaddressed question is – how deep is this change and how much does it now differ from the original one? Also, are there any people who have preserved the original tradition?

There is material evidence that prehistoric people knew and appreciated the medicinal aspect of mushrooms. In 1991, a 5000 year-old mummy, known as the Iceman, was discovered in the Alps, in northern Italy.[15] Interestingly, in a bag with the Iceman were two mushroom species, *Fomitopsis betulina* and *Fomes fomentarius*, which were speculated to be of medicinal importance at that time.[16] In the meantime, the tradition of using medicinal mushrooms and folk remedies was characterized as outdated, unscientific, and unprecise. Only recently have scientists recognized the potential of the kingdom of fungi, especially macrofungi, so the new class of medicine of natural origin was established. Professors Solomon P. Wasser and Shu Ting Chang were pioneers in bringing knowledge of medicinal mushrooms from folk medicine to the international scientific level. They introduced the term *culinary-medicinal mushrooms*. Their immense contribution is evident not only from their research, but also from outreach activities such as founding of the *International Journal of Medicinal Mushrooms*, and the International Medicinal Mushroom Conference at the beginning of 2000s.[17,18] The popularity and business prospects of nutraceuticals spread to mushrooms as well, and a separate branch, called *mycoceuticals*, was formed.

How to bioprospect such a vast group of organisms? The first step is to go back to folk culture and consult the ancient medicinal sources. The oldest medicinal collection of data originates from the ancient Sumer, some 5000 years ago, and mentions 250 plants. The most famous is *Chinese de Materia Medica* which includes several mushroom-based remedies. However, China remains an unknown field when it comes to contemporary ethnomycological studies. Another example of written medicinal botanical knowledge (including mushrooms) is *Chilandar Medical Codex*, which is the only medieval medicinal handbook written in vernacular.[19] This unique European medicinal manuscript comprises eight recipes containing mushrooms. At the same time, people are preserving some of this knowledge through the oral tradition, which is the main way of its transfer in rural areas. Contrary to the statement of Montalto *et al.*[20] who expressed that culture mostly happens in urban areas, the preservation and transfer of mushroom knowledge and culture are the strongest inside the small rural communities.[21,22]

Besides several rare examples, the cultural connection with mushrooms of people from most parts of the world remains sporadic and underexplored. Due to a strong mycophobia among many Western cultures, a significant part of mushroom knowledge was lost. This chapter aims to help preserve and summarize different elements of cultural use of mushrooms across the globe viewed through several cultural aspects – as food, medicine, beliefs, folk tales, superstitions and religion, art, economy, and linguistics. Moreover, it examines the tools of cultural use assessment, their development, and their importance in collecting and protecting knowledge about mushrooms. Finally, we discuss the new avenues of mushroom-connected cultural phenomena that have been developed and spread through digital technologies.

9.2 Food

The choice of food makes a heart of the cultural identity as pointed out by Rubel and Arora.[23] Nai and Meyer wrote that we are an *eating culture*.[24] No wonder so far the most frequent cultural use of mushrooms is for food.[25] In Europe, mushrooms came second as the most often collected non-wood forest products based on a study conducted in 28 European countries.[26] This percentage increases from Western to Eastern Europe. On the individual product level, they even hold the first place (*Boletus edulis* and *Cantharellus cibarius*, 15.8 and 12.8% of households collect these two species, respectively). Peintner *et al.* proved that there is a strong connection between mushroom consumption and culture.[27] The authors examined lists of mushrooms as a part of legislation related to consumption and commerce in 27 European countries. They stated that consumption behaviour strongly relates to geographical location since there are different preferences for the same species across different countries.

Mushrooms are highly valued worldwide, and in some indigenous communities mushrooms are appreciated more than meat.[28] People in rural areas are familiar with more species and, unlike the urban population, collect and use them in various dishes regularly.[29] For most informants across literature, eating mushrooms is not connected with a time of food scarcity, like war or natural disasters. On the contrary, they are prized as highly nutritious, tasty, and healthy.[28] Mushrooms, being a source of non-animal proteins, are quite important ingredients for lean foods for Orthodox Christians in Russia during all fasting periods except Holy Week.[30]

The choice of species and the method of preparation depend on the region, availability, season, or customs. For example, the Ngando people from Congo eat more than 50 species, while the Kumu and Ngelema use only a few.[31] While in most of the world, species like *B. edulis* or *C. cibarius* are among the most favourite because of their taste and exquisite flavour, in Africa *Schizophyllum commune* was reported to be regarded as the tastiest.[31,32] Moreover, in some cultures, some species are considered as a delicacy and are highly prized, expensive, and given as a gift, like *Tricholoma matsutake* in Japan or *Neolentinus lepideus* in Mexico.[33,34] Species that are scarce, big, and tasty are usually prepared and eaten alone, while those that are abundant, small, and have a simple taste are mixed with other species or ingredients.[33] Since mushrooms are highly perishable goods, and cannot last more than 1–3 days unless refrigerated or stored in modified atmosphere packaging, people often preserve them using traditional methods.[35] The most often used methods are sun drying, smoking, chilling, freezing, and pickling.[22,28,29,36] In Australia, Aborigines consumed several species, like *Elderia arenivaga*, *Cyttaria gunnii*, and *Mycoclelandia bulundari*, prepared in a unique way, cooked in sand and ashes, while *Pisolithus* sp. was eaten only when very young and as an emergency food.[37] In ancient Rome mushrooms were highly appreciated and constituted an important part of high-class banquets. They were often consumed in tremendous quantities and even had a special vessel reserved only for cooking mushrooms named the *boletaria*.[38] Fungi were even manipulated and served using amber knives and silver service. How strong the culture of eating mushrooms was in ancient Greece is illustrated by the gastronomic classification system introduced by Discorides. Mushrooms were either edible or poisonous, a classification which remains in practice until today. Almost every field guide for collecting mushrooms presents species from the point of view of edibility.[23] The same culture gave birth to the known Western mycophobia. As reviewed by Buller, ancient Greek and Roman physicians like Pliny, Discorides, and Galen, were rather suspicious of mushrooms, advising people to avoid them.[38]

There is an interesting aspect that the same species of mushrooms can be considered as (conditionally) edible in one country and inedible in another. For example, a lot of acrid species of *Lactarius* (*L. torminosus*, *L. rufus*) and *Lactifluus* (*L. vellereus*) are listed in Western and Central European field guides as inedible, because of the very acrid milky fluid in their flesh.[39–41] However, in Russia, Ukraine, and Belarus, these species are conditionally edible and prepared using long and complex treatments, so called *zasolka* (salting), which is also one of the oldest traditional methods of mushroom preparation in these countries.[40,42,43] This includes two steps: (1) maceration in cold water and salt with daily water changes (cold method) or boiling (hot method) and (2) further dry brining (fermentation) under pressure with coarse salt and seasonings.

Foraging for edible mushrooms has been an important activity in our past and, as a study from 2010 showed, affected sex differences in human cognition expressed as spatial abilities.[44] Women in this study collected similar quantities of mushrooms as men but at a lower cost by means of travel distances, heartbeat rate, and energy consumption. Those results imply that women evolved around the gathering lifestyle. Ethnomycological studies also confirm this hypothesis since in the majority of them women were found to be the principal gatherers and knew more about mushrooms, especially in the food use category. The instructions on how to cook mushrooms are transferred mainly by the oral route from generation to generation in rural communities, while in cities, media like television and the internet are the main sources, supported by famous chefs, celebrities, and cookbooks.[29] Besides, scientific interest in the nutritive and biological value of mushrooms is growing, which translates into innovative food products like flour enriched with mushroom powder, pasta, burgers, chocolate, frankfurters, and sausages with edible and medicinal mushrooms.[45–50] The results of the National Health and Nutrition Examination Survey, conducted for the period from 2001 to 2010, showed a positive association between mushroom consumption and better diet quality and nutrient intake.[51] This goes hand in hand with new cultural tendencies among the younger population toward a healthy diet, sustainability, innovation, and globalization.[52] It is also supported by different international directives and guidelines, such as the 17 UN sustainable development goals.[53] Consumption habits have changed in the direction of plant-based food. The paradigm shift increased international trade of wild edible mushrooms spectacularly, as reported by de Frutos,[54] and even the communities which previously had no *mushroom culture* increased their consumption.[55] An example is Hungary, which has gone through a slow increase in mushroom consumption per capita (around 1.5 kg).[56] This trend is especially important for regions with poorer populations, hunger, or malnutrition, like in sub-Saharan Africa, which previously had a mushroom use culture but neglected it as a consequence of the popularization of Western eating habits.[57]

A true *mycocultural* revolution has happened in Israel in the past 15 years. Before this, a few species of mushrooms were foraged, mainly by traditional Arabs, Druze, and Circassian villagers, who collect only a few species (*e.g.*, *Pleurotus eryngii*, *Volvopluteus gloiocephalus*, *Tirmania nivea*), as well as Russian-speaking repatriates, who collect the same or similar species to those growing in temperate forests of their native countries (*e.g. Suillus* spp. and *Lactarius deliciousus*).[58,59] Recently, interest in mushrooms has exploded in this country following the trends of meatless diets and the love of natural products, so currently thousands of people go foraging various edible fungi in forests, grasslands, and deserts. A few websites and four Facebook groups with more than 50 000 followers in total have appeared. And as a consequence of the acute need for mycological knowledge for the general public, the Mycological Society of Israel was established in 2019.[60]

New tendencies have been hardly explored in the sense of scientific studies. However, market research companies have recognized these changes in eating trends (vegetarians and vegans), as well as the category of superfoods, in which mushrooms perfectly fit, supported by the latest research findings. The rise of the food use of mushrooms among Western countries is also supported by the growing Asian immigrant population implementing its food cultural behaviour in a new setting (mainly the USA). Another factor affecting contemporary Western society's taste is the COVID-19 pandemic. Physical and social distancing, trading disturbances, and the fear of food scarcity have enabled the conditions for the rebirth of mushroom use as a food. People started to explore home/urban mushroom growing or gathering as a form of escapism as well as a tasty and secure diet source.

Still, the scientific information concerning mushrooms used as food is limited to a small number of studies including even fewer countries. Traditional knowledge mainly depends on oral transmission and with declining rural communities this is in danger. In addition, future studies should include the new generations, Alpha and Z, *the customers and parents of the future*, as Bringye, Fekete-Farkas, and Vinogradov vividly called them.[56] Some business journals have already recognized this new food market influence, saying that it "changes the taste of the food industry".[61] Some of the most important characteristics of this largest population group are eating autonomy, willingness to cook, and great spending power. On the other hand, the *Z generation* is known as a food culture disruptor which is upending tradition. It is important to understand whether the mushroom subject will play a part in this novel culture frame, especially in connection with digital technologies and knowledge transfer (including social media, blogs, and podcasts). How is it connected with the old ones and what are the implications concerning other cultural uses?

9.3 Medicine

Before the invention of contemporary synthetic drugs, diseases were treated with the resources people had around them, like plants, animals, minerals, and mushrooms. These first remedies were also used in prevention and prophylaxis. Sometimes, the remedy was applied followed by soothsaying and singing performed by a shaman or in the form of music like with the Thai instrumental folk music.[62,63] The process of curing has been seen as magic supported by natural remedies. From a cultural standpoint, the process of curing an illness cannot be divided from the cultural environment. It is affected by local beliefs, traditional health practices, language, communication, and family structure.[64] Throughout time medicine has transformed, losing most of its transcendental nature. In rural or poor communities or in places where healthcare facilities are not available, people still rely on traditional medicine. On a global scale, the percentage of people using folk remedies is between 70% to 80%, with a declining tendency in more developed countries.[65] With contemporary science recognizing the potency of natural products, folk medical knowledge has gone through a revival. Bioprospecting based on the traditional use of plants and mushrooms has become one of the activities with the highest scientific and business importance, especially with the emergence of antibiotic-resistant microorganisms.

Probably one of the oldest proofs of mushrooms being used for medicinal purposes was that found in a frozen bag carried by Ötzie, the Iceman. At least one of two mushrooms, *F. betulina*, found in the frozen mummy's bag, was thought to have curative value. It is speculated to be a remedy for intestinal parasites, while novel research proved its health benefits going beyond antiparasitic effects.[16] The second one, *F. fomentarius*, was also a medicine in ancient Greece. The great physician of that time, Hippocrates, claimed that this polypore mushroom was known for its anti-inflammatory activity and cauterization of wounds.[15] Based on written artefacts, mushrooms were seen as cures mainly by Slavic countries and the Far East cultures, like Chinese, Japanese, Korean, and Indian.[66] However, ancient Greeks and Romans also used mushrooms to treat diseases. In a study by De Vos, mushrooms appeared in 10 European historical texts covering a 2000 year period.[67] Among the listed cultures practicing healing with mushrooms the best known is traditional Chinese medicine (TCM) with its 5000 years old practice. The majority of Chinese medicinal texts date from the period from 581 to 979 AD.[68] However, the oldest one mentioning mushrooms, particularly *Ganoderma lucidum*, is from around 100 BC, *Shennong Materia Medica*. *G. lucidum* was highly prized, categorized among the highest graded materials without known toxicity. In the case of the Slavs, there are several medical handbooks mentioning mushrooms dating from medieval times, such as the Russian *Lechebniki*, *Travnyki*, and *Vertogrady* (herbal manuscripts), and chronicles dated from the 13th century onward, and the *Chilandar Medical Codex* (15–16th centuries) written and practiced in Serbia.[19,69,70] Mushrooms were not some rare medicine among Eastern Slavs but intensively used, especially during the 16th and 17th centuries.[69] *Polyphore*, *Boletus*, *Amanita*, and *Phallus* species were used to treat illnesses from oedema and headache to cancer. *Vetrograd Prohladnyi* (end of the 17th century), the first Russian medical encyclopaedia and the first medical book officially adopted by authorities, mentions wood ear (*Auricularia auricula-judae*) as a remedy for some throat ailments.[70,71] Based on literature dating from the 19th century, Grzywnowicz reported bracket fungi as the most popular medicinal mushrooms in Polish folk medicine.[72] Among natives populating Siberia, mushrooms were regularly used, although the number of species was narrow. This was probably due to a fatalistic life outlook where spirits played a major role in a person's health and disease. In this case, only a shaman was able to deal with the supernatural forces. As Saar wrote, the Khanty tribe from Siberia used *A. muscaria*, *F. fomentarius*, *Inonotus obliquus*, *Phellinus igniarius*, and several puff-ball species.[73]

In Arabia, desert truffles (*Terfezia* spp. and *Tirmania* spp.) were cherished not only as a delicacy but as a traditional medicine too. It is believed that the prophet Mohammed recommended them as helpful in the case of eye diseases and pain. They were used for over two millennia mainly as a remedy for trachoma, the most common eye disease in the Arabia region.[36] In today's Uzbekistan spores of puff-ball species are used as a folk remedy for itching redness between the toes.[74] Indian Ayurveda considers mushrooms as calming agents (*tamasic*) for the nervous system. Although TCM and Ayurveda share some theoretical aspects, like the five basic elements, they do differ in recommendations concerning mushrooms.[75] This demonstrates how the same subject has a different context in different cultures. However, Vaidya and Rabba concluded that fungi play a major role in Ayurvedic medicine, listing mushrooms like *F. fomentarius*, *I. obliquus*, *Fomitopsis officinalis*, *Trametes flavida*, and species of the genus *Phellinus*, the way they are prepared, and the conditions for which they are used in Ayurvedic medicine.[76] In Pakistan, Yasin *et al.* explored the medicinal use of 50 mushroom species. These authors identified that mushrooms are used for treating 14 ailments.[65] The majority were used to cure sexually transmitted diseases, followed by urinary and nervous disorders. The main method of using them was for internal applications, in the form of an extract or powder, with the same practice being the most common in other parts of the world as well.

Unfortunately, the majority of folk knowledge about the medicinal uses of mushrooms has not been recorded in any form and its maintenance depends solely on rural communities and oral transmission. Inevitably, with the transition of the population to the urban areas, where Western medicine is dominant, this sensitive knowledge has been greatly lost or distorted. Haro-Luna, Ruan-Soto, and Guzmán-Dávalos observed that Mestizos from Mexico (mixed population) do not use mushrooms as medicine, contrary to the indigenous Wixaritari.[28] The older ethnic population complained that younger generations are not interested in learning about mushrooms or the traditional knowledge of their use. In general, new generations do not value natural resources. The deterioration of ethnomycological knowledge was also observed on the Balkan peninsula. Stoyneva-Gärtner and Uzunov reported a significant decline in the ability to recognize mushroom species among the rural population in Bulgaria, from 250 species in the 1950s to about 200 only 50 years later.[29]

Contrary to the decline of the traditional medicinal mushroom use knowledge, contemporary dietary supplements and the new group of food-medicine-in-between products, known as *mycoceuticals*, are going through exponential growth. This trend is especially on the rise among populations of nations without traditional mushroom knowledge, such as Americans. Mushrooms are marketed as magic ingredients, effective against the main maladies of modern times, including cancer, diabetes, and cardiovascular diseases. Although *in vitro* studies largely confirm these claims as undoubtful, randomized, double-blind, clinical studies are scarce. However, they are over-the-counter products, of natural origin, and broad activity, which makes them convenient and highly sought after.

Another specific group of mushroom products making their way into the modern medicinal culture are preparations based on psychedelic mushroom species. Once seen as a treat for society, psychedelic mushrooms and their active compound psilocybin are gaining more and more public interest as the latest research breakthroughs have occurred. As a study from 2020 found, "magic" mushrooms are perceived as being safer than heroin, cocaine, prescription painkillers, gamma-hydroxybutyrate, ecstasy, tobacco, and alcohol, which is in line with data on their actual harm.[77] Based on novel findings that psychedelic mushrooms can help in treating mental health disorders, followed by modern business pressures, especially connected with jobs in the information technology sector, the *microdosing culture* of mushroom use nested in the Silicon Valley and continues to spread. This way of using mushrooms was recently explored by the so far largest placebo-controlled study on psychedelics. Szigeti *et al.* explained the beneficial effects of microdosing by the placebo effect.[78] How the cultural perception of mushroom's medicinal use changed over time and culture is especially visible in the case of psychedelics. While contemporary consumers see it as a new panacea, cultures like the Mazatecs from Oaxaca in Mexico always associated the psychedelic experience with the cure of illness.[79] Mushrooms were their only medicine. The fruit bodies were eaten only when a person had a problem, whether physical, mental, or ethical. However, traditional and contemporary mushroom-as-medicine users share one same trait – belief. For the present, mushrooms are perceived as the answer to modern health issues.

9.4 Linguistic Use

To transfer all different aspects of culture (customs, beliefs, practices, *etc.*), besides nonverbal means a person has to use language too. Thus, language is an integral, inevitable, part of the culture.[80] Hassan also stated that there is a strong connection between culture and language contextualization. The development of language started in nonhuman primates with pre-linguistic tool use, meaning that it was situational.[81] Ancient people up-rise in nature, the most similar with today's rural communities, so, it is logical that this setting served as a context pool. This is in accordance with the findings of Taverna *et al.*[82] They discovered that conceptual and semantic human–nature relations differ among people from urban and rural areas. In rural communities, the connection with nature is stronger and humans are an integral part of nature. In ethnomycological studies, researchers found similar integration; folk names of mushrooms are derived from or connected with plants, animals, myths, natural phenomena, smell, or colour.[21,28,83,84] For example, indigenous communities in Ecuador use the name *anj kijtiutiu* for *Volvariella volvacea*, pointing to its white colour.[21] In some cases, like with the Wixaritari people from Mexico, a combination of words for smell and shape or form and animal are used for mushrooms.[28] Similarly, the association with plant and colour can be found in the Serbian language, *e.g. Hrastova sjajnica* (*G. lucidum*), which is a combination of words meaning *oak's* and *shiny*. Russian folk names of some popular mushrooms have semantic word formation 'animal→mushroom' using diminutive forms, expressing similarity based on form, colour, properties, *e.g. lisichka* (*Cantharellus*) meaning a little fox, and *svinushka* (*Paxilus*, *Tapinella*) formed from the word *svynya* (swine); as well as 'non mushroom→mushroom', on the basis of metaphorical transfer, *e.g. volna* or *volnushka* (*Lactarius torminosus*) meaning wave, and *poplavok* (*Amanita* sect. *Vaginatae*, former *Amanitopsis*) for bobber, *etc.*[85,86]

In parts of India, ethnic names for several mushroom species can overlap. These are mainly based on the specific substrate on which the mushroom is growing.[83] Names of mushrooms also point to different associations in different cultures. *Sparassis crispa* is known as cauliflower mushroom in English, while people from Mexico call it lion's head (Cabeza de leon).[33,87]

The diversity of mushroom shapes has always attracted attention, and many descriptive folk names, connected with morphology, reflect this animosity. Some tribes in the Western Ghats of India use the word *amme* meaning breast-like.[83] Mushrooms with the most unusual appearance seemed to drive more inspiration. It is no wonder that species of puffballs and phallic mushrooms are among those with the highest number of folk names. Extensive research on the folklore of gasteromycetes was published by Spooner and Læssøe, pointing to a very complex etymology of these words.[88] Depending on the culture, vernacular names can be a single word, or consist of binominal, trivernacular words, or a phrase.[21,28] However, almost every ethnomycological research paper lists vernacular names for species with which local people are familiar with only a small number indicating the linguistic implications. The deepening of this type of research requires a multidisciplinary approach and a collaborative investigation. However, it might be worth the effort since it can give novel insights into the cultural aspects of a particular society.

Some cultures from Zambia appear to have been deeply affected by mushrooms which became a part or focus of proverbs, sayings, riddles, and stories.[89] For example, the saying *tuli samfwe tumenena ukumbola* reflects on the elusive, evanescent nature of mushrooms meaning that our own life is short. A riddle from Kenya also includes mushrooms: A European standing on one leg; to which solution is *a mushroom*.[90] As Dienhart mentioned in the same paper 'riddles are means of gaining insight into the nature of the mental relationship which members of a given society perceive among the elements of material and non-material culture'.[90]

Interestingly, in the tradition of the Mazatec Indians, mushrooms had a different though language-connected use. Shamans used psychedelic mushrooms to speak for the ill person. Mushrooms enable the liberation of language in the form most probable to safely arrive and speak to the spirits.[79] The language stimulated by the mushroom presented a medicine. Thus, inexpressible, individual, and painful mental and physical states can be transformed and articulated through words. These Indians also called mushrooms the flesh of language, while the shamans were men of language.

One significant although almost overlooked language use of mushrooms (fungi) has a strong cultural implication – prejudices and phobia. There are two expressions describing the fear of fungi, *fungiphobia* and *mycophobia*, pertaining mainly to the Anglo-Saxon cultures. From special terminology, like the British name for mushrooms, *toadstool*s, to fear and disgust-loaded poems, to everyday school mockery, mushroom's presence in the English language has a negative meaning.[91] As an illustration, there is an old saying "*There are old mushroom hunters and bold mushroom hunters, but there are no old, bold mushroom hunters*". According to the *Urban Thesaurus*, a platform searching for word slang synonyms, there are 507 synonyms for the word *mushroom*.[92] The majority of them are characterized as racist, sexist, offensive, or downright appalling. This goes hand in hand with the recent work of Vráblíková who gave an insight into and speculated on less naive linguistic connections, which mushrooms have in contemporary UK society.[93] She identified the fear of East European immigrants, nationalism, imperial colonialism, the empowerment of antisemitism, and connection of these processes with the Brexit campaign. *Mushroom mobs* is a derogatory term, which appeared in *The Daily Express* article '*Eastern European mushroom mobs: New forest is stripped bare by fungi rustlers*' and used by Brexit supporters.[94] Revival of antisemitism is particularly evident from the recent republishing (2008 and 2020) of the Nazi Germany children's book *Der Giftpilz* from 1938, which translates in English to *Toadstool*. The book exploits Anglo-Saxon phobia of mushrooms, enabling dehumanization of Jews by comparing them with poisonous mushrooms. Furthermore, the author gave a linguistic example of this cultural prejudice, a folk name of *A. auricula-judae* – Jew's ear – which appeared in medieval Europe. To add to this, Kaishan and Djoulakian even compared mycophobia with queerphobia, discovering many similar prejudices existing in these cases. Based on the evidence, we can say that mushrooms are (ab)used in Anglo-Saxon languages.[95]

9.5 Folk Tales, Beliefs, and Superstitions

The collection of stories that reflect human beliefs, customs, values, lifestyle, moral issues, and explanations of natural phenomena passed from generation to generation are considered as the heart of cultural heritage. They give a deep insight into the particular culture as they are condensed manifestations of collective phantasies, worries, knowledge, sets of values, rights, and justice. In parallel, these stories serve as an instrument of cultural preservation and as a source of entertainment. The main route of transfer was oral. It enabled personal interaction between the storyteller and the audience. Body language and vocalization enabled emphasizing important messages and the audience's attentiveness. As Hogue pointed out, by transferring folk stories a strong connection is established between the generations.[96] The importance of folk tales is reflected in the example of African slaves brought to the West Indies, from the 17th to 19th centuries, whose only way to preserve their indigenous knowledge and culture was through oral practice.[97] However, oral transmission is affected by an individual, so the same story undergoes reinterpretation and recontextualization. Each reproduction/retelling contains subtle linguistic features in which we can identify the reflection of various dimensions of cultural, social, conceptual, and cognitive information.[98]

From a global perspective, folk tales and beliefs about mushrooms are numerous. These creatures are divine, they have feelings, and can be sad, happy, or sick. They are autonomous, able to see and feel, and even express their emotions through morphological changes. Besides, mushrooms can communicate with God, overlook and protect their surroundings, and enable further fructification of edible species. These are some of the rich mushroom-connected beliefs of the Wixaritari population of Mexico who are referring to mushrooms in general and not to a specific species.[28] Similarly, in the Democratic Republic of Congo, in the Tshopo province, the beliefs connected with mushrooms are mainly general. They can bring luck but also discord between lovers or friends, and they can be used to chase away birds from rice fields. The exceptions are *Termitomyces robustus* and *S. commune*. The former if eaten by a child can cause it to become an orphan,[31] while the latter is forbidden to young mothers with vaginal infections since there is a belief that it can cause the death of their child. On the contrary, in Ecuador, women who have just given birth are recommended to eat *S. commune* to restore their energy.[21] The same group of authors reports on a species-oriented belief, in which *Agaricus pampeanus* and *A. argyropotamicus* are eaten during the celebration of the *Day of the Dead*. These species are food sent by God. The older population connects mushrooms with spirits wandering in forests. In Zambia, mushrooms are part of children's tales, thus *Termitomyces titanicus* presents a shelter for gazelle, snakes, and green amoebas.[89] The Blackfoot Indians believed that puff-ball species were fallen stars and kept them as items for exorcist rituals.[99] They also practiced painting them on the bases of their wigwams since there was a belief mushrooms would provide fire in the heart.

In China, mushrooms always bear a positive connotation. *G. lucidum* is the most prized among the mushroom species and is believed to bring immortality, luck, fortune, peace, and prosperity. It is known as a heavenly herb. Even the folk names like *Ruizhi*, *Shenzhi*, and *Ziancao* emphasize the belief that this species brings good fortune and power.[100] The special appreciation for *G. lucidum* is reflected in many idioms as reported by Lin, *Ji Xiang Ru Yi* meaning good fortune and happiness or *Guo Tai Ming An* meaning country is prosperous and people live in peace.[101] It also found its place in several tales throughout history, including the *Book of Mountains and Seas*, and poems like *Nine Sons*, *Mountain Ghost*, *The Song of Lingzhi*, *On Lingzhi*, *Praise of the Goddess Luo*, *Flying Dragon*, and *Singing Trip of the Han Dynasty*.[101] In these, the divine mushroom provides actors with magical health benefits. Especially interesting is the story of the *Legend of the White Snake*, in which the snake goddess goes to the Kunlun Mountain to find the herb of immortality to save her beloved.[102] Taoism especially supported the growth of *Lingzhi culture* and the spread of belief in this mushroom's healing powers.[100] Even its shape was seen as an auspicious symbol.

In Europe, particularly in its Western part, mushrooms were considered scary and mysterious. This fear was a result of ignorance. People were disgusted and amazed by fungal sudden appearance, strange and often offensive shape, especially in the Victorian era. They grow from moist dark places, usually soil or dung, which were seen as having a negative connotation. They were connected with decay. This atmosphere also made a fruitful substrate for superstitions. To add to the negative burden placed upon mushrooms, the Western Christian churches were not mycophilic. Those who admired mushrooms were under suspicion, scorned, or even considered idiots. During the medieval times, collecting mushrooms or having them at home could result in an accusation of witchcraft, followed by trial and even the death penalty. The horror of being perceived as a witch in medieval England is observable from the words of the English theologian William Perkins "The most notorious traitor and rebel that can be is the witch, for she renounceth God himself, the king of kings, she leaves the society of his church and people, she bindeth herself in league with the Devil". People were ready to testify against their neighbours and accuse them of a connection with fungi.[103] Another, although less dangerous, belief came from the phenomenon of *fairy rings*. The folk explanations were numerous, from fairies dancing in a circle, playful dwarfs, hedgehogs in love, cows defecating in circle, the devil milking cows, to dragons burning soil by in-circle flight.[104] Some people believed that dew from the fairy ring could bring either beauty or misery by turning a person into an ugly old lady. For more detailed folk stories with mushroom motives, beliefs and superstitions one should look at the detailed paper by Dugan.[105]

9.6 Religious Use and Entheogens

Mushrooms played an important role in many ancient cults, they were worshiped, and some species served as entheogens and elements of rituals that have persisted to the present. One such species is *A. muscaria (fly agaric, mukhomor)*, which is still used in shamanic rituals of various indigenous peoples from northern Europe (Saami, or Lapp), western Siberia (Khanty and Mansi), northeastern Siberia (Chukchi, Evenki, Koryaks, and Itelmens), and North America (Ojibway), as well as Mesoamerican Indian tribes (putatively Mayan).[69,106–108] Gordon Wasson authored the famous and highly influential hypothesis about the widest diffusion of the *A. muscaria* in the archaic cultures of Eurasia, namely in the cultural strata of the Vedas. Wasson's main "test-case" was the mysterious and suggestive *mukhomor*-based image of *Soma*, which is the *god* and the *divine food* of high importance in the hymn-poetry of Rigveda.[107]

Based on Díaz's classification of psychodysleptic magic plants and fungi, *A. muscaria* belongs to the *deliriants*, which induce a delirium characterized by dim and clouded consciousness, stupor, confusion, disorientation, perception distortion, difficulties in recollection, anxiety, irritability, excitation, and behavioural disorganization employed in sorcery, purification, or exorcism rituals.[106,109] These effects are due to the presence of hallucinogenic compounds – *muscimol* and *ibotenic acid*, which are also neurotoxic. However, fly agaric contains also other toxic compounds, *e.g. muscarine* and *muscaridin*.[110] According to Vereschaka, indigenous Siberians were familiar with the peculiarities of the effect of *A. muscaria* on the human body for many centuries and developed the most acceptable methods of its use and dosage.[108] They applied different detoxification procedures and methods to increase its potency, such as drying, sucking in the mouth for a long time, maceration (both dried and fresh mushrooms), boiling (also dried and fresh), and mixing with other foods. There was a peculiar tradition to drink the urine of the shaman or of deer that had consumed the mushrooms in some Siberian tribes in order to get a 'second-hand' stimulus.[110] The consumption of fresh non-treated fly agaric was rather rare.[108]

To summon helping spirits, shamans usually ate fly agarics, which caused hallucinations, and began beating their drums.[14] People unquestioningly obeyed the fly agaric spirits and carried out all their commands, no matter how ridiculous they may have seemed.[108] In order to cure a person, a shaman (or ielta-ku) of the Vakh area in Western Siberia, prepared themselves by smoking a pipe, swallowing the smoke, and eating a few dried fly agaric mushrooms.[14] Similar to this, shamans (or curanderos) of the Valley of Puebla in Mexico added the fly agaric to tobacco and smoked the mixture to perform ritual medicinal diagnoses.[106] Lowy described the probable sacred use of *A. muscaria* among Mayan Indians of Mexico and Guatemala in relation to the Thunderbolt Legend and from the study of several codexes.[111]

Psilocybin mushrooms (*Psilocybe* spp., *Panaeolus* spp., and *Pholiotina smithii*), or so-called *magic mushrooms*, are well-known entheogens, used in long-lasting divination rituals, prophecy, and sacramental practice by various tribes in Latin America, including the Aztecs, Mazatecs, Otomí, and Totonaca.[109,112] Probably the oldest evidence of the use of *magic mushrooms* as entheogens was discovered in northern Africa in the Sahara desert. One of the Neolithic frescos depicts a bee-faced mushroom shaman holding mushrooms, presumably *Psilocybe mairei* Singer, dated from 7000 to 9000 BP.[113,114]

Díaz classified psilocybin mushrooms as *hallucinogens*, because they induce strong perceptual changes, affective intensification, and cognitive enhancement.[106,109] Their active compounds are *psilocybin* and its derivatives. Aztecs and Chichimecas called psilocybin mushrooms '*teonanácatl*', which means 'the flesh of the God'.[115,116] In Latin America, the most important mushrooms with religious uses have been *Psilocybe aztecorum* R. Heim, *P. caerulescens* Murrill, *P. hoogshagenii* R. Heim, *P. mexicana* R. Heim, *P. yungensis* Singer & A. H. Sm., *P. zapotecorum* R. Heim, *P. cubensis* (Earle) Singer, *Panaeolus sphinctrinus* (Fr.) Quél., and *Pa. squamulosus* Velen.[106,115,117]

Indigenous groups in Mexico are known to make aqueous extractions of *Psilocybes* for use in curing ceremonies, even though cooking is not required to achieve the hallucinogenic effect.[118,119] The extraction methods may be a remnant of earlier practices involving the use of *A. muscaria*.[120] The remarkable '*mushroom stones*' (of stone or pottery) of the ancient Mayas and mural frescoes found in central Mexico are striking archaeological evidence, dating from the period 1000–300 B.C. to 200 A.D. These *mushroom stones* are found in Guatemala, southern states of Mexico, and El Salvador.[112,121] The interpretations supported that these sculptures were probably associated in Mayan culture with religious ceremonies in which hallucinogenic fungi played a major role. There is an assumption that at least some of these stones may have been linked to a fertility cult.[111] Mixtec, a modern indigenous group in Mexico, are known to prepare aqueous extractions of *Psilocybe* spp., generally grinding the mushrooms on a metate and mixing them with water. Interestingly, sets of metates and manos (grinding stones) have been found in association with mushroom stones in Guatemala.[121] The association of metates with mushroom stones tends to support the prevailing theory that the mushroom stones are connected with the ritual use of hallucinogenic mushrooms.[30,121]

The similarity between Siberian and American cults of worshipping psychedelic fungi can be explained by the fact that the Americas were inhabited by a single migration wave from Siberia across the Bering Strait at most 23 000 years ago, based on genetics and archaeology results.[122,123] Considering this we can assume that '*mushroom cults*' could be even older than this migration. Apparently, mushrooms may be an integral attribute of the most ancient religious cults, especially taking into account evidence of their ritual use in the 'Cradle of Humankind', Africa.[113,124]

Anthropologist John A. Rush concluded that all Abrahamic religions – Judaism, Christianity, and Islam – are political constructions evolving out of the use of *A. muscaria* and a plethora of other mind-altering substances.[125] Shanon proposed that psychedelic mushrooms expressed their activity upon Moses when he was receiving the Ten Commandments.[126] However, there is no strong material evidence on the territory of ancient Israel supporting this theory. The only finding includes cannabis residue (cannabidiol, Δ^9-tetrahydrocannabinol, and its degradation byproduct cannabinol, identified by gas chromatography with mass spectrometry) from an ancient temple at Tel Arad.[127] Polish anthropologist Sara Benetowa (Sula Benet) argued that cannabis had been used in early Judaism and that the plant *kaneh bosem*, mentioned five times in the Hebrew Bible, and used in the holy anointing oil of the Book of Exodus, was in fact cannabis.[128] Moreover, as Schilling explains, the Hebrew Bible was a product of a compiling and editing process, while Moses is a religious and mythical figure.[129] More probable than Moses being influenced by psychedelic mushrooms, is the theory that some of the people who told the stories were. The author further points out that Judaism cannot be imagined without words (while the psychedelic experience is non-verbal), interpretation, and debate. Still, some contemporary Judaism practitioners see psychedelics as a tool of great value for Jewish mysticism. On that basis, a new religious movement, known as Psychedelic Judaism, was formed. Inevitably, mushrooms if not proved to be part of Judaism from its beginning are becoming a new cultural and religious stream.

There is an assumption that ancient cults worshipping *A. muscaria* also influenced the Christian tradition of Christmas celebration and beliefs. According to the theory of Rush, the legend of Santa derives from shamans in the Siberian and Arctic regions who collect fly agarics, dry them, drop them into *yurts* (the locals' teepee-like homes), with a bag full of hallucinogenic mushrooms as presents in late December on the winter solstice.[125] Because snow is usually blocking the doors, there was an opening in the roof through which people entered and exited, thus the chimney story. Reindeers are common in Siberia, and seek out hallucinogenic fungi, as the local humans have been known to do. Mycologist Donald H. Pfister suggested that a legend about *flying Santa's reindeer* also has origin in shamanism: 'Siberian tribesmen who ingested fly agaric may have hallucinated into thinking that reindeer were flying. Amongst the Siberian shamans, you have an animal spirit you can journey with in your vision quest'.[130] Arthur James pointed out that *A. muscaria* is found throughout the Northern Hemisphere under conifers and birch trees, with which it has symbiotic relationships, and this partially explains the practice of the Christmas tree, and the placement of bright red-and-white presents underneath, which look like fly agaric mushrooms.[131]

9.7 Art and Fashion

Besides being an emotional experience, art has often served as a portal to some other time, mindset, and culture. One of the earliest artefacts supporting the theory that mushrooms were extremely important for early civilizations are works of art, namely cave paintings. A few examples of murals and sculptures depicting mushrooms were mentioned in an earlier section, because the earliest mushroom effigies were closely connected to the religious cults of ancient people. Previously, the oldest known images of mushrooms were considered Epipalaeolithic effigies dating back to 5000–7000 BC, which were found in the Sahara desert in northern and central Africa (Algeria, Libya, Egypt, and Chad).[114,132] However, '*Mushroom head*' depictions of Bradshaw art discovered in the Kimberley region of Australia could be much older. Bradshaw art is very likely to have a Pleistocene origin, based on three independent measures that include thermoluminescence studies of wasp nests overlying the art (17 000–30 000 years). An amazing similarity of these '*Mushroom head*' images made by Australian Bradshaw and African Sandawe (Tanzania) cultures, as well as some other remarkable features, suggest a close relationship between these two cultures.[124]

The oldest fungoid effigies in Europe hypothesized to depict neurotropic fungi, especially *Psilocybe hispanica*, were recognized on the Selva Pascuala mural, a work of post-Palaeolithic rock art in Spain (6000–4000 BC).[133] In northeastern Siberia, Dikov discovered anthropomorphic mushroom petroglyphs on large exposed rocky formations in the Pegtymel River region, apparently of shamanistic significance and likely representing *A. muscaria.*[134] Pegtymel petroglyphs dated to the period from 1000 BC to 1000 AD. Besides a few neolithic findings, mushroom effigies have been identified in the Bronze Age rock art of Scandinavia, England, and southern France.[132]

In the archaeological site of Aryannoor (Kerala, India), the megalithic monuments known as *kuda-kallu* ('umbrella-stone') resemble mushrooms or have a parasol shape. These megalithic monuments belong to the Iron Age of the Indian peninsula.[135] In modern times, these forms were taken up by the German architect Jürgen Mayer, who designed the Metropol Parasol in Seville (Spain).[136]

As for contemporary art, Aleksandr Vyazemsky and Tiffany Bozic should be mentioned as artists, who are famous for their paintings of mushrooms. Vyazemsky is a world-class artist, a founder of watercolour botanical painting in Russia. His works almost exclusively depict mushrooms surrounded by a collection of debris from the fungus's environment, such as pine needles, berries, moss, and blades of grass. To achieve unrivalled realism, he always paints his mushroom portraits from life and to size. Vyazemsky illustrated the *Red Data Books of Russia* and of a few other regions, as well as the *Kola Encyclopaedia* and a few mushroom field guides. His works are featured in the Shirley Sherwood Gallery of Botanical Art in Kew Gardens (UK), Minneapolis Institute of Arts (USA), Farlow Library (Boston, USA), Grand Hotel Europe in Saint Petersburg (Russia), *etc.*[137,138] Tiffany Bozic is an American self-taught painter whose style combines nature illustration with surrealism. The complexities of natural processes like death and decomposition percolate through pictures of otherwise shining life. Mushrooms are popular heroes of her paintings. Bozic is in the forefront of artists today, redefining the aesthetics and imperatives of global change. Her work has been collected in two books, both published by Gingko Press: *Drawn By Instinct* (2012) and *Unnatural Selection* (2019).[139]

The usage of mushroom and lichen pigments as dyes is one of the most interesting and old fields of their application. A renaissance of this handicraft took place in the 1970s–80s and it became popular especially in Scandinavia, Finland, the USA, and the UK. These natural dyes, in combination with some chemical compounds, are mainly used for wool yarn and fabric colouration. Miriam C. Rice, Dorothy Beebee, and Carla and Erik Sundström made a great contribution to the study of old and the development of new colouring methods using mushrooms, as well as to popularization of this craft. The International Mushroom Dye Institute was established by Rice in 1985.[140,141]

Mushroom mycelium is one of the most promising eco-friendly materials for the fashion industry, especially as a leather substitute. A spongy felt-like material called *Amadou*, derived from fruit bodies of *F. fomentarius*, has been used for the production of hats, bags, and other artisanal objects over the centuries in Romania. The processed Amadou was also used as a plaster for wound dressing and haemostasis, before the modern sticking plaster was invented.[142] In past 5 years a true revolution in biofabrication of mushroom-derived materials has taken place. Fungi-derived leather substitutes are an emerging class of ethically and environmentally responsible fabrics that are increasingly meeting consumer aesthetic and functional expectations and winning favour as an alternative to animal and synthetic leathers. These renewable sustainable leather substitutes are obtained through the upcycling of low-cost agricultural and forestry by-products into chitinous polymers and other polysaccharides using a natural and carbon-neutral biological fungal growth process.[143]

There are a number of mushroom-based non-woven textiles that have appeared recently on the market including *Amadou Leather*™, *MuSkin*™, *MycoTEX*®, *Reishi*™, *Mylo*™, *Mylea*™, and *MOGU fabric*.[143–145] Amadou Leather™ is an award-winning 'vegan' leather, produced in commercial scale from mushroom material grown on recycled sawdust, by a London-based company.[146] MuSkin™ is produced from the fruit bodies of subtropical species *Phellinus ellipsoideus* by the Italian corporation Grado Zero Espace.[145] Both Amadou Leather™ and MuSkin™ have a natural brown colour and suede texture.

MycoTEX®, innovated by Dutch textile designer Aniela Hoitink (NEFFA), is a mushroom-based textile shaped on custom-fitted moulds, consisting only of pure mycelium, grown in Petri dishes. Such a garment can be built three-dimensionally and straight onto a model and the length can be changed, or elements can be added. This allows for growth of the right amount of material, eliminating waste.[147]

Mylo is a leather-like material, developed by Bolt Threads in partnership with Ecovative based in the USA.[148] The mycelium is grown in beds of agricultural waste and byproducts. Under certain temperature and humidity conditions, the mycelium grows upward and assembles into an organized mat of interconnected cells. After harvesting, the mycelium mat is tanned with environmentally sound dyes. Then it is compressed and imprinted with a pattern, which gives the final material. However, Mylo is not biodegradable. While Mylo is not petroleum-based, it is also not plastic-free. The Mylo fabric is between 50–85% bio-based. The goal of the Bolt Threads teams is to ultimately have a completely plastic-free product in Mylo. However, this goal has not been achieved yet.[144] Today, Mylo is being used by leading brands such as Adidas, Stella McCartney, and Lululemon.[148]

The Indonesian biotechnology company MycoTech released a range of products including shoes, sandals, handbags, wallets, and watch bands made using its mycelium-derived leather substitute Mylea in 2019.[143] Italy-based Mogu produces mycelium fabric on top of agro-industrial residues. Mogu's company mushroom fabric is non-toxic, fire-resistant, waterproof, 100% plastic-free, biodegradable, and has wide application.[144]

MycoWorks, a biotechnology company founded by Philip Ross and Sophia Wang in San Francisco (USA), grows a leather-like material called Reishi™ from *G. lucidum*.[149,150] Mycelium is grown using the readily available nutritional waste resources of sawdust, corncobs, and pistachio shells. Tissues of mycelia are stored in a jar, and within three to seven days, the mixture expands and forms clumps which are then moulded into a leather sheet, which can be dyed in different colours. The material is then baked to kill the fungal organisms and spores so that new fungal growth will not occur. MycoWorks can control many aspects of the growing conditions to create a faux leather of any shape, thickness, or size. The mycelium can also be manipulated with organic compounds to result in various textures.[151] The company is already catering to designers, customizing the material's characteristics for various consumer products.

9.8 Economic Use

From legendary stories from ancient China where the immortality mushroom (*G. lucidum*) and *Grifola frondosa* were paid for in gold or silver equivalent to their weight to the current international mushroom trade, fungi goods had been the main or additional source of income in a number of worldwide economies. *Ophiocordyceps sinensis* presents one of the most extreme cases. In Tibet, collecting and selling this species brings a significant amount of money to the rural communities, so that it is known as the 'spore money'.[152] On average this medicinal mushroom apprised as an aphrodisiac contributes to 74% of household cash income that is literally transforming the Himalayan people's livelihoods.[153] Another highly valuable species is truffles, whose prices for particularly large specimens can go up to several thousand euros. However, this is also an example of the dark side of the mushroom market. Cases of people committing fraud, kidnapping, shooting, and killing each other or poisoning opponent's dogs for truffles, are not rare in Europe, including in France, Serbia, and Italy.[155] Recently, two brothers in Serbia shot each other over a truffle dispute. The lure of mushroom money, again including truffles, but this time the prospect of their cultivation has attracted many farmers from Serbia from 2010–2014. On this occasion, people paid for truffle-inoculated hazelnut seedlings with a contract from a ghost company from London.[156] The cost of this fraud was estimated at around nine million British pounds. However, not all stories about the mushroom market are not so lurid but rather trivial. In the majority of cases, people are collecting different, usually mycorrhizal, species that are sold as fresh or raw materials for making food and dietary supplements. Lovrić *et al.* discovered that mushrooms are the most collected individual non-wood forest product among European countries.[26] In Serbia, being among the least developed countries in Europe, the mushroom trade has become one of the most important income sources, especially due to its high-quality *Boletus* species.[154] Similarly, Yilmaz and Zencirci reported the importance of wild mushroom species as a source of economic gain for the rural communities in the western Black Sea region of Turkey.[22] They identified 31 species, with 43.8% of them being important as a source of income. The prices are season-dependent, while the economically most important species were *L. deliciosus*, *L. salmonicolor*, *Morchella* spp., and *Amanita caesarea*. In Tanzania's rural communities, the principal mushroom collectors and sellers are women (76.25%), as discovered by researchers.[28,33] This is a well thought of job which brings about 400–900 US$ per year, and women doing it are known as *mothers of mushroom*.[157]

Recently, the trend of urban farming, including small-scale mushroom cultivation, has occurred worldwide. Being the only food not requiring arable land, mushrooms fit well into the novel tendencies of sustainable food production. Even the Food and Agriculture Organization (FAO) has recognized and supported this business by publishing a cultivation booklet.[158] Also, their cultivation can be managed on zero-waste or blue economy principles. This implies the use of waste materials like coffee waste, as well as the transformation of mushroom spent substrate into other products with market value.

9.9 The Emergence of New Mushroom Cultural Phenomena

The end of the 20th and especially the 21st century have brought the most radical factor influencing the contemporary culture, globalization. Once isolated and strongly protected individual cultures of smaller groups represented by individuals interacting in a web-space started to merge and influence each other to a greater degree in no time frame. Society has experienced cultural homogenization. The Internet, social networks, in particular, have enabled communication between individuals from different, in some cases small and isolated, cultures, to connect with the larger cultural groups. This resulted in the exchange and merging of their knowledge. They were able to influence each other's culture (including mushroom-connected culture) in a short time. Although technical, these instruments brought about a cultural revolution as well.[159] Today's culture is practiced in cyberspace, an imaginary area formed by a global connection of personal computers.[160] In this space the dominant language is English. It is no wonder that these new phenomena, concentrated around or including mushrooms, have some of the English 'linguistic flavour'. They are known as *cottagecore* and *goblincore*.

Cottagecore is a sort of lifestyle marked by fashion or aesthetics evoking fantasy and romanticism, the pastoral idyll. It became popular throughout 2020 as a result of the social implications of the COVID-19 pandemic. As pointed out by Fernandes *et al.*, adolescents and young people were under significant stress as the pandemic measures were prolonged.[161] People turned to in-house activities like constructing a small balcony garden, baking or knitting. For the majority of people situated in urban areas, this was the only way of accepting the limited-space reality. These were all traits of escapism, so cottagecore was a logical cyberspace step among digital natives. It is supposed to be the total opposite of the chaotic reality of pandemics. Adjacent to it developed *goblincore*, with several synonyms, like *dirtcore*, *crowcore*, *feralcore*, and *cottagegoth*. This was reserved not only for mushrooms but for all ugly and less appreciated creatures of nature, such as snails, animal skulls, worms, moss, and toads. It appeared on the social media platforms Tumblr, Pinterest, Reddit, and TikTok with more than 100 million auditoriums. It maximizes details and on many occasions resembles the clutter of a witch cottage. The only place for people to be is in their houses and apartments, so why not devote your time decorating it. For some this aesthetic evokes childhood, collecting items from nature, playing in the mud, and amusement with fairy tales or the construction of magic spaces. *Cottagecore* was at the beginning, an aspiration to show a simple but perfect, presentable way of living. However, as the pandemic prolonged showing off your perfect reality became meaningless so people retreated to their inner child without worrying how presentable they are, just like mushrooms emerging overnight through the mud and dirt. *Goblincore* aimed at showing what nature really is or, translated to our culture, who we really are. This trend also goes hand in hand with popular, but still minority, diets, such as veganism, and acceptance of marginal groups of people and groups with restricted rights, such as the LGBT population.[95]

Mobile devices, the Internet, and cell phone applications transformed the way of teaching and learning and a new method, *mobile learning*, emerged.[162] Similar to Internet learning, mobile learning enables fast, even in-the-moment changes, which occur constantly and make this way of education sustainable. This translates into the constant expansion and dramatic increase of the mobile application use rate.[163] Moreover, applications are accessible and attractive for the majority of the population with mobile devices. In mycology, this sort of learning appeared recently and proved suitable, since the previous practice of using field guides was not very practical. However, one of the main features of a good application is reliability, which can hardly be achieved in this instance. Mushroom identification applications are based on machine-learning recognition of photographs, which are uploaded by an individual, and depend upon those being previously used to train the system.[164] Therefore, the application is based only on visual recognition, lacking all other aspects of species identification that one can encounter when foraging in the field. On the other hand, there are applications developed to help modern mushroom farmers. These are known as *smart farming*, which follows the fourth industrial revolution, allowing automation and remote operation.[165,166] These apps offer whole-cycle management and support, from substrate requirements and preparation, pest and disease control, to composing of mushroom spent substrate, and even further mushroom-based product development. In the work of Thapwiroch *et al.* an *Easy-Mushroom* mobile application was developed to assist mushroom farmers. The authors identified the surrounding humidity as the key to successful cultivation.[167] For more accurate results, light intensity was also included. It was concluded that the app would be a useful tool for those growing mushrooms, especially when predicting yield. In another attempt to develop such an application, performed by Mahmud *et al.*, besides humidity, the researchers were able to monitor temperature and CO_2 levels.[165] Sensors were used to collect the mass amount of data, which would serve to form the Internet of Things (IoT) sensing system. Finally, the system was supposed to be trained to automatically regulate the irrigation system in the growing room. The same parameters were used by Shakir *et al.* for developing an android web-based application.[166] Kassim, Mat, and Yusoff claimed to have succeeded in increasing the mushroom yield by almost twice using an IoT-based monitoring and control system.[168] Besides, these authors emphasized that smart mushroom farming enables optimization of resources like water and fertilizer, while at the same time maximizing the quality. Another group of authors reported as having succeeded in using the IoT to enhance the prototype of smart *G. lucidum* farming.[169] However, mushroom cultivation is more complex and includes more parameters than were used in the discussed work. Far from advising restraint from using apps, mushroom growers and other farmers should rely on the fundamental knowledge of crop biology, expertise, and their own experience.

9.10 Quantitative Methods for Assessing the Cultural Use of Mushrooms

Besides the external layers of *culture*, like language and traditions, which are studied mainly qualitatively, contemporary science tends to reach further through applying mathematical models and measurable parameters to assess the importance of each cultural aspect. This *cultural dissection* is important since *culture* is a complex construct with more than 150 definitions and up to 700 000 articles provided by social science electronic databases.[2] Culture is a sociological phenomenon that, at first, seems impossible to measure. However, the 20th century and the development of computational technology offered novel possibilities. As Taras, Rowney, and Steel reported, more than 100 instruments for quantitative assessment of culture have been developed so far.[2] However, before using any of them, precise boundaries should be set up, particularly to avoid vagueness of the term *culture* and to improve the *use quality* of the particular research. In ethnomycological research, measuring instruments are fairly new, and researchers are still adapting the terminology and methods of cultural assessment.

The *cultural significance* index (CS) represents a growing trend in ethnobotanical research and can be considered as the crowning method for the evaluation of biological taxa, mainly plants, and recently of mushrooms too, in a defined cultural context.[33,170,171] It is a valuable ethnobotanical tool, whether it is used to determine the importance of a single species or a family. In a modified form it enables cross-cultural research, as well as the establishment of a protection strategy.[172] CS can be defined as the sum of the use-value of a plant (mushroom) species, represented by different indices, like relative importance, specific reported use, frequency of mention, and others.[31,33] These indices are actually variables or use categories identified to affect the CS. As with other cultural studies the number of variables can go from three to five or seven, depending on the study's scope and the researcher's opinion. Still, too many variables can overwhelm the study and make it imprecise, so the best practice is to choose the most relevant one to a given setting.[2]

The *use quality* variable first appeared in research by Turner and was later elaborated in Pieroni's work as a *compound index* developed for wild food botanicals occurring in Italy.[170] Pieroni's model presented a significant methodological step forward, since the previously used *freelisting method*, suffered from bias, and was not able to clarify the specific role of botanical taxa.[170] Similarly, the consensus use index, which represents the extent of a general agreement in a group of people, might overlook the species, which are less known or with fewer mentions.[173,174] However, the compound index suffers from subjectivity since the researcher, but not the questionnaire respondents, defines the variables and a score-value relation, as pointed out by Thomas *et al.*[175] In addition, Pieroni's method cannot be transferred literally since some of the variables have no practical meaning when applied to mushrooms, for example *part used index*. In most cases, mushrooms are used as a whole, unlike plants. Moreover, some indices like the *food-medicinal role index* are not applicable worldwide, because not every culture recognizes the medicinal value of mushrooms.[33] On the other hand, new and more meaningful indices (*e.g.*, *the economic index*), which fit in the context of a particular community, might be introduced in areas where mushrooms are an important source of income.[22] Another approach to access CS is to allow participants to rate the specific indices by themselves, as done by Stagegaard, Sørensen, and Kvist, or to combine informant consensus and the diversity of reported plant use.[176] Thomas *et al.* observed that the latter method enables no more than one quality assessment per use category, which in some cases can lead to a loss of data if a species has multiple uses within different use categories.[175] Thus, the authors applied the *informant indexing technique* and the *quality use value index*, which are less researcher-biased since they include no primary subjectively assigned categories. This is more objective and incorporates the quality of all individual uses. Respondents are a source of individual plant use information, which then forms the basis of cultural indices. However, ethnomycological studies dealing with quantitative assessments, which have been published so far, are mostly based on Pieroni's method. In addition, the freelisting method, if based on frequency and order of mentioning, enables cross-cultural comparison of those communities, which are ethnically different, but share the same mushroom-gathering site.[25] As noted by Quinlan, the freelisting method can provide high-quality data on people's medical understanding, attitudes, and behaviours when proper attention has been given to the details.[177]

Most researchers used open-ended semi-structured questionnaires or interviews.[28,33,55,178] Although closed questions present many advantages, mainly easiness to collect and process, open questions provide better insight into a respondent's conception of the topic/term of interest. Since the knowledge of mushrooms exists mainly in older populations, some researchers have developed a graphical stimuli model for some of the examined categories such as in the case of *perceived abundance index* or *taste score appreciation index*.[25,33]

The existing literature presents a good insight into the state of quantitative ethnomycology – it is a developing area. The existing tools are still a work in progress although they have prospects to manage big-data sets, see the results, and attract scientific interest. There are several models in use, but they are mainly based on subjectively formed indices, which makes it difficult to compare the existing studies. Interestingly, there is still no research reporting on cross-cultural comparison of ethnomycological knowledge or any of its indices.

In most of the presented studies, researchers used a convenience sample, too small sample, or a not sufficiently based target group. For example, Živković *et al.* interviewed only participants from two mycological societies.[84] Similarly, Stoyneva-Gärtner *et al.* interviewed only professors and students to examine the use of polypore fungi in the production of alcoholic beverages in Bulgaria.[178] Haro-Luna *et al.* questioned only those Wixaritari informants who were perceived by their community to know the most about mushrooms, which, again, does not represent a general culture.[28] In studies reporting on myths and beliefs, there is usually a small number of participants (sometimes only one or few) describing a specific myth, which raises a question: "is it a group phenomenon, existing in one family, class/rank inside of a tribe, or community?"

Although young, the field of quantitative assessment of ethnomycological knowledge has already developed a number of tools, mainly by taking them from botany and modifying them to better suit the mycological field. So far, the indices used in the literature have provided results on traditional cultural uses. With the fast changes in the time of digital technologies and phenomena recently evolved around mushrooms, it would be useful to construct new indices and apply them in analysing contemporary cultural trends and implications they might have for the future.

9.11 Future Cultural Use of Mushrooms

We have become a global society. This means a more homogeneous culture, especially in the domain of cyberspace and digital natives. For certain, we can say that a significant part of our culture has transformed into a technology hybrid. Our language is digitalized, and through computers, it is transformed into a binary code. The development of science has enabled us to communicate with other more silent species, like plants. But this was not enough. We started to look to space, anticipating contact with alien species. We even theorized upon the out-of-this-galaxy origin of fungi.[179] In all these matters, the key is communication or language as an integral part of our culture. But how can two distinct species from which one does not even speak in a traditional sense communicate? Science has proposed biological communication. Although we might think that humans are the pioneers in this sense, mushrooms were already there. The fact that they create symbiotic communities with trees, plants, and bacteria illustrates that mushrooms already have a highly developed interspecies language. Somehow they coded the many processes occurring in the underground/substrate network and made it harmonious. This non-existent language exists after all. Microorganisms and plants do communicate by altering their bodies. The existence of symbiotic communities depends upon biological communication, meaning that there are ways of transferring the untranslatable. Therefore, the future use of mushrooms might go in this direction; we might embrace mushroom language by studying their physiology and biochemistry, experiencing their psychedelic properties, and merging it with the pool of ethnomycological knowledge. Or is this already happening as we can see with the example of bitcoin as proposed by Brandon Quittem?[180]

Acknowledgements

This work was supported by the Ministry of Education, Science and Technological Development, Republic of Serbia (Grant No: Contact No. 200116).

References

1. G. Martinez, S. Costas and Ó. Ferreira, *Adv. Clim. Change Res.*, 2020, **11**, 297–309.
2. V. Taras, J. Rowney and P. Steel, *J. Int. Manage.*, 2009, **15**, 357–373.
3. M. Gradstein, *Rev. Econ. Dyn.*, 2021, **44**, 91–103.
4. M. Iyigun, J. Rubin and A. Seror, *Eur. Econ. Rev.*, 2021, **135**, 103734.
5. M. Salehan, D. J. Kim and J. N. Lee, *Inf. Manage.*, 2018, **55**, 725–745.
6. M. Lityńska-Zając and A. Czekaj-Zastawny, *J. Archaeol. Sci.*, 2021, **35**, 102710.
7. M. E. Biwer and A. M. VanDerwarker, *Ethnobiology Lett.*, 2015, **6**, 28–31.
8. J. Vunduk and S. Veljović, in *Advances in Macrofungi Industrial Avenues and Prospects*, ed. K. R. Sridhar and S. K. Deshmukh, CRC Press

Taylor & Francis Group, Boca Raton, Florida, USA, 1st edn, 2021, ch. 8, pp. 108–141.
9. N. el-Guebaly and A. el-Guebaly, *Int. J. Addictions*, 1981, **16**, 1207–1221.
10. B. P. Akers, J. F. Ruiz, A. Piper and C. A. P. Ruck, *Econ. Bot.*, 2011, **65**, 121–128.
11. D. H. Pfister, *Mycologia*, 1988, **80**, 11–13.
12. R. E. Schultes, *J. Ethnopharmacol.*, 1987, **19**, 117–118.
13. N. Lopez, *J. Conscious Evol.*, 2020, **16**, 6.
14. A. A. Znamenski, *Shamanism in Siberia: Russian records of indigenous spirituality*, Springer, Netherlands, 2003.
15. P. Stamets and H. Zwickey, *Integr. Med.*, 2014, **13**, 46–47.
16. J. Vunduk, A. Klaus, M. Kozarski, P. Petrovic, Z. Zizak, M. Niksic and L. J. L. D. Van Griensven, *Int. J. Med. Mushrooms*, 2015, **17**, 1113–1125.
17. M. A. Bondartseva, I. V. Zmitrovich, S. P. Arefyev and V. I. Kapitonov, *Mikol. Fitopatol.*, 2021, **55**, 382–384.
18. S. T. Chang and S. P. Wasser, *Int. J. Med. Mushrooms*, 2012, **14**, 95–134.
19. S. Jarić, M. Mitrović, L. Djurdjević, O. Kostić, G. Gajić, D. Pavlović and P. Pavlović, *J. Ethnopharmacol.*, 2011, **137**, 601–619.
20. V. Montalto, M. C. J. Tacao, S. Langedijk and M. Saisana, *Cities*, 2019, **89**, 167–185.
21. P. Gamboa-Trujillo, F. Wartchow, C. Cerón-Martinez, D. Andi, P. Uwinjin, G. Grefa, M. Entza, E. Chimbo, J. Payaguaje, N. Piyaguaje, D. Payaguaje, V. Licuy, V. López, M. Mendua, M. Criollo, M. Jati, S. De La Cruz, M. Calazacón, S. Flores, E. Aules, C. Aigaje, M. De Aro, A. Morales, Y. Murillo, G. Farinango and T. B. Gibertoni, *Ethnobotany Res. Appl.*, 2019, **18**, 1–15.
22. H. Yilmaz and N. Zencirci, *Econ. Bot.*, 2016, **70**, 270–284.
23. W. Rubel and D. Arora, *Econ. Bot.*, 2009, **63**, 227.
24. C. Nai and V. Meyer, *Fungal Biol. Biotechnol.*, 2016, **3**, 10.
25. L. Alonso-Aguilar, A. Montoya, A. Kong, A. Estrada-Torres and R. Garibay-Orijel, *J. Ethnobiol. Ethnomed.*, 2014, **10**, 27.
26. M. Lovrić, R. Da Re, E. Vidale, I. Prokofieva, J. Wong, D. Pettenella, P. J. Verkerk and R. Mavsar, *For. Policy Econ.*, 2020, **116**, 102175.
27. U. Peintner, S. Schwarz, A. Mešić, P. A. Moreau, G. Moreno and P. Saviuc, *PLoS One*, 2013, **8**, e63926.
28. M. X. Haro-Luna, F. Ruan-Soto and L. Guzmán-Dávalos, *IMA Fungus*, 2019, **10**, 16.
29. M. P. Stoyneva-Gärtner and B. A. Uzunov, *J. Nutr. Food Sci.*, 2015, **5**, 5.
30. V. P. Wasson and G. R. Wasson, *Mushrooms Russia and history*, Pantheon Books, New York, 1957, vol. 1–2.
31. H. M. Kamalebo, H. N. S. W. Malale, C. M. Ndabaga, J. Degreef and A. De Kesel, *J. Ethnobiol. Ethnomed.*, 2018, **14**, 13.
32. J. Parladé, F. Martínez-Peña and J. Pera, *For. Ecol. Manage.*, 2017, **390**, 73–79.
33. R. Garibay-Orijel, J. Caballero, A. Estrada-Torres and J. Cifuentes, *J. Ethnobiol. Ethnomed.*, 2007, **3**, 4.

34. D. N. Pegler, *Mycologist*, 2000, **14**, 86–87.
35. J. Vunduk, M. Kozarski, I. Djekic, I. Tomašević and A. Klaus, *Eur. Food Res. Technol.*, 2021, **247**, 829–838.
36. Q. A. Mandeel and A. A. A. Al-Laith, *J. Ethnopharmacol.*, 2007, **110**, 118–129.
37. A. Kalotas, *Fungi of Australia*, Australian Biological Resources Study/ CSIRO Publishing, 1996, vol. 1B.
38. A. H. R. Buller, *Trans. Br. Mycol. Soc.*, 1914, **5**, 21–66.
39. G. Ewald, G. Vasas and C. Locsmándi, *Gombászok kézikönyve*, CSER Kiadó, Budapest, 2017.
40. M. M. Sukhomlyn and V. V. Dzhagan, Mushrooms of Ukraine, in *Atlas-handbook*, Km-buks, Kiev, 2014.
41. N. Sitta, P. Davoli, M. Floriani and E. Suriano, *Guida ragionata alla commestibilità dei funghi*, Regione Piemonte, 1st edn, 2021.
42. S. P. Wasser, *Edible and poisonous mushrooms of the Carpathian Mountains*, Karpaty Press, Uzhgorod, Ukraine, 2nd edn, 1990.
43. O. S. Gapienko and Y. A. Shaporova, in *Flora of Belarus*, ed. V. I. Parfenova, Belorurusskaya nauka, Minsk, 2012, vol. 1, vol. 1, p. 199.
44. L. Pacheco-Cobos, M. Rosetti, C. Cuatianquiz and R. Hudson, *Evol. Hum. Behav.*, 2010, **31**, 289–297.
45. M. Kozarski, A. Klaus, J. Vunduk and M. Nikšić, *Food Feed Res.*, 2020, **47**, 23–32.
46. X. Lu, M. A. Brennan, L. Serventi, J. Liu, W. Guan and C. S. Brennan, *Food Chem.*, 2018, **264**, 199–209.
47. S. Novakovic, I. Djekic, A. Klaus, J. Vunduk, V. Djordjevic, V. Tomović, B. Šojić, S. Kocić-Tanackov, J. M. Lorenzo, F. J. Barba and I. Tomasevic, *Foods*, 2019, **8**, 635.
48. S. Novakovic, I. Djekic, A. Klaus, J. Vunduk, V. Đorđević, V. Tomovic, B. Šojić, S. Kocić-Tanackov, J. M. Lorenzo, F. J. Barba and I. Tomasevic, *J. Food Process. Preserv.*, 2020, **44**(8), DOI: 10.1111/jfpp.14556.
49. I. Patinho, E. Saldaña, M. M. Selani, A. C. B. Teixeira, B. S. Menegali, T. C. Merlo, J. D. Rios-Mera, M. D. B. Dargelio, H. Rodrigues and C. J. Contreras-Castillo, *Food Res. Int.*, 2021, **147**, 110551.
50. L. Wang, C. Li, L. Ren, H. Guo and Y. Li, *J. Food Sci.*, 2019, **84**, 3091–3098.
51. T. A. Nicklas, C. E. O'Neil and V. L. Fulgoni, *J. Acad. Nutr. Diet.*, 2013, **113**, A86.
52. G. Sogari, J. Li, Q. Wang, M. Lefebvre, M. I. Gómez and C. Mora, *Food Qual. Preference*, 2020, 1–9.
53. United Nations, https://sdgs.un.org/goals (accessed August, 2021).
54. P. de Frutos, *For. Policy Econ.*, 2020, **112**, 102093.
55. S. S. Lee, Y. S. Chang and M. N. R. Noraswati, *For. Ecol. Manage.*, 2009, **257**, 2062–2065.
56. B. Bringye, M. Fekete-Farkas and S. Vinogradov, *Agriculture*, 2021, **11**, 677.
57. T. Fernandes, C. Garrine, J. Ferrão, V. Bell and T. Varzakas, *App. Sci.*, 2021, **11**, 4221.

58. A. Y. Biketova, *Planet. Gribov*, 2016, **10**, 52–56.
59. O. Godorova and R. Kuznetsov, *Planet. Gribov*, 2016, **9**, 38–45.
60. The Wild Mushroom Association in Israel, https://society.mushrooms.org.il/ (accessed October 2021).
61. C. Mullen, https://www.bizjournals.com/bizwomen/news/latest-news/2019/05/gen-z-changes-the-taste-of-the-food-industry.html?page=all (accessed July 2021).
62. S. Im-oun, P. Kotruchin, P. Thinsug, T. Mitsungnern, P. Techa-atik and C. Pongchaiyakul, *Complementary Ther. Med.*, 2018, **39**, 43–48.
63. L. Roseman, Y. Ron, A. Saca, N. Ginsberg, L. Luan, N. Karkabi, R. Doblin and R. Carhart-Harris, *Front. Pharmacol.*, 2021, **12**, 607529.
64. D. Cora-Bramble, F. Tielman and J. Wright, *Clin. Pediatr. Emerg. Med.*, 2004, **5**, 102–108.
65. H. Yasin, M. Zahoor, Z. Yousaf, A. Aftab, N. Saleh, N. Riaz and B. Shamsheer, *Phytomedicine*, 2019, **54**, 43–55.
66. S. P. Wasser, *Int. J. Med. Mushrooms*, 2005, **7**, 363–364.
67. P. De Vos, *J. Ethnopharmacol.*, 2010, **132**, 28–47.
68. D. Lu, *Int. J. Med. Mushrooms*, 2014, **16**, 189–204.
69. B. P. Vasilkov, *The study of mushrooms in the USSR: Historical and bibliographic essay*, Izdatelstvo Akademii Nauk SSSR, Moscow-Leningrad, 1953, p. 192.
70. I. A. Dudka, *Int. J. Med. Mushrooms*, 2001, **3**, 1.
71. G. F. Lapteva, *Zemskyi Vrach*, 2010, **4**, 44–46.
72. K. Grzywnowicz, *Int. J. Med. Mushrooms*, 2001, **3**, 154.
73. M. Saar, *J. Ethnopharmacol.*, 1991, **31**, 175–179.
74. E. Sezik, E. Yesilada, H. Shadidoyatov, Z. Kulivey, A. M. Nigmatullaev, H. N. Aripov, Y. Takaishi, Y. Takeda and G. Honda, *J. Ethnopharmacol.*, 2004, **92**, 197–207.
75. H. Sharma and H. M. Chandola, in *Bioactive Nutraceuticals and Dietary Supplements in Neurological and Brain Disease*, ed. R. R. Watson and V. R. Preedy, Academic Press, 1st edn, 2015, ch. 18, pp. 173–177.
76. J. Vaidya and A. Rabba, *Mycologist*, 1993, **7**, 131–133.
77. C. A. Roberts, I. Osborne-Miller, J. Cole, S. H. Gage and P. Christiansen, *J. Psychopharmacol.*, 2020, **34**, 999–1007.
78. B. Szigeti, L. Kartner, A. Blemings, F. Rosas, A. Feilding, D. J. Nutt, R. L. Carhart-Harris and D. Erritzoe, *eLife*, 2021, **10**, e62878.
79. H. Munn, in *Hallucinogens and Shamanism*, ed. M. J. Harner, Oxford University Press, 1973, ch. 7, pp. 86–122.
80. Z. M. Hassan, *Procedia-Social and Behavioral Sciences*, Elsevier, 2014, pp. 31–35.
81. A. Kuhle, *Lang. Sci.*, 2018, **67**, 72–88.
82. A. S. Taverna, S. R. Waxman, D. L. Medin, N. Moscoloni and O. A. Peralta, *J. Cognit. Cult.*, 2014, **14**, 213–233.
83. N. C. Karun and K. R. Sridhar, *Data Brief*, 2017, **14**, 320–328.
84. J. Živković, M. Ivanov, D. Stojković and J. Glamočlija, *J. Fungi*, 2021, **7**, 349.

85. G. A. Nikolaev, *Gwary Dziś*, 2019, **8**, 151–157.
86. O. Senk-Szelągiewicz, *Structural and semantic organization and functioning of mushroom names in Russian*, Wydział Neofilologii UAM w Poznaniu, Poznan, 2016.
87. R. Kiyama, Y. Furutani, K. Kawaguchi and T. Nakanishi, *Sci. Rep.*, 2018, **8**, 16053.
88. B. Spooner and T. Laessøe, *Mycologist*, 1994, **8**, 119–123.
89. G. D. Piearce, *Bull. Brit. Mycol. Soc.*, 1981, **15**, 139–142.
90. J. M. Dienhart, *J. Pragmatics*, 1999, **31**, 95–125.
91. D. Arora, in *Fungal Conservation (Issues and Solutions)*, ed. D. Moore, M. M. Nauta, S. E. Evans and M. Rotheroe, Cambridge University Press, 2009, pp. 105–110.
92. Urban Thesaurus – Find Synonyms for Slang Words, https://urbanthesaurus.org/ (accessed July 2021).
93. L. Vráblíková, *Fem. Encounters: J. Crit. Stud. Cult. Polit.*, 2021, **5**, 05.
94. R. Perring, https://www.express.co.uk/news/uk/591570/Gangs-mushrooms-black-market-New-Forest-Eastern-European-fungi (accessed October 2021).
95. P. Kaishian and H. Djoulakian, *Catal.: Fem. Theory Technosci.*, 2020, **6**, 1–26.
96. J. N. Hogue, Encyclopedia of Insects, in *Folk Beliefs and Superstitions*, ed. V. H. Resh and R. T. Cardé, Academic Press, Elsevier, 2nd edn, 2009, pp. 372–376.
97. C. A. Smart, *IFLA J.*, 2019, **45**, 16–25.
98. F. Karsdorp and L. Fonteyn, *Palgrave Commun.*, 2019, **5**, 25.
99. Á. M. Nieves-Rivera and D. A. White, *Mycologist*, 2006, **20**, 22–25.
100. Z. Lin, *Adv. Exp. Med. Biol.*, 2019, **1181**, 1–13.
101. Z. Lin, *Lingzhi from Mistery to Science*, Peking University Medical Press, Beijing, China, 2009.
102. C. Lee, *Psychol. Perspect.*, 2007, **50**, 235–253.
103. E. Gruffydd, *Bull. Brit. Mycol. Soc.*, 1985, **19**, 63–65.
104. G. Rutter, *Field Mycol.*, 2002, **3**, 56–60.
105. F. M. Dugan, *North. Am. Fungi*, 2008, **3**, 23–72.
106. J. L. Díaz, *J. Psychoact. Drugs*, 1979, **11**, 71–101.
107. D. Ioffe, *Sov. Post-Sov. Rev.*, 2020, **47**, 135–184.
108. E. A. Vereschaka, in *Sibirskij sbornik 4: Grani social'nogo: Antropologicheskie perspektivy issledovanija social'nyh otnoshenij i kul'tury*, ed. V. N. Davydov and D. V. Arzjutov, MAJe RAN, Sankt-Peterburg, 2014, pp. 326–339.
109. J. L. Díaz, *Phenomenol. Cognit. Sci.*, 2010, **9**, 159–170.
110. D. Michelot and L. D. Melendez-Howell, *Mycol. Res.*, 2003, **107**, 131–146.
111. B. Lowy, *Mycologia*, 1974, **66**, 188–191.
112. H. Nyberg, *Karstenia*, 1992, **32**, 71–80.
113. G. Guzmán, *Acta Bot. Mex.*, 2012, **100**, 79–106.
114. G. Samorini, *Integration*, 1992, **2**, 69–78.
115. R. E. Schultes, *Bot. Mus. Leafl.*, 1939, 7, 37–56.

116. R. Heim and R. G. Wasson, *Les Chamignons Hallucinogenes du Mexique: Etudes Ethnologiques, Taxonomiques, Biologiques, Physiologiques et Chimiques*, Editions du Muséum National d'Histoire Naturelle, Paris, 1958, vol. 1.
117. R. E. Schultes and A. Hofmann, *The botany and chemistry of hallucinogens*, Charles C. Thomas, Springfield, Ill, 2nd edn, 1980.
118. J. Ott, *J. Psychoact. Plants Compd.*, 1998, **1**, 9–37.
119. R. G. Wasson, *Soma and the Fly Agaric: Mr. Wasson's Rejoinder to Professor Brough*, Botanical Museum of Harvard, 1972.
120. K. Feeney, in *Entheogens and the development of culture: The anthropology and neurobiology of ecstatic experience*, ed. J. Rush, North Atlantic Books, Berkeley, CA, 2013, pp. 279–318.
121. S. Borhegyi, *Am. Antiq.*, 1961, **26**, 498–504.
122. M. R. Bennett, D. Bustos, J. S. Pigati, K. B. Springer, T. M. Urban, V. T. Holliday, S. C. Reynolds, M. Budka, J. S. Honke, A. M. Hudson, B. Fenerty, C. Connelly, P. J. Martinez, V. L. Santucci and D. Odess, *Science*, 2021, **373**, 1528–1531.
123. M. Raghavan, P. Skoglund, K. Graf, M. Metspalu, A. Albrechtsen, I. Moltke, S. Rasmussen, T. W. Stafford Jr, L. Orlando, E. Metspalu, M. Karmin, K. Tambets, S. Rootsi, R. Mägi, P. F. Campos, E. Balanovska, O. Balanovsky, E. Khusnutdinova, S. Litvinov, L. P. Osipova, S. A. Fedorova, M. I. Voevoda, M. DeGiorgio, T. Sicheritz-Ponten, S. Brunak, S. Demeshchenko, T. Kivisild, R. Villems, R. Nielsen, M. Jakobsson and E. Willerslev, *Nature*, 2014, **505**, 87–91.
124. J. Pettigrew, *Clin. Exp. Optom.*, 2011, **94**, 403–417.
125. J. A. Rush, *Failed God: Fractured myth in a fragile world*, Frog Books, Berkeley, CA, 2008.
126. B. Shanon, *Time Mind*, 2008, **1**, 51–74.
127. E. Arie, B. Rosen and D. Namdar, *Tel Aviv*, 2020, **47**, 5–28.
128. R. Robinson, *The great book of hemp: The complete guide to the environmental, commercial, and medicinal uses of the world's most extraordinary plant*, Park Street Press, Rochester, 1995.
129. C. L. Schilling, *Zen Judaism The Case Against a Contemporary American Phenomenon*, Palgrave Macmillan, 2021, pp. 67–111.
130. D. Main, https://www.livescience.com/25731-magic-mushrooms-santa-claus.html (accessed October 2021).
131. A. James, *Mushrooms and Mankind: The Impact of mushrooms on human consciousness and religion*, The Book Tree, US, US, 2003.
132. G. Samorini, *The stone mushrooms of Thrace*, Ekataios, Alexandroupoli, 2012, pp. 16–44.
133. B. P. Akers, J. F. Ruiz, A. Piper and C. A. P. Ruck, *Econ. Bot.*, 2011, **65**, 121–128.
134. N. N. Dikov, *Rock art puzzles of Ancient Chukotka: Pegtymel petroglyphs*, Nauka, Moscow, 1971.
135. G. Samorini, *Integration*, 1995, **6**, 33–40.
136. M. L. Gargano, *Bocconea*, 2019, **28**, 183–188.

137. G. Popov, к-я.рФ/xudozhnik-aleksandr-vyazmenskij/ (accessed October 2021).
138. S. Sherwood, *The Shirley Sherwood collection: Modern masterpieces of botanical art*, Kew Publishing, Richmond, 2019.
139. Tiffany Bozic, http://tiffanybozic.com/about (accessed August 2021).
140. D. M. Beebee, www.mushroomsforcolor.com/HISTORY.htm (accessed October 2021).
141. M. C. Rice, *Mushrooms for dyes, paper, pigments and Myco-Stix™*, Mushrooms for Color Press, 2008.
142. A Legend Reborn, www.revolution-amadou.com/a-legend-reborn (accessed October 2021).
143. M. Jones, A. Gandia, S. John and A. Bismarck, *Nat. Sustainability*, 2021, **4**, 9–16.
144. A. Gomez, www.theuptide.com/mushroom-fabric-clothing-made-from-mushrooms-mycelium-cells/ (accessed July 2021).
145. C. Riccio, www.lifegate.com/muskin-leather-mushrooms (accessed August 2021).
146. www.amadouleather.com (accessed June 2021).
147. C. Nai and V. Meyer, *Fungal Biol. Biotechnol.*, 2016, **3**, 10.
148. https://www.mylo-unleather.com/ (accessed May 2021).
149. P. Ross, *Rot on! State of the World's Fungi Symposium*, Royal Botanic Gardens, Kew, 2018.
150. www.mycoworks.com (accessed June 2021).
151. K. Deeg, Z. Gima, A. Smith, O. Stoica and K. Tran, in PH 290 – Greener Solutions Fall, 2017.
152. D. Winkler, *Field Mycol.*, 2010, **11**, 60–67.
153. P. K. Yadav, S. Saha, A. K. Mishra, M. Kapoor, M. Kaneria, M. Kaneria, S. Dasgupta and U. B. Shrestha, *Oryx*, 2018, **53**, 247–255.
154. R. Mandić, M. Adžemović and Ž. Marjanović, *Nat. Conservation*, 2018, **25**, 31–53.
155. https://www.novosti.rs/drustvo/vesti/942921/pecurke-zbog-kojih-srbi-ubijaju-ovo-najunosnije-lokacije-tartufe.
156. https://www.gov.uk/government/news/court-shuts-down-companies-behind-9m-truffle-scam.
157. D. D. Tibuhwa, *J. Ethnobiology Ethnomed.*, 2013, **9**, 49.
158. E. Marshall and N. G. T. Nair, *Make money by growing mushrooms*, Rural Infrastructure and Agro-Industries Division Food and Agriculture Organization of the United Nations, Rome, 2009.
159. M. Combi, in *Cultural Heritage in a Changing World*, ed. K. Borowiecki, N. Forbes and A. Fresa, Springer, Cham, 2016, pp. 3–15.
160. www.dictionary.cambridge.org (accessed April 2021).
161. B. Fernandes, U. N. Biswas, R. Tan-Mansukhani, A. Vallejo and C. A. Essau, *Rev. Psicol. Clin. Con Ninos Adolesc.*, 2020, **7**, 59–65.
162. K. Demir and E. Akpınar, *Malays. Online J. Educ. Technol.*, 2018, **6**, 48–59.
163. R. Islam, R. Islam and T. A. Mazumder, *Int. J. Eng. Technol.*, 2010, **10**, 104–111.

164. D. Mihail-Vaduva, *Inf. Econ.*, 2019, **23**, 72–88.
165. M. S. A. Mahmud, S. Buyamin, M. M. Mokji and M. S. Z. Abidin, *Indones. J. Electr. Eng. Comput. Sci.*, 2018, **10**, 847.
166. A. A. Shakir, F. Hakim, M. Rasheduzzaman, S. Chakraborty, T. U. Ahmed and S. Hossain, 2019 International Conference on Electrical, Computer and Communication Engineering (ECCE), IEEE, 2019, pp. 1–6.
167. K. Thapwiroch, A. Kumlue, N. Sayong, P. Taprasan and S. Puengsungewan, *Indones. J. Educ. Res. Technol.*, 2021, **1**, 1–4.
168. M. R. M. Kassim, I. Mat and I. M. Yusoff, IEEE, Sydney, NSW, Australia, 2019.
169. O. Chieochan, A. Saokaew and E. Boonchieng, 2017 14th International Joint Conference on Computer Science and Software Engineering (JCSSE), IEEE, 2017.
170. A. Pieroni, *J. Ethnobiol.*, 2001, **21**, 89–104.
171. H. Shaheen, R. Qureshi, M. F. Qaseem, M. S. Amjad and P. Bruschi, *Eur J. Integr. Med.*, 2017, **12**, 27–34.
172. W. Sujarwo and G. Caneva, *J. Cult. Herit.*, 2016, **18**, 342–348.
173. M. O. Faruque, S. B. Uddin, J. W. Barlow, S. Hu, S. Dong, Q. Cai, X. Li and X. Hu, *Front. Pharmacol.*, 2018, **9**, 40.
174. Y. Tsuchiya and N. Hiramoto, *Inf. Sci.*, 2018, **439–440**, 50–60.
175. E. Thomas, I. Vandebroek, S. Sanca and P. Van Damme, *J. Ethnopharmacol.*, 2009, **122**, 60–67.
176. J. Stagegaard, M. Sørensen and L. P. Kvist, *Perspect. Plant Ecol. Evol. Syst.*, 2002, **5**, 103–122.
177. M. B. Quinlan, in *Handbook of Research Methods in Health Social Sciences*, ed. P. Liamputtong, Springer, Singapore, 2019.
178. M. P. Stoyneva-Gärtner, B. A. Uzunov, P. H. Dimitrova and N. Chipev, *Res. J. Food Nutr.*, 2018, **2**, 13–17.
179. A. M. Nieves-Rivera and D. A. White, *Mycologist*, 2006, **20**, 22–25.
180. B. Quittem, https://medium.com/@BrandonQuittem/bitcoin-is-a-decentralized-organism-mycelium-part-1-3-6ec58cdcfaa6 (accessed January 2021).

CHAPTER 10

Applications of Mushrooms in the Food Industry

J. PETROVIĆ,* M. KOSTIĆ, D. STOJKOVIĆ AND J. GLAMOČLIJA

Institute for Biological Research "Siniša Stanković", National Institute of Republic of Serbia, University of Belgrade, Bulevar Despota Stefana 142, 11000 Belgrade, Serbia
*Email: jovana0303@ibiss.bg.ac.rs

10.1 Introduction

Fungi are an untapped resource with both nutritive and medicinal potentials, that have been undervalued and underexplored until recently, but they are now increasingly utilized in the pharmaceutical and food industries. With the re-evaluated concept of using food in prophylaxis of diseases and as therapeutic agents, contemporary research has been strongly focused on this area, leading to scientifically supported facts. Since the nutritive properties of edible mushrooms have been praised in scientific research, an up-to-date literature review on these properties will be presented so as to evaluate the novel prospects and possible applications of edible mushrooms in the food industry.

10.2 Mushroom Products as Functional Food

In the age of health and food crises, some of the forgotten concepts have been re-evaluated – such as the consumption of certain foods so as to achieve and attain optimal health and balance. This claim refers to the consumption of food ingredients which exert disease-preventing and/or health-promoting properties in addition to their nutritional value, which is the very definition

Food Chemistry, Function and Analysis No. 36
Edible Fungi: Chemical Composition, Nutrition and Health Effects
Edited by Dejan Stojković and Lillian Barros

Published by the Royal Society of Chemistry, www.rsc.org

of functional food.[1] Recent studies have revealed that wild-growing and cultivated mushrooms have immense potential to be regarded as functional foods, due to their favourable nutritional profile, the presence of bioactive molecules and several demonstrated medicinal effects. For most of the species analysed so far, the macronutrient content usually follows the same pattern: high content of water (>90% in some species), followed by a high content of carbohydrates and proteins, with a low share of total lipids, all of which suggest desirable consumption of mushrooms on a daily basis.

A more in-depth chemical profile reveals that carbohydrates may be divided into: (i) available – which contribute to our energy level and are digested and (ii) unavailable – which do not contribute to our energy level, passing through our gastrointestinal (GI) tract undigested, and posing a fermentation substrate for gut microbiota. Available carbohydrates include various monosaccharides, disaccharides, and trisaccharides, whereas unavailable ones include different types of polysaccharides, also known as dietary fibres. As for the protein level, it varies somewhere between those of leguminous plants and meat products.[2] With a rather high level in some wild-growing species, mushrooms are often referred to as 'meat of poverty' – providing an adequate meat substitute in nations with low levels of meat consumption (due to religion issues, low-income nations, *etc.*). Regarding the content of lipids, the prevalence of nutritionally desirable polyunsaturated fatty acids (PUFAs) over monounsaturated fatty acids (MUFAs) as well as nutritionally undesirable saturated fatty acids (SFAs), makes them a desirable food in several tailored diets and lifestyles. Multitude of minerals, provitamins, and vitamins, along with a low total calorie value, increases the potential applications of mushrooms in circumstances where controlled nutrient intake is necessary.

Aside from the nutrients, fruiting bodies of mushrooms are also rich in other groups of compounds which do not contribute to nutritional value, but are often associated with the bioactive properties of mushrooms. These mycochemicals go beyond our nutritional needs and include, among many: phenolics (mainly phenolic acids), terpenoids, lectins, *etc.*[3,4]

The presented data on their chemical profile imply that mushrooms have immense potential to be used as functional food and/or to be added as functional ingredients to foods already present in the market. Since consumers are increasingly interested in purchasing nutritious, yet widely available food products with health-beneficial attributes, the functional food industry has become one of the most profitable ones. With a gross market size of nearly 200 billion dollars annually worldwide and a fast growing rate, it is expected to double its value in the next five years.[5] However, the placement of food products with functional properties on the market is often paved with difficulties as this type of product must meet certain criteria. On the one hand, quality and safety of the product come first, and on the other, its health-beneficial properties must be validated. The latter is generally when things become tricky – even though several breakthroughs have been achieved, there is still missing information on how these bioactive/nutritive compounds benefit our health after consumption. Many model systems have been developed, but precise translation of the obtained knowledge into practice remains delayed. What we do know is that compounds ingested via food suffer numerous chemical modifications in our GI tract, but the resulting outcomes are still a matter of debate. These changes may decrease bioactivity, but practice has undoubtedly shown that people with healthy eating habits are more likely to be in good shape. Since diet preferences proved to have a positive effect on managing symptoms of several diet-related health issues (including high sugar and cholesterol content, high blood pressure), mushrooms may very well be used in prophylaxis of these conditions. Furthermore, their unique organoleptic properties make them adaptable for numerous food products, which allows their application in several aspects of food industry, as elaborated on throughout this chapter.

10.3 Mushrooms as Food Value Enhancers

Multifarious proteins, dietary fibres, and polysaccharides from mushroom mycelia and fruiting bodies contribute to their nutritional value and biological activity, including antioxidant, antimicrobial, antitumour, immunomodulatory effects, *etc.*[6–9] However, due to the limited seasonal and geographical availability of different mushroom species, their consumption has been focused on the distribution of processed products. The use of preserved/processed products in recent years has tremendously increased the application of mushrooms in various areas of the food industry, leading to the global availability and popularity of numerous mushroom products.[10,11] Numerous products, such as sauces, seasonings, extracts, and supplements/nutraceuticals with mushroom added-value are available on the market.[9] These products may be subcategorized according to different criteria as dried (mushroom soup powder, nuggets, chips, biscuits, noodles, mushroom extract flavour), fermented (mushroom flavour sauce), liquid (mushroom tea), extruded (mushroom noodles), preserved/canned (mushroom sauce, pickle, jam, ketchup, candy, beef paste), and miscellaneous products (chitin, chitosan, mushroom patties).[10]

10.3.1 Mushroom-based Proteins as Fortifiers of Food Products

According to Derbyshire,[12] food proteins are currently divided into two rough categories depending on their origin, plant-based and animal-based, with mushroom-based proteins being totally omitted. With the increased demand for meat products and accompanying livestock farming in the last 50 years, it is of utmost importance to exhaust all the possibilities regarding alternative sources of proteins. Moreover, newly adapted sources must fit in with the green sustainable technologies, as livestock production is among the highest greenhouse gas emitters. In line with rising ethical questions associated with the consumption of meat, as well as the connection between the development of several diseases and increased meat utilization, current trends have shifted towards the establishment of novel protein sources.

Mushrooms have emerged as an extraordinary alternative to animal-based proteins for several reasons. Aside from the presence of both essential and non-essential amino acids, they are widely tolerant when exposed to extreme temperature and pH values. Moreover, they exert a wide range of bioactive properties including immunomodulatory, antitumour, antiviral, antimicrobial, *etc.* This makes them applicable in different technological processes that involve the production of food and pharmaceuticals.

The high amount of proteins in edible mushrooms indicates that regular consumption of mushrooms may satisfy required daily intake of proteins, *i.e.* our requirements for both non-essential (glutamine, glutamic acid) and essential amino acids. However, the downside of using mushrooms as a sole source of proteins includes the great variability of their content depending on the species, type of growing substrate, environmental factors, *etc.*[13] This reflects the numbers behind the proteins as well; with them varying from 12.00 g/100 g dry weight in *Sarcodon asparatus* to 59.40 g/100 g dry weight in *Lepista nuda*,[14] whereas the free amino acids content in mushroom fruiting bodies ranges from approximately 7.00 to 12.5 mg per g^1 d.w. This may be overcome with the cultivation of mushrooms: since cultivation of various mushrooms has been established, changes in external factors may increase total protein content in the harvested fruiting bodies. Furthermore, as it is sustainable (using several types of waste) and achieved in shorter cycles than growing animals, mushroom growing is more in compliance with contemporary lifestyles than farming. The resulting low costs, high yield, and short cycles, in combination with increasing population and their growing demand for the consumption of meat and meat products, make mushrooms excellent candidates for adequate meat alternatives in nutrition, or as "fortifiers" of food products.

To "fortify" a food product in the narrowest sense means to add a nutrient to it, regardless of the fact whether the nutrient is already present in it. This strategy has been used for a long time in circumstances when a population was deprived of the nutrient, but recently it has been utilized for numerous ingredients and became a trend of its own. Since mushrooms are a rich source of nutritive/bioactive compounds, they have emerged as a versatile tool to enhance the nutritive value of food products. On several occasions, attempts to incorporate mushroom powder/extract into existing food products to fortify their nutritional value and improve organoleptic properties have been successively made. For example, meat (sausages, meatballs, patties, nuggets, *etc.*), vegetables, grain (flour, cookies, and pastas), and dairy products have been successfully mixed with mushrooms, resulting in products acceptable for consumers.[15] Along with the beneficial properties (enhanced nutritional value and improved sensorial properties), the addition of mushroom powder may also exert food-preserving properties. This is due to the presence of certain compounds in mushrooms (phenolics, organic acids, tocopherols, *etc.*) which may prevent lipid/protein peroxidation and retard bacterial/fungal growth. Since oxidative degradation and microbial growth are the two main causes of food decay, eventually, as a result, the shelf life of a product may be extended. This assumption was validated using several *in situ* food matrices.[16–18]

However, among 14 000 edible mushrooms, only a handful have been intensively used for this purpose. The most extensively used species include *Pleurotus ostreatus* (oyster mushroom), *P. eryngii* (king oyster mushroom), *P. sajor caju* (grey oyster mushroom), *Agaricus bisporus* (white button mushroom), *Agrocybe aegerita* (chestnut mushroom), *Lentinus edodes* (shitake), and occasionally *Tremella fuciformis* (white jelly mushroom), *Boletus edulis* (porcini mushroom), and *Flammulina velutipes* (winter mushroom). Numerous meat and dairy products such as pâtés, burgers, salamis, cream cheese, and cheese have been mixed with mushroom products, usually to increase their protein content and to enhance their organoleptic properties. To name a few, β-glucan-rich fraction (*P. eryngii*) added to pasta increased its protein level and improved the sensorial features.[19] Moreover, according to Doğan *et al.*,[20] the incorporation of *P. ostreatus* powder to the baking mixture resulted in the development of a "healthy snack" – chips fortified with proteins and phenolics, with increased antioxidant properties. As for the processed meat products, several attempts to develop enriched food were successfully accomplished. Thus, beef salami was fortified with dried powder of *P. ostreatus*, which efficiently inhibited lipid oxidation of salami meat and prolonged the shelf life of this ready-to-eat product. Furthermore, sensory evaluation analysis showed that the incorporation of mushroom powder to salami improved its taste, but did little for the texture.[21] As for the application of other species, powdered button mushroom was used to develop meat products with different consistency; porcini mushroom (*Boletus edulis*) prevented oxidative processes and microorganism growth in meat which are held responsible for food deterioration. Flour derived from the mushroom powder of *P. ostreatus* unused lower section was used successfully to develop patties with increased total phenolic compounds content and the accompanying antioxidant activity. The obtained product was acceptable for consumers as well as in terms of appearance and taste, and it fitted very well in the contemporary concept of "from waste to taste", which may be used for advertisement purposes also.

The ability of mushrooms to blend well with different flavours, as well as to improve the sensory and rheological properties (viscoelasticity, higher heat resistance, improved emulsion structure) of food has been covered very well in the review by Das *et al.*[15] Their versatility, structure of their fruiting bodies, as well as unique taste, are the key features that improve the organoleptic attributes and quality of the resulting "muscle" product. Due to their "umami" taste and meaty-like texture, mushrooms fit very well into meat products, which increases their consumption and decreases consumed meat portions.[22] Proteins derived from fungi are currently present on the market as a sole product as well, in the form of powder. They are obtained from both mushrooms and moulds: for example, purified protein powder, known as mycoprotein is produced by *Fusarium venenatum*. Approved back in 1983 by the United Kingdom Ministry of Agriculture, Fisheries and Food, it has been distributed throughout EU markets for more than 40 years.[23]

The presented results indicate that mushroom proteins are unique regarding their potential use in the food industry, since they affect positively several aspects: (i) enhance organoleptic properties, (ii) improve nutritional value, and (iii) prolong the shelf life of food products. Hence, extensive research covering these aspects (chemical characterization, sensory evaluation, and rheological properties) must be conducted to obtain all the necessary information on the acceptability, bioavailability, and physicochemical properties of the newly developed product and its market placement options. Several parameters have to be monitored during these phases, but the quality of the resulting food product determines its acceptability for consumers, and eventually seals the deal. That being said, the amount of added mushroom in the initial stages of production strongly influences the sensorial attributes of the final product. Several reports have indicated that the addition of ~20% mushrooms into a product is sufficient to improve the nutritional, organoleptic, rheological, optical (discoloration of food), and cooking characteristics of the food. Furthermore, due to the presence of various types of compounds (including bioactive proteins) with antioxidant and antimicrobial properties, the addition of mushrooms may decrease to a certain level any rancid odour, repulsive flavour, or change in colour and texture, making them acceptable for consumers.

As deprivation of meat products in some parts of the world results in protein deficiency in nutrition, this may be overcome with the addition of mushroom proteins to different food products. Also, given the growing trend in vegetarian/vegan nutrition in the last decade, deprivation of proteins in such regimens indicates the necessity for their alternative sources. Mushrooms, which are considered a rich source of this macronutrient, may be utilized as an adequate option for those who do not consume products of animal origin. However, in parts of the world where mushrooms are not used extensively in the diet, application of mushrooms as food fortifiers remains delayed.

10.3.2 The Rise of Mushroom-based Food: Upgraded Snacks and Flours, Faux Meat/Dairy Products, and Others

The contemporary lifestyle encourages the consumption of pre-cooked, frozen, and partially prepared food. Moreover, since the number of ailments associated with poor/incorrect nutrition is continuously on the rise, there is growing pressure on both medical practitioners and society for people to acquire healthy eating habits. Thus, ready-to-use mushroom products, rich in proteins, fibres, vitamins, and minerals are an excellent, quick, and acceptable way to improve poor eating habits.

Both sweet and savoury ingredients have been evaluated as candidates for the development of new, upgraded food products with enhanced nutritional value: added macronutrient vitamin, mineral, and fibre content.[24] Biscuits, cakes (made from added mushroom powder into flour), cereals (with added dietary fibres), and nutritional bars are just some of the sweet products that have been used for this purpose, though using different methodologies.

Various mushrooms have been utilized for this purpose including *Lentinus edodes*, *Agaricus bisporus*, *Pleurotus ostreatus*, *etc.*[3,25] Several studies have revealed that the addition of mushroom extract/powder in pastas, flour, milk, cookies, and noodles may improve their nutritional (increase crude fibres, minerals, proteins, and vitamins) and sensorial profile, along with exerting various health-beneficial effects.[26] For this purpose some of the edible species were thoroughly elaborated, including *A. bisporus*, *A. blazei*, *P. ostreatus*, *Calocybe indica*, and *L. edodes*.[27–31] Also, Petrović *et al.*[32] demonstrated that *Agrocybe aegerita* powder increased the organoleptic properties when incorporated into cream cheese. According to Spim *et al.*,[25] flakes of *Lentinula edodes* were used for the development of sweet and savoury bars intended for distribution on the market. The consistency, texture, and taste of the product were evaluated, as well as the shelf life and consumer acceptability. The obtained results showed that consumers were more prone to accept sweet bars, highlighting the potential of bioactive compounds derived from mushrooms to effectively retard growth of foodborne pathogenic microorganisms for 180 days. The addition of mushroom powder to flours/baking products may also increase their level of fibre and decrease the amount of starch, resulting in a food ingredient with low glycaemic index that is desirable in diet regimens of people suffering from type 2 diabetes. Thus, according to Rathore *et al.*,[30] consumption of cookies enriched with *Calocybe indica* powder resulted in a glycaemic index reduction, suggesting it as a suitable ingredient for diabetic patients.

As for faux meat and dairy products, the increasing number of vegetarian/vegan-prone consumers has led to increased demand for these type of products which may substitute animal origin of food in terms of both "meat-like texture" and the daily protein requirements. Filamentous fungi have turned out to be an important biotechnological tool in the production of food product substitutes, as well as flavour enhancers, including some esters, terpenes, *etc.* Indonesian tempeh is an appropriate meat replacement and a good example of it. This product derived from soybeans fermented by *Rhizopus oligosporus* is not lacking in flavour or meat-like texture, as well as having favourable nutritional properties. Consumption of tempeh is known to reduce the risk of cardiovascular diseases, various types of carcinomas, *etc.*, which suggests it should be consumed regularly. Furthermore, Quorn, a well-known cholesterol-free meat substitute with low lipid content is produced by *Fusarium venenatum*. Its production is rather sustainable in terms of gas emissions compared to meat production, making it eco-friendly, which is of great importance, since certain animal-based products have become deficient in this respect. A good example of this substitute is rennet, which is developed using *Rhizomucor miehei* and *R. pusillus*. This affordable product used for milk clotting and production of various milk products is acceptable for vegetarians/vegans as well.[33]

As for the health-promoting effects, it has been demonstrated that bioactive compounds present in mushrooms may help manage several health issues. For example, they may alleviate blood pressure and blood sugar level, maintain a good level of cholesterol, and improve immune system-regulated functions; all of which suggests the intake of mushrooms on a daily basis.[34] With increasing demand from consumers for "ready to eat/drink" products, the type of product which will be developed is mainly determined by its market demand. After its development, consumers will be able to consume quality food with health benefits and at the same time with reduced preparation/cooking time.

Due to their versatile nature, mushrooms should be intensively used in both newly developed food products and the existing ones.[15] Their ability to improve nutritional quality and express health-beneficial effects in a way that is convenient to use may encourage consumers to acquire healthier life habit as well.

10.4 Mushroom Bioactives in the Food Industry

Along with the use of "ready to eat" food products, consumers are also interested in using capsules and pills to supplement their nutrition. Mushrooms have been used extensively in this aspect as well as in various other forms: as extracts/tonics prepared from mushroom's fruiting body and its mycelium, as a single bioactive compound, or part of various mixtures, *etc.*

These preparations are often referred to as "mushroom nutraceuticals", a term which encompasses and highlights both the alimentary and pharmaceutical potentials of mushrooms.[3] Three main segments covering the nutraceutical industry encompass: functional foods, dietary supplements, and herbal/natural products.[11] Strictly speaking, the term "nutraceutical" refers to a product that has health-promoting and disease-preventive effects validated by clinical testing.[35,36] In more detail, the term mushroom nutraceutical refers to refined extracts or dried mushroom biomass (mycelium or fruiting body), which is consumed for potential therapeutic effects in the form of capsules/tablets, or as a dietary supplement.[37,38] Their beneficial activity must be achieved through an appropriate dietary regimen, with regular intake not causing secondary adverse effects.[39] (Scheme 10.1).

Although medicinal properties have been validated for approximately 700 mushroom species,[40] only a small number may be regarded as nutraceuticals. However, due to limitations in the availability of wild-growing species, for commercial production of nutraceuticals only cultivated species are currently being utilized. The most commonly found species in dietary supplements include: *A. bisporus* (button mushroom), *Cordyceps sinensis* (cordyceps), *Ganoderma lingzhi* (reishi), *Grifola frondosa* (maitake), *Hericium erinaceus* (lion's mane), *Lentinus edodes* (shiitake), and *Trametes versicolor* (turkey tail).[41,42] Others that are not exploited on a commercial scale, but have promising bioactive properties include, among many: *Auricularia auricula-judae*, *Calocybe indica*, *Pleurotus* spp., *Schizophyllum commune*, and *Volvariella volvacea*.[11] In the last few decades, studies have shown a broad spectrum of biological activities that may be ascribed to different chemical compounds identified in mushrooms.[43–46] Consequently, this accumulating

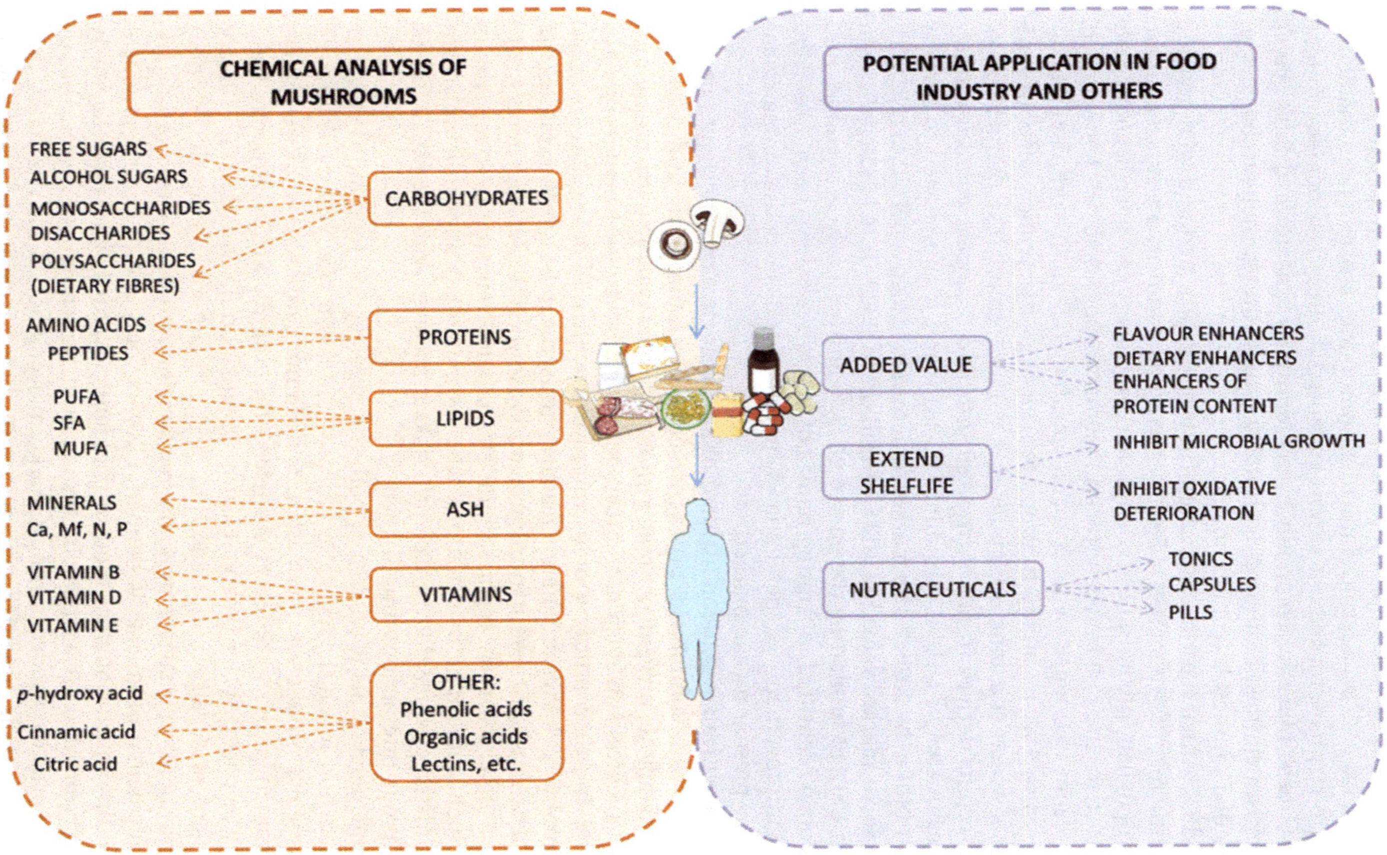

Scheme 10.1 Map of the potential applications in the food and other industries. Reproduced from servier.com, under the terms of the CC BY 3.0 license, https://creativecommons.org/licenses/by/3.0/.

knowledge has intensified the interest of the non-scientific community to pursue, collect, and exploit indigenous mushroom resources and their traditional use.[47,48] As a result, numerous compounds isolated from medicinal mushrooms have turned out to have potential in drug development. Among many, these bioactive compounds include lipids (unsaturated fatty acids), proteins, peptides, amino acids, lectins, carbohydrates (polysaccharides in particular), minerals (copper, manganese, and iron), and vitamins (complex B and C), which have been isolated from both cultivated and wild-growing mushrooms.[5,49] To name a few, cordycepin, coriolan, krestin, lentinan, schizophyllan, pleuran, hericenones, and amyloban, *etc.* have been recently explored for their pharmaceutical and medicinal potential.[50–52] For instance, cordycepin is extensively used for its antidiabetic, antihyperlipidaemic activity as well as its ability to promote beneficial metabolic effects in individuals which consume a high-fat diet.[53] Furthermore, the market size of cordycepins derived from *Cordyceps sinensis* and *Cordyceps militaris* extracts was estimated at USD 473.4 million in 2018, and it is predicted to surpass USD 1 billion by 2026.[54] The market for lentinan is also expected to rise, with the global market valued at USD 10 million in 2020. It is predicted it will reach USD 12 million by the end of 2027.[55] Furthermore, hericenones and amyloban identified in fruiting bodies of *H. erinaceus* have been associated with improvement of dementia symptoms and cognitive impairment (already present on the market).

Nutraceuticals with demonstrated pharmacological activity should not be confused with commercial drugs (prescribed by a physician) and used as a sole therapy to prevent or treat a disease.[49] However, if used properly or under medical supervision, they can be very effective in the management of several health conditions or to improve overall well-being and to boost immunity.[56–58] Precautionary measures must be taken regardless of the fact that most of these natural preparations are considered generally safe to consume, with few reported side effects and contraindications. This is extremely important in situations where other medications/therapies are administered, since there are certain risks associated with the possibility of dangerous interactions among them.[59,60]

Nonetheless, this is an area of both the medical and food industries that reflects how much the legislation inconsistency can affect the availability of a particular product on the market. For example, different countries follow recommendations of different agencies – for example EFSA (European Food Safety Agency), WHO (World Health Organization), and DSHEA (Dietary Supplement Health and Education Act) are authorities in the Europe. According to their suggestions, the registration and distribution of "botanical medicinal products" is banned, allowing only use of natural products (including mushrooms) as food supplements. No further acknowledgement of their potential use as medicines is required, nor comprehensive studies in clinical practice needed for their market placement. On the one hand, this diminishes their potential applications in the prevention and treatment of diet-related health conditions, and on the other, it allows placement of food products in populations with no background supporting evidence on their safe consumption. This enables re-selling and re-packaging of raw materials of unknown and potentially unreliable origin, which is an issue that immediately has to be addressed. Having this in mind, one must be very cautious when consuming these products with "alleged" medicinal properties, since reliability of their chemical composition as well as their safety precautions are often lacking. Most of the available studies rely on *in vitro* model systems, without appropriate animal model systems, let alone clinical trials in humans. Therefore, self-treatment with natural products should be limited only to ailments that cause mild discomfort, with serious precaution regarding the amount of consumed products in order to avoid potential contraindications and damage to health.[33] However, it seems that a potential solution is approaching. Recent data indicate that herbal/fungal preparations may acquire licences to be used as "medicines" if they have been used traditionally for a certain condition, over a period of 30 years. Moreover, a safety sheet granted by an expert in the relevant field will be required. This will possibly enable the inconsistencies in the quality of the raw material to be overcome, which is nowadays mainly defined by the distributor. This is an issue that has to be addressed as soon as possible since this is one of the most profitable industries on the market. According to PMMI Business Intelligence,[61] the global nutraceutical market was worth approximately USD 241 billion in 2019, and it is continuously growing and expected to reach as much as USD 373 billion by the year 2025. In line with this, emerging food technology has to be directed towards the development of novel delivery systems that will enable increased bioavailability of bioactive compounds belonging to natural matrices.

10.5 Mushrooms as a Source of Pre- and Probiotics in Food

The balance of intestinal microbiota is crucial for maintaining human health and disease prevention. Hence several dietary regimens, including different food ingredients, have been developed in order to tackle the underlying disease and improve the health status of consumers. For example, a diet rich in dietary fibres is used to reduce levels of triglycerides and cholesterol in the blood; this is achieved indirectly, through the activity of the intestinal microbiome. To be more precise, fibres represent a substrate for gut microorganisms which enables their proliferation. Subsequently, products of microbial metabolism enter the systemic circulation, which further leads to the establishment of energy homeostasis.[24] In cases of intestinal microbial dysbiosis – an imbalance in the number/presence of gut bacteria – the consumption of fibres is highly recommended for the re-establishment of the number and composition of gut bacteria. Nowadays, three approaches are available for the restoration of microbiota homeostasis in the gut, *i.e.* for this purpose we can use: prebiotics, probiotics, and synbiotics.[62] Their mutual differences are elaborated in the following section.

Prebiotics are defined as "non-digestible food ingredients that beneficially affect the host via selective stimulation of bacterial growth and/or their activity in the colon".[62,63] Studies on prebiotics have demonstrated their implications in several areas of health, including: maintenance of gut homeostasis, prevention of colitis, cardiovascular, and malignant diseases, enhancement of immune functions, cholesterol removal, prevention of obesity and type II diabetes, restoration of vaginal microbiota, production of bacteriocin (toxins produced by bacteria to inhibit growth of other bacteria), *etc.* Aside from this, they have applications in the food industry, as starter culture formulations, as well as in the production of animal feed.[62,64] Recent studies have shown that fruiting bodies of mushrooms are significant sources of dietary fibres (various types of complex polysaccharides), which have prebiotic activity, indicating their potential in the pharmaceutical and food industries.[24] Some of the compounds with promising prebiotic potential have already been identified in mushrooms, such as pleuran, lentinan, and schizophyllan, derived from *P. ostreatus*, *Lentinus edodes*, and *Schizophyllum commune*, respectively. Others include complex polysaccharides, α- and β-glucans, such as mannans, xylans, galactans, chitin, and inulin as well as hemicellulose.[65] As for their activities, polysaccharides obtained from mushrooms such as *Pleurotus* sp.,[66] *Lentinus edodes*, *Tremella fuciformis*,[67] and *Agaricus bisporus*[68] can significantly alter the composition of the intestinal microbiota. This modification may be achieved through metabolic changes and increased proliferation of beneficial bacteria (Lactobacilli and Bifidobacteria), as well as a reduced number of pathogenic bacteria (*E. coli, Clostridium* sp., and *Salmonella* sp.).[69] According to Saman *et al.*,[70] increased numbers of *Bifidobacterium bifidum* TISTR 2129, *B. breve* TISTR 2130, *B. animalis* TISTR 2195, and *B. longum* TISTR 2194 have already been achieved with glucans extracted from *Pleurotus ostreatus*, *P. sajor-caju*, and *P. abalonus*. Similar results with glucans derived from *P. ostreatus* and *P. eryngii* were obtained in the study conducted by Stachowak *et al.*[71] Our literature survey also showed that pleuran from *Pleurotus ostreatus* and lentinan from *L. edodes* are currently the most frequently used β-glucans as prebiotics. Aside from them, β-glucans from *Pleurotus tuber-regium*, *Polyporus rhinocerus*, and *Wolfiporia cocos* have been exploited for their prebiotic potential as well, since they were shown to affect gut homeostasis by selective modulation, *i.e.* by increasing the number of probiotic bacteria.[72] Another mushroom with probiotic potential is *Ganoderma lucidum*, which reduced obesity (decreased body weight, inflammation, and insulin resistance) *in vivo*, by modulating the composition of the gut microbiota.[73] Furthermore, consumption of *A. bisporus* stimulated growth of desirable gut microorganisms, which consequently improved GI tract health,[74] and a similar GI effect was achieved with polysaccharopeptides isolated from *Trametes versicolor* and *L. edodes* as well. As for the potential of *H. erinaceus* to be used as a prebiotic, studies have shown that its polysaccharides also positively affect the composition and activity of the GI tract microbiota of the host.[73]

Probiotics are defined as "products that contain sufficient amount of microorganisms to alter the microbiota in the host compartment, thus exerting beneficial health effects in this host".[62] To be qualified as a probiotic, a microorganism has to be safe for human consumption, without any pathogenic or toxic effects. Preferably, it should originate from the intestinal tract of healthy persons, which ensures its safer use in humans. Furthermore, these microorganisms have to be tolerant to gastric and bile acids, resistant to digestive enzymes and, lastly, they have to act beneficially on the intestinal microbiota.[62] Various calculations have shown that microorganisms have to be consumed at a level of at least 10^7 CFU per mL to demonstrate a probiotic effect.[62]

From numerous tested potential probiotics, only a few microorganisms have been taken into consideration for use as efficient probiotics, including fungi (mushroom and yeast), bacteria, or mixed combinations. These include *Lactobacillus* and *Bifidobacterium* among bacteria and *Saccharomyces boulardii* among yeasts.[62] As for the mushrooms, Kumar *et al.*[24] have shown that several species (*A. bisporus*, *Auricularia auricular*, *Pleurotus sajor-caju*, *P. eryngii*, *Cordiceps militaris*, and *Lentinus edodes*) have the capacity to be referred to as probiotics (when used in powdered form or as an extract). According to Synytsya *et al.*,[75] glucans isolated from *Pleurotus ostreatus* and *P. eryngii* were able to stimulate growth of probiotic bacteria, including *Lactobacillus* sp. (Lac A–D), *Bifidobacterium* sp. (Bifi A–C), and *Enterococcus faecium* (Ent A and B). Additionally, the synergistic effects of mushroom polysaccharides have been demonstrated by several researchers in probiotic growth: for example, three different crude polysaccharide fractions from *Agaricus blazei* had better growth-promoting activity on lactic acid bacteria when used simultaneously than individually.[76] Furthermore, Yamin *et al.*[77] demonstrated the prebiotic effect of crude polysaccharides extracted from *Ganoderma lucidum*: growth promotion of *Bifidobacterium* sp. and *Lactobacillus* sp. and growth inhibition of *Salmonella* sp. Recently, Chou *et al.*[78] studied the prebiotic activity of crude polysaccharides from *Lentinula edodes* stipe, *Pleurotus eryngii* base, and *Flammulina velutipes* base, which enhanced the survival rate of probiotic cultures, such as *Lactobacillus acidophilus*, *Lactobacillus casei*, and *Bifidobacterium longum* subsp. *longum* during cold storage, thus demonstrating the prebiotic potential of the polysaccharide in question.

As for the term synbiotics, this refers to "products containing both probiotics and prebiotics". Synbiotics have their own beneficial properties, which may be attributed to possible synergistic effect between pro- and prebiotics.[62] The applications of synbiotics in integrative medicine and human health cover a broad spectrum of health-beneficial effects, including restoration of the immune function, reduction in systemic inflammation, alleviation of irritable bowel disease, and acute pancreatitis, *etc.*[79]

It has been demonstrated that these new products have great potential as functional food ingredients.[80]

10.6 Mushrooms in Biotechnology (Encapsulation, Emulsification)

Microencapsulation is a technique used in the food industry to ensure higher stability of sensitive compounds, that is, to protect them from humidity, oxygen, and other stressors. Furthermore, this technology provides protection to the unstable bioactive components, which allows them to be incorporated into designed food products and to effectively deliver the active material at the targeted site at a controlled rate.[81]

Currently available encapsulation techniques employed in food industry include coating, spray drying/cooling/chilling, air suspension, liposome entrapment, among others.[82] Coating of an active component by a solution of shell material is the simplest method of encapsulation used in the food industry. This involves pan, spray, or fluidized bed coating followed by drying. Other methods involve the emulsification of a core material (*e.g.* oil or enzyme) with a solution of a wall material (usually a polysaccharide, a protein, or their synergies) followed by a drying process to obtain the encapsulated material in a dry powdered form for convenient handling and storage.[82]

Encapsulating agents used in spray drying include proteins, gums, and carbohydrates.[83] Few natural materials are suitable as encapsulating agents due to their chemical nature. This may be of practical importance, given the importance of a "clean label" in food production these days. Hence replacement of any synthetic additives with natural ingredients is appreciated.[84] Consequently, the search for new possible natural compounds that can be used freely in the food industry as materials and/or emulsifiers that can replace synthetics has increased tremendously. Several natural macromolecules have emerged as suitable alternatives to the synthetic emulsifiers: β-glucans, plant proteins, and pectins to name a few.[85] Since mushrooms are considered as an important source of several bioactive compounds, such as polyphenols, polysaccharides (*i.e.* β-glucans), and sterols (*i.e.* ergosterol),[86] they may be used on a wider scale for this purpose. Ergosterol has high commercial value because it acts as a precursor of vitamin D_2 and it has also been related to immune activity and antioxidant capacity.[86–88] As for the β-glucans, the complexity of their chemical structure enables their use as prebiotics, enhancers of the immune system, with antitumour and antioxidant capacities among others.[89] For example, β-glucans derived from barley and *Pleurotus ostreatus* were able to prevent oxidation in the encapsulation of fish and sunflower oil.[90]

However, for their use on a wider scale, several issues must be addressed. To start with, it is necessary to gain insights into how these materials affect the characteristics of the microcapsules. Furthermore, we have to evaluate the encapsulation efficiency, particle size, solubility, density, colour, *etc.*

10.7 Wild-growing Mushrooms and Their Perspective in Commercialization

People have been foraging for wild-growing mushrooms since ancient times, even before recorded history. Once they discovered which were the most delicious, it made sense to try to grow them and have a more reliable supply. Ethnic communities inhabiting different regions of the globe regularly consume edible mushrooms collected from the wild based on their traditional knowledge of mushroom identities and their nutritional benefits.[91] Wild edible mushrooms have received significant scientific and socio-economic attention in the last few decades, as they have become the subject of a booming trade. The largest number of these species belongs to the Basidiomycota, while a smaller number of representatives are from the Ascomycota.

Inventive approaches have been applied in order to formulate and stimulate environmentally sustainable and economically feasible mechanisms for the cultivation of wild mushrooms. These have been tested over millennia with the participation of the local community.[92]

Mushroom-growing techniques have been developed to a high degree of technical efficiency. Nowadays, many well-known wood-decomposing species can be easily grown on hardwood logs (oak, maple, elm, cottonwood, and alder) including Reishi (*Ganoderma lucidum*), shiitake (*Lentinula edodes*), hen of the woods (*Grifola frondosa*), lion's mane (*Hericium erinaceus*), and oyster mushrooms (*Pleurotus* spp.). Over the last few decades, cultivation has taken place using a synthetic log production system with substrate consisting of sawdust enriched with organic and inorganic additives. This approach has been used for cultivation of sulphur polypore or sulphur shelf (*Laetiporus sulphureus*),[93] black poplar mushroom (*A. aegerita*; syn. *A. cylindracea*),[94] stubble rosegill (*Volvariella speciosa*), *Termitomyces* sp., and other saprobic mushroom species.[92] Commercial quantities of mushrooms are grown using large-scale production methods. The classic commercial button mushroom (*Agaricus bisporus*) is grown in tremendous quantities on mushroom farms around the world. Since the natural habitat of its wild predecessors was in pastures, this species is grown on a substrate of horse manure compost. Many other species from the genus *Agaricus* that are not widely cultivated, *A. bitorquis*, *A. campestris*, *A. macrosporus*, and *A. silvaticus*, show evident strong biological activity[95,96] and so their introduction into mass cultivation would contribute to wider consumption of these species. Other species cultivated on compost include *Coprinus comatus* (shaggy mane) and *Lepista nuda* (blewit), which may increase their consumption worldwide.

Recently, many different inventive techniques have been used for mushroom cultivation attempts, which include sustainable technologies and waste recycle. For example, in the study by Huang and Ohga,[97] ceramic beads were examined as a reusable material for cultivation of 20 species of edible mushrooms. The control groups were several types of sawdust, and the results indicated that 11 species responded well to the ceramic bead substrate as compared to the traditional sawdust substrate. This opens up new prospects for mushroom cultivation, in line with contemporary needs. In contrast to saprobic mushrooms, the cultivation of ectomycorrhizal species is more difficult, due to their dependence on several biotic and abiotic factors such as climate, humidity, soil conditions, and the surrounding flora and fauna.[98] Even so, their cultivation began centuries ago, and achieved accomplishments that are still under the veil of secrecy. As for the cultivation of *Ascomycota* species, data indicate that morels (*Morchella importuna*, *M. sextelata*, and *M. eximia*) are currently cultivated in China. These mushrooms, with immense repute, are still largely collected in the wild.[99]

10.8 Conclusion

Bearing in mind that the food production chain is accompanied by an immense waste problem, contemporary research has focused on developing sustainable and green strategies that will decrease the amounts of underutilized food products as well as enable a circular economy. Mushroom cultivation fits very well into this concept. Furthermore, due to the presence of bioactive compounds, mushrooms have wide applications in the pharmaceutical industry and drug development. With this in mind, the resulting question can be raised: can we resolve the increasing challenges in the food and pharmaceutical industries with achievements we obtain from mushroom fundamental research? Bearing in mind that mushrooms meet top trends in the agrifood industry with a broad spectrum of applications in these industries, it is not difficult to imagine that we are approaching the possible solution. Mushrooms cover several requirements in the food industry that other resources such as plants cannot meet. Flavour, consistency, functionality, and versatility, are just some of the features that expand the range of potential uses of edible mushrooms, resulting in designing food and beverages with functional properties. This may possibly open up new prospects on how to unlock underexplored and underutilized potential of mushrooms, making the mushroom nutrition accessible to everyone.

References

1. K. Arihara, in *Encyclopedia of Meat Sciences*, ed. C. Devine and M. Dikeman, Elsevier, 2nd edn, 2014, DOI: 10.1016/B978-0-12-384731-7.00172-0, ISBN: 9780123847317, eBook ISBN: 9780123847348.
2. G. A. Martinez-Medina, M. L. Chávez-González, D. K. Verma, L. A. Prado-Barragán, J. L. Martínez-Hernández, A. C. Flores-Gallegos,

M. Thakur, P. P. Srivastav and C. N. Aguilar, Bio-funcional components in mushrooms, a health opportunity: Ergothionine and huitlacohe as recent trends, *J. Funct. Foods*, 2021, 77, 104326.
3. F. Cateni, M. L. Gargano, G. Procida, G. Venturella, F. Cirlincione and V. Ferraro, Mycochemicals in wild and cultivated mushrooms: nutrition and health, *Phytochem. Rev.*, 2021, **21**, 339–383.
4. J. Petrović, J. Glamočlija, T. Ilić Tomić, M. Soković, D. Robajac, O. Nedić and A. Pavić, Lectin from *Laetiporus sulphureus* effectively inhibits angiogenesis and tumor development in the zebrafish xenograft model of melanoma and colorectal carcinoma, *Int. J. Biol. Macromol.*, 2020, **148**, 129–139.
5. J. R. Al-Obaidi, N. N. Jambari and E. I. Ahmad-Kamil, Mycopharmaceuticals and Nutraceuticals: Promising Agents to Improve Human Well-Being and Life Quality, *J. Fungi*, 2021, 7(7), 503.
6. M. Kostić, M. Smiljković, J. Petrović, J. Glamočilija, L. Barros, I. C. F. R. Ferreira, A. Ćirić and M. Soković, Chemical, nutritive composition and wide-broad bioactive properties of honey mushroom *Armillaria mellea* (Vahl: Fr.) Kummer, *Food Funct.*, 2017, **8**(9), 3239–3249.
7. J. Glamočlija, M. Kostić and M. Soković, Antimicrobial and hepatoprotective activities of edible mushrooms, in *Biology of macrofungi*, ed. B. Pratap Singh, L. Chhakchhuak and A. K. Passari, Springer International Publishing AG, 2019, ch. 4, pp. 81–113, eBook ISBN 978-3-030-02622-6, Hardcover ISBN 978-3-030-02621-9.
8. M. Kostić, M. Ivanov, Â. Fernandes, J. Pinela, R. C. Calhelha, J. Glamočlija, L. Barros, I. C. F. R. Ferreira, M. Soković and A. Ćirić, Antioxidant Extracts of three Russula genus species express diverse biological activity, *Molecules*, 2020, **25**, 4336.
9. D. Yadav and P. Singh Negi, Bioactive components of mushrooms: Processing effects and health benefits, *Food Res. Int.*, 2021, **148**, 110599.
10. S. Singhal, P. Rasane, S. Kaur, U. Garba, J. Singh, N. Raj and N. Gupta, Mushroom cultivation, processing and value-added products: a patent based review, *Recent Pat. Food, Nutr. Agric.*, 2019, **10**(1), 3–19.
11. A. G. Niego, S. Rapior, N. Thongklang, O. Raspé, W. Jaidee, S. Lumyong and K. D. Hyde, Macrofungi as a nutraceutical source: promising bioactive compounds and market value, *J. Fungi*, 2021, 7, 397.
12. E. J. Derbyshire, Is There Scope for a Novel Mycelium Category of Proteins alongside Animals and Plants?, *Foods*, 2020, **9**(9), 1151.
13. P. Kalač, Chemical Composition and Nutritional Value of European Species of Wild growing Mushrooms, in *Mushrooms: Types, Properties and Nutrition*, ed. S. Andres and N. Baumann, Nova Science Publishers, 2012, pp. 129–152.
14. C. Barroetaveña and C. V. Toledo, The Nutritional Benefits of Mushrooms, in *Wild Plants, Mushroom and Nuts. Functional Food Properties and Applications*, ed. I. C. F. R. Ferreira, P. Morales and L. Barros, Wiley Blackwell, 2017, pp. 65–81.

15. A. K. Das, P. K. Nanda, P. Dandapat, S. Bandyopadhyay, P. Gullón, G. K. Sivaraman, D. J. McClements, B. Gullón and J. M. Lorenzo, Edible Mushrooms as Functional Ingredients for Development of Healthier and More Sustainable Muscle Foods: A Flexitarian Approach, *Molecules*, 2021, **26**(9), 2463.
16. J. Petrović, J. Glamočlija, D. Stojković, A. Ćirić, M. Nikolić, D. Bukvički, M. E. Guerzoni and M. Soković, *Laetiporus sulphureus*, edible mushroom from Serbia: Investigation on volatile compounds, in vitro antimicrobial activity and in situ control of *A. flavus* in tomato paste, *Food Chem. Toxicol.*, 2013, **59**, 297–302.
17. D. Stojković, F. S. Reis, I. C. F. R. Ferreira, L. Barros, J. Glamočlija, A. Ćirić, M. Nikolić, T. Stević, A. Giveli and M. Soković, *Tirmania pinoyi*: Chemical composition, in vitro antioxidant and antibacterial activities and in situ control of Staphylococcus aureus in chicken soup, *Food Res. Int.*, 2013, **53**(1), 56–62.
18. D. Stojković, F. S. Reis, A. Ćirić, L. Barros, J. Glamočlija, I. C. F. R. Ferreira and M. Soković, *Boletus aereus* growing wild in Serbia: chemical profile, in vitro biological activities, inactivation and growth control of food-poisoning bacteria in meat, *J. Food Sci. Technol.*, 2015, **52**(11), 7385–7392.
19. S. Kim, J.-W. Lee, Y. Heo and B. Moon, Effect of *Pleurotus eryngii* Mushroom β-Glucan on Quality Characteristics of Common Wheat Pasta, *J. Food Sci.*, 2016, **81**, C835–C840.
20. N. Doğan, C. Doğan, M. Çam and İ. Hayoğlu, Optimization and comparison of three cooking methods for wheat flour-oyster mushroom (*P. ostreatus*) powder composite chips, *J. Food Process. Preserv.*, 2020, **44**, e14873.
21. O. Özünlü and H. Ergezer, Possibilities of Using Dried Oyster Mushroom (Pleurotus ostreaus) in Production of Salami, *J. Food Process. Preserv.*, 2020, **45**(2), e15117.
22. W. A. A. Q. I. Wan-Mohtar, S. A. Halim-Lim, N. Z. Kamarudin, Y. Rukayadi, M. H. Abd Rahim, A. A. Jamaludin and Z. Ilham, Fruiting-body-base flour from an Oyster mushroom waste in the development of antioxidative chicken patty, *J. Food Sci.*, 2020, **85**, 3124–3133.
23. M. Wiebe, Myco-protein from fusarium venenatum: A well-established product for human consumption, *Appl. Microbiol. Biotechnol.*, 2002, **58**, 421–427.
24. H. Kumar, K. Bhardwaj, R. Sharma, E. Nepovimova, N. Cruz-Martins, D. S. Dhanjal, R. Singh, C. Chopra, R. Verma and K. A. Abd-Elsalam, *et al.*, Potential usage of edible mushrooms and their residues to retrieve valuable supplies for industrial applications, *J. Fungi*, 2021, **7**, 427.
25. S. Spim, N. Castanho, A. Pistila, A. Jozala, J. Oliveira Jr and D. Grotto, Lentinula edodes mushroom as an ingredient to enhance the nutritional and functional properties of cereal bars, *J. Food Sci. Technol.*, 2020, **58**(4), 1349–1357.

26. F. Salehi, Characterization of different mushrooms powder and its application in bakery products: A review, *Int. J. Food Prop.*, 2019, **22**, 1375–1385.
27. X. Lu, M. A. Brennan, L. Serventi, J. Liu, W. Guan and C. S. Brennan, Addition of mushroom powder to pasta enhances the antioxidant content and modulates the predictive glycaemic response of pasta, *Food Chem.*, 2018, **264**, 199–209.
28. J. R. Ishara, D. N. Sila, G. M. Kenji and A. K. Buzera, Nutritional and functional properties of mushroom (*Agaricus bisporus* & *Pleurotus ostreatus*) and their blends with maize flour, *Am. J. Food Sci. Technol.*, 2018, **6**(1), 33–41.
29. A. C. P. Vital, C. Croge, S. M. Gomes-da-Costa and P. T. Matumoto-Pintro, Effect of addition of *Agaricus blazei* mushroom residue to milk enriched with Omega-3 on the prevention of lipid oxidation and bioavailability of bioactive compounds after *in vitro* gastrointestinal digestion, *Int. J. Food Sci. Technol.*, 2017, **52**(6), 1483–1490.
30. H. Rathore, S. Sehwag, S. Prasad and S. Sharma, Technological, nutritional, functional and sensorial attributes of the cookies fortified with *Calocybe indica* mushroom, *J. Food Meas. Charact.*, 2019, **13**, 976–987.
31. L. Wang, H. Zhao, M. Brennan, W. Guan, J. Liu, M. Wang and C. Brennan, *In vitro* gastric digestion antioxidant and cellular radical scavenging activities of wheat-shiitake noodles, *Food Chem.*, 2020, **330**, 127214.
32. J. Petrović, J. Glamočlija and D. Stojković, *et al.*, Nutritional value, chemical composition, antioxidant activity and enrichment of cream cheese with chestnut mushroom *Agrocybe aegerita* (Brig.) Sing, *J. Food Sci. Technol.*, 2015, **52**, 6711–6718.
33. K. D. Hyde, J. Xu and S. Rapior, *et al.*, The amazing potential of fungi: 50 ways we can exploit fungi industrially, *Fungal Diversity*, 2019, **97**, 1–136.
34. S. H. Ng, S. D. Robert, W. A. N. W. Ahmad and W. R. W. Ishak, Incorporation of dietary fiber-rich oyster mushroom (*Pleurotus sajor-caju*) powder improves postprandial glycaemic response by interfering with starch granule structure and starch digestibility of biscuit, *Food Chem.*, 2017, **227**, 358–368.
35. P. Dudeja and R. K. Gupta, Nutraceuticals, in *Food Safety in the 21st Century*, ed. P. Dudeja, R. Kumar, G. Amarjeet and S. Minhas, Elsevier, Amsterdam, The Netherlands, 2017, pp. 491–496, ISBN 9780128017739.
36. P. Daliu, A. Santini and E. Novellino, A decade of nutraceutical patents: Where are we now in 2018?, *Expert Opin. Ther. Pat.*, 2018, **28**, 875–882.
37. I. Giavasis, Bioactive fungal polysaccharides as potential functional ingredients in food and nutraceuticals, *Curr. Opin. Biotechnol*, 2014, **26**, 162e173.
38. S. P. Wasser, Medicinal mushroom science: Current perspectives, advances, evidences, and challenges, *Biomed. J.*, 2014, **37**, 345e356.
39. E. K. Kalra, Nutraceutical-definition and introduction, *AAPS PharmSci.*, 2003, **5**(3), E25.

40. S. T. Chang and S. P. Wasser, The cultivation and environmental impact of mushrooms, in *Oxford research encyclopedia of environmental science – agriculture and the environment*, 2017.
41. M. E. Valverde, T. Hernández-Pérez and O. Paredes-López, Edible mushrooms: Improving human health and promoting quality life, *Int. J. Microbiol.*, 2015, **2015**, 1–14.
42. D. J. Royse, J. Baars and Q. Tan, Current Overview of Mushroom Production in the World, in *Edible and Medicinal Mushrooms*, 2017, ed. C. Z. Diego and A. Pardo-Giménez, DOI: 10.1002/9781119149446.ch2.
43. D. Stojković, N. Kovačević-Grujičić, F. Reis, S. Davidović, L. Barros, J. Popović, I. Petrović, A. Pavić, J. Glamočlija, A. Ćirić, M. Stevanović, I. C. F. R. Ferreira and M. Soković, Chemical composition of the mushroom *Meripilus giganteus* Karst. and bioactive properties of its methanolic extract, *LWT–Food Sci. Technol.*, 2017, **79**, 454–462.
44. M. Soković, A. Ćirić, J. Glamočlija and D. Stojković, The Bioactive Properties of Mushrooms, in *Wild Plants, Mushrooms and Nuts: Functional Food Properties and Applications*, ed. I. C. F. R. Ferreira, P. Morales and L. Barros, Wiley-Blackwell, 2017, ch. 4, pp. 83–122, 496, ISBN: 978-1-118-94462-2.
45. D. Stojković, M. Smiljković, J. Petrović, M. Kostić, A. Ćirić, J. Glamočlija and M. Soković, Edible and medicinal fungi inhibit enzymes linked to diabetes type-2 therapy, The 2nd Balkans-China Mini-symposium on Natural Products and Drug Discovery, 11–13 April, 2019 Belgrade, Serbia, Book of abstracts, 2019, p. 37, ISBN 978-86-80335-10-0.
46. J. Petrović, A. Fernandes, D. Stojković, M. Soković, L. Barros, I. C. F. R. Ferreira, A. Shekhar and J. Glamočlija, A step forward towards exploring nutritional and biological potential of mushrooms: a case study of *Calocybe gambosa* (Fr.) Donk wild growing in Serbia, *Pol. J. Food*, 2022, 72(1), 17–26.
47. S. Patel and A. Goyal, The current trends and future perspectives of prebiotics research: a review, *Biotechnology*, 2012, **2**(2), 115–125.
48. A. Ćirić, I. Kruljević, D. Stojković, Â. Fernandes, L. Barros, R. C. Calhelha, I. C. F. R. Ferreira, M. Soković and J. Glamočlija, Comparative investigation on edible mushrooms *Macrolepiota mastoidea*, M. rhacodes and M. procera: Functional food with diverse biological activities, *Food Funct.*, 2019, **10**, 7678–7686.
49. F. Reis, A. Martins, H. Vasconcelos, P. Morales and I. C. F. R. Ferreira, Functional foods based on extracts or compounds derived from mushrooms, *Trends Food Sci. Technol.*, 2017, **66**, 48–62.
50. S. M. Badalyan, Potential of mushroom bioactive molecules to develop healthcare biotech products, in Proceedings of the 8th International Conference on Mushroom Biology and Mushroom Products, New Delhi, India, 19–22 November 2014, pp. 373–378.
51. D.-W. Qin and C. Han, Medicinal and edible fungi as an alternative medicine for treating age-related disease, *Evidence-Based Complementary Altern. Med.*, 2014, **2014**, 638561.

52. C. Sánchez, Bioactives from Mushroom and Their Application, in *Food Bioactives: Extraction and Biotechnology Applications*, ed. M. Puri, Springer International Publishing, Cham, Switzerland, 2017, pp. 23–57.
53. S. A. Ashraf, A. Elkhalifa, A. J. Siddiqui, M. Patel, A. M. Awadelkareem, M. Snoussi, M. S. Ashraf, M. Adnan and S. Hadi, Cordycepin for Health and Wellbeing: A Potent Bioactive Metabolite of an Entomopathogenic *Cordyceps* Medicinal Fungus and Its Nutraceutical and Therapeutic Potential, *Molecules*, 2020, **25**(12), 2735.
54. Coherent Market Insights Cordyceps sinensis and militaris Extract Market Analysis. Available online: https://www.coherentmarketinsights.com/market-insight/cordyceps-sinensis-and-militaris-extract-market-2578 (accessed on 18 March 2021).
55. MarketWatch Global Lentinan Market 2021 Analysis with Key Players, Applications, Trends and Forecasts by 2027. Available online: https://www.marketwatch.com/press-release/global-lentinan-market-2021-analysis-with-key-players-applicationstrends-and-forecasts-by-2027-2021-03-09 (accessed on 18 March 2021).
56. L. Das, E. Bhaumik, U. Raychaudhuri and R. Chakraborty, Role of nutraceuticals in human health, *J. Food Sci. Technol.*, 2012, **49**, 173–183.
57. G. Elhassan, M. Khan, H. Khalilullah and E. Eid, Nutraceuticals–an opportunity of healthcare: A review, *Trop. J. Pharm. Res.*, 2016, **5**, 229–239.
58. A. Santini, S. M. Cammarata, G. Capone, A. Ianaro, G. C. Tenore, L. Pani and E. Novellino, Nutraceuticals: Opening the debate for a regulatory framework, *Braz. J. Clin. Pharmacol.*, 2018, **84**, 659–672.
59. J. Zhao, Nutraceuticals, nutritional therapy, phytonutrients, and phytotherapy for improvement of human health: A perspective on plant biotechnology application, *Recent Pat. Biotechnol.*, 2007, **1**, 75–97.
60. I. G. Télessy, Nutraceuticals, in *The Role of Functional Food Security in Global Health*, ed. R. B. Singh, R. R. Watson and T. Takahashi, Elsevier, Amsterdam, The Netherlands, 2019, pp. 409–421. ISBN 978-0-12-813148-0.
61. PMMI Business Intelligence 2019 Nutraceuticals Market Assessment. Available online: https://www.pmmi.org/report/2019nutraceuticals-market-assessment (accessed on 10 March 2021).
62. K. L. Lam and P. C. K. Cheung, Nondigestible long chain beta-glucans as novel prebiotics, *Bioact. Carbohydr. Diet. Fibre*, 2013, **2**(1), 45–64.
63. G. R. Gibson and M. B. Roberfroid, Dietary modulation of the human colonic microbiota: Introducing the concept of probiotic, *J. Nutr.*, 1995, **125**, 1401–1412.
64. S. Patel and A. Goyal, Recent developments in mushrooms as anti-cancer therapeutics: A review, *3 Biotech*, 2012, **2**, 1–15.
65. S. K. Singdevsachan, P. Auroshree, J. Mishra, B. Baliyarsingh, K. Tayung and H. Thatoi, Mushroom polysaccharides as potential prebiotics with their antitumor and immunomodulating properties: A review, *Bioact. Carbohydr. Diet. Fibre*, 2016, 7(1), 1–14.
66. A. Daneshmand, G. H. Sadeghi, A. Karimi and A. Vaziry, Effect of oyster mushroom (*Pleurotus ostreatus*) with and without probiotic on growth

performance and some blood parameters of male broilers, *Anim. Feed Sci. Technol.*, 2011, **170**, 91–96.

67. F. C. Guo, R. P. Kwakkel, B. A. Williams, W. K. Li, H. S. Li, J. Y. Luo, X. P. Li, Y. X. Wei, Z. T. Yan and M. W. A. Verstegen, Effects of mushroom and herb polysaccharides, as alternatives for an antibiotic, on growth performance of broilers, *Br. Poult. Sci.*, 2004, **45**(5), 684–694.
68. I. Giannenas, E. Tsalie, E. F. Chronis, S. Mavridis, D. Tontis and I. Kyriazakis, Consumption of *Agaricus bisporus* mushroom affects the performance, intestinal microbiota composition and morphology, and antioxidant status of turkey poultry, *Anim. Feed Sci. Technol.*, 2011, **165**(3), 218–229.
69. B.-L. Zhou, Q. Liang, Y. Zou, J. Ming and G.-H. Zhao, Research progress in prebiotic properties of edible mushroom, *Food Sci.*, 2011, **32**(15), 303–307.
70. P. Saman, A. Chaiongkarn, S. Moonmangmee, J. Sukcharoen, C. Kuancha and B. Fungsin, Evaluation of prebiotic property in edible mushrooms, *Biol. Chem. Res.*, 2016, **3**, 75–85.
71. B. Stachowiak and J. Reguła, Health-promoting potential of edible macromycetes under special consideration of polysaccharides: a review, *Eur. Food Res. Technol.*, 2012, **234**, 369–380.
72. K. H. Wong, K. Y. Wong, H. S. Kwan and P. C. Cheung, Dietary fibers from mushrooms clerotia: 3. *In vitro* fermentability using human fecal microflora, *J. Agric. Food Chem.*, 2005, **53**(24), 9407–9412.
73. M. Jayachandran, J. Xiao and B. Xu, A critical review on health promoting benefits of edible mushrooms through gut microbiota, *Int. J. Mol. Sci.*, 2017, **18**(9), 1934.
74. J. Varshney, J. H. Ooi and B. M. Jayarao, White button mushrooms increase microbial diversity and accelerate the resolution of *Citrobacterrodentium* infection in mice, *J. Nutr.*, 2013, **143**, 526–532.
75. A. Synytsya, K. Míčková, A. Synytsya, I. Ablonský, J. Spěváček, V. Erban, E. Kovářiková and J. Čopíková, Glucans from fruit bodies of cultivated mushrooms *Pleurotus ostreatus* and *Pleurotus eryngii*: Structure and potential prebiotic activity, *Carbohydr. Polym.*, 2009, **76**(4), 548–556.
76. Y. Lili and H. Jianchun, Effect of *Agarcus blazei* murill polysaccharides on lactic acid bacteria, *China Dairy Ind.*, 2008, **12**, 7–10.
77. S. Yamin, M. Shuhaimi, A. Arbakariya, A. B. Fatimah, A. K. Khalilah, O. Anas and A. M. Yazid, Effect of *Ganoderma lucidum* polysaccharides on the growth of *Bifidobacterium* spp. as assessed using Real-time PCR, *Int. Food Res. J.*, 2012, **19**(3), 1199–1205.
78. W. T. Chou, I. C. Sheih and T. J. Fang, The applications of polysaccharides from various mushroom wastes as prebiotics in different systems, *J. Food Sci.*, 2013, **78**(7), M1041–M1048.
79. S. Bengmark, Integrative medicine and human health – the role of pre-, pro- and synbiotics, *Clin. Transl. Med.*, 2012, **1**(1), 6.
80. B. Bigliardi and F. Galati, Innovation trends in the food industry: The case of functional foods, *Trends Food Sci. Technol.*, 2013, **31**(2), 118–129.

81. M. Umaña, C. Turchiuli, C. Rosselló and S. Simal, Addition of a mushroom by-product in oil-in-water emulsions for the microencapsulation of sunflower oil by spray drying, *Food Chem.*, 2021, **343**, 128429.
82. Y. P. Timilsena, Md. A. Haque and B. Adhikari, Encapsulationin the Food Industry: A Brief Historical Overview to Recent Developments, *Food Nutr. Sci.*, 2020, **11**, 481–508.
83. C. Turchiuli, M. T. Jimenez Munguia, M. Hernandez Sanchez, H. Cortes Ferre and E. Dumoulin, Use of different supports for oil encapsulation in powder by spray drying, *Powder Technol.*, 2014, **255**, 103–108.
84. C. Berton-Carabin and K. Schroen, Towards new food emulsions: Designing the interface and beyond, *Curr. Opin. Food Sci.*, 2019, **27**, 74–81.
85. M. Umaña, C. Turchiuli, V. Eim, C. Rosselló and S. Simal, Stabilization of oil-in-water emulsions with a mushroom (*Agaricus bisporus*) by-product, *J. Food Eng.*, 2021, **307**, 110667.
86. K. Papoutsis, S. Grasso, A. Menon, N. P. Brunton, J. G. Lyng, J. C. Jacquier and D. J. Bhuyan, Recovery of ergosterol and vitamin D2 from mushroom waste – potential valorization by food and pharmaceutical industries, *Trends Food Sci. Technol.*, 2020, **99**, 351–366.
87. S. Shao, M. Hernandez, J. K. G. Kramer, D. L. Rinker and R. Tsao, Ergosterol profiles, fatty acid composition, and antioxidant activities of button mushrooms as affected by tissue part and developmental stage, *J. Agric. Food Chem.*, 2010, **58**, 11616–11625.
88. W. Guan, J. Zhang, R. Yan, S. Shao, T. Zhou, J. Lei and Z. Wang, Effects of UV-C treatment and cold storage on ergosterol and Vitamin D2 contents in different parts of white and brown mushroom (*Agaricus bisporus*), *Food Chem.*, 2016, **210**, 129–134.
89. A. A. Khan, A. Gani, F. A. Khanday and F. A. Masoodi, Biological and pharmaceutical activities of mushroom β-glucan discussed as a potential functional food ingredient, *Bioact. Carbohydr. Diet. Fibre*, 2017, **16**, 1–13.
90. F. Gallotti, V. Lavelli and C. Turchiuli, Application of *Pleurotus ostreatus* β-glucans for oil-in-water emulsions encapsulation in powder, *Food Hydrocolloids*, 2020, **105**, 105841.
91. J. Živković, M. Ivanov, D. Stojković and J. Glamočilja, Ethnomycological Investigation in Serbia: Astonishing Realm of Mycomedicines and Mycofood, *J. Fungi*, 2021, 7, 349.
92. D. Wendiro, A. P. Wacoo and G. Wise, Identifying indigenous practices for cultivation of wild saprophytic mushrooms: responding to the need for sustainable utilization of natural resources, *J. Ethnobiol. Ethnomed.*, 2019, **15**, 64.
93. M. Pleszczynska, A. Wiater, M. Siwulski and J. Szczodrak, Successful large-scale production of fruiting bodies of *Laetiporus sulphureus* (Bull.: Fr.) Murrill on an artificial substrate, *World J. Microbiol. Biotechnol.*, 2013, **29**, 753–758.

94. V. Kleofas, L. Sommer, M. A. Fraatz, H. Zorn and M. Rühl, Fruiting Body Production and Aroma Profile Analysis of *Agrocybe aegerita* Cultivated on Different Substrates, *Nat. Resour.*, 2014, **5**, 233–240.
95. J. Glamočlija, D. Stojković, M. Nikolić, A. Ćirić, L. Barros, I. C. F. R. Ferreira and M. Soković, Comparative study on edible Agaricus mushrooms as functional foods, *Food Funct.*, 2015, **6**, 1900–1910.
96. A. Pavić, T. Ilic-Tomic and J. Glamočlija, Unravelling Anti-Melanogenic Potency of Edible Mushrooms *Laetiporus sulphureus* and *Agaricus silvaticus* In Vivo Using the Zebrafish Model, *J. Fungi*, 2021, 7, 834.
97. P. Y. Huang and S. Ohga, Utilization of Ceramic Beads for Edible Mushrooms Cultivation, *Adv. Microbiol.*, 2017, 7, 853–862.
98. H. Lee, K. Nam, Z. Zahra and M. Q. U. Farooqi, Potentials of truffles in nutritional and medicinal applications: a review, *Fungal Biol. Biotechnol.*, 2020, 7, 9.
99. Q. Liu, H. Ma, Y. Zhang and C. Dong, Artificial cultivation of true morels: current state, issues and perspectives, *Crit. Rev. Biotechnol.*, 2018, **38**(2), 259–271.

CHAPTER 11

The Global Market for Mushrooms, Their Uses as Dietary Supplements and Associated Safety Issues

VINÍCIUS MATEUS SALVATORI CHEUTE,[a] EMANUELI BACKES,[b] RÚBIA CARVALHO GOMES CORRÊA,[c] VANESA GESSER CORRÊA,[b] ADELAR BRACHT[a,b,d] AND ROSANE MARINA PERALTA*[a,b,d]

[a] Post-Graduate Program in Biochemistry, State University of Maringá, Maringá, Paraná, Brazil; [b] Post-Graduate Program in Food Sciences, State University of Maringa, Maringá, Paraná, Brazil; [c] Program of Master in Clean Technologies, Cesumar Institute of Science Technology and Innovation (ICETI), Cesumar University – UniCesumar, Maringá, Paraná, Brazil; [d] Department of Biochemistry, State University of Maringá, Maringá, Paraná, Brazil

*Email: rmperalta@uem.br

11.1 Introduction

Mushrooms have formed part of the human diet for thousands of years. They are regarded as valuable foods, not only because of their attractive sensory characteristics and culinary features but also because of their nutritional properties. Mushrooms are rich in carbohydrates, proteins, fibers, vitamins, and minerals, but their fat content is relatively low.[1]

Food Chemistry, Function and Analysis No. 36
Edible Fungi: Chemical Composition, Nutrition and Health Effects
Edited by Dejan Stojković and Lillian Barros

Published by the Royal Society of Chemistry, www.rsc.org

Moreover, mushrooms have been described as functional foods, potentially beneficial for human health by virtue of their contents in numerous biologically and physiologically active substances. These substances have been reported to exert antitumor, antioxidant, antibacterial, hepatoprotective, anti-inflammatory and antidiabetic effects.[2] Also, the consumption of mushrooms in replacement of other foods is likely to reduce the risk of high blood pressure and cardiovascular diseases, as they are low in sodium and high in potassium.[3] Because of all these potential benefits, mushrooms are steadily gaining popularity and are gradually approaching the status of "super foods".

More than 2000 species of mushrooms exist in nature, but only around 25 are widely accepted as food. Among these, just a few are effectively cultivated for commercial purposes. The most common mushrooms in the market are *Agaricus bisporus* (Champignon), *Lentinus edodes* (shiitake), and species of *Pleurotus* genus: *P. citrinopileatus* (golden oystermushroom), *P. djamor* (pink oystermushroom), *P. eryngii* (king oyster mushroom), *P. florida* (Florida oyster mushroom), *P. ostreatus*, and *P. pulmonarius*, the last two known as oyster mushrooms. In turn, the major medicinal mushrooms that are commercialized worldwide are *Ganoderma lucidum* (lingzhi or reishi), *Grifola frondosa* (maitake), *Inonotus obliquus* (chaga), *Cordyceps militaris* (caterpillar fungus), *Trametes versicolor* (turkey tail), *Hericium erinaceus* (lion's mane), and *Agaricus blazei* (sun mushroom). Frequently, shiitake and some *Pleurotus* species are also considered as medicinal.[2,4]

The objective of this chapter is to evaluate the global market for mushrooms, their uses as dietary supplements and the associated safety issues. Due to the prominence of the above-mentioned species, evidently, facts and analysis will be focused on them in as much as they are also shaping the most important characteristics of the market.

11.2 The Global Market for Mushrooms

The global mushroom production and consumption chain is projected to witness significant growth in the next years or decades, as there are several factors that are likely to increase the worldwide per capita mushroom consumption. The mushroom market will certainly benefit from the growing consumer awareness of the multiple health benefits of these foods, also meeting the rising demand for vegan and natural foods in the diet, an irreversible movement around the globe. As a kind of side effect, the mushroom market is also expected to be boosted by the global increase in the consumption of processed food, which involves continuously improving technologies to extend the mushrooms shelf-life. In view of the ascendant demand, mushroom cultivation will be widely adopted by new growers as a promising and profitable business. The lack of technical knowledge for spawning production in developing countries, however, will certainly represent a bottleneck, limiting the market. Table 11.1 shows some of the main mushroom-producing and -trading companies in different countries.

Table 11.1 Representative list of key players in the global edible mushroom market.

Company	Web site[a]	Country of origin
Agro Dutch Industry Ltd	www.agro-dutch.com	India
Okechamp S.A.	www.en.okechamp.pl	Poland
Ganoherb Technology (Fujian) Corporation	www.organicreishi.en.made-in-china.com	China
Zhejiang Fangge Pharmaceutical Industry Co.	www.fgmushroom.en.made-in-china.com	China
Hangzhou Molai Biotech Co. Ltd	www.mushroomsextracts.com	China
Biosan Biotech Co Ltd	www.qualitymushroom.com	China
Weikfield Foods Pvt. Ltd	https://weikfield.com/	India
The Mushroom Company	www.themushroomcompany.com	USA
Monaghan Company	www.monaghan.eu	Ireland
Walsh mushrooms.com	www.walshmushrooms.com	United Kingdom
Isle of Wight Mushroom Culture	https://www.iwmushrooms.co.uk	United Kingdom
South Mill Mushrooms Sales Inc.	www.southmill.com	USA and Canada
Smithy Mushrooms	www.smithymushrooms.com	United Kingdom
Rheinische Pilz Zentrale GmbH	https://www.rhpz.de	Germany
Mushroom Sas di Pozza Fernando & C. Società Agricola	www.mushroomfem.com	Italy

[a]Web sites were accessed June 11, 2021.

Statistics on mushroom production per country vary considerably according to the sources. A good example of the disparity between the available data is the production of China, which is agreed to be (by far) the greatest mushroom producer in the world. A Western source attributes to China an annual production of 7.8 million tons per year.[5–7] The figures presented by the China Business Research Institute,[8] however, are different. According to this source China's annual output was 38.42 million tons in 2017. This corresponds to approximately 75% of the whole global production. In the same year, the CFNA[9] (China Chamber of Commerce of Foodstuffs and Native Produce) Edible Mushroom Branch reported that edible mushroom exports were valued at US$3.8 billion. From other sources it can be inferred that mushroom cultivation in China, valued at US$24.0 billion, can be considered as the fifth-largest agricultural sector in the country. Despite the discrepant data, there is no doubt that mushrooms have become one of the most important crops in Chinese agriculture, and that China absolutely leads the international mushroom production.[10,11] Considering, thus, China's production and that of its neighbors, there can be little doubt that the Asia Pacific region leads the global mushroom production market. This reflects also the annual per capita consumption in China, which presumably reaches 10 kg per person. This greatly exceeds the consumption of any other country, the only exception being perhaps the Netherlands. Besides, in Asian countries such as India, Japan and others, the consumption of mushrooms is experiencing substantial annual increases. This is credited to increasing production, which in turn was boosted by both the increasing vegan 'population and consumers' trend in choosing more nutrient-rich food.

China's huge mushroom production is followed by that of the United States, Italy and the Netherlands, each one of which currently yields less than 1.0 million tons per year. In Europe, the per capita mean mushroom consumption is around 3.5 kg per person per year, whereas in the Netherlands it surpasses 10 kg per person per year.[10,11] Over the last decade, there has been a marked increase in the consumption and commerce of medicinal mushrooms over the whole European trading space. The European medicinal mushroom market is projected to expand at the highest compound annual growth rate (CAGR) of 8.07% during the period 2017–2023. A more pronounced expansion is expected during the same period for the dried medicinal mushroom segment, namely 14.04% (Kenneth Research, https://www.kennethresearch.com/, New York validated by Global Information Inc., West Hartford, USA).[12]

Considering the divergent information on China's mushroom production, market characteristics and size, the 'market analysts' reports in the Western area should be regarded with some caution. Zion Market Research's analysts have reported that the worldwide mushroom market was capitalized at more than US$35.08 billion in 2015. According to them, the mushroom market was predicted to attain more than US$59.48 billion in 2021 and was expected to increase at a CAGR of more than 9.2% from 2016 to 2021.[13] Furthermore, for 2026, great heights are projected, with a global mushroom consumption of 20.84 million tons, thanks to a CAGR of 6.41%.

The Fortune Business Insights report, "Mushroom Market Size, Share & Industry Analysis, By Type (Button Mushroom, Shiitake Mushroom, Oyster Mushroom, & Others), Form (Fresh Mushroom, Frozen Mushroom, Dried Mushroom, and Canned Mushroom), and Regional Forecast, 2019–2026", informed of a global mushroom consumption of 12.74 million tons in 2018.[14] The report also displayed a detailed rating of the various factors that are likely to shape the market during the forecast period as well as a detailed analysis of its dynamics. The view is expressed that the turning point of the global dietary shift toward vegetarianism and the increasing awareness of the health benefits of mushrooms will be key drivers for the worldwide growth of the mushroom market. In 2019 the button mushroom (*Agaricus bisporus*) branch accounted for the largest market share. In fact, its contribution exceeded 40% of the global production. In all the most important mushroom producer countries, such as China, Japan, the US, the UK, Germany, and Poland, this is the most cultivated mushroom type for commercial purposes. The intense research on white button mushrooms aiming at unveiling their potential to prevent cancer and to promote their availability at affordable prices as compared to special mushroom varieties is projected to escalate the demand during the forecast period.

Of all the main properties attributed to medicinal mushrooms (see Table 11.2), those corresponding to the antioxidant segment have received greater attention and represent a large share of the mushroom market.[12] The increasing use of medicinal mushrooms in the food and nutraceutical industries suggests that the global market for medicinal mushrooms could reach approximately 7250 kilotons by the end of 2023, with a CAGR of 12.60%.[15] The Asia-Pacific area is expected to dominate the market, followed by Europe. The market also includes products deriving from medicinal mushrooms and used as food commodities, for personal care and as cosmetics. In addition, fresh and dried medicinal mushrooms are processed

Table 11.2 Overview of the most common mushroom supplements and their potential uses.

Mushroom species, popular name and reference	Potential uses	Commercial products[a]
Agaricus blazei, sun mushroom[24]	Antioxidant, antiallergic, antitumoral activities	Capsules of pure *Agaricus blazei* https://alohamedicinals.com
Cordyceps sinensis, Chinese caterpillar fungus[25]	Antiaging, antibacterial, antitumor, immunomodulatory activities	Capsules of *Cordycepis*™ https://alohamedicinals.com
Ganoderma lucidum, reishi, lingzhi[26]	Antioxidant, antibacterial, immunomodulatory, anti tumoral activities	Reishi powder and Reishi extract https://naturalin.com
Grifola frondosa, maitake[27]	Antioxidant, antiviral, antitumoral activities	Maitake mushroom extract https://vnibioscience.com
Hericium erinaceus, lion's mane[28]	Antihyperlipidemic, Antioxidant, hepatoprotective activities	Pure Lion's Mane™ capsules https://alohamedicinals.com
Inonotus obliquus, chaga mushroom[29]	Antihyperglycemic, antioxidant, antimicrobial, antimutagenic, immunomodulatory activities	Organic Chaga mushroom powder https://phytoextractum.com
Lentinula edodes, shiitake[30]	Immunomodulatory, antitumoral, cholesterol lowering, anti-inflammatory activities, among others	Shiitake Mushroom capsules https://natureslab.com
Pleurotus ostreatus, oyster mushroom[21]	Antioxidant, anti-inflammatory, antibacterial, immunomodulatory activities	Oyster mushroom capsules https://terezia.eu
Trametes versicolor, turkey tail[31]	Antitumoral activity, regulation of immune system	Turkey tail capsules https://fungi.com

[a]Web sites accessed June 11, 2021.

and incorporated into pizzas, burgers and soups. Of importance are also various skin-care products containing selenium, zinc and/or kojic acid obtained from medicinal mushrooms.

11.3 Mushroom Dietary Supplements

The definition of dietary supplement given by the United States (US) Food and Drug Administration (FDA)[16] has gained wide acceptance. According to the FDA a dietary supplement is a product taken in orally and which contains a dietary ingredient that will supplement the diet. Dietary ingredients can be concentrates or extracts derived from plants and herbs. They can also contain vitamins, minerals, amino acids and related substances (including enzymes, organ tissues and glandular extracts).

When mushrooms are used in dietary supplements, they are often in the form of an extract, capsule or powder. The part usually visible above ground, the "fruiting body" or basidium, is typically eaten as food (Figure 11.1). The mycelium is the "root" of the mushroom, as illustrated in Figure 11.1. In general, the mycelium grows faster and easier than the fruiting body. Both parts, however, contain beneficial chemical components (macronutrients as proteins, polysaccharides and glucans, and small molecules such as phenolics, triterpenes, *etc.*). Up to date, it remains unclear if mushrooms

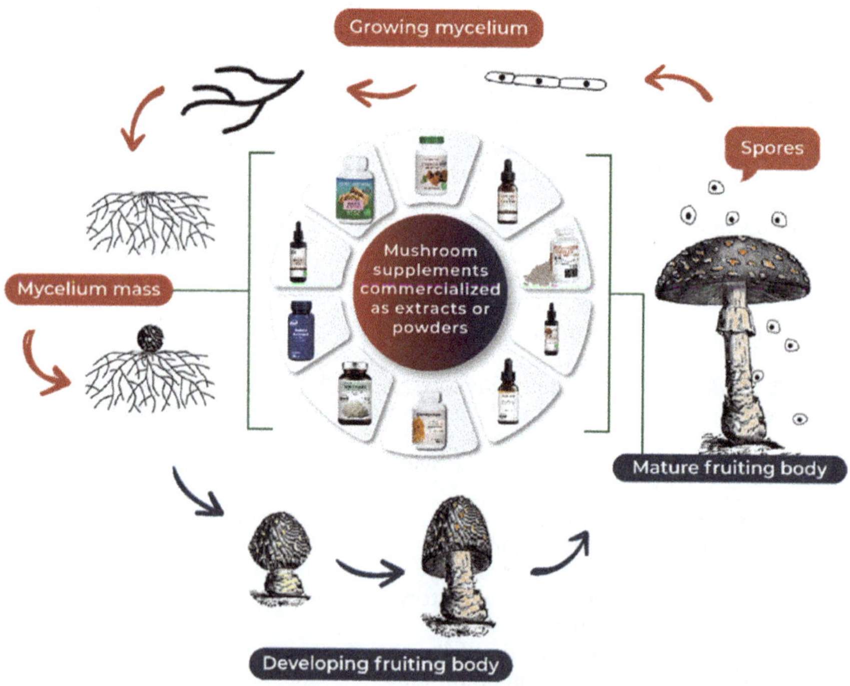

Figure 11.1 The fungi life cycle and sources for the preparation of dietary supplements (mycelium mass and fruiting bodies).

ingested in the form of dietary supplements deliver the same components or biological effects as when eaten as food.

About 15 years ago, approximately 80–85% of all edible-medicinal mushroom products were derived from the fruiting bodies and only 15% were from mycelia extracts.[17] However, the process of producing fruiting bodies is effortful and time-consuming, as it demands large volumes of substrate, space, growing conditions and qualified labor, factors that hinder research in the laboratory. Cultivations that are performed in the vegetative phase are much more interesting and functional for research considering that they can be kept in the laboratory and performed on a small and medium scale. Furthermore, important parameters, such as temperature, humidity, pH and aeration can be more easily controlled.[18,19] Thus, submerged cultivation is a promising and still under-explored alternative for the extraction of bioactive molecules in a short time, which also allows for mycelia storage for a long period without genetic alterations, benefiting the conservation of biodiversity.[20] However, for using the mycelial biomasses, it is necessary to prove that they are similar to fruiting bodies.[21,22] A number of mushroom-producing companies have invested in technologies to obtain and commercialize mycelial biomass as well as for the extraction of bioactives from these materials. Biosan Biotec Co. Ltd (www.qualitymushroom.com) and Hangzhou Molai Biotech Co. Ltd (www.mushroomsextracts.com) are two examples.

There are several ways to prepare mushroom powders and extracts (Figure 11.1). Extracts can be prepared by boiling fruiting bodies or mycelial biomass in alcohol or hot water. Powders can be made by grinding either the dried fruiting body or the dried mycelial biomass. Some products contain mixtures of different mushrooms. Control and standardization of the chemical composition of mushroom-based ingredients are difficult because of the diversity of preparation methods and growth conditions (in logs, sawdust, liquid, *etc.*). These are characteristics that are likely to compromise both quality and consistency among batches and products.

Table 11.2 lists the most common mushroom supplements commercialized in the world. A scientific reference that describes the potential uses of each mushroom is also given. More detailed lists of marketed mushroom supplements can be found in previous works which explored the functional properties of mushroom extracts.[15,23]

11.4 Safety Issues Associated with the Consumption of Mushroom Supplements

For a long time – thousands of years – mushroom tinctures have been explored by the traditional medicine sector as a primary means of treatment due to their alleged long list of health benefits. Asia is the continent where these beneficial mushrooms have been used more extensively.[32] Nowadays, however, their use is spreading to all parts of the world. In most cases, mushrooms as food are safe to consume, excepting of course those that are poisonous (toxic). The consumption of toxic mushrooms can produce severe side effects including nausea, diarrhea, vomiting and liver failure. In some cases, it can even cause death. In a general way, wild edible mushrooms can be contaminated from many sources, from the environment, processing, degradation and spoilage, besides food contact materials. The most common are toxic metals, pesticides, process contaminants, radioactive isotopes, microbial contamination, inherited toxins (*e.g.*, biogenic amines and mycotoxins), allergens and natural toxic substances in wild mushrooms. Since the European Union (EU) Rapid Alert System for Food and Feed (RASFF) was established, only three significant problems have been singled out in the trade of edible mushrooms: unsuitable microbial contamination, unlawful chemical contamination (*e.g.*, high levels of cadmium or radioactive cesium) and misidentification of inedible or poisonous mushrooms as edible species (*e.g.*, edible *Amanita caesarea* was found mixed with deadly *Amanita phalloides*).

The growing popularity of the medicinal mushrooms in dietary supplements has not been paralleled by a consistent evaluation of their safety for human consumption, at least not by methods that use modern techniques. Although the long history of use can be considered as self-provided evidence for safe consumption, many toxicological questions remain obscure. Consensus is widespread that there is not enough reliable information available supporting the safety of medicinal mushrooms and of the corresponding dietary supplements. The same can be said about their effectiveness or appropriateness. In some labels of dietary supplements that contain mushroom ingredients, warning statements have been added. The warning "avoid using if you're pregnant, breastfeeding, or anticipate surgery", is a good example. Mentions of side effects, such as nausea, vomiting, diarrhea, dizziness, skin sensitivity, headache, gastrointestinal discomfort and allergic reactions, can also be found.

Contributing to making the fungi safety evaluation more difficult, modern cultivation practices and preparations are rarely consistent with traditional medicinal uses. Whereas fruiting bodies are the matrices most often used in traditional Chinese medicine,[33] the most typical commercial raw material is the mycelial biomass, which grows faster and is also less expensive to produce. It is important to note that the growth conditions alter the secondary metabolite profile of the fungi.[34] This has an impact on the ability of investigators in bridging data generated in the analysis of different raw materials or different suppliers (even of the same species).

The safety and effectiveness of mushroom supplements are less investigated because for dietary supplements the regulations are much less abundant and detailed when compared to food and drug products. However, very important points to be considered regarding the safety of using mushroom supplements are the quality and quantity of their bioactive components. Each species holds a unique chemical composition that, in turn, accounts for the mushroom nutritional, functional and pharmacological features. However, the conditions of cultivation and the extraction procedures can influence the chemical composition of the extracts and powders. Ergothioneine, lovastatin and γ-aminobutyric acid (GABA) are secondary metabolites of several edible mushrooms with proven beneficial actions for human health.[35–37] Lin *et al.*[38] evaluated the presence of these molecules in both fruiting bodies and mycelia of different culinary-medicinal mushrooms from 18 genera, including *Agaricus*, *Cordyceps*, *Ganoderma*, *Grifola*, *Pleurotus* and *Trametes.* These molecules were found at different levels in both fruiting bodies and mycelia, suggesting that in addition to the choice of the best species, a careful search for the best culture conditions must be carried out to obtain mycelia rich in target bioactives.

More recently, VanderMolen *et al.*[39] performed an interesting safety investigation of dietary supplements of chaga, turkey tail, shiitake, maitake and reishi mushrooms. Their study combined modern analytical data with *in silico* toxicological evaluations of several commercial raw materials. According to the authors, this is an approach that could provide an effective alternative to conventional toxicological animal studies for the safety assessment of fungal dietary ingredients. Confidence that materials are safe for consumption at doses consistent with dietary intakes of culinary mushrooms can be derived from their apparent prevalence in the marketplace, the lack of reported adverse events and the very high degree of similarity between their mycelium growths (the raw materials investigated) and the culinary fruiting bodies to which they were compared.

Ganoderma lucidum is among the most popular medicinal mushrooms. In this species, the three most important physiologically active constituents are polysaccharides, peptido-glycans and triterpenes.[40–45] However, the amount and percentage of each component can differ in commercial and natural products. Chang and Buswell[46] evaluated the triterpenes and polysaccharides of 11 randomly selected samples of commercial lingzhi products purchased in Hong Kong. They found that the triterpene content ranged from undetectable to 7.8%, and the polysaccharide content varied from 1.1% to 5.8%. This kind of variation can have several causes, including differences in the species or strains that were used, as well as differences in the production methods. Wu *et al.*[47] used modern methodologies, including chromatographic methods and saccharide mapping, for the evaluation of the triterpene and polysaccharide contents of 19 different *Ganoderma lucidum* supplements commercialized in the United States. Unfortunately, the authors concluded that only five tested samples (26.3%) were in accordance with their labels, evidencing that the quality of the *G. lucidum* dietary supplements in the US market is poor, an issue that certainly demands careful investigation.

A very important point to be considered about the safety of the consumption of mushroom supplements is related to their heavy metal content. The fruiting bodies of mushrooms not only have an effective mechanism to accumulate macro- and trace elements, but also the capability to accumulate heavy metals such as Cd, Cr, Hg and Pb.[48–52] The accumulation of heavy metals in wild-growing mushrooms is well documented in the literature.[53,54] A study conducted in Nigeria revealed elevated levels of metals in several wild mushrooms.[55] In another study, conducted in the Yunnan Province of China, the authors reported the presence of heavy metal in wild edible mushrooms.[56] On the other hand, in a more recent study, wild-growing *Boletus edulis* from Yunnan, China, was considered safe according to the provisionally tolerable weekly intake (PTWI) of heavy metals recommended by the United Nations Food and Agriculture Organization and World Health Organization (FAO/WHO).[57] Similar results were obtained in a study carried out in Iran, where ten samples of *Agaricus bisporus* (between wild and cultivated samples) were scrutinized for the presence of heavy metals, and most samples were considered safe for consumption according to the criteria of the *Codex Alimentarius* (CODEX) and FAO/WHO.[58]

As experimentally proven, mushrooms cultivated on artificially contaminated substrates can also take up and accumulate health-threatening concentrations of toxic elements such as mercury[50] and arsenic.[59] Studies conducted with the royal sun mushroom (*Agaricus brasiliensis*), cultured on a substrate that had been artificially contaminated with heavy metals, found the bioaccumulation of cadmium, lead and chromium.[60,61] Consequently, it is crucial for food and health safety that the substrates used for mushroom cultivation have the lowest possible content of toxic elements. There is a recent report about the presence of lead, mercury, arsenic and cadmium in *Ganoderma lucidum*, *Phellinus linteus*, *Umbilicaria esculenta*, *Trametes versicolor*, *Agaricus blazei*, *Hericium erinaciium*, *Paecilomyces japonica*, *Fuscoporia obliqua*, *Sarcodon imbricatu* and *Hypsizygus marmoreus* mushroom samples. These samples were from the South Korean market.[62] The authors concluded that the consumption of these 10 mushrooms did not present any significant health risks for the general public in Korea. Therefore, the heavy metals and metalloid contents of mushrooms should be studied for effective safety management.

11.5 Conclusion

There is no doubt that the global mushroom market offers many favorable perspectives in terms of production and marketing. This conclusion applies to both the food section *per se* as well as to the various dietary supplement branches. Likewise, mushrooms have great potential for pharmaceutical application, and many scientific contributions and case studies in traditional medicine and/or nutritional studies endorse such ability. However, to fully exploit these potentialities, not only in the form of dietary supplements but also as regular drugs, continuous and defined production processes, standardized quality and methods for its control, together with preclinical and clinical trials according to the regulations, and legal authorization, are mandatory. These processes require time, money and close interdisciplinary collaboration, which is reason why large companies will make such investment only when the developments are protected by patents. In this sense, new ideas are necessary to allow patent protection or to make the development profitable without patent protection. These ideas could include special formulations, new fields of application and indications, new chemical entities or new production technologies.

Acknowledgements

The authors are grateful for the financial support of Coordenação de Aperfeiçoamento de Pessoal de Nível Superior (CAPES), Conselho Nacional de Desenvolvimento Científico e Tecnologia (CNPq), Instituto Cesumar de Ciência Tecnologia e Inovação (ICETI) and Universidade Estadual de Maringá. V. M. S. Cheute is a scholarship recipient of CAPES. E. Backes, V. G. Correa, R. M. Peralta and A. Bracht are scholarship recipients of CNPq.

References

1. J.-H. Choi, Biologically functional molecules from mushroom-forming fungi, *Biosci., Biotechnol., Biochem.*, 2018, **82**(3), 372–382.
2. M. L. Gargano, L. J. L. D. van Griensven, O. S. Isikhuemhen, U. Lindequist, G. Venturella and S. P. Wasser, *et al.*, Medicinal mushrooms: Valuable biological resources of high exploitation potential, *Plant Biosyst.*, 2017, **151**(3), 548–565.
3. M. E. Valverde, T. Hernández-Pérez and O. Paredes-López, Edible Mushrooms: Improving Human Health and Promoting Quality Life, *Int. J. Microbiol.*, 2015, **2015**, 1–14.
4. G. Ma, W. Yang, L. Zhao, F. Pei, D. Fang and Q. Hu, A critical review on the health promoting effects of mushrooms nutraceuticals, *Food Sci. Hum. Wellness*, 2018, 7(2), 125–133.
5. Atlas big, *World's top Mushroom and Truffle Producing Countries*, Atlas big, 2018. Available from: https://www.atlasbig.com/en-us/countries-mushroom-truffle-production (accessed 2021 Jun 20).
6. D. J. Royse, J. Baars and Q. Tan, Current overview of mushroom production in the world, in *Edible and medicinal mushrooms: technology and applications*, ed. D. C. Zied and A. Pardo-Giminez, John Wiley & Sons Ltd., Hoboken, 2017, pp. 5–13.
7. G. C. Wakchaureand, Production and Marketing of Mushrooms: Global and National Scenario, in *Mushrooms-cultivation, marketing and consumption*, ed. Singh M., Vijay B., Kamal S. and Wakchaure G. C., Indian Council of Agricultural Research, New Delhi, 2011, pp. 15–22.
8. Research and Markets Global edible Mushrooms Market—Industry Trends, Opportunities and Forecasts to 2023. Available from: https://www.researchandmarkets.com/reports/4451952/global-edible-mushrooms-market-industry-trends (accessed 20 june 2021).
9. China Chamber of Commerce of Foodstuffs, Native Produce and Animal Byproducts (CFNA). Available from: https://en.cccfna.org.cn/ (accessed Jul 2021).

10. R. Fang, P. Kirk, J. C. Wei, Y. Li and L. Cai, *et al.*, Country focus: China, in *State of the World's Fungi 2018, Report*, ed. K. J. Willis, Kew Publishing, Richmond, 2018, pp. 48–55.
11. Y. Zhang, W. Geng, Y. Shen, Y. Wang and Y.-C. Dai, Edible mushroom cultivation for food security and rural development in China: bio-innovation, technological dissemination and marketing, *Sustainability*, 2014, **6**, 2961–2973.
12. M. Kozarski, A. Klaus, D. Jakovlievic, N. Todorovic, J. Vunduk, P. Petrovic, M. Niksic, M. M. Vrvic and L. van Griesven, Antioxidants of edible mushrooms, *Molecules*, 2015, **20**(10), 19489–19525.
13. Zion Market Research, Mushroom Market To Witness Fabulous Growth In The With A CAGR Of 9.2%, 2018. Available from: https://www.zionmarketresearch.com/news/global-mushroom-market (accessed 2021 Jun 13).
14. Fortune Business Insights, Mushroom Market to Grow at a Steady CAGR of 6.41% from 2019 to 2026, Public Sector Investment in Commercial Cultivation of Mushroom to Boost the Market, 2019. Available from: https://www.fortunebusinessinsights.com/press-release/mushroom-market-9301 (accessed 2021 Jun 13).
15. A. G. Niego, S. Rapior, N. Thongklang, O. Raspé, W. Jaidee, S. Lumyong and K. D. Hyde, Macrofungi as a nutraceutical source: promising bioactive compounds and market value, *J. Fungi*, 2021, 7, 397–431.
16. US Food and Drug Administration (FDA), FDA 101: Dietary Supplements, 2015. Available from: https://www.fda.gov/consumers/consumer-updates/fda-101-dietary-supplements (accessed 2021 May 10).
17. U. Lindequist, H. Timo, J. Niedermeyer and W. D. Julich, The pharmacological potential of mushrooms, *Evidence-Based Complementary Altern. Med.*, 2005, **2**(3), 285–299.
18. F. N. Anike, O. S. Isikhuemhen, D. Blum and H. Neda, Nutrient requirements and fermentation conditions for mycelia and crude exo-polysaccharides production by *Lentinus squarrosulus*, *Adv. Biosci. Biotechnol.*, 2015, **6**(8), 526–536.
19. V. Elisashvili, Submerged cultivation of medicinal mushrooms: bioprocesses and products (review), *Int. J. Med. Mushrooms*, 2012, **14**(3), 211–230.
20. K. F. Benson, P. Stamets, R. Davis, R. Nally, A. Taylor and S. Slater, *et al.*, The mycelium of the *Trametes versicolor* (Turkey tail) mushroom and its fermented substrate each show potent and complementary immune activating properties *in vitro*, *BMC Complementary Altern. Med.*, 2019, **19**(1), 1–14.
21. R. C. G. Corrêa, T. Brugnari, A. Bracht, R. M. Peralta and I. C. F. R. Ferreira, Biotechnological, nutritional and therapeutic uses of *Pleurotus* spp. (Oyster mushroom) related with its chemical composition: A review on the past decade findings, *Trends Food Sci. Technol.*, 2016, **50**, 103–117.
22. A. E. S. S. Carvajal, E. A. Koehnlein, A. A. Soares, G. J. Eler, A. T. A. Nakashima and A. Bracht, *et al.*, Bioactives of fruiting bodies and

submerged culture mycelia of Agaricus brasiliensis (A. blazei) and their antioxidant properties, *LWT–Food Sci. Technol.*, 2012, **46**(2), 493–499.

23. F. S. Reis, A. Martins, M. H. Vasconcelos, P. Morales and I. C. F. R. Ferreira, Functional foods based on extracts or compounds derived from mushrooms, *Trends Food Sci. Technol.*, 2017, **66**, 48–62.
24. A. C. S. Souza, V. G. Correa, G. A. Gonálves, A. A. Soares, A. Bracht and R. M. Peralta, *Agaricus blazei* bioactive compounds and their effects on human health: benefits and controversies, *Curr. Pharm. Des.*, 2017, **23**(19), 2807–2834.
25. P. X. Chen, S. Wang, S. Nie and M. Marcone, Properties of Cordyceps Sinensis: A review, *J Funct. Foods.*, 2013, **5**(2), 550–569.
26. Y. Yang, H. Zhang, J. Zuo, X. Gong, F. Yi and W. Zhu, *et al.*, Advances in research on the active constituents and physiological effects of *Ganoderma lucidum*, *Biomed. Dermatol.*, 2019, **3**, 1–17.
27. X. He, X. Wang, J. Fang, Y. Chang, N. Ning and H. Guo, *et al.*, Polysaccharides in *Grifola frondosa* mushroom and their health promoting properties: A review, *Int. J. Biol. Macromol.*, 2017, **101**, 910–921.
28. M. Friedman, Chemistry, nutrition, and health-promoting properties of *Hericium erinaceus* (Lion's mane) mushroom fruiting bodies and mycelia and their bioactive compounds, *J. Agric. Food Chem.*, 2015, **63**(32), 7108–7123.
29. M. E. Balandaykin and I. V. Zmitrovich, Review on Chaga medicinal mushroom, *Inonotus obliquus* (Higher Basidiomycetes): Realm of Medicinal Applications and Approaches on Estimating its Resource Potential, *Int. J. Med. Mushrooms*, 2015, **17**(2), 95–104.
30. R. Gaitán-Hernández, D. López-Pena, M. Esqueda and A. Gutiérrez, Review of bioactive molecules production, biomass and Basidiomata of shiitake culinary-medicinal mushrooms Lentinus edodes (Agaromycetes), *Int. J. Med. Mushrooms*, 2019, **21**(9), 841–850.
31. C. Hobbs, Medicinal Value of Turkey Tail Fungus Trametes versicolor (L.:Fr.) Pilát (Aphyllophoromycetideae). A Literature Review, *Int. J. Med. Mushrooms*, 2004, **6**(3), 195–218.
32. K.-H. Lee, S. L. Morris-Natschke, X. Yan, R. Huang, T. Zhou and S.-F. Wu, *et al.*, Recent progress of research on medicinal mushrooms, foods, and other herbal products used in traditional Chinese medicine, *J. Tradit. Complementary Med.*, 2012, **2**(2), 84–95.
33. D. Bensky, S. Clavey and E. Stöger, *Chinese Herbal Medicine: Materia Medica*, Eastland Press, Seattle (WA), 2004.
34. K. M. VanderMolen, H. A. Raja, T. El-Elimat and N. H. Oberlies, Evaluation of culture media for the production of secondary metabolites in a natural products screening program, *AMB Express*, 2013, **3**, 1–7.
35. S. Nachimuthu, R. Kandasamy, R. Ponnusamy, J. Deruiter, M. Dhanasekaran and S. Thilagar, L-Ergothioneine: A Potential Bioactive Compound from Edible Mushrooms, in *Medicinal Mushrooms*, ed. D. Agrawal and M. Dhanasekaran, Springer, Singapore, 2019, pp. 391–407.

36. K. Kala, A. Kryczyk-Poprawa, A. Rzewinska and B. Muszynska, Fruiting bodies of selected edible mushrooms as a potential source of lovastatin, *Eur. Food Res. Technol.*, 2020, **246**(4), 713–722.
37. Y. J. Park, T. S. Oh and M. J. Jang, Effect of adding amino acids on the production of gamma-aminobutyric acid (GABA) by mycelium of Lentinula edodes, *Int. J. Eng. Res. Appl.*, 2019, **15**(9), 20180287.
38. S. Y. Lin, Y. K. Chen, H. T. Yu, G. S. Barseghyan, M. D. Asatiani and S. P. Wasser, *et al.*, Comparative study of contents of several bioactive components in fruiting bodies and mycelia of culinary-medicinal mushrooms, *Int. J. Med. Mushrooms*, 2013, **15**(3), 315–323.
39. K. M. VanderMolen, J. G. Little, P. Sica, T. El-Elimat, H. A. Raja and N. H. Oberlies, Safety assessment of mushrooms in dietary supplements by combining analytical data with *in silico* toxicology evaluation, *Food Chem. Toxicol.*, 2017, **103**, 133–147.
40. C. Dong and Q. Han, *Ganoderma lucidum* (Lingzhi, Ganoderma): Fungi, algae, and other materials, in *Dietary Chinese Herbs Chemistry: Pharmacology and Clinical Evidence*, ed. Y. Liu, Z. Wang and J. Zhang, Springer, London, 2015, pp. 759–765.
41. K. K. Hapuarachchi, W. A. Elkhateeb, S. C. Karunarathna, C. R. Cheng, A. R. Bandara and P. Kakumyan, *et al.*, Current status of Global *Ganoderma* cultivation, products, industry and market, *Mycosphere*, 2018, **9**(5), 1025–1052.
42. T. Lai, Y. Gao and S. F. Zhou, Global marketing of medicinal Ling Zhi mushroom *Ganoderma lucidum* (W.Curt:Fr.) Lloyd (Aphyllophoromycetideae) products and safety concerns, *Int. J. Med. Mushrooms*, 2004, **6**(2), 189–194.
43. J. Li, J. Zhang, H. Chen, X. Chen, J. Lan and C. Liu, Complete Mitochondrial Genome of the Medicinal Mushroom *Ganoderma lucidum*, *PLoS One*, 2013, **8**(8), e72038.
44. Z. T. Xu, X. U. Chen, Z. F. Zhong, L. D. Chen and Y. T. Wang, *Ganoderma lucidum* polysaccharides: Immunomodulation and potential anti-tumor activities, *Am. J. Chin. Med.*, 2011, **39**, 15–27.
45. R. Russell and M. Paterson, *Ganoderma* – A therapeutic fungal biofactory, *Phytochemistry*, 2006, **67**(18), 1985–2001.
46. S. T. Chang and J. A. Buswell, Safety, quality control and regulational aspects relating to mushroom nutriceuticals, in *Proceedings of the 6th International Conference in Mushroom Biology and Mushroom Products*, GAMU GmbH, Institut für Pilzforschung, Krefeld, Germany, 2008, pp. 188–195.
47. D. T. Wu, Y. Deng, L. X. Chen, J. Zhao, A. Bzhelyansky and S. P. Li, Evaluation on quality consistency of *Ganoderma lucidum* dietary supplements collected in the United States, *Sci. Rep.*, 2017, 7, 7792–7801.
48. Q. Huang, Y. Jia, Y. Wan, H. Li and R. Jiang, Market survey and risk assessment for trace metals in edible fungi and the substrate role in accumulation of heavy metals, *J. Food Sci.*, 2015, **80**(7), 1612–1618.

49. S. R. Koyyalamudi, S. Jeong, S. Manavalan, B. Vysetti and G. Pang, Micronutrient mineral content of the fruiting bodies of Australian cultivated *Agaricus bisporus* white button mushrooms, *J. Food Compost Anal.*, 2013, **31**, 109–114.
50. P. Rzymski, M. Mleczek, M. Siwulski, M. Gasecka and P. Niedzielski, The risk of high mercury accumulation in edible mushrooms cultivated on contaminated substrates, *J. Food Compost. Anal.*, 2016, **51**, 55–60.
51. J. Vetter, Chemical composition of fresh and conserved *Agaricus bisporus* mushroom, *Eur. Food Res. Technol.*, 2003, **217**, 10–12.
52. J. Vetter, J. Hajdú, J. Györfi and P. Maszlavér, Mineral composition of the cultivated mushrooms *Agaricus bisporus*, *Pleurotus ostreatus* and *Lentinula edodes*, *Acta Aliment.*, 2005, **34**, 441–451.
53. P. Palac, Trace element contents in European species of wild growing edible mushrooms: A review for the period 2000–2009, *Food Chem.*, 2010, **122**, 2–15.
54. M. Dowlati, H. R. Sobhi, A. Esrafili, M. Farzadkia and M. Yeganeh, Heavy metals content in edible mushrooms: a systematic review meta-analysis and health risk assessment, *Trends Food Sci. Technol.*, 2021, **109**, 527–535.
55. C. C. Ndimele, P. E. Ndimele and K. S. Chukwuka, Accumulation of heavy metals by wild mushrooms in Ibadan, Nigeria, *J. Health Pollut.*, 2017, 7(16), 26–30.
56. B. Liu, Q. Huang, H. Cai, X. Guo, T. Wang and M. J. F. C. Gui, Study of heavy metal concentrations in wild edible mushrooms in Yunnan Province, China, *Food Chem.*, 2015, **188**, 294–300.
57. J. Su, J. Zhang, J. Li, T. Li, H. Liu and Y. Wang, Determination of mineral contents of wild *Boletus edulis* mushroom and its edible safety assessment, *J. Environ. Sci. Health, Part B*, 2018, **53**(7), 454–463.
58. H. Karami, N. Shariatifar, S. Nazmara, M. Moazzen and B. Mahmoodi, The concentration and probabilistic healthy risk of potentially toxic elements (PTEs) in edible mushrooms (wild and cultivated) samples collected from different cities of Iran, *Biol. Trace Elem. Res.*, 2021, **199**(1), 389–400.
59. M. Mlecze, M. Niedzielski, P. Rzymski, M. Gasecka, P. Golinskil and L. Kozak, Importance of low substrate arsenic content in mushroom cultivation and safety of final food product, *Eur. Food Res. Technol.*, 2016, **242**(3), 355–362.
60. E. M. Morales, B. L. Baldissera and D. F. de Angelis, Monitoring heavy metals in royal sun medicinal mushroom, *Agaricus brasiliensis* (Agaricomycetes), *Int. J. Med. Mushrooms*, 2019, **21**(2), 169–179.
61. L. Liu, H. Chen, J. Yuan, Y. Wang, B. Weng and P. Liu, Effects of cadmium stress on physiological indexes and fruiting body nutrition of *Agaricus brasiliensis*, *Sci. Rep.*, 2021, **11**, 1–12.
62. J.-M An, S.-Y Gu, D.-J Kim, H.-G Shin, K.-S Hong and Y.-K. Kim, Arsenic, cadmium, lead, and mercury contents of mushroom species in Korea and associated healthy risk, *Int. J. Food Prop.*, 2020, **23**(1), 992–998.

Subject Index